Proceedings of the International Coastal Congress ICC-Kiel '92
Interdisciplinary Discussion of Coastal Research and
Coastal Management Issues and Problems

Horst Sterr
Jakobus Hofstede
Hans-Peter Plag
(Hrsg.)

Proceedings of the International Coastal Congress ICC - Kiel '92

Interdisciplinary Discussion of
Coastal Research
and Coastal Management
Issues and Problems

PETER LANG
Frankfurt am Main · Berlin · Bern · New York · Paris · Wien

Die Deutsche Bibliothek - CIP-Einheitsaufnahme

Interdisciplinary discussion of coastal research and coastal management issues and problems : proceedings of the International Coastal Congress, ICC - Kiel '92 / Horst Sterr... (Hrsg.). - Frankfurt am Main ; Berlin ; Bern ; New York ; Paris ; Wien : Lang, 1993
  ISBN 3-631-45906-8

NE: Sterr, Horst [Hrsg.] ; International Coastal Congress >1992, Kiel<

ISBN 3-631-45906-8
© Verlag Peter Lang GmbH, Frankfurt am Main 1993
All rights reserved.

All parts of this publication are protected by copyright. Any utilisation outside the strict limits of the copyright law, without the permission of the publisher, is forbidden and liable to prosecution. This applies in particular to reproductions, translations, microfilming, and storage and processing in electronic retrieval systems.

Printed in Germany 1 2 3 4 5 6 7

PREFACE

More than most other places the coastal zones are affected by an abundance of - often conflicting - interests between further economic development, protection of people or values and nature conservation. Furthermore, the coasts are seriously threatened by impacts of climate changes such as sea level rise, increasing storminess etc. from which additionally high risks (for instance flooding, erosion or salt water intrusion) can evolve.

On the one hand coastal managers are often not prepared to examine these newly arising issues without assistance from the scientific community. Coastal scientists on the other hand, are generally specialists in one particular discipline and are depending on the knowledge and experience of their collegues for an overall appreciation of complex problems. For these reasons it is essential that an intensive dialogue is held between both groups in order to archieve an ecologically and economically compatible legal concept for a sustainable development of the vulnerable coastal zones.

It is to this end that the German Organization for the Advancement of Interdisciplinary Coastal Science (GEO-COAST) organized the International Coastal Congress - ICC - Kiel '92. The main goal of GEO-COAST is to support multidisciplinary coastal research with special emphasis on current important issues (e.g. climate impact, tourism, conservation aspects).

The congress was organized at Kiel. As a university and seaboard town Kiel has long had a strong tradition towards coastal and marine sciences and activities. In addition its position on the German-Danish Peninsula makes it a link between the North Sea and the Baltic coastal areas. Finally the political opening of all of eastern Europe now gives it an added integrative function for bridging the interests of this region while ties to the southern European countries have also been strengthened. Of a total of 112 participants from 24 countries 11 came from eastern Europe. Allthough this figure seems to be relatively low, the financial situation in the newly democratized eastern European states should be kept in mind. Considering this it becomes clear that especially the Baltic states share a great interest in commencing closer contacts with western Europe.

The Kiel Congress was initially intended to be held as the Second International Symposium of the Association EUROCOAST. Because of basic legal, financial and management problems within EUROCOAST, the congress organizing committee decided to draw support for ICC - Kiel '92 from other sources (see below).

During the congress four regional sessions as well as a thematic session on Statistics, Modelling and GIS were held. Besides, over 25 posters diplaying various coastal issues were presented during the congress. Finally, two excursions to the Wadden Sea coast and to the Baltic coast of the Federal State of Schleswig-Holstein were held.

Parallel to the regular sessions two workshops, one on Climate Impact Related Coastal Problems and one on Coastal Conservation Aspects with emphasis on the Baltic Area were held. These workshops were organized by the Climate Impact Research Unit (CIRU) and jointly by the European Union for Coastal Conservation (EUCC) and the World Wide Fund for Nature (WWF), respectively (table 1). Especially the former workshop was met with great interest by the audience; it dealt with both organizational and scientific aspects of this important future challenge.

sessions	oral presentations	manuscripts
Mediterranean	10	17
North sea	11	13
Baltic	9	11
Atlantic	9	11
Statistics, GIS and modelling	7	11
Sedimentology	0	3
Various topics	7	4
Workshop EUCC/WWF	4	4
Workshop CIRU	4	4
Total	60	78

Table 1: Number of lectures held at ICC - Kiel '92 and number of contributions to the proceedings.

The discrepancy between the number of papers and the number of contributions to the proceedings represents a current problem which all persons organizing scientific meetings are increasingly facing. Although their scientific results will surely be reliable, scientists seem to become increasingly unreliable where it concerns the meeting of obligations.
A number of scientists who submitted a manuscript to the organizing committee did not present their contribution to the audience. However, since they had payed their contribution it was decided to incorporate their papers into the proceedings. On the other hand 12 scientists who announced papers did not appear at the congress at all and without any form of cancellation or paying their contribution. As a result not only organizational problems occured but the attendants to the congress found themselves confronted with a number of unplanned extra coffee breaks as well. Let us all hope that this phenomenon will turn out to be nothing more than a temporary dissonance.

The 60 papers presented during ICC - Kiel '92 concerned four main topics, covering all major coastal regions of Europe:
- historic coastal development (8 papers);
- modern coastal processes (20);
- biology/ecology (9);
- coastal protection and planning (23).

As can been seen from this overview the interdisciplinary approach proved to be successful and urgently needed. During the congress coastal managers were able to present their specific problems to a highly specialized audience. Coastal scientists on the other hand were able to discuss their results with more practical-oriented coastal engineers. However, it should be mentioned that in a number of papers the results were presented in a way not easily understood by non-specialists.

This stresses the need for a specifically coastal-oriented and interdisciplnary organisation as a basis for cooperation among coastal scientists and managers in the future. Association EUROCOAST, which was started in 1989 by an EC workgroup of coastal experts, was up to now met with great interest from the scientific community but has not yet operated efficiently enough to fulfill the tasks expected by its members. At about the same time the European Union for Coastal Conservation EUCC was founded as an organisation with similar goals for research and management cooperation but focusing primarily on coastal conservational aspects. One sensible solution might be that these existing international and also national organisations which share the same interests join forces.

Due to these problems with EUROCOAST the organizing committee provided a forum for the discussion of a satisfactory structure for a future European coastal organization at ICC - Kiel '92. At this general assembly two possible structural concepts for such an organization were presented. Due to the importance and complexity of the issue an ad-hoc agreement could not be reached during the congress. Consequently it was decided to set up an advisory committee under the leadership of Prof. G. Fierro, one of the former vice-presidents of EUROCOAST. This committee shall prepare an organizational concept which can be adopted as a legal constitution for EUROCOAST and as such be discussed with its current president, Dr. Quellenec.

ACKNOWLEDGEMENTS

We wish to thank the following institutions who sponsored or supported ICC - Kiel '92:

- the Federal Ministry of Research & Technology through its Climate Impact Research Unit at Oldenburg University
- German Research Foundation DFG
- the state of Schleswig-Holstein through its Ministry of Environment
- University of Kiel/FTZ Büsum
- the European Union for Coastal Conservation EUCC
- the World Wide Fund for Nature WWF
- The EUROCOAST National Committee of Great Britain

With the support of the sponsoring institutions it was possible to keep the registration fees relatively low. Moreover with their financial subsidies 11 scientists from eastern European countries were enabled to attend the congress.

The local organizing committee of GEO-COAST consisted of the following persons (in alphabetical order): D. Boedeker, Dr. J.L.A. Hofstede, Dr. H.-P. Plag, U.-H. Schauser and Dr. H. Sterr.

ICC - KIEL '92 PROGRAM (06.- 11.09.92)

	General Events	Block I	Block II	Block III
Sunday	Registration			
Monday	-Registration -Welcoming -Reception	Mediterranean	North Sea	Poster presentation
Tuesday	- WWF-EUCC Workshop on Coastal Conservation	Mediterranean	- Statistics - Modelling - GIS	- Ecology/ Conservation Baltic
Wednesday	Excursions			
Thursday	General assembly - open workshop climatic impact	Atlantic	Baltic	Poster presentation
Friday		- Various topics	-Sedimentology	

During the congress 4 regional sessions were held: Mediterranean, Atlantic, North Sea and Baltic. Moreover, there were sessions on statistics, modelling, GIS, sedimentology and various topics.

	page
Preface	I - III
Contents	IV - XVIII
Papers	1 - 797
List of adresses	798 - 808

Contents according to the sessions

MEDITERRANEAN
page

Holocene coastline shifts at the Mediterranean coast of Andalucia (SE Spain), **G. Hoffmann** ... 1

Geomorphological evolution of the cliffs in the central Adriatic Sea (Italy), **Cicco, L., Elmi, C., Fanucci, F. & Nesci, O.**
- **no manuscript received** - ... --

Morphological types of rocky coast on southeastern Apulia, **Mastronuzzi, G., Palmentola G. & P.Sansò** ... 784
- **manuscript belated received** -

Spatial and temporal changes at Alexandria beaches, Egypt, **Frihy, O.E., Nasr, S.M., Dewidar, K. & Raey, M.E.** ... 13

Drift rates on accreted and eroded beaches by fluorescent sand, Nile delta coast, **El-Fishawi, N. & Badr, A.** ... 22

Estimation of sediment budget and longshore transport rates from coastal response. Application in the Ebro delta, **Jimenez, J.A. et al.** ... 38

Preliminary investigations of gravel barrier evolution in the Baie des Anges, French Riviera, **Anthony, E.** ... 48

Barrier breaching in microtidal environments: the Trabucador Bar case (Ebro delta), **Jimenez, J.A.** ... 58

Coastal erosion in the Eastern Black Sea coast of Turkey - A critical review, **Gokce, K., Dakoglu, A., Koc, S. & M. Gurhan** ... 68

Erosion along the Black Sea coast of Turkey and proposed protective measures, **Bilgin, R.** ... 74

Relationship between the port structures and coastal dynamics in the Gulf of Gela (Sicily - Italy), **Brambati, A., Amore, C., Giuffrida, E. & G. Randazzo** ... 773
- **manuscript belated received** -

Channel maintenance by fluidation, a first experiment in France, **Larcher, M.**
- **no manuscript received** - ... --

Artificial beach nourishment on the Mediterranean Nile delta coast, **Frihy, O.E. & R. Dean** ... 84

Integrated planning as a new paradigm of a coastal zone management, **Simunovic, I.** ... 96

Coast-Hinterland socio-economic relations as an essential element of integrated planning in the coastal zones, **Filipic, P.** 107

Climatical conditions and tourism through the Turkish Aegean coastal zone, **Kocman, A.** 126

Proposals and program for the organization of the natural park on the coast of Arbus (Sardinia), **Castelli, P. & F. Di Gregorio** 136

ATLANTIC

Morpho-climatic evolution of the Aveiro region littoral (NW of Portugal) during Tertiary and Quaternary, **Rocha, F. & Gomes, C.** 157

Recent evolution of Obidos and Albufeira coastal lagoons, **Freitas, M., Andrade, C. & F. Jones** 167

Dune erosion and shoreline retreat between Aveiro and Cape Mondego (Portugal). Prediction of future evolution, **Ferreira, O. & Alveirinho Dias, J.** 187

Recent dynamics in the Alto Minho Coast (NW Portugal), **Alves, A.M.C.** 201

Determination of the evolutionary condition of coastal cliffs on the basis of geological and geomorphological parameters, **Rivas, V. & Cendrero, A.** 214

The turbulent resuspension of cohesive intertidal muds: some new concepts and ideas, **Black, K.S.** 223

Persistant marine debris along the Glamorgan Heritage coast, UK: a management problem, **Simmons, S.L. & Williams, A.T.** 240

Beach aesthetic values: the south west peninsula, UK, **Williams, A.T., Leatherman, S.P. & Simmons, S.L.** 251

Coastal Protection by biological means, **Haiden, W.**
- **no manuscript received** - --

An approach for the rehabilitation of coastal wetlands and estuarine areas in northern Spain, **Frances, E., Rivas, V. & Cendrero, A.** 264

Planning and design for the leisure craft within a traditional fishing harbour, **Teixeira, A.T. & Gamito, T.** 274

VARIOUS TOPICS

Environmental development among the TENMILE CREEK coastal dunes in Oregon, USA, **Heikkinen, O.** — 283

Fluidization and beachface dewatering: recent progress and case histories, **Parks, J.** — 296

Chemical and biological characteristics of different water masses in the German Bight **Gerdes, D. & K.-H. Hesse** — 305

Intertidal zone of Spitsbergen and Franz Josef Land, **Weslawsky, J.M., M. Zajaczkowsky & T. Suryn** — 322

NORTH SEA

Coastal zone management: tools for initial design; **Wind, H.G. & E.B. Peerbolte** — 332

Middelkerke Bank; sedimentology and hydrodynamics, **Stolk, A.**
- **no manuscript received** - — --

Morphodynamics of inter-tidal areas - the Wadden Sea, **Noorbergen, H.H.S** — 342

Simulation of morphodynamics of tidal inlets in the Wadden Sea, **Van Overeem, J., Steijn, R.C., Van Banning, G.K.F.M.** — 351

Morphological structures in German tidal flat areas, **Dieckmann, R.** — 365

Dynamics of the Eider channel. An analysis of the precision of profiles, **Gönnert, G.** — 377

The Nösse-Peninsula of the island Sylt/ North Frisia: hydrology and sedimentology of storm surges; **Stieve, B. & J. Ehlers** — 387

Dynamic preservation of the coastline of the Netherlands. How does it work, **Hillen, R.** — 397

Measurement of tidal water transport with an acoustic doppler current profiler (ADCP) and comparision with calculations, **Kolb, M. & Lobmeyr M.** — 410

In situ erosion flume (erosf): Determination of critical bedshear stress and erosion of a kaolinite bed and natural cohesive sediments, **Houwing, E.-J. & L.C. van Rijn** — 420

Determination of benthic Wadden Sea habitats by hydrodynamics, **Damm-Böcker, S., Kaiser, R. & Niemeyer, H.D.** — 430

Towards sustainable development of coastal zones, **Hillen, R., Bijlsman, L. & Misdorp, R.** 442

Mechanisms for recurrent nuisance algal blooms in the coastal zones: Resting cyst formation as lifestrategy of Dinoflagellates, **Nehring, S.** 454

STATISTICS, MODELLING and GIS

Slope of the equilibrium range in the frequency spectra of wind-generated waves in Greek seas, **Moutzouris, C.I.**
- **no manuscript received** - --

Interdisciplinary investigation of aquatic communities using size spectra, **Kamenir, Y.G.** 468

Description of changes in a pocket beach using empirical eigenfunctions, **Kypraios, N.G. & I.C. Moutzouris** 477

Beach dynamics of barred nearshores: Gold Coast (South Pacific) and island Sylt (North Sea), **Boczar-Karakiewicz, B. et al.** 487

Comparison of measured and calculated nearshore bottom profiles, **Szmytkiewicz, M.** 495

GIS for shoreline management, **Townend, I.H. & D. Legget** 505

Geographical information system for the inventory of coastal characteristics in Finland, **Granö, O. & M. Roto** 519

A GIS-supported sensitivity analysis. Implementation of results from ecosystem research, **Stock, M. et al.** 528

Thematic mapping and sensitivity study of mud flat areas in the German Wadden Sea, **Van Bernem, K.-H. et al.** 542

GIS-application on Wadden Sea areas, **Liebig, W.** 550

Shoreline change simulation and equilibrium shore-arc analysis of embayed muddy coast, **Qui Jianli** 559

BALTIC

Coastal and near-shore erosion at Vejrö, Denmark: combined effects of a changing wind-climate and near-shore dredging, **Christiansen, C., Christoffersen, H. & Binderup, M.** 566

Short and long term variations in vertical flux and sediment accumulation rates in a semi-enclosed bay at the frontal zone between the North Sea and the Baltic Sea, **Lund-Hansen, L.C. et al.**	576
Measurement of morphological changes by means of a sand-surface-meter, **Straube, J.**	586
Erosion-accretion system of south Baltic coast during the last 100 years, **Zawadzka-Kahlau, E.**	595
Near-shore sedimentation and pollution influence on the ecosystem of the Baltic sea in Lithuanian boundaries, **Pustelnikovas, O.**	608
Classification and characterization of contemporary coasts of Estonia, **Orviku, K.**	615
The present state of Estonian coasts and evolution tendencies, **Orviku, K.**	621
Environment and dynamics of the Kaliningrad coast, the Baltic Sea, **Boldyrev, V.**	625
Coastal management in Poland, **Cieslak, A.**	628
Experience from shore protection of the Hel peninsula, **Basinski, T.**	639
Environmental aspects of mining clastic material from the sea bottom, **Uscinowicz, S.**	648
Coastal anchoring, **Larsen, O.F.** - **no manuscript received** -	--

SEDIMENTOLOGY

Sediment variability - an important element in the evolution of tidal flats, **Eitner, V. & G. Ragutzky**	651
Beach cusp granulometry: a study of beach cusp sediment grain-size statistics, **Kristensen, P.J., Ghionis, G. & C. Christiansen**	661
Sedimentological and geochemical characteristics of the carbonatic beaches of the Gulf of Orosei (east-central Sardinia), **Cristini, A., Di Grigorio, F. & C. Ferrara**	671

SESSIONS ON ECOLOGY AND CONSERVATION IN THE BALTIC REGION (WWF & EUCC)

Describing the coastline of Europe, **Doody, J.P., Davidson, N. & F. van der Meulen**   683

Coastal woodlands, forestry and nature conservation, **Tekke, R.M.H. & A.H.P.M. Salman**   694

Distribution of epigaeic anthropods in dune-habitats and salt-meadows along the Baltic coast of Schleswig-Holstein, **Hoerschelmann, C.**   703

Biochemistry of semi-closed bays of the Baltic Sea and human factor of sedimentation, **Pustelnikovas, O.**   709

WORKSHOP ON CLIMATE-RELATED IMPACTS AND PROBLEMS IN THE COASTAL ZONE

The "sea level rise" problem: An assessment of methods and data, **Plag, H.-P.**   714

Vulnerability of the coasts of Germany due to the impacts of climate change: analyses and research demands, **Sterr, H.**   733

Detailed Holocene sea level curve, Northern Denmark, **Tanner, W.F.**   748

Modelling Holocene sea levels in the Irish and Celtic Seas, **Wingfield, R.**   760

Contents in alphabethic order

	page
Alves, A.M.C.	
PRESENT DYNAMICS IN THE ALTO MINHO COAST (NW PORTUGAL)	201
Anthony, E.	
PRELIMINARY INVESTIGATIONS OF GRAVEL BARRIER EVOLUTION IN THE BAIE DES ANGES, FRENCH RIVIERA.	48
Basinski, T.	
EXPERIENCE FROM SHORE PROTECTION OF THE HEL PENINSULA	639
Bilgin, R.	
EROSION ALONG THE BLACK SEA COAST OF TURKEY AND PROPOSED PROTECTIVE MEASURES	74
Black, K.S.	
THE TURBULENT RESUSPENSION OF COHESIVE INTERTIDAL MUDS: SOME NEW CONCEPTS AND IDEAS	223
Boczar-Karakiewicz, B.; Jackson, L.A.; Kohlhase, S. & A. Naguszewski	
BEACH DYNAMICS OF BARRED NEARSHORES: GOLD COAST, SOUTH PACIFIC AND ISLAND SYLT, NORTH SEA	487
Boldyrev, V.	
ENVIRONMENT AND DYNAMICS OF THE KALININGRAD COAST THE BALTIC SEA	625
Brambati, A.; Amore, C.; Giuffrida, E. & G. Randazzo	
RELATIONCHIP BETWEEN THE PORT STRUCTURES AND COASTAL DYNAMICS IN THE GULF OF GELA (SICILY - ITALY)	773
Castelli, P.; Di Gregorio, F. & C. Ferrara	
PROPOSALS AND PROGRAM FOR THE ORGANIZATION OF THE NATURALPARK ON THE COAST OF ARBUS (SARDINIA)	136

Christiansen, C.; Christoffersen, H. & M. Binderup

    COASTAL AND NEAR-SHORE EROSION AT VEJRÖ, DENMARK: COMBINED EFFECTS OF A CHANGING WIND-CLIMATE AND NEAR-SHORE DREDGING   566

Cieslak, A.

    COASTAL MANAGEMENT IN POLAND   628

Cristini, A.; Di Gregorio, F. & C. Ferrara

    SEDIMENTOLOGICAL AND GEOCHEMICAL CHARACTERISTICS OF THE CARBONATIC BEACHES OF THE GULF OF OROSEI (EAST-CENTRAL SARDINIA)   671

Damm-Böcker, S., Kaiser, R. & Niemeyer, H.D.

    DETERMINATION OF BENTHIC WADDEN SEA HABITATS BY HYDRODYNAMICS   430

Dieckmann, R.

    MORPHOLOGICAL STRUCTURES IN GERMAN TIDAL FLAT AREAS   365

Doody, J. P., Davidson, N.C. & F. van der Meulen

    DESCRIBING THE COASTLINE OF EUROPE   683

Eitner, V. & G. Ragutzki

    SEDIMENT VARIABILITY - AN IMPORTANT ELEMENT IN THE EVOLUTION OF TIDAL FLATS   651

El-Fishawi, N. & A. Badr

    DRIFT RATES ON ACCRETED AND ERODED BEACHES BY FLUORESCENT SAND, NILE DELTA COAST   22

Ferreira, O. & J. Alveirinho Dias

    DUNE EROSION AND SHORELINE RETREAT BETWEEN AVEIRO AND CAPE MONDEGO (PORTUGAL). PREDICTION OF FUTURE EVOLUTION   187

Filipić, P.

> COAST-HINTERLAND SOCIO-ECONOMIC RELATIONS AS AN ESSENTIAL ELEMENT OF INTEGRATED PLANNING OF COASTAL ZONES (THE MEDITERRANEAN CASE)
>
> APPENDIX: CASE STUDY: COAST-HINTERLAND SOCIO-ECONOMIC INTERRELATIONS IN A SELECTED AREA OF CROATIA (SUMMARY)   107

Frances, E.; Rivas, V. & A. Cendrero

> AN APPROACH FOR THE REHABILITATION OF COASTAL WETLANDS AND ESTUARINE AREAS IN NORTHERN SPAIN   264

Freitas, M.; C. Andrade & F. Jones

> RECENT EVOLUTION OF ÓBIDOS AND ALBUFEIRA COASTAL LAGOONS   167

Frihy, O.E.; Nasr, S.M.; Dewidar, Kh. & M. El. Raey

> SPATIAL AND TEMPORAL CHANGES OF ALEXANDRIA BEACHES, EGYPT   13

Frihy, O.E. & R. Dean

> ARTIFICIAL BEACH NOURISHMENT ON THE MEDITERRANEAN NILE-DELTA COAST   84

Gerdes, D. & K.-J. Hesse

> CHEMICAL AND BIOLOGICAL CHARACTERISTICS OF DIFFERENT WATER MASSES IN THE GERMAN BIGHT   305

Gönnert, G.

> DYNAMICS OF THE EIDER CHANNEL. AN ANALYSIS OF THE PRECISION OF PROFILES   377

Gokce, K., Dakoglu, A., Koc, S. & M. Gurhan

> COASTAL EROSION IN THE EASTERN BLACK SEA COAST OF TURKEY - A CRITICAL REVIEW   68

Granö, O. & M. Roto

> GEOGRAPHICAL INFORMATION SYSTEM FOR THE INVENTORY OF COASTAL CHARACTERISTICS IN FINLAND   519

Heikkinen, O.

    ENVIRONMENTAL DEVELOPMENT AMONG THE TENMILE CREEK
    COASTAL DUNES IN OREGON, USA   283

Hillen, R.

    DYNAMIC PRESERVATION OF THE COASTLINE IN THE NETHERLANDS.
    HOW DOES IT WORK OUT?   397

Hillen, R., Bijlsman, L. & R. Misdorp

    TOWARDS SUSTAINABLE DEVELOPMENT OF COASTAL ZONES   442

Hoerschelmann, C.

    DISTRIBUTION OF EPIGAEIC ARTHROPODS IN DUNE-HABITATS AND
    SALT-MEADOWS ALONG THE BALTIC COAST OF SCHLESWIG-
    HOLSTEIN   703

Hoffmann, G.

    HOLOCENE COASTLINE SHIFTS AT THE MEDITERRANEAN COAST OF
    ANDALUCIA (SE SPAIN)   1

Houwing, E.J. & L.C. van Rijn

    IN SITU EROSION FLUME (EROSF): DETERMINATION OF CRITICAL
    BEDSHEAR STRESS AND EROSION OF A KAOLINITE BED AND NATURAL
    COHESIVE SEDIMENTS   420

Jiménez, J.A.; Sánchez-Arcilla, A. & M.A. Garcia

    BARRIER BREACHING IN MICROTIDAL ENVIRONMENTS: THE
    TRABUCADOR BAR CASE (EBRO DELTA)   58

Jiménez, J.A.; Garcia, V.; Valdemoro, H.; Garcia, M.A. & A. Sánchez-Arcilla

    ESTIMATION OF SEDIMENT BUDGET AND LONGSHORE TRANSPORT
    RATES FROM COASTAL RESPONSE. APPLICATION TO THE EBRO
    DELTA   38

Kamenir, Y.G.

    INTERDISCIPLINARY INVESTIGATION OF AQUATIC COMMUNITIES
    USING SIZE SPECTRA   468

Kocman, A.

    CLIMATICAL CONDITIONS AND TOURISM THROUGH THE TURKISH
AEGEAN COASTAL ZONE   126

Kolb, M. & M. Lobmeyr

    MEASUREMENT OF TIDAL WATER TRANSPORT WITH AN ACOUSTIC
DOPPLER CURRENT PROFILER AND COMPARISION WITH
CALCULATIONS   410

Kristensen, P.J.; Ghionis, G. & C. Christiansen

    BEACH CUSP GRANULOMETRY: A STUDY OF BEACH CUSP SEDIMENT
GRAIN-SIZE STATISTICS   661

Kypraios, N.G. & C.I. Moutzouris

    DESCRIPTION OF CHANGES IN A POCKET BEACH USING EMPIRICAL
EIGENFUNCTIONS   477

Liebig, W.

    GIS - APPLICATION ON WADDEN SEA AREAS   550

Lund-Hansen, L.C.; Floderus, S.; Pejrup, M.; Valeur, J. & A. Jensen

    SHORT AND LONG TERM VARIATIONS IN VERTICAL FLUX AND
SEDIMENT ACCUMULATION RATES IN A SEMI-ENCLOSED BAY AT THE
FRONTAL ZONE BETWEEN THE NORTH SEA AND THE BALTIC SEA   576

Mastronuzzi, G.; Plamentola G. & P. Sansò

    MORPHOLOGICAL TYPES OF ROCKY COAST ON SOUTHEASTERN
APULIA   784

Nehring, S.

    MECHANISMS FOR RECURRENT NUISANCE ALGAL BLOOMS IN THE
COASTAL ZONES: RESTING CYST FORMATION AS LIFE-STRATEGY OF
DINOFLAGELLATES   454

Noorbergen, H.H.S.

    MORPHODYNAMICS OF INTER-TIDAL AREAS - THE WADDEN SEA   342

Orviku, K.

    CLASSIFICATION AND CHARACTERIZATION OF CONTEMPORARY COASTS OF ESTONIA   615

Orviku, K.

    THE PRESENT STATE OF ESTONIAN COASTS AND EVOLUTION TENDENCIES   621

Parks, J.

    FLUIDIZATION AND BEACHFACE DEWATERING:RECENT PROGRESS AND CASE HISTORIES   296

Plag, H.-P.

    THE "SEA LEVEL RISE" PROBLEM: AN ASSESSMENT OF METHODS AND DATA   714

Pustelnikovas, O.

    NEAR-SHORE SEDIMENTATION AND POLLUTION INFLUENCE ON THE ECOSYSTEM OF THE BALTIC SEA IN LITHUANIAN BOUNDARIES   608

Pustelnikovas, O.

    BIOGEOCHEMISTRY OF SEMI-CLOSED BAYS OF THE BALTIC SEA AND HUMAN FACTOR OF SEDIMENTATION   709

Qiu Jianli

    SHORELINE CHANGE SIMULATION AND EQULIBRIUM SHORE-ARC ANALYSIS OF EMBAYED MUDDY COAST   559

Rivas, V. & A. Cendrero

    DETERMINATION OF THE EVOLUTIONARY CONDITION OF COASTAL CLIFFS ON THE BASIS OF GEOLOGICAL AND GEOMORPHOLOGICAL PARAMETERS   214

Rocha, F. & C. Gomes

    MORPHO-CLIMATIC EVOLUTION OF THE AVEIRO REGION LITTORAL (NW OF PORTUGAL) DURING TERTIARY AND QUATERNARY   157

Simmons, S.L. & A.T. Williams

    PERSISTANT MARINE DEBRIS ALONG THE GLAMORGAN HERITAGE COAST, UK: A MANAGEMENT PROBLEM   240

Šimunović, I.

    INTEGRATED PLANNING AS A NEW PARADIGM OF A COASTAL ZONE MANAGEMENT   96

Sterr, H.

    VULNERABILITY OF THE COASTS OF GERMANY DUE TO SEA LEVEL RISE AND CLIMATE CHAMGE: ANALYSES AND RESEARCH DEMANDS   733

Stieve, B. & J.Ehlers

    THE NÖSSE-PENINSULA OF THE ISLAND SYLT / NORTH FRISIA: HYDROLOGY AND SEDIMENTOLOGY OF STORM SURGES   387

Stock, M.; Boedeker,D.; Schauser, U.-H. & R. Schulz

    A GIS-SUPPORTED SENSITIVITY ANALYSIS. IMPLEMENTATION OF RESULTS FROM ECOSYSTEM RESEARCH   528

Straube, J.

    MEASUREMENTS OF MORPHOLOGICAL CHANGES BY MEANS OF A SAND-SURFACE-METER   586

Szmytkiewicz, M.

    COMPARISON OF MEASURED AND CALCULATED NEARSHORE BOTTOM PROFILES   495

Tanner, W.F.

    DETAILED HOLOCENE SEA LEVEL CURVE, NORTHERN DENMARK   748

Teixeira, A.T. & T. Gamito

    PLANNING AND DESIGN FOR THE LEISURE CRAFT WITHIN A TRADITIONAL FISHING HARBOUR   274

Tekke, R.M.H. & Salman, A.H.P.M.

    COASTAL WOODLANDS, FORESTRY AND NATURE CONSERVATION   694

Townend, I.H. & D. Leggett

    GIS FOR SHORELINE MANAGEMENT   505

Uścinowicz, S.

    ENVIRONMENTAL ASPECTS OF MINING CLASTIC MATERIAL FROM THE SEA BOTTOM   648

Van Bernem, K.H.; Müller, A.; Grotjahn, M.; Knüpling, J.; Krasemann, H.L.; Neugebohrn, L.; Patzig, S.; Ramm, G.; Riethmüller, R.; Sach, G. & S. Suchrow

    THEMATIC MAPPING AND SENSITIVITY STUDY OF MUD FLAT AREAS IN THE GERMAN WADDEN SEA   542

Van Overeem, J.; Steijn, R.C. & G.K.F.M. Van Banning

    SIMULATION OF MORPHODYNAMICS OF TIDAL INLETS IN THE WADDEN SEA   351

Węslawski, J.M.; Zajączkowski M. & Suryn T.

    INTERTIDAL ZONE OF SPITSBERGEN AND FRANZ JOSEF LAND   322

Williams, A.T.; Leatherman, S.P. & S.L. Simmons

    BEACH AESTHETIC VALUES: THE SOUTH WEST PENINSULA, UK   251

Wind, H.G. & E.B. Peerbolte

    COASTAL ZONE MANAGEMENT; TOOLS FOR INITIAL DESIGN   332

Wingfield, R.

    MODELLING HOLOCENE SEA LEVELS IN THE IRISH AND CELTIC SEAS   760

Zawadzka-Kahlau, E.

    EROSION ACCRETION SYSTEM OF SOUTH BALTIC COAST DURING THE LAST 100 YEARS   595

Holocene Coastline Shifts at the Mediterranean Coast of Andalucia (SE Spain)

Dr. Gerd Hoffmann, Institute for Baltic Sea Research (IOW), Seestr. 15, O-2530 Warnemünde

Abstract

The stratigraphy of Holocene sediments in the valleys of the Mediterranean coast of Andalucia was investigated in a joint program with the German Archaeological Institute Madrid,

The results of the investigations showed a marine transgression into the valleys of this coast in the early Holocene. The marine sediments of the valleys prove a relatively constant sedimentation with a rate of 1 to 2 m per millenium until the end of the 15th century A.D.. After the period of Reconquista, when the catholic Kings of Spain expelled the Arabs from the southern part of the Iberian Peninsula, an enormous increase in the accumulation rate was caused by the deforestation of the hinterland. These sediments of the 16th and 17th centuries mostly have a thickness of up to 20 meters. During this time the valleys were filled up with the eroded soils of the hinterland.

The dominant geomorphological factor for this evolution of landscape was man, supported by a wetter and cooler climate and an increase of torrential rainfalls during the Little Ice Age (1500 - 1750 A.D.).

During this time the rivers of Rio Andarax, Rio Adra, Rio Guadalfeo and Rio Velez also developed deltas and coastal plains, where nowerdays erosional processes can be observed. They are originated by the coastal, W-E directed "Gibraltar-current". Other reasons for this erosion also are coastal constructions like the harbour of Motril and the missing sediment supply from the hinterland due to the canalisation of the river beds and due to the construction of artificial lakes and dams.

Introduction

In an interdisziplinary project with the German Archaeological Institute Madrid the stratigraphy of Holocene sediments in the valleys of the Mediterranean coast of Andalucia (Fig. 1) was investigated.

The aim of this cooperation between Geology and Archaeology was the investigation of human settlements and the reconstruction of the ecological environment of this area, especially in historical times. A particular interest of this project is the evolution of coastlines during the Holocene period (Arteaga, Hoffmann, Schubart & Schulz, 1988).

Study area

The Mediterranean coast of Andalucia is situated south of the Betic Cordillere, the westernmost alpidic mountains of Europe. The highest elevation of the Iberian Peninsula, the Mulhacen with an altitude of 3481 m height is only 35 km away from this coast. Hence there are steep gradients particularly in the central part of the investigated area.

Climatically there are arid until extremely arid conditions close to the coast and humid conditions in the hinterland. The torrential rainfalls are of particular interest for erosional processes. In the easternmost part of the area with normally 220 mm/year rainfall have been measured 300 mm in 36 hours in October 1973.

The area of investigation is situated between the Rio Almanzora (Province of Almeria) in the east and the Rio Guadarranque (Province of Cadiz) in the west (Figs. 1 and 2). The results presented here are of the valleys of Rio Andarax and of Rio Guadalfeo (Figs. 2, 5 and 6).

Investigation of Holocene sediments

For the investigation of the sediments more than 450 drillings with a hand auger of Eijkelkamp Company (Netherlands, Fig. 9) were carried out. With the mostly used auger bits of 10 cm diameter a maximal depth of 19,50 m was reached (Fig. 7).

The grain size of the Holocene sediments in the investigation area was mostly a sandy silt or a clayish silt. Only a small quantity of the drilled sediment represented fine until coarse sand, gravel or stones.

The most striking macroscopic feature of the sediments is that at a certain depth close to the recent water level the colour of the sediments changes from brown to bluish grey. The colour shows the oxidation state of the iron: the brown colour testifies an oxidated, terrestrial environment and the bluish grey colour an aquatic environment.

Some of the drilling profiles, drawn in relation to the recent sea level, are shown in Fig. 6.. Most of the bluish grey sediments have been investigated micropalaeontologically. Many marine and brackish microfossils have been found. These sediments are proof of marine or brackish bays of Holocene age, which had been filled up and covered by brownish flood plain deposits of the different rivers (Figs. 5-8; Hoffmann, 1988). Also the still submarine parts of the pleistocene valleys are filled up with Holocene sediments. Beside of Rio Antas only the submarine valley of Rio Guadalfeo can still be followed by the isobaths (Fig. 4).

Dating methods

Two methods of dating the Holocene sediments were used:
- the Radiocarbon method and
- the dating with ceramic fragments

The importance of dating the Holocene sediments by artefacts must be emphasized, especially in connection with the reconstruction of the ancient landscape. Dating with this method is usually not only more exact than Radiocarbon dating, but the lack of organic material in the sediments of the Mediterranean valleys in this area underlines its importance. For saving the ceramic fragments it was necessary to use the auger bits with the diameter of 10 cm.

For better understanding of the landscape evolution also results of archaeological excavations and historical maps (Figs. 2 and 5) have been used. Comparing Figs. 5 and 6 it is striking, that in 1759 Rio Andarax had not yet built a delta.

Summary

Brackish until marine bays of the Mediterranean coast of Andalucia, flooded about 6000 y.b.P. (Fig. 3), still existed in the late Middle Ages, The origin of coastline change are erosional processes in the hinterland and accumulation of the ancient soils in the coastal zone.

This enormous erosion in the late Middle Ages has the following origins (Hoffmann, 1988):
- deforestation of the hinterland during the expulsation of the Arabs by the Catholic Kings of Spain
- installation of a non-adapted agriculture particularly in the terraces of the Arabs
- increasing shipbuilding- (discovery of America) and mining -activities
- foundation of the "mesta"-organisation of goat breeders
- little glaciation with a colder and wetter climate and a higher amount of torrential rainfall (Lopez Vera, 1986)
- the influence of anthropogenic activities in the mediterranean subtropics is increased by its border situation between the evergreen forests and the steppe (ROTHER 1984).

To separate anthropogeneous and natural factors the peat profiles, found during this project, should be investigated. Up to now there are only very sparse climatic informations of this coastal region, mostly deriving from botanical investigations of archeological excavations.

Recently erosional processes are noted on all the deltas and coastal plains of the mediterranean coast of Andalucia:
- Rio Andarax and Campo de Dalias (Almeria)
- Rio Grande de Adra and Campo de Dalias (Almeria)
- Rio Guadalfeo and Llanos de Carchuna
- Rio Velez and coastal plain of Torre del Mar (Malaga)

The origins of the coastal retreat are the following (Drescher & Hoffmann, 1991, Drescher, 1988)):
- nearly all the soils of the hinterland have been eroded at the end of Middle Ages, hence a change in the water-household of the rivers took place. Today the soil erosion is accelerated by modern terracing activities with big machines without any protection of the slopes.
- nearly all bigger rivers are canalized and in their upper course have been built artificial lakes, dams and irrigation canals to avoid catastrophes during torrential rainfalls. Hence the suspended material of the rivers, eroded specially during these torrential rainfalls, sedimentate in the arteficial lakes upstream or in the Mediterranean, but not on the flood plains. Here it has to be pointed out, that the support of silty and clayish flood plain deposits are also the origin of the fertility of the coastal plains.
- the West-East directed coastal current, so called "Gibraltar -current", is the active erosional factor, because Erosion on the above mentioned deltas and coastal plains is noted only on their western side. In former times the erosion by this coastal current was balanced by the accumulation of the old soils of the hinterland.
- and by man's constructions at the coastline like the harbour of Motril (Fig. 4).

References

Arteaga, O., Hoffmann, G., Schubart, H. & Schulz, H.D. (1988): Geologisch-Archäologische Forschungen zum Verlauf der andalusischen Mittelmeerküste.- Madrider Beiträge 14, Mainz.

Drescher, A. (1988): Untersuchungen eines Agrarökosystems in den Winterregensubtropen Spaniens: Naturpotential und Auswirkungen innovativer Entwicklungen in der Landwirtschaft.- Dissertation University of Freiburg/Breisgau.

Drescher, A. & Hoffmann, G. (1991): Soil erosion in the Mediterranean - the impact of agriculture and consequences on the coastal landuse planning; in: 1992 Europe and the Mediterranean countries. Center for Mediterranean studies, University of Ankara, 1991.

Hoffmann, G. (1984): Holozänstratigraphie und Küstenlinienverlagerung an der andalusischen Mittelmeerküste; Berichte aus dem Fachbereich Geowisenschaften der Universität Bremen Nr. 2, 173 S, Bremen 1988.

Hopley, D. (1978): Sea-level change on the Great Barrier reef: an introduction, London 1978.

Lopez Vera, F. (1986): Quaternary Climate in Western Mediterranean. Proceedings of the symposium on climatic fluctuations during the Quaternary in the western mediterranean region - Madrid 1986.

Acknowledgements
I like to thank all colleagues of the Geological Institutes in Kiel and Bremen, of the German Archeological Institute Madrid and of the different Institutes in Sevilla, Granada, Malaga, Almeria and Madrid for their help.
The investigation was supported by the Volkswagen Foundation.

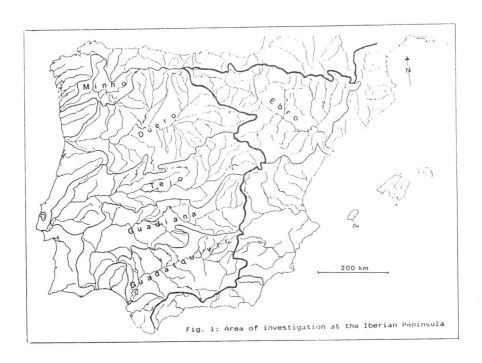

Fig. 1: Area of investigation at the Iberian Peninsula

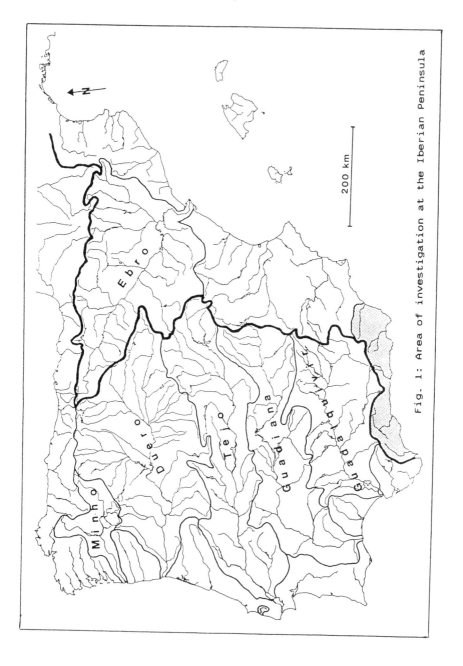

Fig. 1: Area of investigation at the Iberian Peninsula

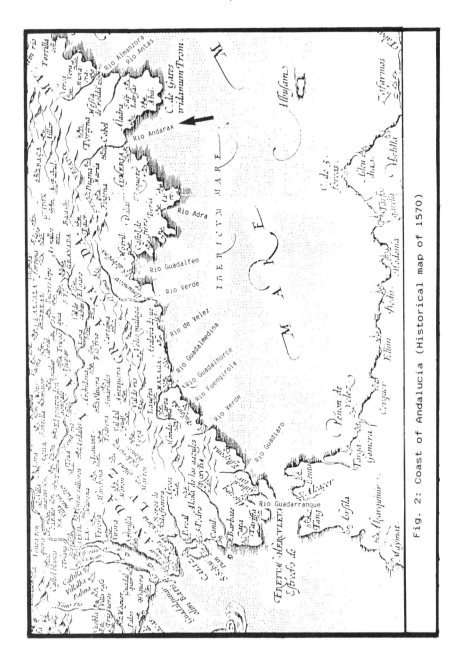

Fig. 2: Coast of Andalucia (Historical map of 1570)

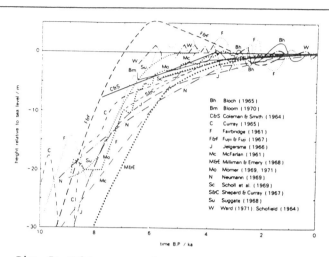

Fig. 3: Holocene sea-level curve (Hopley 1978)

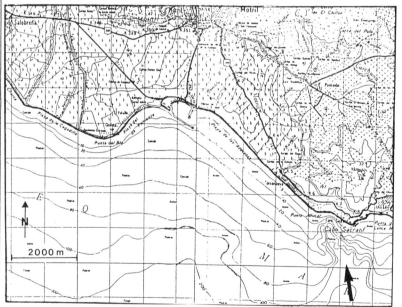

Fig. 4: Valley of Rio Guadalfeo (Fig. 2). The arrow marks the submarine part of the valley.

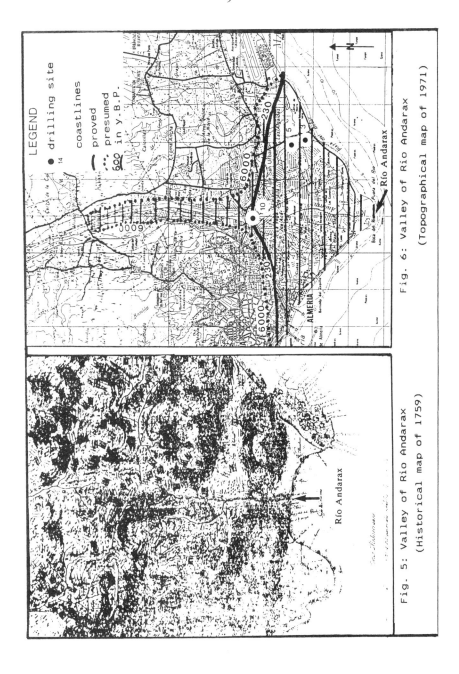

Fig. 5: Valley of Rio Andarax
(Historical map of 1759)

Fig. 6: Valley of Rio Andarax
(Topographical map of 1971)

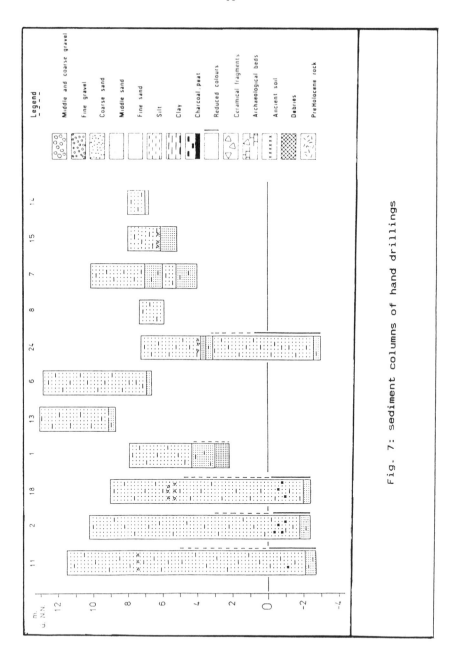

Fig. 7: sediment columns of hand drillings

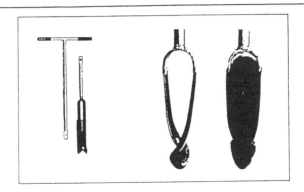

Fig. 9: Hand auger (Eijkelkamp company)

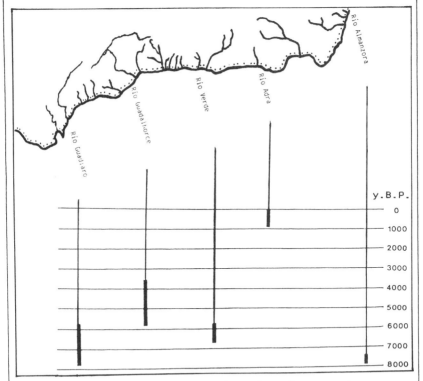

Fig. 10: Peat profiles (age in y.B.P.)

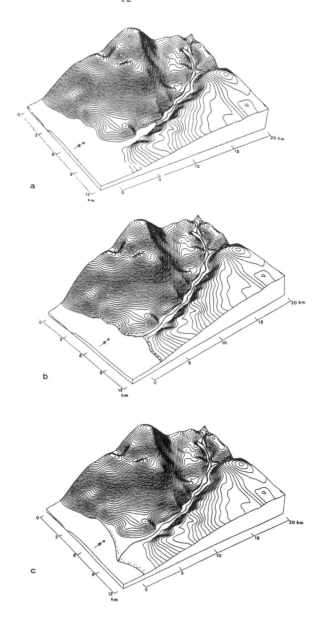

Fig. 8: Landscape evolution of Rio Andarax valley
(a: 6000 y.B.P., b: 230 y.B.P, c: recent situation)

SPATIAL AND TEMPORAL CHANGES OF ALEXANDRIA BEACHES, EGYPT

Omran E. Frihy*, Samir M. Nasr**, Khalid Dewidar* and Mohamed El Raey**

* Coastal Research Intitute, Alexandria, Egypt
** Institute of Graduate Studies and Research, Alexandria, Egypt

ABSTRACT

Long-term variations in beach width of Alexandria Governerate have been analyzed using two sets of aerial photos taken in 1955 and 1983. The analysis reveals that a major part of most beaches has been subjected to coastal erosion except for only beaches. The estimated long-term rate of erosion is approxemitly 0.20 m/yr.

El Maamoura beach, located east of Alexandria, is selected to evaluate seasonal variations. Sand volume losses are found to be 450 m^3/m/yr for the entire beach length (1.2 km). The annual sand transport by wind is estimated as about 37.7 m^3. Subtracting this amount of sand from the net sand loss (450 m^3/m/yr) yields 412.3 m^3/m/yr of eroded sand caused by the action of currents and waves.

INTRODUCTION

The rate of shoreline change resulted from seasonal conditions is one of the most common measurements used by coastal scientists, engineers, and land planners to indicate the dynamics and hazards of the coast. Storms usually erode sediments from the beach, leaving the beach with concave upward profile, and deposit them in the nearshore to form a bar. The swells drive the bar system landward to the beach. With additional sediments added to the beach, a berm builds, and the shape of the beach becomes convex upward.

The phases of beach erosion and beach reconstruction are often cyclic. Studies conducted in various parts of the world show that the beach cycle is often associated with seasons; winter storms erode beaches while summer swells rebuild them (Shepard, 1950; Darling, 1964; Harrison and Wagner, 1964; Dolan, 1965; Psuty, 1965; Owens, 1977; Fox and Davis, 1978; Aubrey, 1979; Short, 1980; Eliot and Clarke, 1982; and Clarke and Eliot, 1983). On the other hand, some beaches show no significant change in their form on a yearly basis simply, because the wave climate regime changes very little during a year (Pilky and Richter, 1964). Other beaches show little changes because they are composed of coarse-grain sediments that absorb wave energy without changing beach form (Dingler, 1981).

El Maamoura beach is selected to test seasonal variations. The study is based on an analysis of repeated beach profiles over one year period. The profile analysis includes average shoreline position, beach volume changes and beach sediment characteristics.

The purpose of this paper is to carry out a quantitative analysis of seasonal coastal changes and long-term changes along the waterfront of Alexandria.

Materials and Methods

Alexandria is located on the western limit of the Nile delta coast. It is one of the most important summer resorts along the Mediterranean coast of Egypt. El Maamoura beach is located on the eastern part of Alexandria waterfront. It lies on a large embayment typical of Alexandria beaches and is considered as the largest touristic and most famous recreational beach of Alexandria (Fig.1).

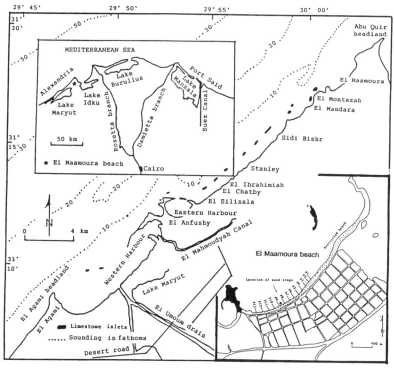

Fig.1. Alexandria waterfront and El Maamoura beach, (case study).

Information on long-term changes along Alexandria beach are obtained from analysis of two vertical aerial photos taken in 1955 and 1983 scaling 1:20000 and 1:25000, respectively. The photo mosaics are overlapped and the entire two shorelines are matched to the same scale and subsequently traced. Comparison between shorelines during a time interval of 28 years is carried out by overlaying these mosaics based on fixed points such as permanent structures, roads and canals. Beach widths are measured for each year for all Alexandria beaches at 82 transects. The difference between measurments from the two successive aerial photographs at each transect is used to calculate the amount of erosion or deposition at each beach.

Short-term changes are studied based on beach profile surveys and are taken at 13 profile lines along El Maamoura beach covering a total of 1.2 km length (Fig.1). The profiles are monitored over the time interval from November 1989 to November 1990 using standard-surveying techniques. Profile lines are taken perpendicular to coastline, spaced 60-100 m apart and extended seaward up to -2.0 m depth. The leveling and sounding data are adjusted to the MSL datum using a beach mark of 2.77 m elevation, located behind the beach area. Distances and corresponding elevations and soundings along profiles are measured at 3.0 m interval, to an approximate depth of 2.0 m below MSL. Beach samples are also collected at the end of each profile line. Grain size parameters are determined using the technique of Folk and Ward (1957).

Sand transport by wind is determined by establishing five sand traps placed on the backshore. A simple box-type sand trap is chosen. It is made of wood except for the open side (inlet) and the back side of the trap. The inlet is kept open while a permeable cap made of silk screening covers the back side of the trap. The collector base is 10 x 50 cm in dimension.

DISCUSSION

Seasonal Variations

The seasonal changes in beach sand of El Maamoura beach are determined by analyzing thirteen beach profiles. The beach profiles are categorized as non winter and winter profiles. The winter profiles are those from November 1989 to March 1990 and the non winter profiles are those from April 1990 to September 1990. The winter beach profiles are characterized by narrow beach width with steep slope and sligthly sand bars at 1 to 2 m below MSL. Non winter beach profiles have wide beach widths with gentle slope and sand bars at 1 to 2 m below MSL.

Three parameters are used to detict and evaluate coastal changes: 1) variations in the shoreline positions (ΔS) (beach width), 2) changes in sand volume change (ΔV), and 3) variations in sediment size of beach sand. Unit volume is defined as the cross-sectional area under the profile multiplied by a unit length of beach in the alongshore direction. The beach width of El Maamoura beach was establised from beach profiles by measuring the horizontal distance between baseline and shoreline at MSL.

Volume Changes in Beach Sand (ΔV)

The cumulative volumetric changes for the surveys in sand volume above MSL along El Maamoura beach is shown in Fig.2a. Most of the winter curves lie on the lower part (erosion side), while the non winter curves lie on the upper part of the figure (accretion side). This pattern reflects the effectiveness of winter and non winter conditions on sand volumes. Erosion prevailed during winter and accretion during non winter conditions. This Figure shows that sediment transport, as evidenced by volumetric changes is from west to east. Severe erosion occurred between profiles P12 and P7, followed by deposition up to profile P1. The net change in sand volume for El Maamoura beach is calculated based on the total summation of erosion (negative values) and deposition (positive values). Therefore, the calculated net result of volume changes is found to be -450 m^3/m for the entire beach length (1.2 km) during the period of study i.e -0.4 m^3/m.

Beach Width Changes (ΔS)

The change in beach width reflects the variation of shoreline positions along the beach. Figure 2b. represents the beach width variability during the period of study. The western side of the beach at profile P13 shows frequent changes in beach width. In winter beach width it attains less than 30 m, while it reaches 40 m in non winter condition. The annual average shoreline fluctuation along the beach is about 15 meters. The net sediment transport is also evidenced from the shoreline changes to be in the eastward direction, where maximum erosion occurs along the western side and maximum deposition at the eastern side. Annual fluctuation in shoreline position is attributable to seasonal changes in coastal weather conditions, particularly in periods of storms characterized by changes in the nearshore wave regime.

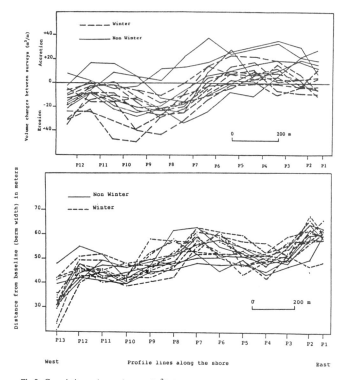

Fig.2. Cumulative volume changes (m^3/m) up to 2 m depth (A) and changes in the beach width of El Maamoura beach (B).

Sediment Transport Rate (Q)

In order to calculate the rate of longshore sediment transport at El Maamoura beach, the net sand volume is calculated for the entire area for each survey time. These data are used to calculate the annual sediment transport rate (Q) employing least squares techniques (Fig.3). The slope of the line is an estimate of the (Q) yielding 0.25×10^6 m^3/yr.

Textural Changes of Beach Samples

The beach sand is composed of biogenic quartz grains with very minor amounts of heavy minerals and shell fragments. The mean grain size and grain sorting for the beach samples collected during winter and non winter conditions are distributed alongshore following the positions of the profile lines. The variation of sediment texture along the shoreline has been attributed to processes of selective grain transport during the longshore movement of sand from eroding to accreting areas (Firhy and Komar, 1990).

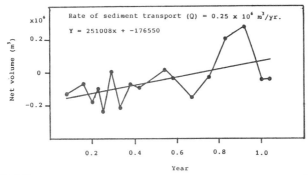

Fig.3. Time series of changes in the net sand volume versus time. The plotted regression line is the least-square fit of the net sand volume a over time period of 13 months.

In order to test the monthly variation in sediment texture of beach sand, the mean grain size and grain sorting are plotted versus time in months. The mean grain size fluctuated between 1.3 ϕ to 1.7ϕ around the average mean grain size for winter months (Fig.4a). This trend progressively decreases through winter and reaches minimum (0.9ϕ) in May and then increases up to September. This trend is similar to the sorting pattern (Fig.4b) where it fluctuates between 0.7ϕ and 0.8ϕ and rapidly decreases during non winter months particularly in May. This means that winter months have been associated with finer and better sorted sand than non winter months.

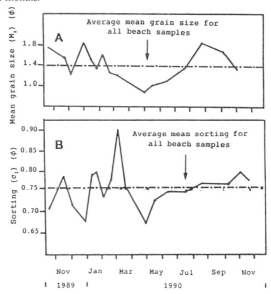

Fig.4. Temporal variation in mean grain size (A) and grain sorting (B) at El Maamoura beach.

Sand Losses by Wind

Eolian transport of sand is one of the processes that control the beach dynamics at shoreline. Eolian transport of sand depends on many factors, such as wind speed and direction, granulometric composition of material, air humidity, water contents on beach face, beach slope, distribution of wind over a dry surface, insolation, cohesion of the surface layer and others (Niespodzinska, 1980). The sand movement by wind action has been treated by several investigators, among others Bagnold, 1954; Horikawa and Shen, 1960; Willetts, 1983; Sarre, 1988; Werner and Izak, 1988.

At El Maamoura beach, five sand traps are located at different levels from the beach surface, being 0.0, 0.10, 0.25, 0.50, and 1.0 m. The inlets of the traps are situated in the plane perpendicular to the observed wind direction. Each trap accumulated sand on its bottom of rectangular dimension of 10 x 50 cm. The traps were left for five months (from November 1989 to March 1990). The amount of sand at each time period in each trap is weighted and corrected for that dimension. They yielded the total sand transport per centimeter. This amount of sand transported is calculated along the beach length to estimate total sand volume transported by the action of wind.

Long-Term Shoreline Changes

Temporal beach changes are studied based on aerial photos of 1955 and 1983, covering Alexandria beaches from Abu Quir headland in the east to El Silsila in the west. Aerial photographic analysis was used to detect erosion and accretionary changes along Alexandria beaches. A total of 82 reference transects along the shoreline are selected on the two photograph sets. The landward and seaward variation in the beach width is measured. The estimated annual rates of erosion and accretion along Alexandria beaches are depicted in Fig.5. Most of Alexandria beaches are subjected to coastal erosion except few ones. The maximum long-term erosion rate of 0.70 m/yr is estimated at El Chatby, Sidi Gaber, and west part of El Maamoura beach, whereas maximum accretion of 0.70 m/yr occurs at Sidi Bishr and the eastern part of El Maamoura beach. Aerial photos analysis shows that El Maamoura beach is exposed to erosion. Its average beach width in 1955 was 65 m and became 53 m in 1983. The calculated rate of erosion is about 0.40 m/yr. The average beach width for Alexandria beaches in 1955 was 37.6 m became 32.4 m in 1983. About 5.2 m were lost from the beach width of Alexandria during 28 years. Therefore, the calculated annual rate of erosion for Alexandria beaches is about 0.20 m/yr.

SUMMARY AND CONCLUSIONS

Temporal beach changes are studied using aerial photos of 1955 and 1983. The analysis reveals that all beaches are subjected to coastal erosion except few ones. The average long-term erosion rate for Alexandria beaches is 0.20 m/yr.

The seasonal changes in beach sand of El Maamoura beach are detected. The calculated net result of sand volume loss is found to be -450 m^3/m for the entire beach length (1.2 km) or -0.4 m^3/m during the period of study (13 months). The present work revealed that, the net sediment transport direction is to the east. Maximum erosion occurs near profile P13 (on the western side) and maximum deposition occurs at the eastern end near profile P1.

Average beach width is less than 30 m in winter while it reaches about 40 m in non winter condition. The annual average shoreline fluctuation in landward and seaward directions along the beach is about 15 meters. The present investigation showed that the winter beach profiles are characterized by narrow beach width with steep slopes and few sand bars at 1 to 2 m below MSL, while non winter beach profiles have wide beach widths with gentle slopes and frequent sand bars at 1 to 2 m below MSL.

Mechanical analysis of investigated sediments showed that, the mean grain size has a slightly coarsing trend to the east, where the sand is coarse near profile P1, close to the accreted side of the beach. Winter months have been associated with finer and better sorted sand than non winter months.

Annual sand transport by wind is found to be about 37.7 m^3 during the period of study (five months). Subtracting this amount of sand from the net sand volume losses (450 m^3/m), we calculate that beach profiles yield 412.5 m^3/m of eroded sand caused by the action of currents and waves.

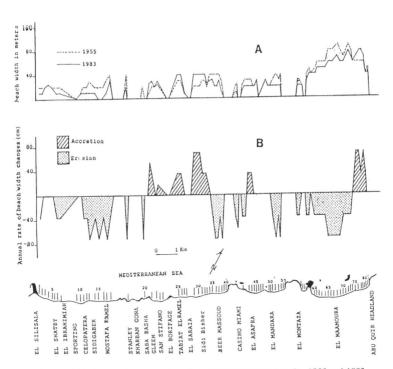

Fig.5. Beach width changes determined from two sets of aerial photographs, 1955 and 1983 (A), annual rate of beach width changes during 28 years (B).

REFERENCES

Aubrey, D.G., 1979. Seasonal patterns of onshore/offshore sediment movement. J. Geophys. R. Res., 84: 6347-6354.

Bagnold, R.A., 1941. The physics of blown sand and desert dunes, Methuen. London, 265p.p.

Clarke, D.J. and Eliot, I.G., 1983. Mean sea-level and beach-width variation at Scarborough, Western Australian. Mar. Geol., 51:251-267.

Darling, J.M., 1964. Seasonal changes in beaches of the north Atlantic coast of the United States. Proc. coastal Eng. Conf., 9th, Am. Soc. Civ. Eng., 236-248p.

Dingler, J.R., 1981. Stability of a very coarse-graind beach at Carmel, California. Mar. Geol., 44:241-252.

Dolan, R., 1965. Seasonal variations in beach profiles along the outer Banks of North Carolina. Shore and Beach, 33:22-26.

Eliot, I.G. and Clarke, D.J., 1982. Seasonal and biennial fluctuation in subaerial beach sediment volume on Warilla Beach, New South Wales. Mar. Geol., 48:89-103.

Folk, R.L. and Ward, W.O., 1957. Brazos River bar, a study in the significance of grain size parameters. J. sed. Petrology, 27:3-27.

Fox, W.T. and Davis, R.A., 1978. Seasonal variation in beach erosion and sedimentation on the Oregon coast. Geol. Soc. Am. Bull., 89:1541-1549.

Frihy, O.E. and Komar, P.D., 1990. Patterns of beach sand sorting and shoreline erosion on the Nile Delta. J. sed. petrology, 61:544-550.

Harrison, W. and Wagner, K.A., 1964. Beach changes at Virginia beach, Virginia. U.S. Army Crops Eng. Coastal Eng. Res. Cent., Misc., 25 pp.

Horikawa, K. and Shen, H.W., 1960. Sand movement by wind (on the characteristics of sand traps). U.S. Army Corps of Engineers Beach Erosion Board. Tech. Memo., 119 pp.

Niespodzinska, L., 1980. Eolian transport of beach material, Lubiatowo 1976. Polish Academy of Sciences, Institute of Hydroengineering, Hydrotechnical Transactions, 41:213-115.

Owens, E.H., 1977. Temporal variations in beach and nearshore dynamics.J. Sed. Petrol., 47:168-190.

Pilkey, O.H. and Richter, D.M., 1964. Beach profiles of a Georiga barrier island. Southeast. Geol., 6:11-20.

Psuty, N.P., 1965. Beach ridge development in Tabasco, Mexico Assoc. Am. Geogr. Ann., 55:112-124.

Sarre, R.D., 1988. Evalution of aeolian sand transport equations using intertidal zone measurements, Saunton sands, England. Sedimentology, 35: 671-679.

Shepard, F.P., 1950. Beach cycles in southern California. U.S. Army Corps Eng. Beach Erosion Board, Tech. Memo. No.20, 26 pp.

Short, A. D., and Wrigh, L. D., 1981. Beach systems of the Sydney Region. Aust. Geog. 15:8-16.

Werner, K.I. and Izak, C.R., 1988. A sand budget for the Alexandria coastal dune field, South Africa. Sedimentology. 35:513-521.

Willetts, B., 1983. Transport by wind of granular materials of different grain shapes and densities. Sedimentology, 30:669-679.

DRIFT RATES IN ACCRETED AND ERODED BEACHES BY FLUORESCENT SAND, NILE DELTA COAST

Nabil M. El-Fishawi and Abdelmoneim A. Badr

Institute of Coastal Research, Geology Depr.,
15 Faraana St., El-Shalallat,
21514 Alexandria, Egypt

ABSTRACT

The Fluorescent sand grains were used to predict the sediment movement quantitatively and qualitatively. This approach was necessitated by the release and sampling methods utilized. The present study was performed at Burullus area to investigate the sediment movement on natural beaches, near man-made structures and through the silted part of the Burullus outlet.

The estimated velocity for sand grains in motion ranges between 1.59 m/min and 2.38 m/min. Such low velocities indicate that the bulk of the sand load travels at a much slower rates along the beach. West of Burullus outlet, the drift rate was estimated to be 0.92-1.20 x 10^6 m^3/yr where continued accretion was found at the up-drift side of the jetty. At the eastern side of the area and where the waves attack the foot of the coastal dunes, high drift rate was occured and estimated to be 2.1 x 10^6 m^3/yr. The amount of offshore drift 1.2 x 10^6 m^3/yr is responsible for severe erosion of the coastal dune area.

A significant portion of the sediments in motion and estimated 0.37 x 10^6 m^3/yr is directed from the eastern tip of the outlet to the west and resulting siltation of the outlet and creating difficulties to the navigation.

INTRODUCTION

Wide consideration has been given to tracing sediment movement over the past 30 years. Some of the studies have included the use of artifical material such as pulverized coal (Shinohara et al., 1958), broken brick (Kidson and Carr, 1961) and magnetic concrete (Pantin, 1961). Grains dyed nonfluorescent hues (King, 1951; Scott, 1954; Luneburg, 1960) and painted cobbles (Phillips, 1963) have also been employed. In addition, sediments have been traced using radio isotopes (Caldwell, 1960; Cummins and Ingram, 1963; Courtois and Monaco, 1969). Naturally occuring radioactive minerals in beach sands have also been used to trace the movement and source area (Kamel, 1962).

Tracing technique employing grains marked with a fluorescent compound was made successfully by Zenkovitch (1960), Aibulatov (1961), Russell (1961), Russell et al. (1963), Inman and Bowen (1963), Griesseier and Voigt (1964), Ingle Jr. (1966), Bruun (1979) and Vries (1971). The use of fluorescent grains holds a number of advantages: 1. natural grains of silt to pebble size are easily, rapidly and cheaply dyed, 2. the fluorescent dyes present no serious legal and health hazard, 3. different fluorescent hues can be used to differentiate between different size fractions, and 4. dyes do not alter the hydrodynamic properties of natural grains.

In attempting to overcome the coastal problems on the Nile Delta presented by siltation of the outlets and severe coastal erosion, the designers are handicapped by the lack of a suitable method for determining the sediment movement characteristics of the region in question. In fact, sand movement studies using tracer grains have been scarce in Egypt and results are mostly depended upon refraction and empirical models. Therefore, attempts have increased lately to use fluorescent grains for sediment movement studies (Kadib, 1968 and 1972; Khafagy et al. 1987; El-Fishawi, 1987).

SCOPE OF THE AREA

Lake Burullus is located in the central part of the Nile Delta coast (Fig. 1). It is a very large lagoon joining the Mediterranean Sea through a narrow outlet at Burg El-Burullus Village. The people of the village earn their living from fishing.

Between 1937 and 1940, five groins were erected at a distance of 180 m to protect the eastern beach of the Burullus outlet. The groins had lengths between 30 m and 90 m. Due to severe erosion between the groins, a concrete wall was constructed between them in 1950 with a length of 600 m. The wall was sunk in many places and repair work mainly consisted of dumping mass concrete behind the blocks for closing the gaps. In 1983 a new sea wall was constructed with tetrapodes parallel to the shore. The sea bottom in front of the sea wall has been eroded by wave action causing the depth to the 3 m compared to less than 1 m some distance away. Furthermore, eastward of the sea wall erosion occured over a distance of about 300 m causing destruction of many houses and coastal dunes. But in fact, the sea wall succeeded to protect the village from the advance of sea waves. Recently, around 1983, a jetty more than 200 m long was erected on the eastern side of the outlet.

In 1972 a jetty was constructed on the western tip of the outlet to keep it open and to widen the narrow western shore. The jetty was 240 m long; 200 m into the sea and 40 m as tie to the beach. The beach west of the jetty has built and there is a proposal to elongate the jetty seawards.

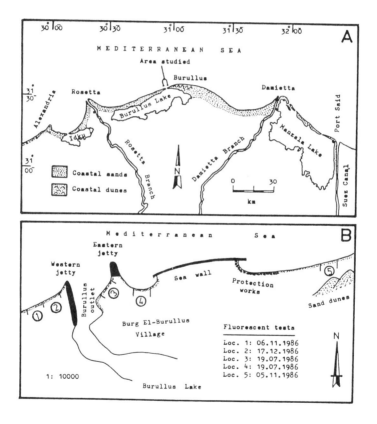

Fig. 1: A- Location map for the Nile Delta coast.
B- Studied area.

PROBLEM AND PURPOSE OF STUDY

Burg El-Burullus village is said to be transferred about 2 km to the south during the last century. Many coastal forts, built in 1880, located along Burullus beaches are at present under the sea. The coastal barrier on both sides of lake Burullus outlet reduced in width (Fig. 2). The sea advanced about 100 m between 1935 and 1947 and more than 100 m during 1947-1964. Till 1987, erosion still occured with high rate on the eastern side while accretion became a general phenomena on the western side due to the construction of the jetty.

Till recently, the outlet of lake Burullus used to be silted up creating difficulties to the local inhabitants who are mainly fishermen.

So, the aim of the present study is to identify the problems of coastal erosion and siltation of the Burullus outlet from the point of view of sediment movement by using fluorescent sand. This study is a part of the operations of the Coastal Research Institute to be used in the coastal protection works.

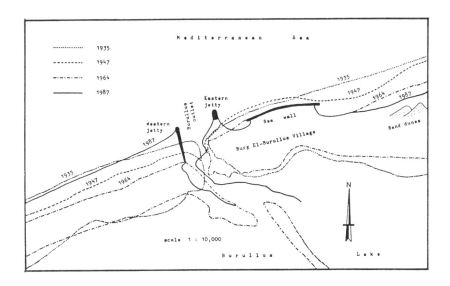

Fig. 2:   Shoreline changes at Burullus area.

TECHNIQUES AND METHOD OF STUDY

Five field experiments on five locations using fluorescent-dyed grains were performed along the western and eastern beaches of Burullus outlet (Fig. 1B). The experiments were carried out during the period from July to Dec. 1986 to trace the sand movement between the foreshore and inshore zones. Such period represents different sea states during summer and winter seasons. The length of each location along the shorelines was about 70 m and the width of the foreshore-inshore was about 50-60 m (up to depth 2 m below sea level).

The field and laboratory techniques were made according to Ingle Jr. (1966). Generally 4 release points were located and 40 kg tracer sand were released during each test. Tracer sand was usually released at the upcurrent of sample stations distributed on a rectangular grid system.

Samples were collected by pressing 10 x 10 cm bentonite-coated wood plank onto the sea bottom surface. Two or three series of samples were collected from each location after release. A total of 560 samples were collected during the 5 tests. In the laboratory, each sample was reviewed under short-wave ultraviolet light and the number of fluorescent grains was tabulated. Values of tracer concentration were then plotted at their respective stations and isopleths were constructed resulting in contoured patterns of tracer dispersion with time.

LITTORAL CURRENT

The direction and velocity of the littoral current were measured at 2 stations astride Burullus outlet. Measurements were made by float and a stop watch to observe the time by float traversing a known distance of 20 m. Measurements were made twice daily.

Littoral currents measured during January and July, 1985 by the shoreline of the Burullus beaches were analyzed. Figure 3 illustrates the velocity, direction and percentage of occurence for winter and summer months. It is indicated that the eastward current of the summer month (100% to east in July) is partially replaced by westward current during the storm month (54-62% to east and 38-46% to west in January). The maximum velocity does not exceed 82 cm/sec for the eastward currents and 66 cm/sec for the westward ones.

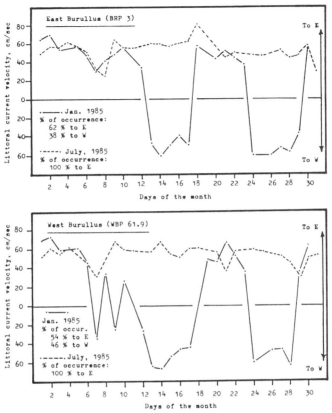

Fig. 3: Littoral current east and west of Burullus during winter and summer seasons.

FLUORESCENT SAND DISPERSION

Average grain velocity:

The technique of the present study depends upon at what rate fluorescent sand left the respective sample grids. After establishing the approximate number of tracer grains released during each field test, a planimeter analysis was made of each individual tracer dispersion pattern constructed from adjust tracer concentration values (e.g. Figs. 4 and 5).

In this manner the approximate number of tracer grains remaining within a sample grid at any elapsed time can be calculated. For instance, 7100×10^6 tracer grains were released during the 19 July 1986 at just east of Burullus outlet and about 59.3×10^6 tracer grains remained within the sample grid at 20 min after release.

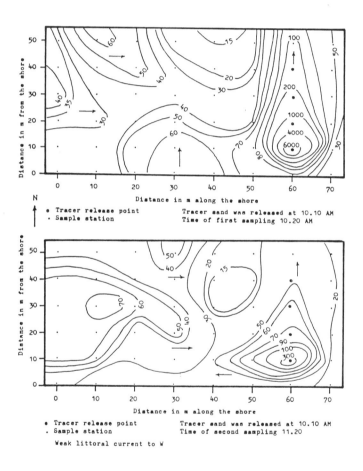

Fig. 4: Fluorescent sand dispersion at location 2 (just west of Burullus outlet).

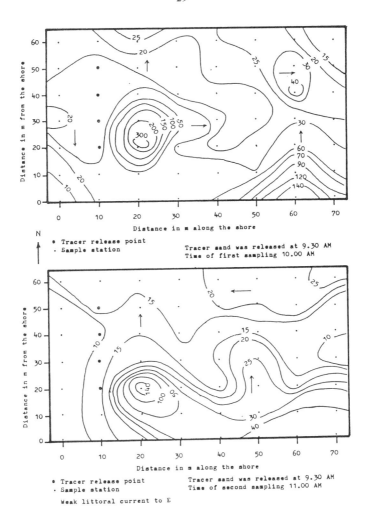

Fig. 5: Fluorescent sand dispersion at location 5 (the eastern side of the sea wall).

The depletion rate, number of tracer grains leaving the sample grid per minute, can be drawn for all tracer tests. The equilibrium depletion rate was found by averaging depletion rate values falling below the inflection point of a depletion curve. Thus it is able to compute the grain velocity regards to the expressions 1 and 2:

$$t_{50} = 1/2\ (G)\ /\ De \quad \text{.. (1)}$$

Where t_{50} = time for one half the total tracer grains to leave the sample grid (minutes);
G = total number of tracer grains released;
De = average equilibrium depletion rate (grain/minute).

An estimate of the average velocity of grains was then calculated using expression 2:

$$Ug = l\ /\ t_{50} \quad \text{.. (2)}$$

where Ug = average grain velocity (meter/minute);
l = average distance of tracer travel (meter); the distance between points of tracer release and the down-drift edge of sample-grid

The calculated average grain velocity ranged between 1.59 m/min and 2.38 m/min. The mean grain velocity for all tests was about 1.91 m/min (3.2 cm/sec). Thus the mean velocity of grains on all test beaches was approximately 1/17 the mean littoral current velocity (52 cm/sec) measured along Burullus beaches. In fact, these velocities do not represent the range of absolute velocities of individual grains. Although the absolute velocities of individual grains may exceed 55 cm/sec, Bruun (1962) and Seibold (1963) found that the ghrain velocity ranging between 1-2 m/min. It is likely that the bulk of the sand load travels along the beach at a much slower rate.

The relatively low average velocities calculated suggest that most sand grains were traveling in traction or within the mobile layer rather than in suspension (Ingle Jr., 1966). He added that the mobile layer calculated will be thicker than the actual depth of this layer. In the present study, the thickness of the mobile layer was determined by using core sampler. It ranges between 1.5 cm and 3.8 cm with an average of 2.74 cm. In fact, it is close to that which found by Ingle Jr. (1966) and far from that calculated at east and west of Damietta (9 cm) by Kadib (1972).

Rate of sand drift:

A unit volume of sand movement is calculated for each fluorescent test at each of the 5 examined beaches. The expression used was as follows:

$$V = K \times W \times B \quad \quad \quad \quad \quad (3)$$

where V = unit volume of sand movement (m^3);
K = constant beach length of 1 m;
W = width of the forehore-inshore zone (m);
B = thickness of the mobile bed layer (m).

Using the computed unit volume of sand movement for a particular test, the rate of sand drift was calculated from the expression:

$$Qi = V \times Ug \times 1440 \quad \quad \quad \quad \quad (4)$$

where Qi = rate of sand movement or drift (m^3/day);
V = unit volume of sand movement (m^3);
Ug = average grain velocity (m/min).

The knowledge from expressions 3 and 4 yielded a wide range of drift rates due to variety of locations and surf conditions prevailing during tracer tests. Table 1 illustrates the mean grain velocity and rate of drift for each location.

Table 1: Summary of grain velocity and drfit rate at Burullus.

Location	Date	Grain velocity (m/min)	driftrate m^3/day	m^3/day
loc. 1	17.12.1986	1.59	2500	0.92×10^6
loc. 2	17.12.1986	1.71	3370	1.20×10^6
loc. 3	19.07.1996	1.74	4100	1.50×10^6
loc. 4	19.07.1986	2.38	4700	1.70×10^6
loc. 5	05.11.1986	2.11	5830	2.10×10^6

The information available from the present investigation reflected an explanation about the behaviour of the sediments astride Burullus outlet. As the area west of Burullus outlet subjected to continued accretion after the construction of the western jetty, calculated rates of sand movement (rate of drift) increased from 0.92×10^6 m^3/yr at loc. 1 to 1.20×10^6 m^3/yr at loc. 2 and up to 1.50×10^6 m^3/yr at loc. 3 where the eastern jetty present. On the other hand, the areas located at the western and eastern side of the sea wall subjected to continued erosion and the drift rate increased from 1.70×10^6 m^3/yr at loc. 4 to 2.10×10^6 m^3/yr at loc. 5 where the maximum destruction of the beach and coastal dunes is found.

By using fluorescent sand tracers near Damietta mouth, Kadib (1969) reported that annual drift rate was about 1.16×10^6 m^3/yr. Comparisons of photographs of the entrances to Bardawill lagoon at Sinai show that easterly sediment transport occurred with rates as high as 0.50×10^6 m^3/yr. (Inman et al., 1976). The estimation of longshore transport along the Nile Delta coast during 1972-1973 was made by using refraction and empirical models (Quelennec and Manohar, 1977). Their easterly drift is estimated to be 3.2×10^6 m^3/yr at Rosetta, 1.5×10^6 m^3/yr at Burullus headland and 2.6×10^6 m^3/yr at Damietta. In fact, the wide range of drift rates between the present investigation and the other ones may reflect the different techniques used and the wide spectrum of surf conditions prevailing during each study.

INTERPRETATION OF TRACER DISPERSION

Tracer concentration values were plotted at their respective stations and contoured with isopleths representing equal concentrations of fluorescent grains per 100 square cm of surface area (e.g. Figs. 4 and 5). Contour maps constructed for each sample series during a test depicted fluorescent grain concentrations within the sample grid at increasing time intervals after tracer release. The fluorescent transport maps were then examined to determine the direction of movement. This method depends upon the decreasing of concentration and presence of elongate tongue in the direction of movement. Percentage of occurrence for each direction of sediment movement was calculated and then used to determine the rate of drift at each direction by using the data in Table 1. The result is summarized in Table 2 and Fig. 6.

Table 2: Percentage of occurrence and drift rates (x 10^6 m^3/yr) for each direction of movement astride Burullus outlet

location	to E %	drift rate	to W %	drift rate	offshore %	drift rate	landward %	drift rate	total annual drift
loc. 1	33	0.31	22	0.20	12	0.10	33	0.31	0.92
loc. 2	50	0.60	12	0.14	38	0.46	-	-	1.20
loc. 3	17	0.26	25	0.37	33	0.50	25	0.37	1.50
loc. 4	25	0.42	25	0.42	12	0.20	38	0.65	1.70
loc. 5	17	0.36	17	0.36	58	1.20	8	0.17	2.10

Patterns of fluorescent sand dispersion at west Burullus outlet (loc. 1 and 2) indicate that the most prominent portion of tracer grains (33-50%) was moved eastwards. Beside the landward drift (0.31 x 10^6 m^3/yr), the western jetty plays an effective role in obstructing the eastward drift which estimated 0.30-0.60 x 10^6 m^3/yr. As a result, a high rate of accretion occurred up-drift of the jetty.

Near the eastern tip of the outlet (loc. 3) a significant portion of tracer grains (25% and estimated to be 0.37 x 10^6 m^3/yr) was directed to the west and resulting siltation of the outlet and creating difficulties to the fishermen. On the other side, this location subjected to accretion after the construction of the eastern jetty due to landward drift.

The beach at the western side of the sea wall (loc. 4) shows that equal amounts of drift rates (0.42 x 10^6 m^3/yr) were moved to east and to west. But in fact, a significant amount of sediments (0.65 x 10^6 m^3/yr) was directed landward.

A contrast situation was found at the eastern side of the sea wall (loc. 5) where the waves attack the foot of the coastal dunes. The most prominent portion of the sand movement (58% and estimated to be 1.2 x 10^6 m^3/yr) was moved offshore. Although eastward and westward drifts were represented (0.36 x 10^6 m^3/yr), the high rate of beach erosion and destruction of the coastal dunes may be related to the high rate of offshore drift.

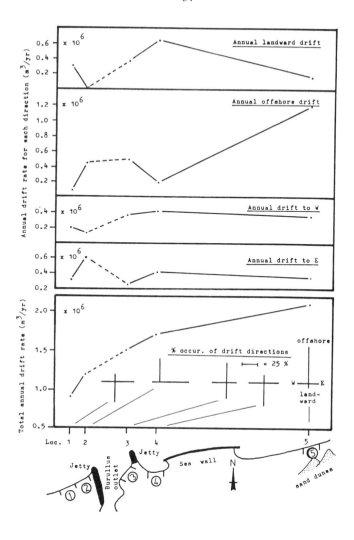

Fig.6 : The annual rate of sediment movement astride Burullus outlet.

CONCLUSIONS

The present investigation illustrates a method for estimating the rate of sediment movement at Burullus beaches of the Nile Delta by using fluorescent tracer technique. Beside qualitative observation of tracer movement, estimations of tracer loss from a unit area in unit time comprised the basis for quantitative statements. The following conclusion can be summarized:

(1) Although the eastward littoral currents predominate, westward currents are common features at Burullus beaches. The maximum velocity does not exceed 82 cm/sec for the eastward currents and 66 cm/sec for the westward ones. On the other hand, the calculated velocity for sand grains in motion ranged between 1.59 m/min and 2.38 m/min with an average of 1.91 m/min (3.2 cm/sec). Thus the average velocity of sand grains on all test beaches was approximately 1/17 the average littoral current velocity (52 cm/sec) measured at Burullus beaches. It is likely that the bulk of the sand load travels along the beaches at a much slower rate.

(2) The estimation of the sediment movement within each test beach yielded a wide range of drift rates due to variety of beaches (natural or with manmade structure) and surf conditions prevailing during each test. Where the accretion predominates at the western side of the constructed jetties, the rate of drift ranges between 0.92×10^6 m^3/yr and 1.50×10^6 m^3/yr. West of the sea wall and eastward where coastal dunes are subjected to severe erosion due to sea attack, the rate of drift increased from 1.7×10^6 m^3/yr to 2.1×10^6 m^3/yr.

(3) The western jetty plays an effective role in obstructing eastward drift. Patterns of fluorescent and dispersion indicate that the most prominent portion of sediments moving to east (33-50% and estimated to be 0.3-0.6×10^6 m^3/yr) is responsible for the accretion processes at the western side of the Burullus outlet. The landward drift (0.31×10^6 m^3/yr) is also effective in encouraging beach accretion.

(4) At the eastern tip of the Burullus outlet and near the eastern jetty, a significant portion of sediments in motion is directed westward across the outlet (25% and estimated to be 0.37×10^6 m^3/yr). At the same time, a similar amount of sediments is moving landwards. As a result, the Burullus outlet used to be silted up creating difficulties to the navigation and inhabitants who are mainly fishermen.

(5) The sediment movement at the eastern side of the sea wall shows contrast situation where the waves attack the coastal dunes. The most prominent portion of the sediment in motion (58% and estimated to be 1.2×10^6 m^3/yr) is moved offshore. Although eastward and westward drifts are represented (0.36×10^6 m^3/yr), high rates of erosion and destruction of the coastal dunes may be related to the high rate of drift towards offshore.

REFERENCES

Aiboulatov, N.A. 1961: Quelques donnees sur le transfer des sediments sableux le long d'un littoral, obtenus a l'aide de luminophores. Cahiers Oceanog., V. 13, p. 292-300.

Bruun, P., 1962: Tracing of material movement on seashores. Shore & Beach, V. 30, p. 10-15.

Bruun, P., 1970: Use of tracers in harbour, coastal and ocean engineering. Eng. Geol., No. 4, p. 73-88

Caldwell, J.M. 1960: Development and tests of a radioactive sediment density probe. U.S. Army Beach Erosion Board, Techn. Mem., No. 121, 29 p.

Courtois, G. and Monaco, A., 1969: Radioactive methods for the quantitative determination of coastal drift rate. Mar. Geol., V. 7, p. 183-206.

Cummins, R.S. and Ingram, L.F., 1963: Tracing sediments movement with radioisotopes. Military Eng., V. 55, p. 161-164.

El-Fishawi, N.M., 1987: Sediment movement at retreating coasts of the Nile Delta, Egypt. European Workshop on Interrelated Bioclimatic and Land Use Changes. 17-21 Oct., 1987, The Netherlands, 15 p.

Griesseier, H. and Voigt, G., 1964: Zur Markierung von Sanden mit Luimineszenten Farbstoffen. Monatsber. Deut. Akad. Wiss. Berlin, V. 6, p. 819-826.

Ingle Jr., G.C., 1966: The movement of beach sand. Developments in Sedimentology 5, Elsevier Publishing Company, 221 p.

Inman, D.L. and Bowen, A.J., 1963: Flume experiments on sand transport by waves and currents. In: J.W. Johnson (Editor), Proc. 8th Conf. Coastal Eng. Council Wave Res., Univ. Calif., Berkely, Calif., p. 137-150.

Inman, D.L., Aubrey, D.G. and Pawka, S.S., 1976: Application of nearshore processes to the Nile Delta. Proc. Sem. Nile Delta Sedimentology, UNDP/UNESCO/ASRT, 25-29 Oct. 1975, Alexandria, p. 205-237.

Kadib, A.A., 1969: Study of the littoral drift at Ras El-Bas using fluorescent tracer. Suez Canal Auth., Res. Center, Tech. Rep., No. 42, 34 p.

Kadib, A.A., 1972: Rate of sediment motion using fluorescent tracer. Proc. 13th Conf. Coastal Eng., p. 985-1003.

Kamel, A., 1962: Littoral studies near San Francisco using tracer techniques. U.S. Army Beach Erosion Board, Techn. Mem., No. 131, 86 p.

Khafagy, A., El-Fishaw, N.M. and Badr, A.A., 1987: The movement of beach sand as indicated by fluorescent sand and grain size at southwest Rosetta. Coastal Res. Ins., Alex., Tech. Rep. 1/87, 38 p.

Kidson, C. and Carr, A.P., 1961: Beach drift experiments at Bridewater Bay, Somerset. Proc. Bristol Nature Soc., V. 30, p. 163-180.

King, C. A. M., 1951: Depth of disturbance of sand on sea beaches by waves. Jour. Sed. Petr., V. 21, p. 121-140.

Luneburg, H., 1960: Sediment transport, sedimentation and erosion on theouter sands of the Weser Estuary. Inter, Geol. Congr., 21st, Copenhagen, 1960, Abstract, p. 247.

Pantin, H.M., 1961: Magnetic concrete as an artificial tracer material. New Zealand Jour. Geol. Geophys., V. 4, p. 424-433.

Phillips, A.W., 1963: Tracer experiments at Spurn Head, Yorkshire, England. Shore & Beach, V. 31, p. 30-35.

Quelennec, R.E. and Manohar, M., 1977: Numerical wave refraction and computer estimation of littoral drift; application to the Nile Delta coast. Proc. Sem. Nile Delta Processes, UNDP/UNESCO/ASRT, 2-9 Oct. 1976, Alexandria, p. 408-433.

Russel, R.C.H., 1961: The use of fluorescent tracers for the measurement of littoral drift. Proc. Conf. Coastal Eng., 7th, p. 418-444.

Russel, R.C.H., Newman, D.E. and Tomlinson, K.W., 1963: Sediment discharges measured by continuous injection of tracer from a point. Int. Assoc. Hydr. Res., Congr. 10th, London, V. 1, p. 69-76.

Scott, T., 1954: Sand movement by waves. U.S. Army Beach Erosion Board, Tech. Mem., No. 48, 37 p.

Seibold, E., 1963: Geological investigation of nearshore sand transport. In: M. Sears (Editor), Progress in Oceanography. Pergamon Press. New York, V. 1, p. 3-70

Shinohara, K., Tsubaki, T., Yoshikata, M. and Agemori, C., 1958: Sand transport along a model sandy beach by wave action. Coastal Eng. Japan, V. 1, p. 111-129.

Vries, M. de, 1971: On the applicability of fluorescent tracers in sedimentology. Delft Hydr. Lab., Pub. No. 94, 17 p.

Zenkovitch, V.P., 1960: Fluorescent substances as tracers for studying the movement of sand on the sea bed; experiments conducted in the U.S.S.R., Dock Harbour Authority, V. 40, p. 280-283.

ESTIMATION OF SEDIMENT BUDGET AND LONGSHORE TRANSPORT RATES FROM COASTAL RESPONSE. APPLICATION TO THE EBRO DELTA.

Jiménez,J.A,; García,M.A.; Valdemoro,H.; Gracia,V. and S-Arcilla,A. Laboratori d'Enginyeria Marítima. Universitat Politécnica de Catalunya. Gran Capitá s/n. Barcelona 08034. Spain.

ABSTRACT

Measurements of shoreline displacements along the Ebro delta coast have been used to estimate the medium-term sediment budget and the associated net longshore transport scheme. The data have been treated using the method of linear regression in order to filter out the short-term components. This method gives promisimg results, allowing thus a reasonable prediction of the delta coast evolution at medium term.

INTRODUCTION

Obtaining the sediment budget and the associated transport scheme is necessary to predict how a coast will behave. There are several methods to do it: using coastal response data (shoreline displacements and/or beach profiles evolution), predicting the transport rates from formulae using wave data, etc...

The first step to model the coastal behaviour is to study its past evolution. From this starting point, it is possible to calibrate an evolution model in order to predict future scenarios.

In coasts where defense structures are absent, shoreline changes and beach profile evolution are common tools to quantify the transport rates. The use of shoreline displacements has the advantage that they can be compared in a coherent manner with historical coastlines. This comparison allows building up a time history of the coastal evolution.

Several years of measurements of shoreline positions have been obtained along the Ebro delta coast to monitor the coastal evolution. These measurements are integrated into a large-scale monitoring programme focusing on the local morphological changes and the driving factors affecting the delta coast.

AREA STUDY

The Ebro delta is located in the Spanish Mediterranean coast (figure 1). The delta coastline has an approximate length of 50 km. The total emerged area is 350 km^2.

After several centuries of growth, the delta evolution trend changed a few decades ago (Maldonado, 1986) in such a way that the delta is no longer an intermediate river-wave dominated system (Wright and Coleman, 1973) but a wave dominated coast. This is mainly due to the decrease of solid discharges by the construction of large dams in the upper course of the river. The initial status of dynamic equilibrium between river discharges and wave transport capacity has evolved to a situation where the second factor is dominant.

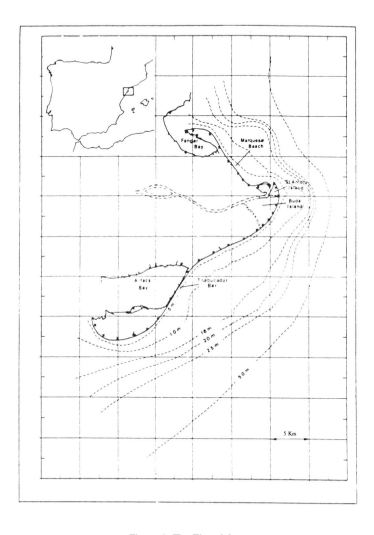

Figure 1. The Ebro delta coast.

At present, the annual average solid discharge of the river is about 32,000 m^3/yr of "useful sand", ranging between 22 and 64,000 m^3/yr as estimated by Jiménez *et al.* (1990a). The minimum flow required to start the movement of sand at the river bed is around 400 m^3/s, which is exceeded only few times per year. Under these conditions, the most important part of the sediment supply consists of wash-load (mainly mud, silt and clay), which does not contribute to the evolution of the inner coast.

The sediment distribution pattern in the Ebro delta coast is shown in figure 2 (Guillén and Maldonado, 1992). Three zones can be distinguished: an inner zone, characterized by the presence of medium-fine sands ($d_{50} \geq 125\mu m$), an outer zone with muddy sediments ($d_{50} \leq 63\mu m$) and a transitional zone in between of very fine sand. There is no evidence of sand migrating out of the delta system through the mud belt.

The wave climate of the zone is characterized by dominant east waves, with an average significant wave height of 1 m and an average wave period of 4 s (Gracia *et al.*, 1988). Waves approach the coast in such a way that they generate a net longshore transport pattern directed towards the south in the southern hemidelta, and towards the north in the northern hemidelta. This transport pattern is consistent with the present coastal configuration, where a big spit has been developed in each hemidelta.

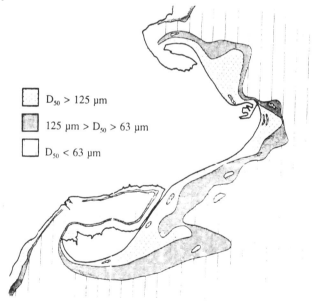

Figure 2. Sediment zonation in the Ebro delta coast (Guillén and Maldonado, 1992).

DATA ANALYSIS

To characterize the evolution of the Ebro delta coast, a collection of shoreline measurements, obtained from a set of beach profiles taken along the coast, has been used (see figure 1). These measurements cover a period of four years, and data have been collected three to four times per year.

The use of shoreline displacements is one of the most common tools to estimate the behaviour of a given coast. These changes are the result of the action of all agents on the coast. In this way, such changes can be considered as formed by two contributions: a medium/large-term component and a short-term component. The former is generally associated to changes due to longshore transport gradients, sea level changes, decrease in the input of sediment to the coast, etc.... The time scale of these changes is of several years. The short-term changes are associated to variations in the driving terms acting on the coast in short-time periods. Examples of these changes are those due to cyclic changes in wave conditions, the action of storm waves on the coast, etc... In this case, the time scale ranges from days to seasons.

If the medium/large-term behaviour is to be considered, the short-term component must be removed. There are several methods to analyse shoreline displacements. Each has its own advantages and disadvantages (see Dolan *et al*, 1991) depending on the quality and the characteristics of the available data. In our case, the least-square linear regression method has been chosen. Since all data have a similar time span, the results will not be polluted by "noisy points" due to big differences in the recording time (*i.e.* there are not clusters of data).

Data from each new survey have been added to the total data set in such a way that the trend has been continuously updated. At first stage, since the time period was short and the cyclic changes dominated we obtained sharp trend changes. For increasing periods, the obtained trend tended to stabilize and reached an assymptotic value, which we identify as the medium/large-term trend (see figure 3).

Figure 4 shows the evolution trends derived for the period 1988/92. In the same figure, two additional trends are also presented: the first one corresponding to the period 1957/73 and the second one to the period 1973/89. These two trends have been obtained by comparing aerial photographs using the method of the end-point rate (*i.e.* only two shoreline positions have been compared). It can be seen that the behaviour is similar for the three covered periods from the qualitative point of view, but the values show some important differences. The first period rates (1957/73) are larger because, at this time the delta coast began to experience the decrease in the river solid discharge. The subsequent trends show a similar trend, but with lower rates, because the shoreline slowly approaches an equilibrium configuration.

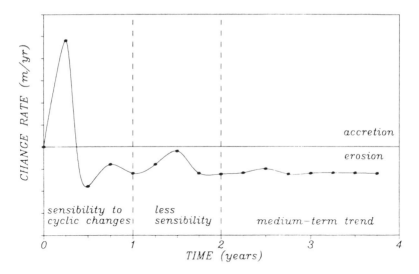

Figure 3. Time history of the evolution trend of a typical profile.

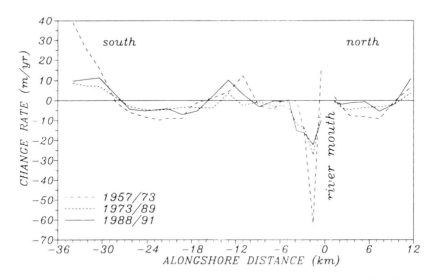

Figure 4. Rates of shoreline changes along the delta coast during the last decades.

If all the recorded shoreline positions had been used, a single trend for the total period since 1957 would have been obtained, thus masking the differences in the evolution over the time. It is, therefore, recommended to segregate the data in groups trying to identify discrepancies in the behaviour.

SEDIMENT BUDGET

From the shoreline evolution trends obtained using the linear regression analysis, volume changes along the Ebro delta coast have been calculated. To transform shoreline displacements into volume changes, a closure depth of 7 m has been adopted (Jiménez et al, 1990b; Jiménez and García, 1991). This depth has been selected upon integrating the information obtained about the evolution and the morphology of beach profiles with empiric criteria such as Hallermeier (1981) or Birkemeier (1985).

The volume changes have been calculated considering a wedge-shaped change over the closure depth. This choice is consistent with the fact that we are mainly concerned with changes due to longshore transport and these changes are more important in the inner part of the profile, where longshore current and longshore transport are dominant.

The so-obtained volume changes are different to those presented by Jiménez et al (1991) because, in this case, a larger number of surveys have been used. This updated sediment budget may, therefore, be considered a more representative value for medium-term scale.

Southern hemidelta

The volume changes in the southern hemidelta are depicted in figure 5. These changes show a pattern of alternating erosion and deposition zones, in such a way that the global change for the hemidelta is nil. Unfortunately, erosive zones are located in areas of special interest (from the agricultural and recreative point of view), whereas the accretive zones have less economic interest.

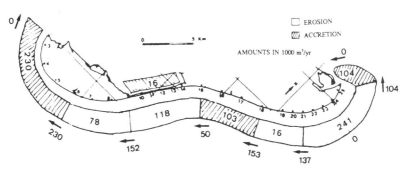

Figure 5. Sediment budget and net longshore transport scheme in the southern hemidelta.

The main erosive zones are located in the areas of Cap Tortosa and the Trabucador Bar (overall erosion rates of 241,000 m^3/yr and 211,000 m^3/yr, respectively). In the second case, this amount of sediment is eroded from the bar itself (133,000 m^3/yr), and the remaining amount (78,000 m^3/yr) is eroded further to the south. It has been assumed that all this material is eroded due to longshore transport gradients, except a part of material which is transported into the Alfacs Bay due to overwash processes (16,000 m^3/yr).

Three deposition zones have also been identified: the southern spit, where 230,000 m^3/yr of sand are deposited; the coastal stretch on the northern side of the Trabucador Bar, where an amount of 103,000 m^3/yr is settled, and the apex of the St. Antoni Island, where the material is deposited (104,000 m^3/yr) as a spit (Jiménez et al, 1991).

Northern hemidelta

Figure 6 shows the volume changes in the northern hemidelta. Two main zones can be distinguished: an area extending from the mouth area to the northern spit, which is eroding at a rate of 70,000 m^3/yr. And the second one at the spit, where part of this material is deposited in the outer coast, whereas the remaining sand is deposited in the inner bay after being transported by locally generated waves from the NW. This deposition pattern originates the expansion of the spit towards the coast and also siltation of the dry surface in the direction of the inner bay.

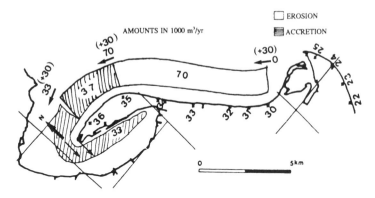

Figure 6. Sediment budget and net longshore transport scheme in the northern hemidelta.

NET LONGSHORE TRANSPORT PATTERN

The net longshore transport scheme has been elucidated from the volume changes calculated in the sediment budget. This transport system must be considered as the resultant of the integrated action of waves approaching the coast. The basic assumption we make is that the volume changes are mainly due to longshore transport gradients. This is quite acceptable since they have been derived from the medium-term trend obtained by linear regression.

Southern Hemidelta

The assumed boundary condition for this hemidelta is zero transport beyond the apex of the southern spit. This condition has been selected because there is no evidence of existing sand transport in the south of this section. The resulting transport scheme is composed by two cells (see figure 5). The transport scheme shows (as expected) a divergence and a change in the transport direction at the Cap Tortosa zone. Thus, in this zone the net longshore transport is directed towards the north at a rate of 104,000 m^3/yr and decreases to zero in the apex of the island due to wave diffraction effects.

In the south of this zone the longshore transport is southwards. The main characteristic of this sector is that two transport maxima exist. Transport quickly increases from zero up to a value of 135,000 m^3/yr south of Buda Island. From this point towards the south, the transport slowly rises up to a first maximum value of 150,000 m^3/yr. Beyond this point, the transport rapidly decreases down to a minimum value attained at the north part of the Trabucador Bar.

Along the external coast of the Trabucador Bar, the longshore transport rate increases towards the spit, where the highest transport rate is reached (230,000 m^3/yr). South of this point, the transport rates decrease again due to wave diffraction, and a zero value is reached in the apex of the spit.

Northern hemidelta

The boundary condition for this hemidelta must be selected as a function of the amount of sediment which is naturally by-passed from the mouth area (solid discharge of the river plus sediments from the St. Antoni Island).

As mentioned before, the transport capacity of the river during a "normal year" is quite low (in the order of 32,000 m^3/yr). In the other hand, most of the sediment eroded from the external coast of the St. Antoni Island is deposited in the apex of the island as a little spit. Only during storm conditions, this spit collapses and part of its sand is transported towards the north. Moreover, a little amount of sand can be transported along the spit towards the hemidelta during low-energy conditions.

If these two facts are taken into account, it is reasonable to assume an upper limit of 30,000 m^3/yr of sediment is by-passed to the north.

The St. Antoni Island acts as a diffraction focus for the dominant waves from the E, which generate northwards longshore transport. This effect causes sheltering in the leeside of the island, and hence gradient currents directed towards the mouth. The present configuration of the coast, similar to a logarithmic spiral, suggests that the shoreline is close to the equilibrium configuration.

All these considerations suggest that flexible boundary condition should be adopted for the northern hemidelta. Thus, it is assumed that the transport in the mouth area has an indetermined value between 0 and 30,000 m^3/yr and is directed towards the north. This assumption brings uncertainty to the longshore transport curve, but not to transport gradients, which remain unaltered (figure 6). North of this point, transport increases up to a maximum value of 70,000 (+30,000) m^3/yr, and beyond it, decreases due to diffraction effects around the spit.

SUMMARY AND DISCUSSION

- Shoreline measurements collected during three years, along the Ebro delta coast, have been used to obtain the sediment budget and the net transport scheme.
- To obtain the coastal medium-term behaviour, the available data have been processed using the least-square linear regression method to remove the short-term component.
- The observed trend has been compared with old trends, and a similar response pattern is observed. Moreover, the recent history of the delta coastline evolution has been reconstructed. If all data had been used, the differences in the evolution over the time would have been masked.
- These shoreline changes have been converted into volume changes assuming a wedge-shaped change over the closure depth. A 7m depth has been estimated for changes due to longshore transport for a normal year.
- The sediment budget shows a quasi-closed system for the sand in both hemideltas, the spits being the most important accretive areas (see figures 5 and 6).
- For a given sediment budget and proper boundary conditions for each hemidelta, the net longshore transport pattern has been estimated. This transport pattern may be considered as the resultant of the integrated action of waves approaching the coast.

ACKNOWLEDGEMENTS

This study has been promoted and funded by Generalitat de Catalunya, Direcció General de Port i Costes (contract no.06890SP175), whose support is duly appreciated.

We also acknowledge Jan van Overeem, whose comments and suggestions have largely enriched this work. We extend thanks to Francisco Rivero for review of this manuscript.

REFERENCES

Birkemeier, A.W. 1985. Field data of seaward limit of profile changes. *Journal of Waterways, Port, Coastal and Ocean Engineering*, 111, 3, 598-602.

Dolan, R.; Fenster, M.S. and Holme, S.J. 1991. Temporal analysis of shoreline recession and accretion. *Journal of Coastal Research*, 7, 3, 723-744.

Gracia, V.; Collado, F.; García, M. and Monsó, J.L. 1989. *Analysis and proposal of solutions to stabilize the Ebro delta: Wave climate II*. Technical Report LT-2/5. Direcció General de Port i Costes, Generalitat de Catalunya (in spanish).

Guillén, J. and Maldonado, A. 1992. *Analysis and proposal of solutions to stabilize the Ebro Delta: Morphology and sediment cover in the internal and medium continental shelf in the Ebro delta*. Technical Report. Direcció General de Port i Costes, Generalitat de Catalunya (in spanish).

Hallermeier, R.J. 1981. Seaward limit of significant sand transport by waves: an annual zonation for seasonal profiles. *Coastal Engineering Technical Aid No.81-2*. U.S. Army Corps of Engineers, CERC.

Jiménez, J.A. and García, M.A. 1991. *Analysis and proposal of solutions to stabilize the Ebro Delta: Sediment budget from shoreline evolution trends*. Technical Report LT-3/3. Direcció General de Port i Costes, Generalitat de Catalunya (in spanish).

Jiménez, J.A.; García, M.A. and S-Arcilla, A. 1990a. *Analysis and proposal of solutions to stabilize the Ebro Delta: Estimation of the sediment transport in the Ebro river. Contribution to the coastal evolution*. Technical Report LT-2/7. Direcció General de Port i Costes, Generalitat de Catalunya (in spanish).

Jiménez, J.A.; García, M.A. and S-Arcilla, A. 1990b. *Analysis and proposal of solutions to stabilize the Ebro Delta: sediment budget*. Technical Report LT-2/8. Direcció General de Port i Costes, Generalitat de Catalunya (in spanish).

Jiménez, J.A.; S-Arcilla, A.; García, M.A.; Overeem, J.van and Maldonado, A. 1991. The Ebro Delta Project: A first sediment budget. *In*: Kraus,N.C., Gingerich,K.J. and Kriebel,D.L.(ed.) *Coastal Sediments '91*, ASCE, 2323-2335.

Maldonado, A. 1986. Sedimentary dynamic and recent littoral evolution in the Ebro delta (western Mediterranean). *In*: Mariño,M.(ed.) *The integral Ebro system: basin, delta and marine environment*, 33-60, Hermes, Madrid (in spanish).

Wright, L.D. and Coleman, J.M. 1973. Variations in morphology of major river deltas as functions of ocean wave and river discharge regimes. *The American Association of Petroleum Geologists Bulletin*, 57, 2, 370-398.

PRELIMINARY INVESTIGATIONS OF GRAVEL BARRIER EVOLUTION IN

THE BAIE DES ANGES, FRENCH RIVIERA.

E.J. ANTHONY, *Département de Géographie, Université de Nice-Sophia Antipolis, 98 Bld Edouard Herriot, B.P. 209, F-06204 NICE CEDEX 3, FRANCE.*

ABSTRACT

The gravel barriers discussed in this paper fringe two embayments in an area of dense touristic development. The morphology and stratigraphy of these deposits suggest that they started as a series of post-Middle Holocene, drift-aligned fringing beaches and flanking spits that extended across a dissected coastline comprising a Gilbert delta and low hilly outcrops of conglomerates, rich in rounded gravel, alternating with narrow re-entrants, all locked between two bounding resistant rock headlands. Two barrier systems were thus formed, separated by the central Var delta. Distal extension of adjacent spits on a shallow substrate of inner shoreface sands and muddy sands was assured by changing longshore drift termini, with sediment supplied by the delta and the soft sedimentary outcrops. The system south of the Var started as fringing beaches and barriers that became organized into a single concave to rectilinear barrier across small, narrow re-entrants clogged with terrigenous sediments. East of the delta, link-up of opposed spits extending from fringing beaches led to the formation of a concave barrier that grew across a small embayment filled with marine and fluvio-deltaic sediments. The barriers in the study area appear to have undergone little roll-over and have been stable features up to very recently, in association with a relatively stable to slightly recessive sea-level after 4600 B.P., a low wave-energy regime punctuated by fetch-limited, low-frequency storms, and moderate but sustained sediment inputs. Man-induced modifications of the morphology and sediment dynamics of these barriers in the last hundred years have resulted in erosion, initiating a phase of breakdown that presently threatens seafront infrastructure.

INTRODUCTION

The northwestern corner of the Mediterranean coast forms the maritime rim of the French and Italian Alps along which occur beaches niched in embayments between headlands. These beaches are dominated by coarse clasts and are either stable pocket beaches fringing steep coastal slopes or ancient bay-barriers bounding aggraded fluviatile plains or, less commonly, now reclaimed freshwater wetlands. These barrier beaches are best developed in the Baie des Anges (Fig.1), the most important bay on the French Riviera, and have been considerably affected by man-induced modification of their sediment budget and dynamics over the last century. With the exception of very short stretches, these beaches, which are important in terms of both coastal protection and recreation on this developed coast, are presently prone to erosion. Apart from loss of beach recreational surface area, this process threatens seafront property and various infrastructure on the upper beach. Understanding the evolution and dynamics of these barrier beaches has therefore become urgent, especially in view of implementing appropriate solutions, one of which is likely to be more important nourishment in the future. This paper presents the results of preliminary investigations specifically addressing the problems of barrier initiation, establishment, and recent erosion.

SETTING

The present gravel beaches of the Baie des Anges form two bay-barriers separated by the central Var delta bulge and bounded by bedrock headlands. These barrier beaches bound small coastal plains or are directly flanked by steep hills consisting of a variety of Mesozoic and Tertiary rocks ranging from compact limestones and dolomites to marls and puddingstones (Fig.1). These extend seaward below the Holocene coastal sediments onto a steep, very narrow shelf cut by several canyons, some of which form submarine prolongations of steep coastal streams with a torrential regime.

The beaches in the study area experience a very low tide range regime (<0.7 m at spring tides) and are exposed to fetch-limited wind waves, 60 % of which come from the east, the rest embracing a wide sector stretching from NE to SW. Wave period ranges from 3 to 5 s, peaking at 7 s, and the mean wave height from a deepwater bouy recorder is 0.6 m, while the significant wave height is 0.96 m. The low wave-energy regime is punctuated by storm conditions during which wave height may exceed 2 m a few days in the year. Mean sea level shows short-term fluctuations of up to 80 cm resulting from changes in atmospheric pressure associated mainly with the Genoa Depression and, less commonly, with northerly winds blowing offshore.

MORPHO-SEDIMENTARY CHARACTERISTICS OF THE GRAVEL BARRIER BEACHES

The beach clasts in the Baie des Anges are very largely dominated by limestone pebbles representing sediments reworked essentially from coastal outcrops of Pliocene deltaic puddingstones which also occupy the lower catchments of the Var, the major river on the Riviera, the Paillon and the other short, steep coastal

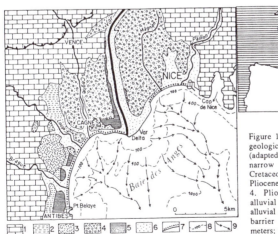

Figure 1. Geographical location and simplified geological sketch of the Baie des Anges area (adapted from Julian, 1980). Note the extremely narrow shelf. 1. Undifferentiated Triassic to Cretaceous limestones, dolomites and marls; 2. Pliocene marls; 3. Miocene volcanic sediments; 4. Pliocene puddingstones; 5. Pleistocene alluvial terraces; 6. Undifferentiated Holocene alluvial and marly sediments; 7. Holocene gravel barrier beaches; 8. Bathymetric contours in meters; 9. Submarine canyons.

streams. Additional sediment sources have been Pleistocene coastal terrace deposits and Miocene volcanic sediments (Fig.1). The morphology and sedimentary characteristics of these beaches have been considerably modified by man through the implantation of recreational, transport and residential infrastructure. Historical data show that the pre-modification morphology is that of massive, single-crested features, around 4 m above the mean sea level and 50 to 100 m-wide. These beaches directly fringe Pleistocene terraces or older cliffed formations or are backed by Holocene alluvial fill and, less commonly, open freshwater bodies that have now been largely reclaimed by infilling. Apart from the Var delta mouth, the barriers are cut by four open stream mouths that have now been stabilized by groynes, but several other coastal streams drain by seepage, periodically opening up in times of very heavy discharge.

Where visible, the sedimentary structure is characterized by horizontal to landward-dipping interbedded units of pebbles, granules and sand. The beach face consists of seaward-dipping beds and is seasonally reworked into a series of low, narrow berms depending on wave energy and runup variations. These berms may be cut by small 1-10 m-wide swash cusps. Beach face reworking is only really important during the few days in the year of stormy waves, which mainly occur between late autumn and early spring. The rest of the time, the beach profile remains basically inert, mobility being limited to the very narrow swash zone. Gravel drift is generally very weak in spite of the often markedly oblique incidence of the short, little-refracted waves crossing the deep shoreface. Drift only becomes significant during storms.

The barrier beaches form a seafront unit commonly 3-15 m-thick. Thickness increases over topographic lows, and may exceptionally attain 30 m. The barriers thin progressively seaward, giving way to shoreface sands, muds or seagrass. They generally rest on old sedimentary formations or on terrigenous Holocene sands and muddy sands interpreted as shoreface facies (Fig.2). These shoreface deposits form the seaward extension of a wedge of heterogeneous Holocene fluviatile, estuarine and deltaic muds and sands that rest in places on marine marls. The present gravel barriers appear to be the youngest of this Holocene suite. Sub-barrier sediments have yielded ^{14}C ages that are no older than 4600 B.P. (Dubar et al., in prep.). Marls have been identified essentially in bore-holes in the Nice bay depression, where they form a transgressive facies that may attain a thickness of up to 30 m and whose summit, dated at 5900 B.P., lies at around -5 m below the present mean sea level (Dubar et al., 1986). The overlying fluviatile and deltaic muds and sandy muds contain peat beds whose ages range up to 5000 B.P. at -2 m, rising to 4200 B.P. at +2 m above present sea level (Dubar et al., 1986). These fluvio-deltaic deposits include intraformational palaeosoils rich in freshwater molluscan assemblages that indicate periodic river flooding.

BEACH BARRIER DEVELOPMENT

Several significant recent studies have provided a conceptual framework for the morpho-sedimentary interpretation of coarse clastic barriers (eg., Carter et al., 1987, 1989; Orford et al., 1991b). Like most meso-scale coastal forms, the formation and evolution of gravel barriers reflect the interplay of various factors, including the sediment supply, the wave regime, the basement control, the need to maintain cross-

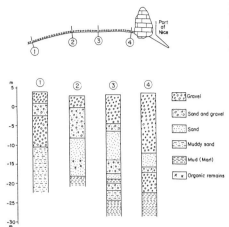

Figure 2. Gravel barrier stratigraphy along part of the Nice seafront. The present barriers rest on a shallow inner shoreface unit of fine sands and muddy sands representing off-lapping terrigenous sediments. These form fluviatile and fluvio-deltaic back-barrier deposits. Shoreface sediments, sometimes exhibiting patches of *Posidonia* sea grass remains, may in turn locally overlie gravel deposits probably reworked from outcropping conglomerates. The basal Holocene unit consists of muds (marl) that form a transgressive landward-thickening wedge within Nice bay.

shore drainage and the rate of relative sea-level change. Orford et al. (1991b), synthesizing much of the work carried out to date on gravel barriers in Britain and North America, have proposed a convenient tripartite framework of barrier evolution that allows for the identification of key factors and processes that successively control the *initiation*, the *establishment* and the *breakdown* of these forms. The formation and evolution of gravel barriers bounding the Baie des Anges are discussed within this framework.

The Initiation Phase

The relatively recent development of gravel barriers has been deduced from their stratigraphic relationship vis-à-vis radiometrically dated sub- and back-barrier deposits. Critical analysis of available ^{14}C data from studies to date, including on-going work (Dubar et al., in prep.) suggests a rapid Holocene sea-level rise of the order of 0.6 cm.a^{-1} up to 6000 B.P. when the sea stood at around -5 m below present, slowing down to 0.3 cm.a^{-1} between 6000 and 5000 B.P. (Dubar, 1987). Stratigraphic sequences from bore-holes show the common presence of spatially discontinuous layers of beach gravel deposits of variable thickness and at variable depths, underlying Holocene shoreface deposits (Fig.2). The pre-5000 yrs B.P. period was probably marked by a weakly erosional front migrating too rapidly landward, under the low wave-energy conditions in this area, to have allowed for the massive liberation and organization of gravels into coherent barriers. Carter et al. (1989) have shown that rapid sea-level rise in a much higher wave-energy environent in Nova Scotia may be associated with both barrier growth and decay. Rapid landward translation of the weakly erosional front with sea-level rise, especially against a steep and irregular inner shelf slope, may have favoured the formation of minor and discontinuous gravel barriers in the Baie des Anges, subsequently overlain by aggradational shoreface sands and muddy sands probably emplaced between 6000 and 4500 B.P. Fine terrigenous sediment inputs at this time may have been sufficiently abundant as to outstrip sea-level rise and to force regressive bay sedimentation and the facies changes in the absence of barriers. The bay facies changes dated by Dubar et al. (1986) between 5900 B.P. and 5000 B.P. from basal marine marls to brackish and finally freshwater sedimentation, relayed seaward by shoreface sands and muds, do not therefore appear to directly correlate with the sea-level history of this area as they coincided with a still slowly rising sea level.

The present barriers started forming after 5000 B.P. in phase with a sea level that was above its present position at ca. 4600 B.P., falling to stabilize since ca. 2600 B.P. (Dubar et al., in prep.). Relative sea-level stability has favoured a more efficient erosional front that reworked poorly consolidated outcropping coastal Pliocene and Pleistocene formations, sometimes cut into cliffs. Barrier development has occurred within the confines of two compact limestone headlands that have rendered the Baie des Anges 'sediment-tight' by preventing gravel leakage into adjacent bays. The Var delta and the less consolidated sedimentary formations between these headlands have served as sediment sources and as hinge points from which have extended the fringing beaches and proto-spits. The Var delta protuberance, which has aggraded on its Pliocene deltaic base and prograded across the narrow shelf into a Gilbert delta by capturing most of the sediments brought down by this river during the Holocene, divided the Baie des Anges into two embayment units, and its bounding gravel spits provided sediment transport corridors allowing for gravel drift both eastwards and westwards. Barrier morphology and stratigraphic relationship to sub- and back-barrier deposits suggest that these barriers were formed by the linking up of drift-aligned fringing beaches and distally growing spits gradually extending alongshore on the more or less shallow, aggraded shoreface, barring small, and generally narrow low-energy re-entrants choked with terrigenous sediments (Fig.3).

In spite of the drift-alignment of these barriers, their final plan-view morphology has been similar to the *swash-aligned Type IIb* barriers of Orford et al. (1991b). These authors have described barrier initiation and consolidation under drift-aligned conditions. The Baie des Anges barriers provide an excellent example of this process. Efficient sediment redistribution from the delta and from intermediate outcropping conglomerates towards the east, and from the Paillon River towards the west, was assured by bi-directional drift. In both sectors, variability of wave approach associated with the wide ranging fetch favoured migration of down-drift sediment termini, leading to the progressive organization of coherent barrier systems, probably over a fairly long time span (10^2-10^3 yrs). The establishment of through-transport corridors formed by the continuous gravel barriers had the positive feedback effect of leading to more efficient longshore redistribution of gravels, thus consolidating the barriers. South of the Var delta, the absence of major re-entrants has favoured the development of a concave to rectilinear linking feature under predominantly southerly drift, while the development pattern east of this delta has been characterized by the link-up of fringing beaches and opposed spits that grew across an open bay which shows facies changes from marine marls to overlying fluvio-deltaic deposits (Fig.3).

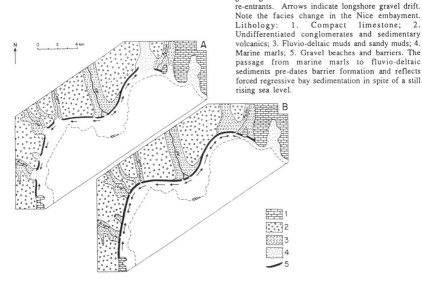

Figure 3. The formation of coherent drift-aligned barriers in the Baie des Anges through link-up of fringing beaches and distally extending spits growing on a shallow shoreface substrate and across re-entrants. Arrows indicate longshore gravel drift. Note the facies change in the Nice embayment. Lithology: 1. Compact limestone; 2. Undifferentiated conglomerates and sedimentary volcanics; 3. Fluvio-deltaic muds and sandy muds; 4. Marine marls; 5. Gravel beaches and barriers. The passage from marine marls to fluvio-deltaic sediments pre-dates barrier formation and reflects forced regressive bay sedimentation in spite of a still rising sea level.

The Established Barrier Phase

The initiation phase over, the overall impression generated by the morphology, stratigraphy and setting of the gravel barrier beaches in the Baie des Anges is that of past stabilization in the form of low but massive, immobile structures, characterized, over the past three millenia, by crestal growth to beyond the limits of normal maximum seasonal swash runup, with wave reworking being limited to the beach face. Conditions for the limited barrier mobility have included relative sea-level stability, a low wave-energy and fetch-limited storm wave regime, and an at least balanced sediment budget. Barrier consolidation, in particular, may have been assured by a 1-2 m fall in sea level between 4600 and 2800 B.P. (Dubar et al., in prep.), with apparent relative stabilization since. The low wave-energy regime has implied a limited capacity for longshore mobility of gravels once the barriers had grown and become consolidated to form inert features that showed little response to minor changes in wave energy and angle of approach. Moreover, because of the low-frequency, fetch-limited storms that have characterized this environment, it seems plausible that barrier roll-over has been very limited, essentially restricted to the initially more open re-entrants, such as the central part of the Nice embayment. The relationship between overwashing, which leads to barrier rollover, and the frequency and magnitude of storms have been highlighted by Orford et al. (1991a). Where sea-level is stationary or very slowly rising (< 0.1 cm.a^{-1}), the rollover rate must become increasingly spasmodic, depending on the increasingly low-frequency, high-magnitude storm events. With time, barrier migration ceases as the fetch-limiting storm event is approached asymptotically.

Examples of these types of barriers include those of southeast Ireland (Orford and Carter, 1982). In the Baie des Anges, sea-level stability and the very limited fetch (Fig.1) may have led to very early cessation of barrier migration. Crestal buildup related to low wave runup, rather than overwash and rollover caused by high runup, may have also been favoured by the predominance of obliquely incident waves.

Finally, sediment inputs, in conjunction with the local sea-level trend and low-energy hydrodynamic conditions, have been such as to maintain barrier stability. Once sea level more or less stabilized, shoreface sediments became exhausted, and cliff outcrops isolated from the transport system, protected by the fringing, aggraded beaches. Gravels brought down by the coastal streams or supplied by the Var, the only available sediment sources other than the barriers themselves, compensated for losses to the offshore zone.

The Breakdown Phase

In spite of the possible existence in the past of longshore variations in barrier dynamics related to sediment source proximity, drift cell dynamics, backbarrier morphology, incident wave energy and clast characteristics, overall barrier evolution on either side of the convex Var delta during the preceding phase was such as to maintain a coherent, essentially concave, plan-view shoreline configuration characterized by sporadic, essentially storm-controlled and slow up-and-down longshore movement of sediments within these "closed" bay-beaches, with the two bounding hard rock headlands allowing for no leakage into adjacent embayments.

The present phase corresponds to one of barrier destabilization, not as a direct result of internal barrier dynamics but as a consequence of human interference in the last hundred years. The erosion and fragile status of the present gravel beaches have been highlighted by Anthony et al. (1992). These are due to a combination of several interacting factors that include past direct beach aggregate extraction, barrage construction across rivers, drastic reduction of beach width and destabilization of the beach face-backbarrier water table as a result of the construction of roads, a major rail link and various other infrastructure, including reflective sea walls to protect routes and infrastructure on the upper beach. The implantation of artificial shorelines associated with leisure ports and reclamation fill, and the emplacement of groyne fields by local government authorities not working in concert and showing no concern for the downdrift consequences, have also been an important factor. Ironically, some of these structures were constructed to stave off beach erosion. The net effects include zero natural sediment inputs and erosion, as well as modification of the sediment drift structure (Fig.4) and the beach morphodynamics. The conjunction of a sediment deficit, of an increasingly polycellular drift system induced by artificial shoreline development, and of barrier constriction through infrastructural implantations on the upper beach has led to beach narrowing and steepening, enhancing the already highly reflective morphodynamic conditions prevailing on these beaches and on the steep inner shoreface. The sediment deficit is due essentially to the fact that losses offshore or in new permanent downdrift sinks are no longer made good by fresh inputs. The construction of groynes and especially breakwaters, including the armouring of the Var stream mouth, have led to a polycellular system comprising both drift and swash alignment. Swash-aligned cells, generally ranging from a few meters to a few tens of meters, are essentially due to changes in local wave refraction patterns engendered by major man-made structures such as the Nice Airport reclamation fill which has taken up the Var delta, while drift-aligned cells vary from a few meters to several hundreds of meters in length, depending essentially on distances between groynes or breakwaters. Short drift-aligned and swash-aligned cells are dominated by seaward directed flows. In the former, effective drift does not attain the necessary thrust to lead to significant downdrift accumulation. Longer drift-aligned cells have tended to become dominated by permanent downdrift accumulation fed by localized reworking of certain hitherto stable updrift sectors that are now being "cannibalized" as a result of new wave refraction-diffraction patterns on either side of marina breakwaters (100-200 m long). Localized beach progradation due to such accumulation of gravel against some of these structures may constitute a mechanism for further sediment loss seaward as the inner shoreface slope increases to attain the critical angle that facilitates offshore sediment movement. Changes in clast organization and sedimentation are also observed locally. The most dramatic example is the immediate vicinity of the west bank of the Var, where a drop in incident wave energy resulting from pronounced diffraction has led to change from gravel to sandy-muddy beach sedimentation and shoreface shallowing as fine sediments from the Var become trapped between the numerous groynes and breakwaters (Fig.4).

The solutions to the problem of gravel loss seem to reside in beach nourishment and reprofiling, to restore the sediment budget and to lessen beach face steepness and reflection. Preliminary analysis of the results of a successful beach nourishment programme in the Nice bay area over the last 15 years shows that annual nourishment, of the general order of 2.5 to 3 $m^3.m^{-1}$ of beach over the entire 4.5 km stretch of beach is just sufficient to maintain the initial beach width.

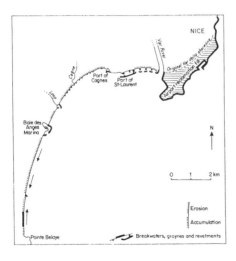

Figure 4. Artificial shoreline structures, present drift cells and instability of gravel barrier beaches south of the Var. Much of this stretch of barrier is affected by gravel loss offshore.

SUMMARY AND CONCLUSION

The preliminary investigations reported here show that barrier development and evolution on the French Riviera have been influenced by sea-level history, antecedent topography, the nature of the wave climate, sediment supply, stream mouths and sediment cell dynamics. These investigations have also provided insight into changes in back-barrier sedimentation. The present barrier beaches appear to be relatively recent, post-Middle Holocene features and their formation has been assured by predominantly longshore gravel redistribution, while roll-over appears to have been minimal. The role of various other factors still needs to be elucidated. This includes the influence of drainage outlets and clast organization, both of which have been shown to be important factors controlling gravel barrier evolution (Carter, 1988; Carter et al., 1989; Orford et al., 1991b).

While being immobile forms, the gravel barriers have tended to break down in places in recent years as a result of heavy man-induced modification of their sediment budget and dynamics. This *in situ* breakdown process involves sediment redistribution alongshore and especially loss seaward to an immediately adjacent steep inner shoreface through stripping of the beach face, leading to barrier thinning and steepening. Interestingly, the breakdown process of anthropogenic origin is similar in certain ways to that affecting naturally disintegrating drift-aligned barriers under sediment deficit stress (Orford et al., 1991b), except that in the present case, the barriers are provisionally wired together by lower- and upper-beach structural defenses that maintain a semblance of coherency, while aggravating the process in places. With the exception of the renourished Nice sector, this breakdown process, although admittedly slow, in response to the prevailing low wave-energy climate, must, if unchecked, lead inexorably to future total barrier beach breakdown. It thus poses a serious socio-economic threat on this developed shoreline.

REFERENCES

ANTHONY, E.J., JAGOUDET, P and PEREZ, S., 1992. Gravel beach stabilization measures in the Baie des Anges, Côte d'Azur. *Cahiers Nantais*. In press.

CARTER, R.W.G. *Coastal Environments*. Academic Press, London, 617 p.

CARTER, R.W.G., FORBES, D.L., JENNINGS, S.C., ORFORD, J.D., SHAW, J. and TAYLOR, R.B., 1989. Barrier and lagoon coast evolution under differing sea-level regimes: examples from Ireland and Nova Scotia. *Marine Geology*, 88: 221-242.

CARTER, R.W.G., ORFORD, J.D., FORBES, D.L. and TAYLOR, R.B., 1987. Gravel barriers, headlands and lagoons: an evolutionary model. *Proceedings, Coastal Sediments '87*, New Orleans, 1776-1792.

DUBAR, M., 1987. Données nouvelles sur la transgression holocène dans la région de Nice (France). *Bulletin de la Société Géologique de France*, 8, 195-198.

DUBAR, M., 1988. La série transgressive côtière holocène de la région de Nice, un modèle sédimentaire. *Bulletin de l'Association Française pour l'Etude du Quaternaire*, 1: 11-15.

DUBAR, M., DAMBLON, F., NICOL-PICHARD, S., VERNET, J.L., CHAIX, L., CATALIOTTI, J., IRR, F. and BABINOT, J.F., 1987. L'environnement côtier des Alpes-Maritimes à la fin de la transgression versilienne d'après l'étude biostratigraphique du site de l'Etoile à Nice (France). *Revue de Paéobiologie*, 5: 289-310.

DUBAR, M., ANTHONY, E.J. and NICOL-PICHARD, S. (*in prep.*). Holocene sea level and sedimentation in the Baie des Anges, French Riviera.

JULIAN, M., 1980. *Les Alpes-Maritimes Franco-Italiennes - Etude Géomorphologique*. Atelier Reproduction des Thèses, Lille, 831 p.

ORFORD, J.D. and CARTER, R.W.G., 1982. Crestal overtopping and washover sedimentation on a sandy-gravel barrier coast, Carnsore Point, southeast Ireland. *Journal of Sedimentary Petrology*, 52: 265-278.

ORFORD, J.D., CARTER, R.W.G. and FORBES, D.L., 1991a. Gravel barrier migration and sea level rise: some observations from Story Head, Nova Scotia, Canada. *Journal of Coastal Research*, 7: 477-488.

ORFORD, J.D., CARTER, R.W.G. and JENNINGS, S.C., 1991b. Coarse clastic barrier environments: evolution and implications for Quaternary sea level interpretation. *Quaternary International*, 9: 87-104.

BARRIER BREACHING IN MICROTIDAL ENVIRONMENTS: THE TRABUCADOR BAR CASE (EBRO DELTA).

Jiménez, J.A.; S-Arcilla, A. and García, M.A. Laboratori d'Enginyeria Marítima. Universitat Politécnica de Catalunya. Gran Capitá s/n. Barcelona 08034. Spain.

ABSTRACT

During the second week of october 1990 a heavy storm hit the Ebro delta (Spanish Mediterranean coast). The coast was seriously eroded, and specially the Trabucador Bar was one of the most damaged zones. A breach of 800 m long and a maximum depth of 40 cm was opened, removing approximately 70,000 m^3 in few hours. Three processes seem to be responsible: longshore transport gradient, offshore transport and overwash/ overtopping. The most of the eroded sediment was transported towards the inner bay due to overwash processes. The recovery of the beach was inefficient, and an artificial dune was constructed to close the breach.

INTRODUCTION

The barrier coasts, and generally the low-profile coasts, are the most sensible coastal areas to storm action. These coasts will experience two kind of response to the storm action. The first one is the same that for sandy beaches, offshore transport and shoreline recession. But the second one, overwash and breach is unique for this kind of coast.

Barrier breaching process has been described by Carter (1982), Carter *et al.* (1987), Penland and Suter (1984) and Penland *et al* (1989) among others. These descriptions may not be considered as valid for all the coasts. Thus, barrier coasts will respond in a different manner depending on their own morphology and storm conditions.

During the second week of october 1990 a heavy storm hit the Ebro delta (Northwest Mediterranean coast). This storm produced important erosive processes. One of the most damaged areas was the Trabucador Bar. This coastal stretch may be identified as a barrier beach, and under storm action it was breached.

The aim of this study is to describe the process, to evaluate the relative importance of each acting factor and to evaluate the coastal response.

PHYSICAL SETTING

The Trabucador Bar is the coastal stretch linking the main body of the Ebro delta (Spain) with the end of the southern spit (figure 1). Its length is about 5 Km, its average width 200 m and the maximum height above mean water level is about 1.5 m. Both width and height are variable with space and time. Figure 2 shows a typical section of the bar.

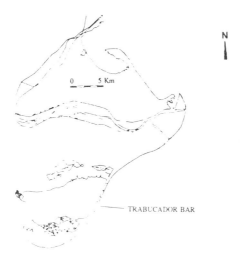

Figure 1. The Ebro delta coast.

In the leeside of the bar, a shallow shelf extending to 1.5 m depth is present. Its length is about 200 m. A continuous submerged longshore bar can be observed along the Trabucador coast. Additional inner bars may appear depending on energy conditions. The main bar system is stable, with only minor movements around its cross-shore position.

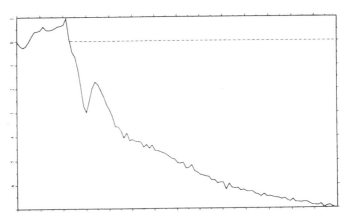

Figure 2. Typical section of the Trabucador Bar.

A positive longshore transport gradient, directed towards the south, exists along the Trabucador (Jiménez et al., 1992). This gradient produces a continuous erosion in its south end, diminishing its width. Moreover, the Trabucador slowly moves landward due to overwash processes. This movement, as in other barrier islands, may be considered as a time-averaged phenomenon, due to storm events over the time (Leatherman, 1983). Due to both processes, the Trabucador Bar has turned clockwise 2 degrees around its north extreme (see figure 3).

The Trabucador is overwashed several times per year. Under these conditions, water flow through overwash channels carries sand from the beach to backshore or to inner bay, depending on its intensity. An important amount of this sediment is recovered by Mestral wind (NW) action, developing little dunes on the top of the bar. The dynamic of these dunes, as expected, is governed by overwash and wind action (e.g. Ritchie and Penland, 1988).

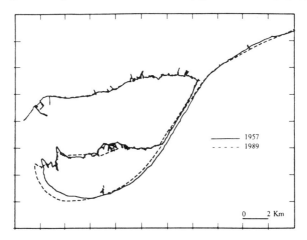

Figure 3. Shoreline evolution between 1957 and 1989.

As all the Mediterranean coast, the Ebro delta is a microtidal environment. The maximum tidal range is around 25 cm. The wave climate of the zone is characterized by dominant east waves, with an average significant wave of 1 m and an average wave period of 4 s (Gracia et al, 1988).

STORM DESCRIPTION

Wave conditions

During the storm, wave characteristics were recorded by a directional waverider at deep waters. Figure 4 shows significant and maximum wave heights recorded during october 1990.

Two main peaks can be identified (october, 8th and 10th), whit significant wave heights of 4.4 m and 3.5 m, and maximum wave heights of 8.8 m and 7.1 m, respectively. Significant wave heights was lower than 2 m after the first peak.

Wave periods were in the highest range, in comparison with usual wave periods for the delta coast. Thus, peak wave periods ranged from 8 to 10 s. This fact is important because it will determine how waves will propagate to the coast.

Waves from E and ENE were predominant during storm, with little directional spreading.

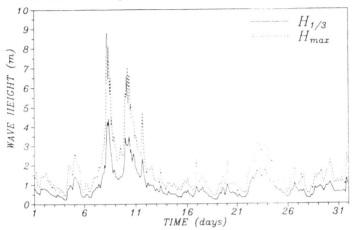

Figure 4. Significant and maximum wave height during october 1990.

Waterlevel

During the storm, the water level rised up flooding the Trabucador Bar. Several factors contribute to water level rise: storm-surge and wave set-up. The former is due to a synergic effect of atmospheric pressure lowering and wind action. Figure 5 shows the water level recorded during the early twenty days of october. The record was obtained by a tide gauge placed at a fishing port near the zone. The surge associated to this factor was around 40 cm.

Additionally, the mean sea level will surge due to wave action. This effect, called wave set-up, depends on wave characteristics. An approximate linear relation between wave set-up and offshore significant wave height was obtained (Jiménez *et al*, 1990), in such a way that, during highest waves action, a set-up around 25 cm was estimated. Taking into account all effects, the surge was estimated around 65 cm.

COASTAL RESPONSE

The storm impact on the catalonian coast produced important erosive processes, being the

Ebro delta one of the most damaged zones. In the Trabucador bar extreme beach erosion occurred.

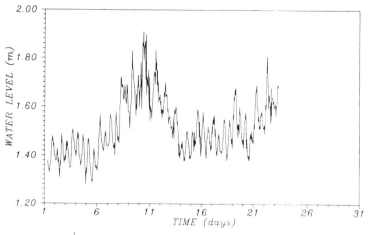

Figure 5. Water level record during october 1990.

The storm produced landward-directed overwash channels and a breach of 800 m long and a maximum depth of 40 cm (figure 6), removing around 70,000 m^3 of sand.

The breached stretch was located, as expected, in the segment where the lowest and narrowest profile existed (70 cm above mean sea level -MSL- and 100 m width). Moreover, the wave energy was highest in that zone, because it presents the highest refraction coefficients along the bar (Jiménez, 1991). Thus, applying the *storm wave susceptibility* concept (Morgan and Stone, 1985), using local wave characteristics, the breached stretch was the most susceptible zone to be hardest eroded.

Trabucador bar breaching may be considered as the result of several processes acting on it. Mainly outer coast erosion due to longshore transport gradient, offshore transport and overwash.

Erosion due to longshore transport gradient

Waves arriving to the coast from E and ENE, generate a longshore transport increasing towards the south along the Trabucador bar (Jiménez et al, 1992). During the storm, high energetic waves arriving from those directions hit the coast. As longshore transport increases with wave energy, stormy waves would produce larger longshore transport gradients in short time periods. This large gradient would erode the outer coast of the bar, narrowing its south part. The narrower the bar, higher the probability to be overwashed and breached.

Offshore transport

During the storm, beach profile was eroded by cross-shore transport. The profile response under these conditions is well known (*e.g.* Vellinga, 1986; Steetzel, 1990). Breaking waves result in an increase in the concentration of suspended sediment in the water column, which is offshore transported by undertow. This process results in the beachface erosion and dune scarping. The morphological result was the narrowing and lowering of the bar.

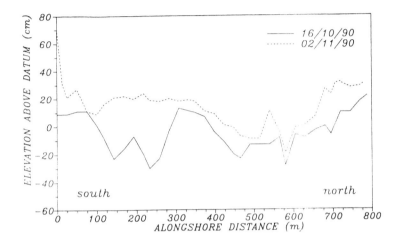

Figure 6. Alongshore profile of the breach.

Overwash

Under sea level rised and storm wave action the bar was overwashed. The overwash intensity was increased by the processes mentioned before, due to the narrowing and lowering of the profile. Under these conditions, during the main storm peak, a water level gradient was stablished between the outer coast and the inner bay. This gradient produced a current directed towards the bay, which intensity was capable to erode significantly the backshore. The removed sediment was transported towards the bay, where it was deposited as a semi-submerged bar along the breach.

The main problem we have to face is to quantify the importance of each process. Potential sediment loss due to longshore transport gradient and offshore transport was estimated around the 30% of the total loss. Thus, under the storm described the most of the eroded sediment was transported towards the inner bay. By that, under those conditions, overwash processes were dominant. Approximately 50,000 m^3 of sand were deposited in the shelf behind the Trabucador forming a bar.

Post-storm recovery

After the wave storm passed, beach recovery processes began. The recovery processes of eroded barrier coast depends on several factors: period of non-storm conditions, availability of sediment, coastal morphology after storm, etc... (*e.g.* see Debusschere *et al*, 1991).

After the storm passed, lower waves arrived to the coast, but sea-level remained rised during one week. At this time, wave direction changed from E to S. Wave action on the coast allowed beach recovery by two processes: longshore transport and onshore transport.

Longshore transport produced swash bars welding on breach ends. This effect can be observed in figure 6.

Lowering in the wave energy allowed onshore sediment transport. This mechanism was no efficient to recover beach profile, due to two reasons:

- The lack of sediment supply, because most of the eroded sediment was deposited in the inner bay. This sediment is only able to be recovered by the beach under the action of waves generated in the bay. These waves with low transport capacity were unable to carry, in an efficient manner, sand towards the beach.
- Since breached stretch was below the MSL, waves broke on beachface reforming and crossing the breach as a bore. Under these conditions, a flow directed towards the bay was established and part of the onshore transported sediment crossed the breach.

The final result of this recovery process was a slow reconstruction of the beachface and sediment trapping by the breach. The breach was not closed during the following two months.

Breach closure

An artificial dune was designed to prevent sediment sinking and winter storms action (Jiménez *et al.*, 1990). Once closed the breach, the beachface was naturally reconstructed, because it acts as a physical barrier to intercept the onshore transported sand. Nowadays, the bar is closed although the breached stretch is narrower than before.

To prevent new breaches, a dune with a longer time of life has been designed (CEDEX, 1988; Steetzel *et al*, 1991). This coastal protection measure is one of the best method to restore and preserve barrier coasts (*e.g.* Penland and Suter, 1988). Moreover, artificial beach and dune nourishment are a good mean to combat the loss of sediment during storm surges (van de Graaff and Koster, 1990).

SUMMARY AND DISCUSSION

During the second week of october 1990, a heavy storm hit the Ebro delta coast. During the storm, high waves arrived from the E and sea level surged. Under these conditions the coast was seriously eroded, being the Trabucador Bar the most damaged zone.

A breach of 800 m long and a maximum depth of 40 cm was opened along the barrier, removing around 70,000 m³ of sand. The most of this sediment (approximately 50,000 m³) was transported towards the bay where it was deposited as a semi-submerged bar.

Three joined processes produced barrier erosion: longshore transport gradient, offshore transport and overwash. The last one was intensified by the others, and it may be considered as the most important agent to breach the bar.

Significance in the sediment budget

The formation of a breach in the Trabucador altered the sediment budget and transport system. At the first stage, during breaching, a large amount of sediment was removed from the outer coast and was deposited in the inner bay. Waves generated in the bay were unable to transport, in an efficient manner, this sediment towards the beach. Thus, breaching must be identified as a punctual sink of sediment. After storm passed, under low-energy waves and high sea-level conditions, constructive processes began. The sediment was transported onshore, but a relative large amount of material crossed the breach towards the inner bay. Under these conditions, onshore transport acted as a sinking process being inefficient to close the breach.

Taking into account the processes described, breaching may be considered as an important agent in the barrier coasts evolution. This process alters coastal morphology and modifies the original sediment budget (*e.g.* McBride *et al.*, 1989).

To avoid this effect adequate protective measures must be taken. In this way, the breach was closed by an artificial dune. The perfomance of this solution was well allowing the beachface recovery.

ACKNOWLEDGEMENTS

This study has been promoted and funded by Generalitat de Catalunya, Direcció General de Port i Costes (contract no.06890SP175), whose support is duly appreciated.

We also acknowledge Oswaldo López for review this manuscript.

REFERENCES

Carter, R.W.G. Barrier Breaching. *Quaternary Newsletter*, 38, 29-32.

Carter, R.W.G.; Orford, J.D.; Forbes, D.L. and Taylor, R.B. 1987. Gravel barriers, headlands and lagoons: an evolutionary model. *Coastal Sediments' 87*, 1776-1792.

CEDEX, 1988. *Study on the Trabucador beach nourishment (Ebro Delta, Tarragona)*. Ministery of Public Works and Transportation (in spanish).

Debusschere, K.; Penland, S.; Westphal, K.A.; McBride, R.A. and Reimer, P.D. 1991.

Morphodynamics of the Isles Dernieres barrier shoreline, Louisiana: 1984-1989. *Coastal Sediments' 91*, ASCE, 1137-1151.

Graaff, J. van de and Koster, M.J. 1990. Dune and beach erosion and nourishment. *In*: Pilarczyk, K.W. (ed.) *Coastal Protection*, 99-120. A.A.Balkema, Rotterdam.

Gracia, V.; Collado, F.; García, M. and Monsó, J.L. 1989. *Analysis and proposal of solutions to stabilize the Ebro delta: wave climate II*. Technical Report LT-2/5. Direcció General de Port i Costes, Generalitat de Catalunya (in spanish).

Jiménez, J.A. 1991. *Analysis and proposal of solutions to stabilize the Ebro delta: study on the wave propagation in the Ebro delta coast*. Technical Report LT-3/1. Direcció General de Port i Costes, Generalitat de Catalunya (in spanish).

Jiménez, J.A.; S-Arcilla, A.; García, M.A.; Overeem, J.van and Steetzel, H. 1990. *Analysis and proposal of solutions to stabilize the Ebro delta: Trabucador Bar erosion during the storm 08/10/90-10/10/90. Analysis and proposal of solution*. Technical Report. Direcció General de Port i Costes, Generalitat de Catalunya (in press).

Jiménez, J.A.; S-Arcilla, A.; García, M.A.; Overeem, J.van and Steetzel, H. 1991. Trabucador Bar erosion during the storm of october 1990. *Revista de Obras Públicas*, Febrero de 1991, 23-30.

Jiménez, J.A.; Gracia, V.; García, M.A.; Valdemoro, H. and S-Arcilla, A. 1992. *Analysis and proposal of solutions to stabilize the Ebro delta: a final sediment budget*. Direcció General de Port i Costes, Generalitat de Catalunya (in spanish).

Leatherman, S.P. 1983. Barrier dynamics and landward migration with Holocene sea-level rise. *Nature*, 301, 3, 415-417.

McBride, R.A.; Penland, S.; Jaffe, B.; Williams, S.J.; Sallenger, A.H. and Wetsphal, K.A. 1989. Erosion and deterioration of the Isles Derniers barrier island arc, Louisiana, U.S.A.: 1853 to 1988. *Gulf Coast Association of Geological Societies Transactions*, 39, 431-444.

Morgan, J.P. and Stone, G.W. 1985. A technique for quantifying the coastal geomorphology of Florida's barrier islands and sandy beaches. *Shore and Beach*, 53, 1, 19-26.

Penland, S. and Suter, J.R. 1984. Low-profile barrier island overwash and breaching in the Gulf of Mexico. *Proc. of the 19th Coastal Engineering Conference*, ASCE, 2339-2345.

Penland, S. and Suter, J.R. 1988. Barrier island erosion and protection in Louisiana: a coastal geomorphological perspective. *Gulf Coast Association of Geological Societies Transactions*, 38, 331-342.

Penland, S.; Suter, J.R.; Sallenger, A.H.; Williams, S.J.; McBride, R.A.; Westphal, K.E.; Reimer, P.D. and Jaffe, B.E. 1989. Morphodynamic signature of the 1985 hurricane impacts on the northern Gulf of Mexico. *Proc. of the 6th Symposium on Coastal and Ocean Management*, ASCE, 4220-4234.

Ritchie, W. and Penland, S. 1988. Rapid dune changes associated with overwash processes on the deltaic coast of South Louisiana. *Marine Geology*, 81, 97-122.

Steetzel, H.J. 1990. Cross-shore transport during storm surges. *Proc. 22th Coastal Engineering*

Conference, ASCE, 1922-1934.

Steetzel, H.J.; Overeem, J.van; García, M.A.; Jiménez, J.A. and S-Arcilla, A. 1991. *Analysis and proposal of solutions to stabilize the Ebro Delta: Artificial dune design to prevent overtopping at the Trabucador Bar.* Technical Report LT-4/6. Direcció General de Port i Costes, Generalitat de Catalunya (in spanish).

Vellinga, P. 1986. *Beach and dune erosion during storm surges*, Comm. 372, Delft Hydraulics Laboratory.

COASTAL EROSION IN THE EASTERN BLACK SEA COAST OF TURKEY
A CRITICAL REVIEW

Gokce K.T.[1], Dakoglu A.[2], Koc S.[3], Gurhan M.[4]
TURKEY

ABSTRACT

Along its 500 km coastline Eastern Black Sea of Turkey is increasingly faced with erosion problems especially during last 10 years both due to natural causes (wave action, land slides) and due to human interventions (marine engineering works, sand mining activities, river basin management). The progressive erosion of the Black Sea coastline causes extensive damage to properties and creates social and economic hardships for the inhabitants at these places. The integral overall management of the coast is necessary to overcome intensive erosion faced today as the essence of the problem is man caused imbalance on coastal system.

1. INTRODUCTION

The relevant part of the coast along the Eastern Black Sea of Turkey stretches over 500 km from city of Samsun to Hopa as shown in Figure 1.

The region has a rough topography and narrow band of flat land. The land in this flat area is very fertile and has a very high commercial value as more than 70% of economical and recreational activity is in this zone. Transportation is mainly by sea-ways as it is considered more economical compared to others. This, together with the strong tradition of coastal fishery led to the construction of numerous coastal works. There are 4 commercial and more than 20 fishery harbors along the coast.

---

1) Dr.Coastal Engineer, Middle East Technical University
2) Environmental Engineer, Trabzon Municipality
3) Chief Engineer, General Dir. of State Hydraulic Works
4) Chief Engineer, General Dir. of State Highways

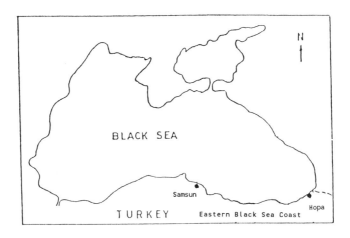

Figure 1. Eastern Black Sea coast of Turkey

State highway runs parallel and very close to the shoreline which is constructed in 1960's, partially by filling the sea. Construction of the highway had both environmental, social and economical impacts on the area. Before the highway the settlement was not heavily concentrated along the shore. Only the central town which was mainly composed of market place, harbor and civic center was situated by the sea. Settlement was distributed to a wide area along the sides of mountains with low density (Kaptan, 1983). The highway has considerably increased the demand to settlement near it, thus along the shore. Also during the construction of the highway some of the inhabitant that used to live on the route of highway have to be moved and where topography permits they preferred to move towards shoreline. New buildings closer to the shoreline, sometimes even on the beach altered the natural equilibrium. Some of these people build some simple coastal protection works just to protect their houses, without being aware of their impact to adjacent coastline. Photograph 1 shows houses along the shore and seawalls constructed to protect them.

The present problems of coastal erosion are therefore primarily caused by man through the construction of highway, several harbors and deterioration of coastal budget in other ways. In the following these will be discussed in more detail.

Photograph 1. Houses and protections on the shore

2. EFFECTIVE MECHANISMS FOR EROSION

The relative stability of a shoreline is depended on the material and energy available to the shore. Any man made changes in the shoreline will progressively change the shoreline configuration until a new condition of equilibrium is reached.

The reasons attributing to coastal erosion may be outline due to loss of littoral material from a specific beach, accretion against littoral barrier, removal of sand for construction purposes, wind action, abrasion by wave action etc. which are also the major causes of coastal erosion in the Black Sea.

2.1 Littoral barriers (harbors, groins and other shore perpendicular structures)

Major rivers and hundreds of smaller ones discharge huge amounts of material to the sea and together with dominant north-western winds a strong littoral drift from west to east is formed. Harbors and groins block the littoral drift resulting in progressive accretion at the west side of breakwaters. Silting the entrance of the harbors causes navigation difficulties which necessitates continuous dredging. Shoreline along the downdrift is heavily effected by the presence of these structures. Usually seawalls are constructed to protect the shoreline but the result is enhanced erosion in further downstream. In several places beaches are lost and houses collapsed due to the presence of groin type structures, which are constructed and used by local people as shelter for their fishing boats. But, unfortunatelly, there is no groin system built along this 500 km coastline for the purpose of protecting a specific coastline.

2.2 Highway and structures to protect the highway

In the Eastern Black Sea region under consideration design wave heights with 50 year return period vary between 5.15 m to 8.60 meter from west to east (Bilgin and Ertas, 1988). Shore parallel highway experiences vital damage due to storm waves and more than 30% of this highway or highway protection are damaged each year. Photograph 2 shows a section of coastal highway with poor protection. As a general measure for the protection of these vital areas, permanent seawall systems have been selected and currently being constructed. Figure 2 shows a typical cross section of shore protection system (Bilgin and Ertas, 1988). However, in areas where such protections are applied, the erosion spreaded along the coasts adjacent to the protected areas and steepening of the coastal profile occurred especially in front of protected areas, which in the long term endanger the stability of coastal works.

Photograph 2. Coastal highway exposed to waves

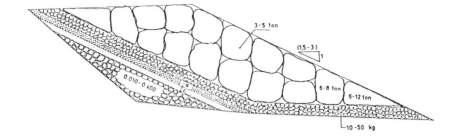

Figure 2. A typical section for highway protection

In principle any coastal defence measure should be designed such that it does not give any harm to the neighbouring sides. Generally the system under being construction aims just to protect the highway and not enough attention was paid on their impact on adjacent coasts.

2.3 Sea filling and uncontrolled sand excavation

It become a common practice to fill the sea especially by municipal authorities to enlarge the land. Possible long term impacts of such constructions have not been given enough importance as short term economical benefits easily overhide them. In the city of Trabzon more than 200.000 m2 of land was filled in recent years to gain land especially for the construction of governmental buildings and recreational purposes.

State highway being so close to the shoreline also increase the uncontrolled excavation of sand and gravel from the beaches. Uncontrolled sand excavation made from the surf zone for the commercial use has also altered the equilibrium of littoral sediment transport.

3. CONCLUSION

The coast erosion effect of these works was in most cases predicted, but apparently had to be materialised before being taken serious. The influence of man on east Black Sea coast is today so large that natural erosion trends are masked. There is however much evidence for the view that coast in the region were in dynamic equilibrium prior to the man made works. The influence of man to these large scale coastal system have broken the continuity of the coastal process.

The essence of the problem is man caused imbalance and the solutions to problems of these nature require interagency co-operation. Therefore national and local priorities have to be prepared jointly with the administrations concerned leading to short term actions and long term planning for coast. Essential is the integral overall management of the coast.

It will thus be required to define a coast protection policy, taking socio-economic aspects into account and determine a set of national priorities. This could be succesfull only if support from the regional inhabitants could be obtained in protecting and developing the coastal region.

ACKNOWLEDGEMENT

Authors would like to express their gratitude to Turkish Chamber of Civil Engineers, Trabzon Branch for supplying relevant material and for their support.

REFERENCES

Bilgin, R. and Ertas, B. (1988) " Eastern Black Sea Coastal Defence Projects " Black Sea Technical University (KTU), Trabzon, Turkey

Kaptan, H. (1983) "A General Overview to the Problems Encountered in the Eastern Black Sea Coastal Zone", 7 th World Cities Day Coast Conference, Trabzon, Turkey.

EROSION ALONG THE BLACK SEA COAST OF TURKEY AND PROPOSED PROTECTIVE MEASURES

Recai BİLGİN

Civil Engineering Department, Karadeniz Technical University,
61080 Trabzon, Turkey

ABSTRACT

The coast of the Black Sea has long been undergoing severe erosion, of which the worst effected sites are in the eastern Black Sea region of Turkey. The use of beach deposits as construction material, construction of various passive protective measures and fishery harbours built without proper design criteria and suitable site conditions had seriously depleted the sediment supply. After construction a new coastal highway during 1960's, erosion rate has accelerated and the highway is now exposed to direct wave action. In 1986, The Highway Department put forward a research project to protect the highway. The extent and the conclusions reached in the project, namely the developed new rubble-mound shore protective structures are described in this paper.

1. INTRODUCTION

The coast of the Black Sea is 4431 km in lenght, nearly 50% of which is actively eroding. The worst effected region by wave erosion with human impact is in the north-east part of Turkey, total length about 500 km. The region has a rough topograhy and a narrow band of flat land.

Tens of groins and fishery harbours constructed in the past three decades without considering the existing wave characteristics and the site conditions have halted the uniform distribution of sediment load transported by the unregulated rivers. The use of beach deposits as construction material for cities, summer resorts and roads is the main cause of coastal erosion in the region. It is estimated that more than 100 million cubic meter of sand and gravel were removed from the beaches (Fig.1) during the thirty-year period between 1960-1990. This unfortunate removal of sediment has been going on continiously each year that resulted in very extensive narrowing of the beaches; and in some cases complete coastal loss of the beach. Coastal recession at rates up to 2-5 m/yr has been recorded with the process extending to depths of 10 m (Bilgin, 1988,1991,1992).

When the new highway extending from Samsun to Hopa was constructed during 1960's, it was not directly exposed to waves. During the past 30 years severe erosion due to the effects of reflected waves has occured and the coastal highway, now in many sites is exposed to direct wave action which necessiates heavy maintenance cost each year. A variety of passive coastal

protection measures were widely used in the past. These included mainly
vertical concrete walls and rubble-mound structures that consist of random
shaped and random placed stones. But these measures have not accomplished
their aims.

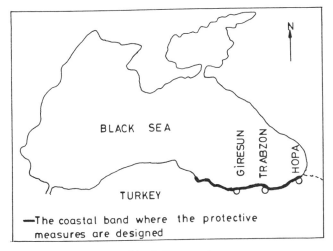

Fig.1. The Eastern Black Sea region

In 1986, The Turkish Highway Deparment started a research project in
cooperation with the Hydraulic Laboratory of Karadeniz (Black Sea) Technical
University to protect the highway from wave attack. The project took three
years and new design criteria for the rubble-mound protective structures
were developed. This is a proper decision. Because rubble-mound structures
require low maintenance and repair costs, their construction is easy and
fast and stones of heavy weight are easily and economically available along
the Black Sea coast. Some other protective measures such as T-groins,
offshore breakwaters and artificial nourishments were also proposed at some
eroded sites. Only the rubble-mound protective measures are under
construction at the sites where urgent measures are needed.

2. THE PROPOSED COASTAL PROTECTIVE STRUCTURES

Before designing the protective structures, following preliminary studies
were undertaken for eastern Black Sea coastal region:

- Damage assessments due to the wave attacks and determination the sites in
urgent need for measures.

- Measurements/estimations of waves, currents, maximum water levels ond
design waves.

- Surveying the sea-buttom topography.

- Application of filter theory.
- Hydraulic model tests.
- Data processing and analysis

Some important items of these studies are briefly explained below.

2.1. Estimation of Design Waves

Around the coasts of Turkey wave measurements are sparse, especially at the coasts of Black Sea are almost absent. The availability in the same region of reliable meteorological data during the past 50 years allowed us to make use of the hindcasting technique. Wind data are derived from two different sources: Namely the meteorological stations situated in the coastal area and the synoptic weather charts drawn in 6-hour period for each day. Thousands of synoptic charts had to be examined to obtain candidate storms providing the highest waves in each year.

Using the fitted line on the Gumbel probability paper, which is chosen as standard plotting paper, design wave heights and the associated wave period for the return periods of 25 years are estimated for each sub-region where the protective structures are to be designed urgently. As an example, design waves estimated for Hopa region are given in Table 1.

Table 1. Estimated design waves for Hopa

return period R_p (year)	synoptic chart		meteorological station	
	$H_{1/3}$ (m)	$T_{1/3}$ (sec)	$H_{1/3}$ (m)	$T_{1/3}$ (sec)
10	6.55	10.24	3.00	6.93
25	7.75	11.13	3.65	7.65
50	8.60	11.73	4.10	8.10

2.2. Hydraulic Model Tests

Depending on the damage characteristics, their cross-sections, buttom materials and extreme wave characteristics of eroded sites, various rubble-mound protective structure types which are considered to be effective were tested in a laboratory model canal under varying waves, water levels, beach and structure slopes.

From the experiments, the rubble-mound protective types were found successful in terms of stability and applicability as shown in Figures 2-5.

2.3. Designing the Protective Structures

Having determined the most suitable types of structures, then dimensional features have to be calculated.

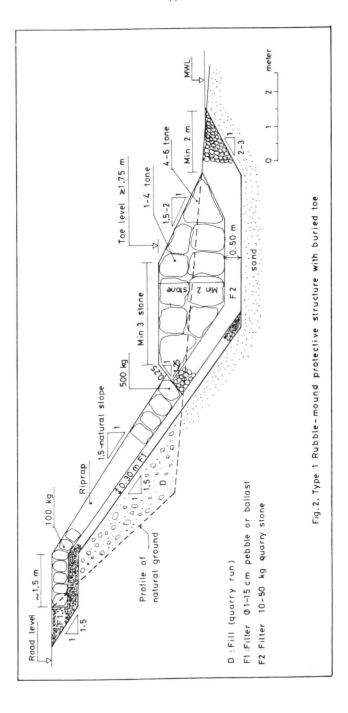

Fig. 2. Type 1 Rubble-mound protective structure with buried toe

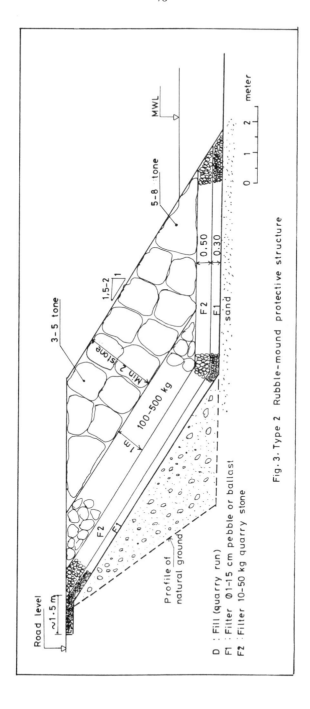

Fig. 3. Type 2 Rubble-mound protective structure

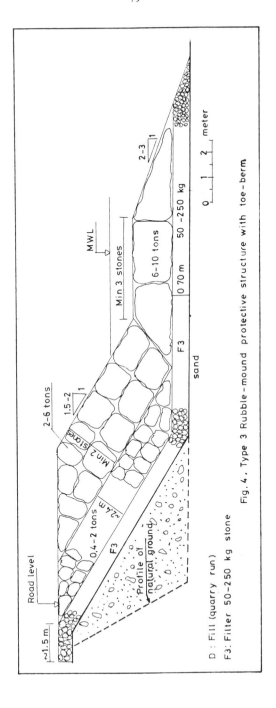

Fig. 4. Type 3 Rubble-mound protective structure with toe-berm

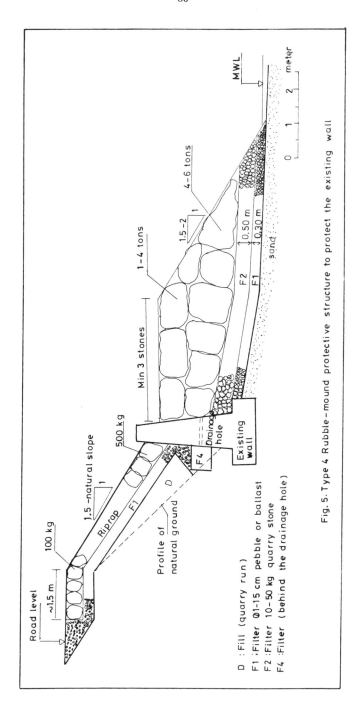

Fig. 5. Type 4 Rubble-mound protective structure to protect the existing wall

A structure can be exposed to wave attack by; broken, breaking and nonbreaking waves depending on the following parameters:

- Water depth at the toe of the structure, ds
- Beach slope in front of the structure,
- Wave height, H.

In designing the rubble-mound protective structures these parameters should be taken into account (CERC, 1984; Bruun, 1985).

Water depth at the toe of the structure, ds, is the maximum measured or estimated MWL, therefore covers the rise of water level due to tide and surge. ds was calculated from the following equation:

$$ds = d + d_g + s \qquad (1)$$

Where;

d = mean water level,
d_g = surge due to tide,
s = surge due to low-pressure and wave characteristics. Besides, wave run-ups and set-ups were considered in the calculations for the Black Sea.

Determination of Stone Weight

For rubble-mound structures, the most important parameter to be determined is the weight of armor units, W_{50}, in the primary cover layer. The other parameters of the structure are determined according to armor unit weight. In this project the armor unit weight for the selected design wave heights (shown in Table 1) were calculated by using Hudson's stability formula:

$$W_{50} = \frac{\gamma_r (H_b)^3_{max}}{K_D (S_r - 1)^3 \cot\alpha} \qquad (2)$$

Where;

W_{50} = weight of armor unit (ton) in primary cover layer,
$(H_b)_{max}$ = Design wave height (m),
α = Angle of surface slope,
γ_r = Unit weight of armor unit (t/m^3),
S_r = Relative spesific gravity which is the specific gravity of armor unit relative to that of the water in which the structure is situated.
K_D = Coefficient that varies primarily with the shape of armor units, roughness of the surface, sharpness of edges and degree of interlocking obtained in placement.

Details of the procedure to calculate the weight is given in CERC (1984).

2.4. Design of Filter System for Rubble-mound Structures

A complete failure of the rubble-mound structure can be expected if the stone gradation is improper or of the filters are not provided (Lee,1972). Many failures have been attributed to internal erosion where in beach materials are removed by percolating water such as that due to water waves, surface runoff and tidal flow. To reduce the danger of piping, either a drainage filter system or an impervious membrane should be considered. A drainage filter system consists of a narrow vertical or sloped layer or layers of graded stone, gravel, and sand behind the rubble-mound structure or underneath the foundation toe.

In view of these facts in mind, it is very easy to understand the reason of failures of the existing rubble-mound structures constructed along the Black Sea coast in the past. Because no filter system is considered at all.

Therefore, three different filters were developed from the conclusions of the laboratory model tests. These are as follows:

Filter F1 : From gravels or broken stones with varying diameters between 1-15 cm and thickness of 30 cm.

Filter F2 : From the quarrystone with varying weights between 10-50 kg and thickness of 50 cm.

Filter F3 : From the quarrystone with varying weights between 50-250 kg and thickness of 70 cm.

3. CONCLUSION

In this study, with the aim of developing new rubble-mound coastal protective structures, especially for sites in urgent need, following conclusions were reached:

1) Coastal protective devices include, but are not limited to, seawalls, groins, offshore (detached) breakwaters, sand bypassing and beach nourishment. These are the "tools". The "rules" for their use come from a knowledge of the beach processes which in many instances are site specific. Systems which work satisfactorily at one location may not necessarily work at another.

2) The rubble-mound sea walls, groins, breakwaters are the most common types of marine structures currently in use. Major reasons are: (a) have good wave energy dissipation characteristics thereby reducing wave run-up and buttom erosion, (b) easy to construct and to repair damages, (c) flexible in response to settlement, (d) economical, since rubble-stones usually are readily available near site.

3) As stated above, the developed devices are urgent measures but the final solution is the artificial nourishments which are very flexible tools in coastal protection policy.

REFERENCES

Bilgin, R. Ertaş, B. and Günbak, A.R. "Prediction of waves and designing the coastal protective structures at eastern Black sea", Proc. of Symp. on "The Role of the Engineering on Turkey's developments", The engineering Faculty of Yıldız University, Istanbul, 1988 (in Turkish).

Bilgin, R. "Coastal and harbour problems and alternative measures at eastern Black Sea of Turkey". Symp. on "the Ecological Problems and Economical Aspects of the Black Sea, Istanbul 1991.

Bilgin, R. "On the calculation of extreme waves and design waves for designing coastal structures", In "Computer Modelling of Seas and Coastal Regions", Computational Mechanics Publications, Elsevier Applied Science, pp. 121-134, London, 1992.

Bruun, P. (Ed) "Design and Construction of Mounds for Breakwaters and Coastal Protection", Elsevire, Amsterdam, 1985.

CERC (U.S. Army Coastal Engineering Research Center),"Shore Protection Manual", Vol 1, Corps of Engineers, Washington D.C., 1984.

Lee, T.T. "Design of filter system for ruble-mound structures", Proc. of Coastal Engineering Conf., pp. 1917-1933, Vancouer, Canada, 1972.

ARTIFICIAL BEACH NOURISHMENT ON THE MEDITERRANEAN NILE-DELTA COAST

Omran E. Frihy* and Robert Dean**

* Coastal Research Institute, 15 El Pharaana St., 21514 Alexandria, EGYPT.
** Department of Coastal and Oceanographic Engineering, University of Florida; Gainesville, Florida, USA

ABSTRACT

Several artificial beach nourishment projects were completed during the last four years on the Nile Delta coast. A beach monitoring program was initiated between 1987 and 1991 at four beaches along the Alexandria waterfront on the western margin of the Nile Delta and at Baltim beach at the central coast of the Delta to evaluate the response of these beaches to coastal processes. These beaches are considered as the primary public summer resort of Egypt on the Mediterranean. Analysis of sand compatibility indicated that inland desert sources of medium and coarse sand are suitable for beach nourishment. Repeated beach profile surveys above and below mean sea level were conducted before and after nourishment. Placement densities ranged from 16 m^3/m to 130 m^3/m and monitoring periods ranged from 2.4 months to 44.2 months. The annual loss rate of sand from the fill was greatest immediately after the placement ranging from 10 to 20% by volume. The calculated sand losses with time within the survey limits were quite variable. The Alexandria projects experienced sand losses ranging between 0% to 70% in a 4 year period, while at Baltim where a high background erosion rate prevails, losses in a one year period exceeded the amount placed.

Most of the Alexandria coasts are rocky and have very little or no beach. The shoreline is generally undulating forming small embayments and pocket beaches. The long-term rate of erosion at Alexandria is fairly small, on the order 20 cm/year. Profile analysis indicates that the small rocky limestone islets and submerged features in the nearshore zone located along the Alexandria shoreline serve to dissipate wave energy in the lower parts of the active profiles, and modify the predominant sediment transport flow to the east. This creates localized shoreline accretion or erosion, especially near the ends of the beach. The central portions are usually either stable or show little erosional and/or accretional changes. Baltim beach is located on a very active shoreline consisting of a sandy arcuate beach, which has experienced a net long-term retreat of about 4 m/yr. The profile analysis of Baltim beach reveals sediment transport consistent with the predominant direction toward the east.

INTRODUCTION

The Alexandria waterfront beach is located along the north western border of the Nile Delta coast (Fig.1). The city is built on a narrow coastal plain separating Maryut lake from the sea. This plain is composed of Pleistocene carbonate sand ridges backed by deltaic deposits to the south. It is 41 km long extending from El Agami to Abu Quir. Alexandria is the second largest city in Egypt after Cairo and considered as the primary public summer resort of Egypt. More than 1.5 million local visitors enjoy the summer season at Alexandria every year. The coastal road of Alexandria (Corniche) is protected from the sea by a concrete bulkhead. The beach is interrupted by two major embayments of the Western and Eastern Harbours. Most of the shoreline is rocky and has very little or no beach fronting the resort facilities constructed along most of the Corniche. The original sand beaches have a slope at the water line which ranges from 7° to 15° and is composed of sediments ranging in diameter from -0.68 to 3.0φ or 1.6 to 0.13 mm (El Wakeel and El Sayed, 1978). They are composed of hydrogenous and biogenic medium to coarse sand varying from loose to fairly well indurated deposits of quartz, shell fragments, heavy minerals and other debris. These sediments are derived from the adjacent Pleistocene limestone ridges located along the western coast of Alexandria and the local rocky limestone outcrops. The shoreline is generally undulating and interrupted by rocky headlands, forming small embayments and pocket beaches ranging from 0.3 to 1.6 km in length. These rocks are

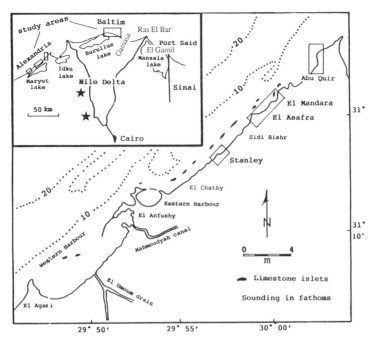

Fig.1. Location of Alexandria and Baltim beach nourishment projects. Asterisks indicate borrow sand quaries.

mainly sandy limestone of Pleistocene formation and beachrock (El Sayed, 1988). The offshore slopes are steeper in Alexandria than the rest of the Nile coast (Nafaa and Frihy, 1992). The inshore and nearshore zones contain outcrops of emerged and submerged rocky islets aligning more or less parallel to the shoreline extending about 300 m into the sea. In most cases, remnants of these rocks form headlands and bottom shoals of the Alexandria beaches. Some beaches, for example Stanley and Abu Quir, are compartmented by rocky limestone headlands.

Most of the Alexandria beaches appear to be experiencing mild erosion with evidence of sand losses, while a few beaches are generally stable. The background erosion rate at Alexandria is fairly small on the order of 20 cm/year (El Sayed, 1988; Frihy et. al., 1992). The existing coastal erosion along the delta was generally caused by the marked reduction in the Nile flow and sediment deposited at the Rosetta and Damietta mouths due to the construction of dams and barrages across the Nile. This sediment deficiency is compounded by the natural reduction of Nile floods due to climatic changes over east Africa. Since closure of the High Aswan Dam in 1964, discharges of sediments at the mouths of the promontories have been reduced to near zero, and subsequently the promontories have been subjected to dramatic erosion. In the case of Alexandria, El Mahmoudyah canal and El Umoum drain were considered as the only source carrying the Nile water and sediments to Alexandria before construction of the Aswan High Dam in 1964 (Fig.1). In spite of the winter storms that attack the beaches, the recession rates at Alexandria are low because of the compartmented nature of the shoreline and the natural protection provided by the rocky sandy limestone islets and shoals. However, the existing erosion may be due in part to the recent reduction of the sediment supply, formerly provided by the El Mahmoudyah Canal and El Umoum drain.

The coastline of the Nile Delta east of Alexandria consists of sandy arcuate beaches, approximately 240 km in total length, interrupted by lagoonal inlets at Idku, Burullus and Manzala lakes (Fig.1). Five recreational beaches lie on the delta coast, including: erosional beaches at Ras El Bar, El Gamil and Baltim and two accretionary beaches at Gamasa and Port Said. Baltim beach lies on the active sector at the central part of the delta, and is experiencing regional retreat (4 to 5 m/yr, Frihy and Komar in preparation). The most significant character of this beach is the cuspate nature of the shoreline along the eastern Burullus coast. Large cusps east of Burullus inlet and Kitchener drain tend to move eastward at a rate of 340 m/month. The fully dissipative characters of the Nile Delta coast and the moderate dissipation of Alexandria have been evaluated by Nafaa and Frihy (1992) based on wave conditions, beach features and sand bar analysis.

COASTAL PROCESSES

Waves and currents associated with the east Mediterranean gyre are the principal driving forces that transport sediment to the east (Inman and Jenkins, 1984). The wave action on the Nile delta is seasonal in nature. Waves reach the shoreline from NW-NNW, this being the direction of major fetch along the delta coast. This in general produces a net longshore sand transport to the east (Manohar, 1981; Inman and Jenkins, 1984). However, local bathymetry particularly at Alexandria reverses that direction to the west. The waves and littoral currents affecting the Nile Delta have been examined by Manohar (1981), Fanos (1986) and Elwany et al. (1988). The wave action is seasonal, with high storm waves approaching from the NW-NNW during the winter (October to March). These are generating eastward-flowing longshore currents and a littoral sediment transport in that direction. Maximum measured current velocities range from 80 to 90 cm /sec. (Fanos, 1986). Swells during the spring and summer are predominantly from NNW-WNW, with a small component from the NNE; this can cause either easterly or westerly sand transport, depending on the local shoreline orientation. The Mediterranean Sea at the Nile delta is almost tideless (range: 30-40 cm) and semidiurnal. Storm surges are significant in coastal erosion and cause overtopping of the corniche and cabins at Alexandria and attack the coastal dunes and delta beaches. Usually, about 14 winter surges per year attack the coast with typical wind speeds ranging to 35 knots (Hamed and El Gindy, 1988).

BEACH PROJECTS

To date, different methods have been used for artificial nourishment on beaches, namely stockpiling, continuous supply, offshore dumping. In Egypt, Tetra Tech (1986) has recommended sand nourishment combined in some cases with hard structures to combat erosion and to create wider beaches for recreational purposes for some eroded beaches at Alexandria and the delta coast. In these projects a direct stockpiling and fill placement method was used. Project dimensions and specifications are listed in Table 1. In Alexandria, for some short projects or long beaches, short groins have been constructed to reduce sand losses. Sand compatibility analysis indicated that the inland desert sources would be suitable. The quality of offshore sources near Alexandria have not been evaluated. The nearest sand quarries are located 120 km from Alexandria, along the desert highway and midway to Cairo.

Table 1. Project Specifications

Operation period	Stanley	Asafra	Mandara	Abu Quir	Baltium
Cost (Egyptian Pounds)	0.1×10^6	0.8×10^6	0.4×10^6	0.3×10^6	0.5×10^6
Mean original beach width (m)	42	30	32	36	55
Beach width after nourish.(m)	63	43	53	62	62
Beach length (m)	250	1100	800	1000	3000
Fill length (m)	172	1100	792	460	2489
Beach elevation (m)	3.3	2.0	2.3	2.4	1.4
Contractor volume (m^3)	22,307	47,777	47,000	21,111	40,000
Distance from borrow area (km)	120	126	126	135	280
Placement Density (m^3/m)	130	43	59	46	16
Monitoring period (months)	42.6	21.2	44.2	42.0	2.4

At Baltim a small fill project was initiated in 1989 as an emergency project and also to serve as a pilot project prior to the final project. This project includes construction of four detatched breakwaters combined with sand fill. The nearest sand quarries are located at El Khatatba, about 280 km south of Baltim, 60 km NW of Cairo (Fig.1). The desert borrow sands are coarser and slightly less well sorted than those from the native beaches of Alexandria and Baltim. Borrow sand was provided by trucking and stockpiled on the native beach. The stockpiled sand was distributed using a Buldozer in accordance with the design dimensions of each beach to minimize losses. Usually the stockpiling operation took place during late winter, generally from February to April, while distribution operations have been carried out during calm conditions of Spring, April and May with waves mostly lower than 1 m. The steep beachface slope of the placed sand is approximately 1:40 is allowed to adjust naturally to an equilibrium slope.

Recently, the Coastal Research Institute (CRI) has carried out marine surveys to develop the coast of the Eastern Harbour, 5 km long, which does not have a natural beach (Fig.1). The specific goal of this project is to create a new beach along this area using sand fill for recreation and also to solve the problem of overtopping during storm conditions (El Fishawy, et al., 1990).

The purpose of the present study is to review the performance of five completed nourishment projects along the coastline of Egypt. The performance of the fill is expressed in terms of sand losses above MSL and above 4 m depth. These projects are located at Alexandria (Stanley, El Asafra, El Mandara and Abu Quir) and at the central delta coast at Baltim. The study is a part of the beach evaluation program of the CRI.

DATA COLLECTION AND ANALYSIS

Monitoring of these projects was conducted using repetitive beach surveys composed of leveling and soundings. The surveys were made between 1987 and 1991. The profile lines were not equally spaced and varied from 25 to 100 m and extended seaward from the baseline to the 4-6 depth contour. The surveys included, pre-existing profiles (before nourishment), post-nourishment profiles (immediately after nourishment) and annual or semiannual profiles. At Alexandria, the pre-nourishment and immediate post-nourishment surveys were carried out by the Egyptian Shore Protection Authority (SPA) and reached only the 4 m depth. Other post-nourishment surveys to 6 m depth were conducted by the CRI. At Baltim beach, beach profile surveys were carried out by the SPA to about -2 m depth during July to September 1989.

Monitoring also included collection and analysis of sand samples in order to characterize the native and nourishment sands, sampling was carried out before nourishment and immediately after flattening the stockpile sand. The survey data were analyzed to obtain the sand volumes. Sand volume changes below the MSL, to a depth of 4 m, at Alexandria and up to 2 m at Baltim were computed by calculating the cross-sectional area under the profile multiplied by distance between profiles in the alongshore direction. Samples were collected from the swash zone of the native beach and during the last survey. Borrow sand samples were also collected along the beach. Repeated photographs of the beach were taken in fixed directions. Selected profiles before and after fill placement are shown superimposed in Figures 3 through 7.

Sediment Texture of the Native and Borrow Sand

The composite grain size distribution of native and desert borrow sand collected from each monitored project is listed in Table 2. and depicted as cumulative curves in Fig. 2. The superimposed cumulative curves reveal significant textural differences between native and borrow sand. It is apparent that for all practical purposes the material placed on the beach from the borrow area was coarser than the natural beach. Field observations indicate that the composition of the fill material has changed in time due to sand sorting and winnowing of the borrow material.

Table 2. Composite grain size distribution of Native and borrow desert sands from Stanley, El Asafra, El Mandara, Abu Qir and Baltim beaches.

LOCALITY		STANLEY		El ASAFRA		El MANDARA		ABU QUIR		BALTIM	
(φ)	(mm)	N(7)	B(14)	N(19)	B(11)	N(15)	B(13)	N(10)	B(10)	N(77)	B(11)
-1	2.000	00.00	08.86	00.00	05.76	00.00	00.00	00.00	00.00	00.00	00.00
0	1.000	10.31	31.67	00.38	02.37	15.83	01.67	02.17	04.40	00.09	00.23
1	0.500	63.35	50.64	04.78	36.14	31.55	21.31	29.30	36.06	03.02	37.17
2	0.250	21.47	08.69	60.06	14.67	33.47	52.47	56.97	36.45	46.79	45.23
3	0.130	04.56	00.14	34.26	20.10	16.33	24.12	11.45	18.39	48.81	15.52
4	0.063	00.26	00.00	00.51	14.55	02.41	00.39	00.11	04.46	01.23	01.83
5	pan	00.05	00.00	00.00	06.69	00.36	00.04	00.00	00.24	00.06	00.02
Phi mean (Dφ)		00.72	00.06	00.83	00.66	01.51	01.11	01.25	01.31	02.05	01.45
Mean (mm)		00.61	00.96	00.56	00.63	00.35	00.46	00.42	00.40	00.24	00.37
Phi sorting (σ_I)		00.61	00.50	00.48	01.50	00.78	00.95	00.72	00.96	00.45	00.55

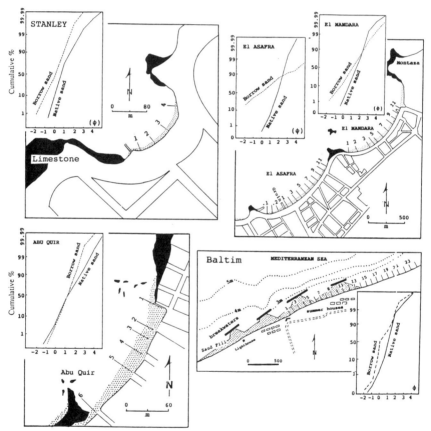

Fig.2. Beach profiles monitored along study projects at Alexandria and Baltim beaches and cumulative composite grain size curves for the borrow and native sands.

Changes Following Beach Nourishment

Changes in the beach berm and nearshore zone were determined by repeated surveys. The performance of the fill in terms of sand losses based on profile comparison before and after completion of the projects is shown in Table 3. The values of sand volume remaining expressed in percentages of the initial volume are plotted for the various surveys in Figs.3 to 7. It can be seen that the volume have systematically decreased with time. These sand losses have primarily been due to littoral drift ("spreading out" losses) and offshore transport beyond the depths surveyed.

Table 3. Loss rates of beach fill in cubic meters at Stanley beach, El Asafra, El Mandara, Abu Quir and Baltim beaches.

Stanley beach (+3.5 to -4.0m).

Period of comparison	Months after project completion	Volume remaining (cumulative)	Volume losses (cumulative)	Volume losses % (cumulative)
Contract volume (22,307 m^3)				
04.04.87 Prenourishment survey				
10.05.87 Post nour.survey	1.0	16,527	5,780	25.61
04.04.87 to 11.08.87	4.0	18,532	3,775	16.92
04.04.87 to 14.10.88	18.0	11,646	11,132	7.79
04.04.87 to 25.09.89	29.0	12,317	10,462	4.78
04.04.87 to 22.10.90	42.0	6,746	15,561	69.76

El Asafra beach (+2.3 to -4.0m).

Contract volume (47,777 m^3)				
10.01.90 Prenourishment survey				
07.05.90 Postnourish.survey	4.0	42,983	04,794	10.03
10.01.90 to 20.10.90	9.0	35,485	12,292	25.73
10.01.90 to 16.10.91	21.2	39,628	08,149	7.06

El Mandara beach (+2.7 to -4.0m).

Contract volume (47,000 m^3)				
11.02.88 Prenourishment survey				
02.05.88 Postnourish.survey	3.0	41,193	05,807	12.36
11.02.88 to 27.09.88	7.0	44,360	02,640	05.62
11.02.88 to 20.03.89	13.0	45,367	01,633	03.47
11.02.88 to 28.09.89	19.0	43,052	03,948	08.40
11.02.88 to 21.10.89	32.0	42,087	04,913	10.45
11.02.88 to 15.10.91	44.2	46,998	00,002	00.00

Abu Quir beach (+2.9 to -2.0m).

Contract volume (21,111 m^3)				
01.05.87 Prenourishment survey				
01.06.87 Post nourish.survey	1.0	19,421	1,690	8.01
01.05.87 to 27.11.87	6.0	15,968	5,143	24.36
01.05.87 to 23.10.88	17.0	17,237	3,874	18.35
01.05.87 to 01.10.89	29.0	12,938	8,173	38.71
01.05.87 to 01.11.90	42.0	8,803	12,308	58.31

Baltim beach (+2.5 to -2.0m).

Contract volume (40,000 m^3)				
11.07.89 Prenourishment survey				
11.08.89 Postnourish.survey	1.0	26,323	13,677	34.19
11.07.89 to 22.09.89	2.4	-13,369	53,369	133.42

Repeated beach profiles are presented for representative beach profiles at each site to demonstrate profile evolution and changes due to nourishment, Figs.3 to 7. The initial beach face slope was steep, approximately 1:40 above MSL. Subsequently this slope was flattened by local current and wave conditions to about 1:50. The final profile appears to represent establishment of a nearly complete dynamic equilibrium profile which could be stable or erosional.

STANLEY BEACH

Stanley beach is a typical small symmetrical pocket beach of approximately 300 m length. The western end of the beach near the swimming pier is characterized by rocky shoals. It had eroded to very narrow width, 10 m on average before nourishment. This erosion threatened the recreational cabins and caused overtopping during storms. The initial volume of nourishment sand was approximately 22 m^3 with medium sand of 0.96 mm. The monitoring profiles cover a total length of 172 meters and included 4 profile lines (Fig.2). Figure 3 shows an example comparison of the pre-nourishment and post-nourishment profiles within the project area. Forty two months after the placement of this amount of sand, approximately 70 % had been lost. The results of the major losses above MSL and up to 4 meters are given in Table 3 and depicted in Fig.3. The net transport, as evidenced by cross section analysis, was reversed from east to west due to the local shoals existing at the western end of the beach near the pier. This end shows remarkable accumulation of sand originating from the eastern end.

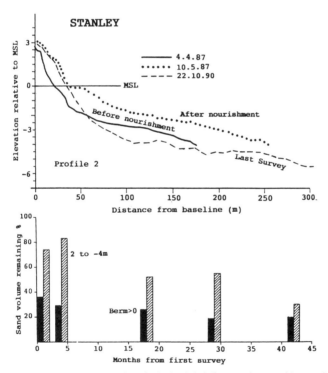

Fig.3. Upper panel: Representative profile at Stanley beach, including pre-and post-nourishment and last survey. Lower panel: Variation in sediment volumes over monitoring period.

EL ASAFRA BEACH

El Asafra beach is an embayment, located approximately 6 km east of Stanley beach. It is 1055 m long with the western and eastern end formed of rocky limestone headlands (Fig.2). El Asafra beach is the largest beach replenishment project on the Alexandria coast. The original sand beach was very narrow or nonexistent in some localities. A sand fill program was executed in conjunction with a short groin of 60 m length. The groin was built to compartment the beach, thereby reducing longshore transport and minimizing sand losses. Approximately 48,000 m^3 of suitable desert sand was placed directly on the beach in 1990. Figure 4 presents a representative beach profile (profile 8) of changes of the pre-nourishment and immediately after project construction to the latest survey in June, 1991. The profile analysis and the natural appearance show an accretion near the east side of the constructed groin. This accretion is due to the local bathymetry which reversed the littoral drift to the west.

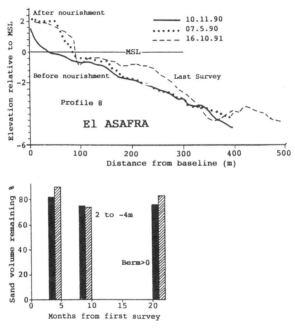

Fig.4. Upper panel: Representative profile at El Asafra beach, including pre-and post-nourishment and last survey. Lower panel: Variation in sediment volumes over monitoring period.

EL MANDARA BEACH

El Mandara beach is located just adjacent to El Asafra beach. The beach is somewhat curved along its 800 m length. Before nourishment, most of the beach had no sand. Limestone shoals exist near the western entrance of the beach while a short groin was constructed on the east side by the Sheraton Hotel in 1982. Approximately 47,000 m^3 sand (medium grain-size $D50= 0.46$ mm) were trucked from a nearby desert and stockpiled on the beach. Beach profiles show the existence of sand bars at about 4 to 5 m depth. This sand was probably eroded from the beach fill in winter and shifted toward the beach in summer. Figure 5 presents representative profile and volume changes over time.

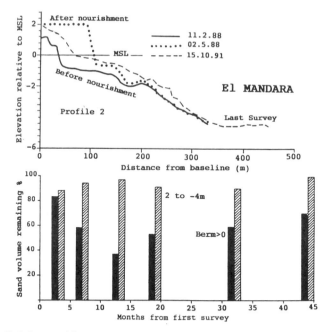

Fig.5. Upper panel: Representative profile at El Mandara beach, including pre-and post-nourishment and last survey. Lower panel: Variation in sediment volumes over monitoring period.

ABU QUIR BEACH

This beach is located farther to the east on the wetern side of Abu Quir Bay. It is a small pocket beach of 100 m length, naturally backed by short rocky subareal headlands at each ends. Rocky islets and shoals front the beach in a water depth of about 2 m. Prior to fill placement of 21,000 m^3 in 1982, there was no natural beach in this area. Figure 6 presents representative profile and volume changes over time.

BALTIM BEACH

The beach fill resort of Baltim is one of the primary public beaches fronting the central sector of the Nile delta coast. It is located almost midway between the Nile promontories and extending from the Baltim lighthouse to 3.0 km east (Fig.1&7). Behind the beach lies a line of Barchan sand dunes. The beach is located on a very active shoreline, which has experienced a net long-term retreat of about 4 to 5 m/yr, based on the CRI monitoring survey.

In order to protect and widen the Baltim beach, four offshore breakwaters in conjunction with sand nourishment have been designed by Tetra Tech (1986). That project, when carried out, will include placement of 0.85 million m^3 of borrow sand between the native beach and the breakwaters (Fig.7). An emergency/ pilot project was initiated in 1989. Between February and August 1989, 90,000 tonnes (40,000 m^3) were placed. Profiling was conducted on a monthly basis to monitor the fill behavior. In less than three months a total of 40,000 m^3 of borrow sand had been removed by waves, whereas a large amount of nourished sand was found just east of the Grand Hotel.

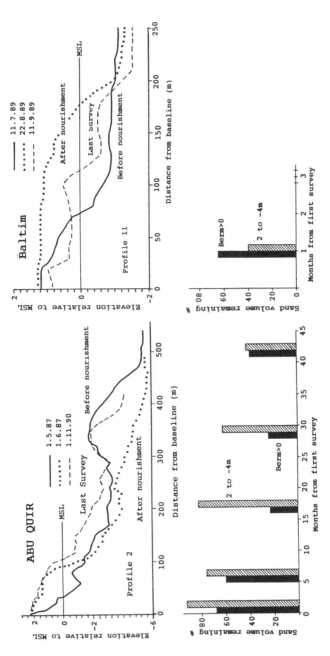

Fig.7. Upper panel: Representative profile at Baltim beach, including pre- and post-nourishment and last survey. Lower panel: Variation in sediment volumes over monitoring period.

Fig.6. Upper panel: Representative profile at Abu Quir beach, including pre- and post-nourishment and last survey. Lower panel: Variation in sediment volumes over monitoring period.

INTERPRETATION

In an attempt to interpret the differences between measured and placed volumes, preliminary calculations were carried out to compare expected volume differences with those measured. Three types of effects were taken into consideration: (1) Profile adjustment to greater depths than surveyed. Sand distributed seaward of the surveyed depths was not lost from the placement area, but simply not documented in the monitoring, (2) "Spreading out" losses which occur at the ends of a project to beaches adjacent to those nourished. In cases of pocket beaches, spreading out losses would be reduced and if the entire beach compartment were nourished and no losses occurred from the compartment to adjacent beaches, the spreading out losses would be zero, and (3) Background erosion losses which are considered to continue at the pre-nourishment rate.

The calculations were necessarily kept simple. It can be shown that the proportion of the fill remaining is given by

$$\frac{V_s}{V_p} = \left(1 - \frac{2}{\sqrt{\pi}} \frac{\sqrt{Gt}}{\ell}\right) \frac{(h_s + B)}{(h_* + B)} - \frac{(ER)(t)}{V_p}(h_s + B) \quad (1)$$

in which V_s and V_p are the volumes surveyed within the placement area and volumes placed, respectively, h_s and h_* are the survey and active depths on the profile, respectively, B is the berm height, ℓ is the project length, G is the longshore diffusivity and represents the capacity of the waves to smooth out the planform anomaly resulting from the nourishment and is defined in detail later, t is the time after nourishment placement, and ER is the erosion rate. Eq(1) considers that complete profile adjustment has occurred, an assumption that is not strictly correct. The longshore diffusivity, G, is defined as

$$G = \frac{KH_b^{2.5}\sqrt{g/\kappa}}{8(S-1)(1-p)(h_s+B)} \quad (2)$$

in which K is a sediment transport coefficient usually taken as 0.77, H_b is the breaking wave height, g is the acceleration of gravity, κ is the ratio of breaking wave height to water depth generally assumed constant at 0.78, S is the ratio of sediment to water densities, and p is the in-place porosity of the sediment.

For a breaking wave height, H_b of 1 m which is considered reasonably representative for this general area of the Egyptian coastline, S= 2.65, p= 0.35, and $(h_* + B)$= 8.5 m, the calculated value of G is 0.037 m^2/s.

In applying the above equations, an attempt was made to the local conditions into consideration. In particular, for the four Alexandria beaches which are of the "pocket" type, the spreading losses were considered to be zero. Table 4 compares the calculated and measured final loss rates. It is seen that there is reasonable agreement in proportions remaining as documented by the surveys at El Mandara (79% predicted vs 83% surveyed) and Abu Quir (58% vs 42%) beaches but the volumes remaining are overpredicted at Stanley (88% vs 30%) and Baltim (13% vs -33%) and underpredicted at El Asafra (74% vs 100%). The spreading losses at Stanley beach were not taken into consideration and, based on the observations, were significant. Also seasonal cross-shore transport exchange beyond the survey depth could cause considerable discrepancies between the measured and calculated volumes and this effect is impossible to taken into consideration with the data available.

CONCLUSIONS

Several artificial beach nourishment projects were completed during the last five years (1987-1992) on the Egyptian coast. A monitoring program has been carried out using repetitive leveling and sounding surveys to evaluate the performance of beach fill projects on the Egyptian coast at Stanley,

El Asafraf, El Mandara, Abu Qir and Baltim resort beaches. The borrow sand was obtained from the desert and transported by means of trucks and subsequently placed directly on the original beaches. It was found that the measured losses from these monitored areas are extremely high. In general, the nourishment projects are effective in protecting and improving the amenity of the eroding beaches. This could be due to the morphologic nature of the compartmented pocket and embayment system of the Alexandria waterfront beach that separated by rocky headlands. The ends of these beaches are either basically erosional or depositional depending on the littoral drift which may be reversed according to the local bathymetry. The central portions are usually either stable or show slow erosional and/or accretional changes. The nourished beaches experienced more rapid erosion than the natural beaches due to losses at the ends of the fill segments and to losses of the fine fraction of the fill material. The texture of the fill material has been changed in time due to sand sorting and winnowing of the borrow material.

During winter, appreciable amounts of sand are eroded from the fill berm and deposited in the breaker zone (2-3 m) as break-point bars. This sand bar mobilizes and shifts landward during summer due to accretionary processes. Sand removed from the beach during storms is moved offshore, where it is temporarily until after the storm when fair-weather swell conditions move it back to the beach.

Table 4. Comparisons of calculated and surveyed volumes for monitoring period

Locality	Volume placed (m3)	Monitoring period (months)	(h_s +B) (m)	Percent volume remaining within survey limits	
				Calculated	Measured
Stanley	22,307	42.6	7.5	88%	30%
El Asafra	47,777	21.2	6.3	74%	100%
El Mandara	47,000	44.2	6.7	79%	83%
Abu Quir	21,111	42.0	4.9	58%	42%
Baltim	40,000	2.4	4.5	13%	-33%

ACKNOWLEDGMENTS

Our appreciation is expressed to Dr. Ahmed A. Khafagy, director of the Coastal Research Institue of Egypt, for providing data and facilities used in this study. Mr. Khalid M. Dewidar for assistance in computer data processing.

REFERENCES

El Fishawi,N.M., Khafagy,A.A., Fanos,M.A. and Frihy,O.E., 1990. Artificial beach project, Eastern Harbor of Alexandria. International Union for Quaternary Research, INQUA, Spain, p.20-25.

Elwany,M.H.; Khafagy,A.A.; Inman,D.L., and Fanos,A.M., 1988. Analysis of waves from arrays at Abu Quir and Ras El Bar, Egypt. Advances in Underwater Technology, Ocean and Offshore Engineering, 16,89-97.

EL Sayed,M.Kh., 1988. Beachrock cementation in Alexandria, Egypt. Marine Geology, 80,29-35.

Fanos,A.M., 1986. Statistical analysis of longshore current data along the Nile Delta coast. Water Science. Journal, Cairo, 1,45-55.

Frihy,O.E, Nasr,S.M., Dewidar,Kh. and El Raey,M., 1992. Spatial and temporal changes at Alexandria beaches, Egypt. International Coastal Congress ICC Kiel' 92, Germany, in press.

Hamed,A.A. and El Gindy,A.A., 1988. Storm surge generation by winter cyclones at Alexandria, Egypt. International Hydrographic Review, Monaco, 36,129-139.

Inman,D.L. and Jenkins,S.A., 1984. The Nile littoral cell and man's impact on the coastal zone of the southeastern Mediterranean. Scripps Institution of Oceanography, Reference Series 84-31, University of California, La Jolla, 43p.

Manohar,M., 1981. Coastal processes at the Nile Delta coast, Shore and Beach. 49,8-15.

Nafaa,M.E. and Frihy,O.E., 1992. Beach and nearshore features along the dissipative coastline on the Nile Delta coastline. Journal of Coastal Research, in press.

Tetra Tech., 1986. Draft final report-shore protection master plan for the Nile Delta coast, V.XI.

INTEGRATED PLANNING AS A NEW PARADIGM OF A COASTAL ZONE MANAGEMENT

Ivo Simunovic, Professor, Faculty of Economics, University of Split
58000 Split, Radovanova 13, Croatia

Abstract

Coastal zones are specific geographical areas, in fact a meeting point of the two different natural phenomena: land and sea. The ecological experience of the coastal zones indicates that these zones are not independent areas unaffected by external influences. On the contrary, the experience has revealed that the populated areas are subject to the causal effect of natural and artificial elements within the immediate or more distant environment, and these influences may have serious effects on the surrounding.

The author uses this fact as a basis for developing the definition of contemporary coastal zone development as an integral concept including the development of nature, man and technology. The coherence of development idea in coastal zones has a further significance, as in these zones interference occurs between continential and maritime life which are two quite different forms of life and nature. That man-generated relation and not nature-generated can be both profitable and damaging. Both the profit and damage jointly require a somewhat different formulation of the development in coastal zones.

The integral type planning is being used while searching for the solution to control the coastal zone development. That means that there has been an attempt to find a corresponding planning type concerning the totality of that development and also of its parts, which runs counter to the concept of the development of overall composite structure. This means: time, space and structure coherence.

In the summarized interpretation, the integral planning respects all the naturally-related life elements of an environment and is relevant for the development and maintenance of a particular environment. From the professional point of view, the discussion concerns the isomorphic sector treated in this case in accordance with the logic of the system applied. An important section of this paper is the part which deals with the interconnections relating to various territorial levels. The author has solved this linking by means of a multilevel matrix system. The solving of a coastal continuum and polycentrism, in fact the development of an integral system from the environmental standpoint, is of utmost importance in developing the structure of the integral approach. The author includes successful achievements and relation possibilities of the suggested system in the context of management of coastal zone development.

1. About the developmental specifics in the coastal zones

From the very beginning till today the most important occupations of people in the coastal zones have been connected with the sea, coast and mild climate. The economic structure of coastal zones is almost the same in each coastal zone all over the world. The slight differences are caused mostly by effects of contemporary littoral processes. However, the very specific feature of life and work in the coastal zones is their particular geographical and other natural prerequisites, which determine nature and structure of human activities in these zones. The main difference between the development process in the coastal and any other geographical zones is to be found in fact that a coastal area itself, as a whole, is the basic resource for its own development.

A coastal zone is a specific geographical area, in fact a meeting point of the two different natural phenomena: land and sea. They represent two different worlds, with different flora and fauna; they can be unified, with the exception of nature, only by man and his activity. That natural phenomenon makes the development process in the coastal zones really specific and unique.

As to better understand the very notion of development process in the coastal zones, some of the principles of ecology will be introduced. Ecology is concerned with the co-existance of living creatures in certain natural stands (oikos).

Ecology as a science had been, for a very long period of time, treated as ecology of sectors, according to the definition of its founder, Ernest Haeckel, and his zoological understanding of ecosystems. Nevertheless, the industrial revolution changed the picture of the world. It was the beginning of the epoch of an exponential growth with an unreached multiplication of production, population and technological achievements. The time of "unnatural products" (substitutes) succeeded that of the "natural products". The occurence of man and its intereference changed everything. Ecosystems, balanced stands, suffered various disturbances.

At the same time, an imbalance in environment became a "normal thing". From the moment when man intervened in nature the ecology as a science has become a controversial and hybrid discipline. The new ecology is no longer the Haeckel's ecology of species, but the ecology of environment, based on the categories and methods taken over from both natural and social sciences.

Ecology as practice is more and more interested in consequences, warning that independent sectoral activities and isolated areas with no mutual influence, either direct or indirect, do not exist any more. In other words, the ecological practice teaches us (by incontestable arguments) about the interrelations within the human environment (oikos); i.e. about the contemporary understanding of development filled with the causal relations among living and not living elements in that natural stand, as well as in all other environmental systems within a cosmic one.

The development in the coastal zones, with all the richness of its contents, takes place on a very narrow coastal strip. This strip today, regardless of country and sea considered, assumes more and more a form of a certain continuum of settlements and economic activities. While the continental zones are mostly characterized by the concentric areals with a high degree of areal cohesion, a particular feature of coastal continuum are the discontinued areals. There is a stronger connection between these areals and continental ones than among the coastal areals themselves. The next characteristic of such a discontinued system of coastal settlements is that they all have similar natural preconditions, structure of economy and are functionally oriented in servicing other economies in the hinterland or overseas. Furthermore, the development process in coastal zones is, to a huge degree, dispersed in the space and therefore, from the standpoint of organization of life and work, extremly uneconomical.

The development in coastal zones possesses enormous natural, economic and social sensitivity. It means that it is very sensitive to the disturbance in natural environment; at the same time, the economy of these zones depends upon the state in other economies, which altogether causes frequent and considerable migrations.

The development in coastal zones relies on a high level of communication, which requires the organization of strong refractional points and traffic regulation (spatial and ecological requirements).

For the development process in coastal zones of the utmost importance is the measure of activities, density and constructions in space; technological and ecological compatibility of activities; measure of natural resources, capability for their exploitation as well as measure of assimilability of particular natural systems.

The development in coastal zones, as a rule, doesn't have a complete developmental structure. In fact, it is highly complementary with the national developmental structure and therefore represents the latter's important part.

The development in coastal zones, due to its intermediary nature, is not based upon its own development poles and does not have the generative power of its own accumulation; hence, it depends on the national driving forces. Nevertheless, once it achieves a certain level of development, it can act independently. When considering the notion and reality of development in coastal areas in the context of previously stated characteristics, the principle of its sustainability regarding its contents and environment, must be especially stressed.

2. About integrated planning of development in coastal zones

At the level of abstaction, a planning process can be described as an intelectual method of transfering the cognition in action or, in other words, as a bridge between human imagination (poesis) and reality (praksis), or, finally, as a guidance on our way from the past to future. By planning, we want to acquire a "knowledge" about the future, i.e. we want to manage the changes.

The planning object is development and its dimensions of space, time and structure. Since the development in the real world manifests in a variety of modalities, the sectoral planning practise is very wide in its range as well.

The main difference between planning and prophecies or other bizzare predictions of future is the fact that planning is based on cognition about the principles and patterns of the development process.

Contemporary planning of today more and more often has an integral concept. In its very notion it represents a system of systems in which the development elements are being coordinated, a balance is created, catastrophies are avoided and rules for all development sectors are made. Naturally, it doesn't imply a perfect system which enables the regulation of each and every occurence in the coastal development system. The main characteristic of an integral planning concept is the necessity of making an overall synthesis at certain stages of the planning process, depending on time, space and contents. A synthesis for determination of means for plan implementation should be done in the same manner. It is specially prominent when the planning process becomes, by its functions, a part of the natural resources and development management system. The term of integral planning appeared in the second half of XX century, when planning was, eventually, accepted as both science and necessity in almost every society. Its real meaning implies the integration of all existing planning systems, such as economic, spatial and social planning, into a new, unique system regarding the same principle which makes the natural stand (oikos) a unity of living creatures and natural environment. Therefore, ecology does not exist as a separate discipline any more. Planning of life in natural environment by its very nature is an ecological activity. This provides a complete meaning of integral planning: it is a process of anticipation of development in certain space, taking into account the values, constraints, sensitivity and potential of all the elements, within the limits of internal sustainability (sustainable development).

None of the sectors in the coastal zones contributes to the development process of these areas either by its generative force or as a development pole. That makes the development, especially in the sectors of economy, infrastructure and environment protection, extremely dependent upon the situation within the immediate and more distant surroundings. The fact of development dependency in coastal zones requires such a planning system adaptable to various levels and times of occurence of particular influences.

The coastal zones are subject to considerable influences coming from different regions and world in general. These influences make part of national development and, in particular local conditions, determine the development paths. From the planning point of view, this can be defined as multilevel planning and the planning with multiple time horizon. The multilevel planning can be defined as a system of matrices, each of them representing different territorial levels, from world to local, with certain vertical flow between different matrices (levels) and horizontal ones within each individual matrix. As far as the coastal zones planning is concerned, the basic matrix is the matrix of coastal zone development. All other matrices are subordinated to this one and being used to identify a planned influence from upper to lower levels. Each of these matrices, analyzed seperately, represent already existing development sectors, which are, in fact, inputs and outputs of the development matrix. Therefore, the causal relations within the system can be understood, as well as interferences and compabilities within the development process (horizontal fluids). In multilevel planning, from the spatial standpoint, the influences are directed from wider to narrower territorial bounds. As far as time dimension is concerned, the multilevel planning can be long- and short-term planning, whereas the long-term planning corresponds to wider territorial levels, while the short-term planning is more suitable for local levels. However, in the multilevel planning system, hereby discussed, long-term anticipation of the future is used in order to enable the definition of relevant world events and processes as a long-term regularities and development patterns. These would represent a determinant of development at lower levels. In such a manner the knowledge frame for creation of short-term and operative plans at local level is obtained.

Naturally, the multilevel planning is not to be understood as a set of separate planning systems at different territorial levels and for different time horizons. It is a unique method of simultaneous planning which is by its objectives and contents oriented toward local coastal areas, in order to provide all necessary prerequisites for operative planning. More precisely, the main goal of this planning system is to manage both the natural resources and the development process in the coastal areas, constantly taking into account the events and processes going on in the system as a whole, as well as changes and new influences and their consequences, as to be able to make crucial decisions for possible redefinition of the development path.

The development in coastal zones is discontinued, i.e. it appears as a great number of smaller units situated along the coast. These units are mutually indifferent, which means that they tend to be better connected with units transversally situated towards the coast than with the other coastal units. That is the reason for the well-know coastal phenomena of the island-coastal-hinterland economic and demographical interrelations as well as of the certain spheres of interest between a continental and overseas economies. The discontinuation can, in fact, be observed as assembling of population and economy around certain points, being the center of these interest fluids. These points are mostly the coastal towns and settlements. Whenever the development process goes on spontaneously, the economic concentrations of a nodal type are being developed along the coastline. Since they have the same, or almost the same, economic and functional structure, they are mutually competitive. In principle, more powerful among these units are always those ones more oriented to the continent and overseas economies than those without these connections.

The most appropriate type of planning for such a development pattern is the polycentric planning. Still, the monocentric planning dominates in some countries due to homogenized interests of the state in their coastal zones. In the great majority of cases, it cuts down the "vertical" economic activities and processes and, simultaneously, makes impossible their integration in a horizontal direction. The monopoly of initiative is created; the development of local capital market and its functions is wrongly shaped and constrained.

The polycentric planning is a natural form of influencing the development of coastal zones, as well as it is the polycentric organization of life and work in coastal zones.

The polycentric planning of development in coastal zones is to be observed as an open planning mechanism. It encompasses the coastal developmental zone and consists of a great number of nodal planning areals. They separately make plans and decisions; they control and execute planned objectives on their own. The polycentric planning in coastal areas is based upon the main principle that each and every plan, both on the horizontal axis (coastal zone plans) and the vertical one (national plan), must take into account all other plans in the whole system. There are several reasons for such a rule, for example the exploitation of natural resources in the case when its repercussions spread out of limits of the particular areal. The other rule is that of hierarchial competency, which implies that determination of certain behaviour types and solution is directed from upper to lower levels.

The misunderstanding of this rule, in addition to its importance in planning practice, leads to the identification of this planning type with the concept of centralistic decentralization. Thus, this rule must be connected with that part of this discussion on development, which dealt with the multilevel system. The development process was successfully defined only when different levels were involved and space (a set of natural systems) treated as an unique and integral natural resource.

The linkage between different levels is established by the means of competition. That is, the choice of activities as well as the extend of their growth in certain areal depend upon the autonomous will of the users from the outside of this areal or in different countries. Since the coastal economy is a dependent system (usually with no independent development pole at the system level), its development is mostly influenced by the including policies and investments, the provision of which is subject to the overall profitability of the entire economic system. According to this planning type a somewhat differently shaped economic systems of individual areals are obtained. They are mutually distinguished by their functions and economic structure. The spontaneous development in coastal zones supports isolated units mostly of the same structure. On the other hand, the polycentric planning establishes the system of units which are both optimally differentiated through their functions and economic structure and, at the same time, mutually connected. In such a way the coastal developmental discontinuity has been broken through.

The development in the coastal zones is integral, composed of various elements interconnected, to a high degree, by the causal relations. Nevertheless, each of those elements has a certain tendency, although confused, toward its own optimal state. As stated above, the influences of development in the coastal zones originate from different levels and different environments, due to their high degree of communication.

Therefore, the development process tends to be discontinued and uneconomical. These facts were basis for the syllogism that the planning activities are the necessity in coastal zones. In order to make the definition of this planning system clearer, let us claim that planning is a permanent activity in coastal zones, unifying the processes of plan preparation, development control and revision of planned proportions.

The continued planning can be described from several standpoints, such as those of planning as a process, preperation of the planning documentation or that of the planning activities.

From the point of view of the planning as a dynamic process, the planning activity is continued because it refers to the entire circle of the planning period, from the moment of a plan definition, through the period of its implementation up to the creation of a new plan.

From the standpoint of planning documentation making, the planning activity is continued for the planning process takes place at various territorial levels and with different time spans, simultaneously and in always new cycles (long-, medium- and short-term time lags). As far as planning activities are concerned, the planning process is continued if the innovations are brought into the development process and planning documents during the planning cycle in such a manner as the planning documentation is to be adapted or changed according to specific demands of the innovation involved (planning - management - revision).

The continued planning, by this definition, is not specific only to the coastal zones development planning. The peculiar feature of the regulation of development in the coastal zones is that it, by its very nature, reqires planning as a continued process.

It is often stated in the literature that a plan is the main result of the planning process, suggesting that planning activity is exhausted in the plan preparation. The more recent planning theory gives the same importance to the activities of the plan realization monitoring, revision of plans and the provision of information for decision making. The integral development, and so the integral planning, are applied to the integral system of heterogeneous elements and of open type. Thus, the activities of observation of certain elements behavior as well as those concerning the preperation of measures applied on the elements tending to digress from the planned path, are of the utmost importance within this system. However, the real reason for introduction of these activities can be found in the fact that the reactions on the higher levels of development remain unknown. These unknown repercusions can not be solved through the plan, but only through the planning activities in the continuity of the development process.

The notion of the continued planning, from the planning activities preperation point of view, is in a high degree complementary to the notion of multilevel and polycentric planning, because it indicates the necessity of permanent planning cycles for different planning periods and for different planning centers. Accordingly, it is necessary to have, at the same time, long-, medium- and short term plans as well as the plans for wider, narrower and local environments.

From the standpoint of planning technology it means that when the short-term plans (sum of their individual cycles equals the medium-term plan cycle) are realized, a new medium-term plan is to be prepared. The same is applied to the preparation of a new long-term plan after the realization of the medium-term plans' cycles. In such a way the territorial scheme can be explained, i.e. it is necessary to have, at the same time, all plans of different cycles at each and every terrritorial level. In this process it is of extreme importance to preserve the continuity, i.e. higher territorial levels must have plans which determine the behaviours at the lower levels.

The continued planning can achieve its purpose only if it goes beyond development, eleminating the obstacles and making the development path always real and dynamic. The obstacles in development are reality, and the innovations and impulse giving the additional energy to development processes. However, these two concepts are unknown, concealed in the moment of the plan preparation. From that point of view, the real meaning of the observation of plan realization is to conceive the problems or the need for innovations in time, in order to create a new planning mechanism for continued and more dynamic development, without threatening the system's internal balance.

All of the three standpoints in fact define in full the notion of continued planning which appears as a necessity in the development in coastal areas.

The characteristics of development planning in coastal zones mentioned above are not all that occur in this process. They just synthesize many other unmentioned details at the level of a general principle.

What is, eventually, the general specific of development planning in coastal zones?

While speaking about the specific features of the planning in coastal zones we often mentioned that some of these specific characteristics can be applied to the planning processes in other geographical zones. Such planning characteristics are the integrity in planning, multilevel planning etc. Nevertheless, when planning of development in coastal zones is concerned, it can be concluded that its general specific is to be found in features of integrity, complexity, polycentrism and continuity. These features represent, altogether and each for itself, an entire nature of the system and thus cannot be separated. They are therefore mutually interconnected in a unique defined entity.

3. The management of natural resources by means of integrated planning

The former practice of the natural resources management had been based on a one-sided definition. The natural resources had been treated as a general good with "unlimited" capacity. Their place in the cost of production was neglected and yet presented a basis for the competition at the commodity market. The expansion of production and, consequently, the natural resources exploitation led to a critical point in the state of natural resources. Hence, an awareness about their limits, price and necessity of control was awaken. That was the reason why the polysectoral understanding of human environment changed the unisectoral one almost everywhere, but especially in the field of management and planning of contemporary development processes.

The general developmental frame in which the integrated planning method was created implies the development and management of natural resources in coastal zones. It focuses on the interactions of various activities going on in the coastal zones as well as on the interactions between these activities and activities in other regions. It can, for instance, signify an integration of environment protection goals in the decision making process regarding some commercial or technical activities, or management of effects of certain agricultural chemical compounds on the coastal sea quality etc.

The integrated planning and coastal zone management, characterized by the interactions of economic activities and natural resources, is in its operational form much simpler in the terms of national than it is in an international jurisdiction. Nevertheless, it can be rather complex even in a national limits in the case when the territory under study encompasses several administrative units.

The integrated management of coastal zones is a continued activity which simultaneously comprises the coordination of the short-term goals and management measures at the local level with the long-term objectives and development policy of national and international level.

The coastal zone management, i.e. its operational activities observe the influence of particular activities as extensively as they effect the natural resources of coastal zones. Therefore, it never replaces a management mechanism within each activity and its integral structure, neither is interested in its internal sectoral reasons.

The essence of integrated management of coastal zones can be simply defined as "management of conflicts and sinergetic effects among the activities in such a way to ensure better use of the coastal zones as a whole, in the accordance with national and international objectives". (The definition is taken from "Integration of environmental considerations into coastal zone management project proposal", OECD, ENV/NRM/89.1.)

4. Summary

In the above chapters a great number of scientifically based opinions about the specific developmental process as well specific development planning in the coastal zones were presented. The intention was to suggest that coastal zones are geographically distinct from other areas, especially because in the coastal zones two completely different naural phenomena meet: land and sea. That natural truth provided evidences about the specific features of development in coastal zones, as much complex and sensitive as its natural environment is. Besides, a set of theses about the development as an integral concept composed of nature, technology and society and also about many other features of development which make it so special, were outlined.

The author's intention was to acquire a knowledge about the development concept as a

whole as to be able to carry out the principles, features and planning methods, in order to establish a scientific basis for definition of development planning in coastal zones.
The third research part of this paper is based on the fact of developmental entropy in coastal zones. It is focused on the field of development and natural resources management as a necessity. The entire notion of development, planning and development management is placed within the frame of space phenomenon, which appears as an element of inspiration and, later on, as a specific phenomenen and resource as a whole, as well as the determining factor of a contemplative and really existing entity.

At the first sight it may be concluded that this paper, as previous ones, suffers the geographical determinism. It may be true in any case where the approach to scientific analysis remains on the principle of strictly divided scientific disciplines. On the contrary, today the marginal contacts and connections among scientific disciplines become more and more evident, as well as facts about the isomorphism of natural, economic and social events. It is almost impossible to find any human activity which is not connected with other activities; it is impossible to imagine that those activities occur somewhere outside the human environment; it is impossible to find completely isolated areas as well as entirely closed limits of space or human activities, i.e. life without any contacts and connections.

When answering the question whether it is possible that all those events and processes, confused and fuzzy by their very nature, assume a sponteaneous and yet regular path, the author indicates that contemporary society enters a phase of strong entropy with overall negative effects.

What is the power of planning as a means for the solution of the entropy problem? The times when some theoreticians glorified planning activity to such an extent that it seemed to be an invisible mechanism making things happen in a certain order in the real world, definitely passed away. The planning activity in the real market economies has from the beginning been defined as an alternative in the hands of politicians.

The contemporary planning paradigm can be understood as a chain within the frame of management mechanisms. The first link of the chain is the planning activity, the second one is monitoring, third is management, the intervention comes in fourth place and, as the fifth link, the planning activity again. The management which directs the development and planning as a crucial means of development management give meaning to entire system.

The intention of this scientific paper was to provoke a need for research of development, planning and management of development in the coastal zones; it was, probably, only partially attained.

References:

1. Enzenberger, H.N., A Critique of Political Ecology, New Left Review, No. 84, 1974.

2. Simunovic, I., Neka suremena nacela ekologije u razvoju privrede i drustva, original scientific paper, Pogledi No. 1/89, Split, 1989.

3. Bicanic, R., O monocentricnom i policentricnom planiranju, article in the book "Policentricni sistem u prostrnoj i tematskoj primjeni", I. Kresic i suradnici, Ekonomski institut Zagreb, Zagreb, 1971.

4. Friedmann, J., Planning in the Public Domain, from Knowledge to Action, Princeton University Press, Princeton-New Jersey, 1987.

5. Kahn, H., On Alternative World Futures: Issues and Themes, Harmon-on-Hudson, 1965.

6. ...... Futures of the Mediterranean Basin, Blue Plan, Sophia Antipolis, 1985.

7. ...... Integration of Environmental Considerations into coastal zone management project proposal, OECD, ENV/NRM/89.1.

8. Stinchcombe, A.L., Ecocnomic sociology, Academic Press, Inc, Orlando, Florida, 1983.

9. Kahn, H., Slijedecih 200 godina, Stvarnost, Zagreb.

10. Pred, A., City-Systems in Advanced Economics, Hutchinson of London, 1977.

PROF. DR. PETAR FILIPIĆ
UNIVERSITY OF SPLIT
CROATIA

COAST-HINTERLAND SOCIO-ECONOMIC RELATIONS AS AN ESSENTIAL ELEMENT OF

INTEGRATED PLANNING OF COASTAL ZONES (THE MEDITERRANEAN CASE)

I General characteristics of the coast-hinterland socio-economic
relations in the Mediterranean

1. Mediterranean is a geographic region and a heterogenous human environment interlinked by joint interest aiming at maintaing equilibrium of all sectors of life and natural systems. The socio-economic reality is however, the only active and mobile element in that realm, being at the same time a developmental goal and origin of problems. Therefore, attempts for studying Mediterranean not only as a natural and geographic unit but as a peculiar and highly sensitive living community appear to be correct.
Mediterranean region makes a triangle enclosed by the coasts of three continents. It is surrounded by countries which differ in size, culture, nationality, as well as in level of development and socio-economic orientation. However, social, political, economic, geographic and other differences do not inhibit Mediterranean countries to undertake and organize common actions for protection and utilization of natural resources.
From that point of view, any attempt for creating generally applicable measures should take into consideration both, common and individual characteristics of area in concern.
Such approach is even more suitable when applied to the coast-hinterland socio-economic relations, an indispensable element of integrated planning process of Mediterranean coastal zones.
This type of socio-economic relations (regarding their inner characteristics) occurs in two ways, as a) common characteristic of the whole Mediterranean area, and b) as specifics of each individual country or regional areas.

2. Common characteristics of the coast-hinterland socio-economic relations in the Mediterranean can be identified in the past, when Mediterranean was, to a great extent, a unique socio-economic system. After its millennial stagnancy, the worldwide litoralization process introduces new activities on coasts of Mediterranean. Their common characteristics are reflected in development of industry, various marine activities, tourism,etc. Location of industry and other activities in the narrow coastal strip also had adverse environmental pollution, excessive urbanization, saturation of natural systems.

Such developmental policy does not take proper consideration of hinterland. People migrate from hinterland to coastal areas, abandoning natural and man-made resources due to which the conversion of land occurs.

Emphasizing common characteristics of the coast-hinterland socioeconomic relations in the Mediterranean has mainly practical value, since it strikes the necessity of a)exchange of experience, b)evaluation of the so far undertaken measures and actions for harmonization of different policies of Mediterranean countries to solving common problems of the coast-hinterland socio-economic relations, and c)undertaking new common actions.

3. Within a global Mediterranean type of development, and under the influence of different natural characteristics, geographic specifics, geopolitical situations and specially different levels of economic development, socio-economic specifics of different parts of Mediterranean have gradually emerged into the first plan.

Identification of such differentials enables definition of specific regional groups of Mediterranean countries with similar characteristics of the coast-hinterland socio-economic relations. This fact makes it possible, to collect and evaluate experiences gained by different countries, to make their regional selection and recommend them in form of measures and actions which should facilitate creation of the socio-economic balance between coast and hinterland, respecting, of course, actual situation and integrity of each individual country.

II Different aspects of the coast-hinterland socio-economic relations

4. The question of the coast-hinterland socio-economic relations in Mediterranean, by its very logic, stems from a triangle: economy - population - human environment. The basic goal of cooperation of Mediterranean countries is to determine new possibilities, through the exchange of experience and use of the existing and new knowledges, for undertaking such socio-economic measures which would contribute to better utilization and more adequate distribution of economic resources and factors, decreased emigration from islands and hinterland and elimination of sources and threats of ecological disturbances.

Each of these basic forms of the coast-hinterland socio-economic relations is in our further work being regarded not as a cause but as consequences which call for seeking adequate measures and actions for their elimination.

5. Processes of concentration of economic activities in the Mediterranean coastlands are very dynamic. They are reflected in increase and concentration of population and economic activities in coastal areas, especially in big cities and ports. Labour force from immediate or distant continental parts (hinterland), attracted by employment possibilities, migrates to coastal towns. These processes are becoming so strong and severe that are very hard to direct and control, and keep within optimum limits. According to available indicators the unemployment, caused by concentration of economic activities on the coast, appears to be the greatest and most severe problem causing a decline in qualitu of living and threat to human environment. At the same time, the concentration process is severely affecting valuable natural resources which in recent developmental approaches are being highly rated (tourism).

Congestion and scarcity of space, traffic congestion, lack of public services and utilities, land speculations, and other added to the previous, reveal the adverse effects produced by such concentration.

The conclusion is that concentration of economic activities in coastal zones appears to be the negative process, both in hinterland and on coasts. In its early phases the concentration process had adverse effects primarily on the socio-economic situation in hinterland. Later on, concentration had, just the same, produced adverse effects in coastal areas; although different in quality and intensity from those in hinterland, those effects had practically similar negative consequences. It proves that
a) coast-hinterland socio-economic relations are related to each other, and that b) they can be neither studied nor treated separately. It is, therefore, necessary to check all what has been learnt so far about those processes, how they have been controled and what conclusions and recommendations they have come to. Such control of the litoralization process would certainly help to relieve the tension and mitigate the concflicts in certain points of concentration of economic activities, as well as to achieve equitable development of the whole area of coast and hinterland.

6. Present socio-economic development in most of the coastal areas of Mediterranean countries is being accompanied with the intensive process of immigration from inland areas. People from immediate hinterland move to urban centres and the relatively narrow coastal strip of the Mediterranean sea.

Excessive litoralization results in depopulation of hinterland and unsufficient utilization of the existing resources. Large concentration of people and their activities on narrow coastal belt severely imperils the environmental balance and in some areas even lessens the possibilities for evaluation of the coast for development of tourism.

Depopulation of hinterland and population movements to coastland have been general processes in Mediterranean so far. Nowadays, those processes are being particularly intensive in less developed countries which go through rapid changes in their socio-economic structure.

The above described process of population movements leads to degradation of social and economic structures and worsening of living conditions in hinterland. Mechanical decrease of population in hinterland affects its age, sex and economic structure. Characteristics of that population are -

deagrarization, decline in the number and domination of aged persons. Generally spoken, we are witnessing lack of human potential, the main factor of the socio-economic development.

Apart from direct adverse effects on demographic structures in hinterland, those occuring in domain of economic development have also been significant; due to lack of labour force some of the important traditional economic activities died out. Since those activities were founded on the primary processing of natural resources, their abandoning caused negative conversions of soil and vegetation.

Being aware of the significance of the problem we strongly believe that all Mediterranean countries will be interested in exchange of experience on measures which are being or could be taken in future to reduce excessive movements of population from inland parts to the coast, and which should result in positive demographic changes in hinterland and improvement of working and living conditions.

7. Recent developmental processes and their rapid growth leave very little time and space for acting upon obtaining a harmonious development. Domination of quantitative over the qualitative aspects of living, which is reflected in increasing concentration of people and goods on coasts of Mediterranean, is transforming into a process of gradual but continous worsening of demographic and physical structures. Consequently, in present days strong emphasize is put on protection of human environment, which is considered to be both, an intervening and a continous discipline for eliminating and checking adverse impacts of development on human environment. Protection of human environment confirms as a sound legal system of actions which are expected to put in accordance different interests of living community, ecosystem and production systems.

Mediterranean area is considered as one space with a high degree of natural sensibility. Therefore, all measures and actions taken to prevent overexploitation of natural systems and deterioration of human environment are given a great importance when dealing with the coast-hinterland socio-economic relations. Any faults that may occur in management of the coast-hinterland development result in a great many adverse effects on living community. Regardless of how these adverse effects are viewed, either individually or as a whole, in their final form they always come out as a worsened state of human environment.

In such circumstances, proper distribution of productive forces, particularly in coastal areas, is of a decisive influence on development and protection of human environment. Mediterranean coasts are the world most significant places of gathering of tourists; those areas are also suitable for intensive agricultural production.

Crisis of energy and transportation requirements attract industries to sea-coasts transportation turns out to be a prerequisite for development of coastal areas. Such constellation of forces and endeavour for new development call for further conquering of coastland, which results in conflicting interests and incompatibility of different activities, as well as in devastation of natural and economic resources of the area.

Ever worsening state of severely spoiled human environment is the result of the inherited philosophy saying that "mistakes are made and mistakes are corrected". Contrary to classical approach ensuring the protection of human environment only by means of juridisction, this attempt should introduce and affirm the protection of human environment as a constitutive part of integrated planning of coastal zones. This approach presupposes exchange of experience between all Mediterranean countries. Different level of development and different approach towards protection of human environment may produce a great variety of measures and actions of the problem concerned.

8. Previously described processes of concentration in Mediterranean coasts are particularly intensive when observed in urban centres on one side and in villages on the other. Movements of population occur much quicker and in much greater quantities than development of functions and physical structures in cities of their immigration. Those processes and their adverse impacts have the economic, demographic and ecological connotations. Municipal authorities and governments of a number of countries are attempting to solve accumulated problems, especially the intensive migrations and uncontrolled growth of cities. Present measures and actions taken in those countries do not apply only to cities but very often to activation of economic resources in hinterland as well. Although those efforts have not produced considerably encouraging results so far, it still stands that such actions are able to add to more adequate distribution of productive forces. There are many examples of the improperly applied methodology, or at one time taken measures incompatible with each other. It is, therefore, necessary to point out to some successful actions, such as the proper allocation of accumulation, seasonal employment, introduction of dual occupations, preparation of integrated plans for coastal zones founded on economic complementarity and interrelationship between coast and hinterland.

III Types of specific regional coast-hinterland socio-economic relations in the Mediterranean

9. Major forms of the coast-hinterland socio-economic relations have been determined in previous chapters, thus defining the frameworks of the case-study analysis, i.e. identification of measures and actions by which socio-economic relations have been defined, at first, and then testing the efficiency of measures and actions taken to check those relations, at second.

Work on identification, analyses and testing will certainly be accompanied with some difficulties of methodological nature. Measures and actions taken by an individual country are being adapted to the situation and actual circumstances in that particular country. The type of the coast-hinterland socio-economic relations and consequently the implemented measures differ to a great extent from one Mediterranean country to another. If a common goal is to be achieved it is, therefore, necessary to identify some regional areas of the typical coast-hinterland socio-economic relations, and then formulate common principles according to which the efficiency of particular measures and actions is to be tested.

Selection of regional areas of the typical socio-economic relations has been done using a simplified method of regionalization based on several indicators. The proposed regionalization is not the only one possible, and does not deny any other regionalization of Mediterranean. This approach is being justified by the necessity for identification of discovered specifics in the initial phase, that representing a basis for starting up the work on case studies.

The results of regionalization reveal the following types of the socio-economic relations:
- of developed Mediterranean countries,
- of moderately developed north Mediterranean countries,
- of the Near East Mediterranean countries,
- of North African Mediterranean countries, and
- of island countries and of the biggest Mediterranean islands.

Selection of principles for definition and regulation of the coast-hinterland socio-economic relations has been done on the assumption that those principles would meet the following requirements:
- parallel development of coast and hinterland,
- integration, i.e. complementarity of coastal and hinterland economies,
- equitable development of coast and hinterland,
- proper and reasonable pressure on natural systems,
- rational allocation of economic and other activities,
- decrease of migrations and activation of people in hinterland,
- assurance of equal living conditions on coast and in hinterland,
- preservation of human environment, and man-made and natural heritage.

IV Description of the model for testing the effectiveness of the taken measures and actions

10. Respecting the selected types of the socio-economic relations, the specific area of Mediterranean and the previously mentioned principles, practical measures and actions for each selected regional area can be tested on a certain sample of population, economy and ecology. This chapter offers the model for testing the effectiveness of the taken socio-economic measures and actions applicable for moderately developed north Mediterranean countries.[1]

It is suggested that testing be done on the following samples:
- distribution of economic activities,
- distribution of population,
- relation between quality of living and economic activities,
- state of saturation of natural systems, and
- state of human environment.

All samples are closely interrelated, well representing the

---

[1] *As seen from the appendix this model is applied to Split region, the central part of coastal Croation area.*

totality of the coast-hinterland socio-economic relations and making part of integrated planning of coastal zones.

A) Testing of measures and actions in relation to the distribution of economic activities

Model for testing of measures and actions is based on the assumption that the implemented measures and actions gave rise to development of new economic activities in the hinterland. Number of questions are put up to asses the effectiveness of such measures and actions - distribution of new economic activities in hinterland, their structure, technology, distribution of population, relation to the coastal economy, development of non-profitable activities, environmental impact of new economic activities.

Measures and actions tested by the model are being discussed also as a certain technique of implementation of the development plans in coastal areas. Therefore, the effectiveness of measures and actions should be measured by different parameters set up by development plans for areas in concern, i.e. from economic, social, physical and environmental aspects (integrative approach).

The reason for testing of measures and actions in relation to the distribution of economic activities is the state in the region before the implementation of measures and actions. Because of the strong processes of littoral concentration of the economy in the region, the population of hinterland, having no choice of employment there, migrated to coastal areas. That, of course, resulted in negative demographic, economic, spatial and environmental processes in the region. These measures and actions were designed to establish the reversibility of such processes, i.e. to induce their positive effects.

Coast-hinterland socio-economic relations greatly depend upon distribution and structure of economic activities. Therefore, measures and actions taken in that domain should meet the following requirements:
a) activate the labour and production potentials of coast and hinterland,
b) prevent congestion of people and goods on coasts,
c) prevent depopulation of hinterland, and
d) prevent devastation of housing stock and productive resources in hinterland.

In order to test the measures and actions taken to ensure a desirable allocation of economic activities the following questions can be put up:
A1 - What new economic activities have been located in the hinterland?
A2 - Was the locating of economic activities prompted by the population distribution?
A3 - What side-effects the locating of economic activities has produced in the hinterland?
A4 - How has the directing of development toward hinterland affected the coastal area?
A5 - Have the measures and actions taken disregarded the human environment protection?
A6 - What are the positive and what the negative effects of the new and

expanded activities in the hinterland?
A7 - Have public utilities and infrastructure been built parallely with new economic activities?

B) Testing of measures and actions in relations to population distribution

In this realm of the coast-hinterland socio-economic relations the test is done for measures and actions expected to stop depopulation of hinterland, prevent the excessive people congestion on coasts and eliminate undesirable consequences caused by those processes. There are two aspects of these migration: one in domain of demography and other in domains of employment, urbanization, deagrarization, ecology and natural systems. Regarding the character and intensity of the process, as well as its consequences, it is to believe that the majority of countries have taken such measures and actions which pertain both, to individual problems and to all of them.
The following questions can be asked in this test:
B1 - Have the measures and actions had an impact on the slowing down of population drain from hinterland?
B2 - Has the process of demographic concentration to a smaller number of larger settlements already begun?
B3 - Has there been abandoning of traditional activities due to the new development?
B4 - Have there been, with the introduction of new economic activities in the hinterland, appearances of double occupation among the inhabitants?

C) Testing of measures and actions in relation to the quality of life and economic activities interaction

This sphere of the coast-hinterland socio-economic relations is the result of policy implemented in distribution of people and goods in a given area; it should agree with the principles of rationality, humanity and quality of living and working environment in a given area. First group of measures and actions to be tested are those which encouraged disperse divelopment on one side and discouraged concentration on the other; those which stimulated parallel and overall coast-hinterland development of unspoiled human environment. Second group of measures and actions consists of those which eliminated conflicts occuring among numerous land users.
Testing of measures and actions in relation to the quality of life and the economic activity interaction is being done either by taking into account their environmental impact or their assistance in the improvement of the quality of life (housing, communal facilities, public services, information).
The test of these measures and actions is done on the basis of the following questions:
C1 - Have the measures and actions taken been at the same time an incentive to a faster economic development and to better quality of life?
C2 - Has economic development been accompanied with corresponding constru-

ction of public services, transportation facilities and infrastructure?
C3 - Have some particular economic activities threatened the quality of life of the population?
C4 - Has life in the hinterland changed because of the new pattern of income earning (housing, health, education, etc.)?

D) Testing the measures and actions in relation to the state of pressure on natural systems

Any development in Mediterranean coastal areas is considered to be happening in a surrounding of very sensitive natural systems, such as: sea, waterways, sources of potable water, seashores, beautiful landscapes, natural and historical sites. Economic and other activities developing on the coast and in hinterland in the past were based primarily on natural systems. Shortage or absence of other factors of development caused excessive exploitation of available natural systems. Utilization of natural systems reveals their strong interdependence, so any improper or faulty utilization threatens to cause disorder in the natural environment. Analyses and tests of measures and actions taken to secure protection of natural systems as prerequisites for development in long run should evaluate the efficiency of reasonable pressure on natural systems.

This particularly applies to natural systems which can be used for various purposes (land), and those which are at one time used by great number of people (air, aquaculture, sea as a waste and effluents recipient, mining, industry, transportation, tourism, etc.).

The following questions can be asked in this test:
D1 - Has the development of new economic activities in the hinterland caused excessive pressure on individual natural systems?
D2 - Are these possible pressures on natural systems the cause of some adverse effects in the development?
D3 - Has there been a concomitant monitoring of the state and pressure on natural systems with the introduction of new economic activities in the hinterland?

E) Testing of measures and actions as to the state of human environment

Human environment is considered to be the qualitative factor in testing of effectiveness of measures and actions. Intensive development on the coast is a severe threat to the human environment. Very often there has been an intention of eliminating the polluting industries from the coast by relocating them to the hinterland. Socio-economic interrelations are expected to secure sound human environment, this criterion is considered to be essential in testing the effectiveness of measures and actions taken to assist the development in the hinterland.

The test of measures and action taken for protection and enhancement of human environment should take into account the already completed work, since effects of previously tested measures and actions were rated by their impacts on human environment. A distinction should be made between measures and actions which had an active attitude towards protection and enhancement of human environment and those which were passive, or restrictive, such as are for example: prohibition, penalty, etc. The efficiency of measures and actions undertaken within this socio-economic relations will, in addition to the previous, depend on the general societal setup, on the state of environmental legislation and awareness of the sound human environment in each country. Testing of measures and actions in this sphere of socio-economic relations always refers to coast-hinterland relations which, in other words, means assessment of human environment that being a result of complex developmental coast-hinterland interrelation.

Within this model, the testing of measures and actions regarding protection of human environment is, in fact, a synthesis, or, in other words, a thorough framework for testifying appropriateness of the whole experience accumulated in the sphere of the coast-hinterland socio-economic relations in Mediterranean.

The test of these measures and actions is done on the basis of the following questions:

E1 - Have the measures and actions undertaken embodied the principle and the human environment?

E2 - Has there been any threat to human environment as the result of the development in the hinterland?

E3 - Has a more "relaxed attitude to environment" been an element of attraction of economic activities from the coast toward the hinterland?

E4 - Which kind of pollution has most frequently occured with the introducing of new economic activities in the hinterland?

11. There are two steps to be taken in implementation of the model for testing the measures and actions. In the first step the measures and actions are being tested for their efficiency and effects produced in one of the three main sectors: economy, population and ecology; in the second step they are being tested with regard to all three sectors at once.

Effects of each measure and action will be considered valuable and useful only if they anticipated the previously mentioned principles, and if they were positive to all three main spheres of the coast-hinterland socio-economic relations.

The testing of measures and actions undertaken is effected in order to evaluate the degree of their contribution to the general objective, i.e. to the more balanced development of coast-hinterland socio-economic interrelations, to the protection and enhancement of the human environment and to more rational use of natural systems.

APPENDIX

CASE STUDY: COAST-HINTERLAND SOCIO-ECONOMIC INTERRELATIONS IN A SELECTED AREA OF CROATIA (SUMMARY)[1]

Presentation of basic socio-economic features of the region of Split

Motives for selecting the area

The coast-hinterland socio-economic interrelation exist in all Croatian coastal regions. They differ from one region to another due to different processes and circumstances under which such relations have been formed.

Motives for selecting the Split region for studying the coast-hinterland socio-economic interrelations are the following:

- the region of Split is, for its territory, economy and demography, the typical Mediterranean area, i.e. it has the inhabited islands, urbanized and densely inhabited littoral and the less developed and geographically isolated interior;

- this region is one of the least connected coastal regions to other parts of the country, both from the aspects of economy and infrastructure, therefore, the influence of the more developed territories on the region is not significant;

- to relieve consequences of such isolation various measures and actions have been taken in this region to encourage the development of the hinterland, based on local models, funds and resources;

- consequences of the uneven development in the region of Split are typical, and may, therefore be used as an appropriate sample in the model for studying the coast-hinterland socio-economic interrelations. In other words, their way of manifestation is very much alike those in many other coastal areas in the country and on the Mediterranean; and,

- the region of Split has long been a very consistent socio-

---

[1] *Filipić, P. and Šimunović I.:"Case study: Coast-hinterland socio-economic interrelations in a selected area of Croatia", UNEP-MAP-PAP, Split, 1987, p. 47.*

economic and physical entity, concentrated around city of Split to which it is connected with strong economic, demographic, cultural and infrastructural links.

Basic geographic characteristics of the region

The region is situated in the central part of the Croatian coast of the Adriatic. It is one of the four regional areas of macroregia Dalmatia. The capital is Split. The region stretches over 4,188 square kilometers, which makes up 7,4% of the territory of the Republic of Croatia.

According to the administrative division, the region is divided into nine communes: Sinj, Imotski, Trogir, Split, Omiš, Makarska, Brač, Hvar and Vis, or into three areas - hinterland, coast and islands according to the geographic-administrative division. Communes of Sinj and Imotski are situated in the interior and they cover the area of 1,683 sq.km.; communes of Trogir, Split, Omiš and Makarska are in the coastal zone and cover the area of 1,726 sq.km. The island communes are Brač, Hvar and Vis with total area of 808 sq.km.

The study of the socio-economic interrelations carried out by the project is focused to the coastal and the hinterland areas. The coastal and hinterland areas of the Split region do not coincide with the boundaries of the coastal and inland communes, therefore the division of the region on its coastal and inland areas being based on geographic criteria of division. The coastal area of the region is, according to geographic criteria, a narrow coastal strip, sharply separated from the inland area by the mountains; it covers the area of 711 sq.km. Interior parts of the region with total area of 2,595 sq.km. lie behind the mountains.

Why is it necessary to change the state?

The uneven development and consequently the uneven distribution of population and of economic activities, as well as the over exploitation of the natural systems have produced many adverse effects on the state of the environment and on development in general in the Split region. All recent development plans suggest that the existing state be changed, primarily for the following reasons:

- population: emigration of population from hinterland to coastal areas; predominance of aged population; weakened biodynamic population structures in hinterland.

- economic development: economic retardation of the hinterland in comparison with the region and other areas; slow and inadequate change of economic structure; old and obsolete technologies; unemployment; poor connections (structural and technological) with economy on the coast and on islands.

- physical development: a) hinterland: deagrarization as a way of neglecting agricultural occupations; sparse population; inproper distribution of economic activities; urbanization of the centres of the communes only; poor quality of public services and public utilities; poor communications with coastal areas; b) on the coast: intensive concentration of the economy and life in general; uninteruped urbanization of the whole littoral

haphazard development; traffic congestion; poor communications with continental part of the country; etc.
- environment: deterioration of natural and man-made resources in hinterland; pollution of rivers, streams; environmental pollution in some urban areas; pollution of the air, water, sea, and the environment in general in coastal areas; haphazard development causes not only developmental disturbancies, but it is also spoiling the landscape.

Many of the stated problems steam from the coast-hinterland socio-economic interrelations which need to be entirely changed. It means that different measures and actions should be taken to encourage faster economic development of the hinterland. Economic development should be accompanied with development of infrastructure, public services, schools, health services, etc. By providing better economic and social preconditions in the hinterland, many negative processes occuring in the spheres of economic, social and urban life would be eliminated. Population pressure on coastal areas would be reduced, thus securing balanced development of the region , rational use of space and of natural resources, and a sound human environment as well.

Testing of measures and actions taken

The procedure of testing measures and actions is quite simple. Descriptive answers to individual questions are being provided, together with some quantitative measure if the required data were being available. This is being compared to the state in the area and to the state after the measures and actions have been implemented.

The procedure for testing measures and actions has started with presenting systematically the state of the coast-hinterland socio-economic interrelations through their consequences for the development. Than the objectives are being described conducive to the change of the state of the socio-economic relations. This is being followed by the description of measures and actions undertaken to change the state, and finally, a systematic presentation of the state after the implementation of these measures is being given.

The testing of measures and actions is being done with relation to the changed state of the coast-hinterland socio-economic interrelations, from five different standpoints.

In evaluating the determined phenomena, efforts have been made to show the distinction between the quality of the intention, the measure and the behaviour of the task performer. Therefore, whenever it has been possible, the repply includes personal appreciation as to whether certain situation is the consequence of an ill-conceived measure or of its inadequate implementation.

The final assessment of the measures and actions adopted has been derived exclusively from the previously given answers. It means that the anewers to the questionnaire, together with the five selected test models, have been the basic for the evaluation of each individual measure and action. For example, an action can be positive as it has raised the level of development in the hinterland and ended the population drain, but it is at the same time negative for its adverse effect on the state of the human environment, on the quality of life and on natural systems. Therefore, in describing its value a synthetic appraisal, either positive or negative, is not given, but rather a detailed description of all the five test models.

In spite of the predominance of the qualitative approach in testing of measures and actions in this case study, some quantitative measure of the state have also been given. The testing model has been designed so as to determine the importance of each question and each reply on the basis of the experts appraisals on the socio-economic relations. Each question and each reply were weinghted by relevant ponders. The influence of each measure and action on the socio-economic relations is graded from -3 to +3, and ponders from 0 to 1.

For assessint the effectiveness of the measures and actions and of ponders the Delphi method is being used, which means that "the truth" about the evaluation and about ponders is not looked for on the level of individual answers, but as a results of the sum of individual opinions.

Evaluations and ponders may be analyzed from different aspects. For example: five groups of criteria may be divided into criteria related to development (distribution of economic activities, distribution of population) and criteria related to environmental protection (interrelation between the quality of life and economic activities, actual state of pressure on natural systems and the state of the human environment). Such division gives us the opportunity of analyzing and testing measures and actions by evaluating the state of the human environment.

The sum of ponders for the developmental criteria in the model is 5.6, and for the environmental criteria it is 5.5. It proves that none of criteria has been prefered in the experts appraisals, but on the contrary, their appraisals have been made on assumptions of the balanced socio-economic development, on the need for protection and enhancement of the human environment, and on more rational pressure on natural systems, or, in other words, on the principle of integrative planning.

Table 1: Socio-economic characteristics of the Split region
(selected features)

	Coast	Hinterland	Islands	Split region	Croatia
Population (1961)	192629	102262	33378	328269	4159696
(1991)	333277	98733	29384	461394	4760344
Population density(1961)	111,6	60,8	41,3	77,8	73,7
(1991)	193,1	58,7	36,4	109,4	84,2
National product					
per person (US$,1985)	2610	732	1756	2110	2277
Economic structure					
(in %, 1985)					
- industry	37,9	41,9	28,5	37,9	
- agriculture	0,8	16,0	6,3	2,2	
- tourism	4,5	2,0	24,3	5,1	
- other	56,8	40,1	40,9	54,8	
Ways of meeting need					
for public services					
(in %, 1985)					
-in place of living	100	25	20		
- out (in Split)	–	75	80		
Length of asphalted roads					
per sq.km. (1985)	0,49	0,38	0,38	0,42	

Table 2: Measures and actions taken to encourage development of under-development areas (by areas and by cronological order)

Measures and actions	Trogir	Split coast	Split hinterland	Omiš	Makarska	Sinj	Imotski
1. Agrarian reform (1946-1948)	xx	xx	xx	xx	xx	xx	xx
2. First industrializat.	x	xx		x		x	
3. First electrification (1946-1955)	x	xx	x	x	x	x	x
4. Federal Loans (1961-65)	x	x	x	x	x	xx	xx
5. Second industrializa. (1965-1972)		x	x	x		xx	xx
6. Croatian Fund (1971-90)						xx	xx
7. Incentive investment in tourism (1969-1972)	x	x			x	xx	
8. Split's Fund for Hinterland (1973-1989)			xx				
9. Dalmatian Fund (1980-1989)			x			xx	xx
10. AIK Programme agriculture (1986-)	x		xx	x		xx	xx
11. Investments of Split's business firm (1965-)	x		xx	x		xx	xx
12. Favourable bank loans (1979-1985)			x			x	x
13. Major infrastructure - regular programmes	x	xx	x	x	xx	xx	xx
14. Middle-range (5-year) economic plans (1955-)	x	xx	x	x	x	x	x
15. Long-range development plans (1986-2000)		xx	x				
16. Physical plans (1969-2000)	xx	xx	xx	xx		x	xx

x = lower intensity   xx = higher intensity

Table 3: Testing model

Question	Sign (+ or -)	Grade (0 to 3)	Ponder (0 to 1)	(3x4)	Possible	Actual in %
1	2	3	4	5	6	7
A.1.	+	2	0,9	1,8	2,7	
A.2.	+	2,5	0,5	1,25	1,5	
A.3.	+	2	0,1	0,2	0,3	
A.4.	+	1	0,6	0,6	1,8	
A.5.	−	2	0,1	−0,2	0,3	
A.6.	−	2	0,5	−1,0	1,5	
A.7.	+	1	0,7	0,7	2,1	
A				3,35	10,2	32,8
B.1.	+	2	0,8	1,6	2,4	
B.2.	+	2	0,5	1,0	1,5	
B.3.	−	1	0,3	−0,3	0,9	
B.4.	+	3	0,6	1,8	1,8	
B				4,1	6,6	62,1
C.1.	−	2	0,8	−1,6	2,4	
C.2.	+	2	0,1	0,2	0,3	
C.3.	+	1	0,6	0,6	1,8	
C.4.	+	2	0,5	1,0	1,5	
C				0,2	6,0	3,3
D.1.	+	1	0,7	0,7	2,1	
D.2.	+	1	0,3	0,3	0,9	
D.3.	−	0	0,4	0	1,2	
D				1,0	4,2	23,8
E.1.	−	2	0,9	−1,8	2,7	
E.2.	−	1	0,6	−0,6	1,8	
E.3.	−	1	0,5	−0,5	1,5	
E.4.	−	2	0,1	−0,2	0,3	
E				−3,1	6,3	0
(A+B+C+D+E)				5,55	33,3	16,7
(A+B)				7,45	16,8	44,3
(C+D+E)				−1,9	16,5	0

Conclusions and recommendations

1. The case study has shown that the coast-hinterland socio-economic interrelations in the Split region are being quite intensive. Before the measures and actions had been taken, the socio-economic relations between the coast and the hinterland were developing spontaneously thus producing both, positive and negative effects in the overall development. The most severe consequences of the spontaneous development are the following: economic underdevelopment and emigration from the hinterland, intensive economic development and concentration of the economy and the population on the narrow coastal strip which resulted in overexploitation of nature systems and in the environmental pollution.
2. Being aware of all negative consequences of such socio-economic relations, the Croatian, the regional and the communal authorities have taken some measures and actions to prevent such adverse effects on the development, on the population and on the human environment altogether.
3. All measures and actions have been initial and incentive, aiming at development of new economic activities, development of infrastructure, schools, medical service and other public services. Funds for assisting the development were limited, therefore they were intended primarily to attract funds from other sources.
4. Measures and actions were at first phase of their implementation expected to alleviate the negative processes, i.e. to decrease the emigration from the hinterland on one side and to raise the level of development in the hinterland on the other side; in the second phase the measures and actions were to create prerequisities for balanced and more intensive socio-economic development of the whole region.
5. Analysis of the effectiveness of the implemented measures and actions has shown that only in their second phase certain changes in the hinterland could have been identified, such as : demographic stability, improvement of the byodinamic structure of population, faster and more qualitative development of economy and of services, increase of urbanization, both of population and economy.
6. The analysis has shown that measures and actions have been properly conceived; objectives they were aiming at have also been very good and realistic. However, these measures and actions were not coordinated with each other and were lacking consistency in organization and in implementation. For that reason they have not produced the full effect in pushing the development in the hinterland and in eliminating development concentration on the coast. The analysis has shown that their effectiveness was only 16,7% of the total possible effectiveness.
7. The study has shown that the coast-hinterland socio-economic developmental process is being very entropic and hard to put to normal again. The on-going process of development of coastal areas proves this fact. It is, therefore, recommended that studying of the coast-hinterland socio-economic interrelations be a continuous activity, i.e. there should be a continuous analysing, planning and monitoring or, in other words, a syistematic and a reasonable management of the socio-economic relations.

8. It is recommended that the Croatian and the regional authorities reorganize the existing measures and actions to meet the present requirements, and that is the direct relation between the coastal and the hinterland economies, the economic criteria and the measures which add to the improvement of the level of development, of the quality of living, to the reasonable use of natural systems and to the protection of the human environment.

9. It is further recommended that by different measures and actions the natural resources in the hinterland be activated. Namely, the study has shown that more effects are to be expected in future, since investment in infrastructure have not produced considerable developmental effects so far. Therefore, further development of economic capacities, of public services and others, would certainly contribute to consolidation of the economy and of population in the interior part of the Split region.

10. Measures and action taken to control the development on the coast by choosing proper technologies, by rational use of natural systems and of space in general should be conceived in the same way. Elaboration of measures and actions aiming at prevention of excessive situations in use of space and the environmental pollution is also recommended.

11. The study has shown that the measures and actions taken on the Croatian level have been less successful in that specific area than those taken by local authorities. It is, therefore, recommended that the measures and acttions be organized regardless the level of authorities by which they have been initiated, and that they be coordinated with other measures and actions taken by local authorities who are most familiar with the problems of that particular area.

12. The study has shown that the implemented measures and actions may serve as a for steering the development in the hinterland, the model should, of course, be updated and ammended with new knowledges and more contemporary approach to the problem. Somewhat transformed, this model could be recommended to other Mediteranean countries, partcularly those experiencing similar socio-economic relations.

13. The study is based on the model previously prepared by the authors of the study. The model consist of elements necessary for analyzing and monitoring of the state of the socio-economic relations, for monitoring and testing of measures and actions intended to direct the socio-economic relations, and elements for decision making.

The study has proved that the model is simple and easy to follow, particularly for experts directly engared either professionally or operatively, in dealing with the problems of the coast-hinterland socio-economic relations. The testing model has an integrative approach, thus making a substiantial contribution to development of the planning methods and to the monitoring of the implementation of the coastal areas development plans.

CLIMATICAL CONDITIONS AND TOURISM
THROUGH THE TURKISH AEGEAN COASTAL ZONE

Dr.Asaf KOÇMAN
Ege University
Department of Geography-İzmir/TÜRKİYE

Unique landscape features, cultural and historical factors and convenient climatic conditions have contributed to birth and progress of tourism. Among these, appropriate climatic conditions are the permanent resource of tourism. In recent years, we observe developments at tourism activities, particularly in our country's coastal regions with an increasing demand. And also Aegean coastal zone is among the places that gets more share from these activities. The climatic conditions and their effects on human health and tourism have increased the usage for recreational purpose at this coastal area.

Aegean coastal zone, in the western part of Türkiye, begins from gulf of Edremit in the north and continuous to the gulf of Marmaris in the south. The formation of the coastal zone, with its more complex tectonic history, is connected to the subsidence movements in the early Quaternary and the rise of Western Anatolia. And the changing of the sea levels have played an important role in the formation of the coastline from Quaternary to Holocene. Narrow bays and gulfs, high cliffs and low beaches have been formed through the coastal zone, and delta alluvions of some streams are still spreading.

The environmental qualities of the Agean coastal zone have affected people since early historical ages. As in the past and still today environmental posibilities and mainly climatic conditions have contributed in development of the settlements and economic and cultural activities. In other words, natural environmental factors or/and possibilites have attracted people and as a result of these, advanced civilizations have been established here. In fact, ancient Greek and Roman cities, such as Adramyttion, Leukai, Smyrna, Teos, Ephesos, Miletos and Knidos, have been constituted on the Turkish Aegean coasts (Fig. 1).

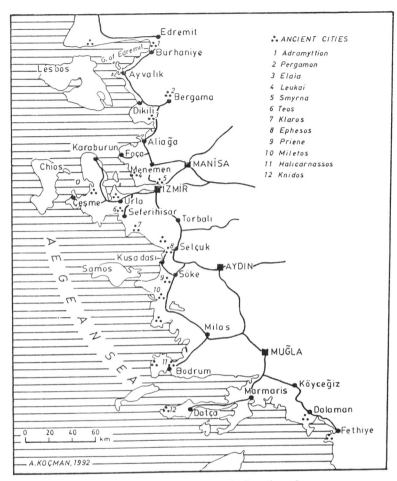

Fig.1- Turkish Aegean coastal zone and location of some ancient cities.

The ancient coastal cities developed depending on convenient climatic conditions and fertile delta plains at the mouth of the streams which have provided suitable environment for agriculture. Without doubt, the possibilities of the natural environment must have played an important part in the foundation, development and importance of the cities which prove their existence.Maritime commerce was encouraged by agriculture and trade founded around bays and gulfs that give shelter to ships (Kayan,1988).

However,the development of the ancient cities have not been unlimited and the human's pressure on the natural environment has brought negative results. Erosion has become a danger in the surroundings where there was a sensitive balance between the climatic conditions, vegetation cover and soils. As a result of erosion soil and sediments filled in the bays and gulfs. With the changing qualities of the environment the ancient cities have declined and lost their importance.

However, the social , economic and cultural conditions have changed today. The climatic features and some environmental possibilities in the Turkish Aegean coastal zone are still dominant. In fact, unique landscape features, rich history and culture and a great number of archaeological sites are highly desirable for the development of tourism in this coastal area. And climatic conditions is one of the most basic resources of tourism. In other words, the effect of environmental factors and climatic conditions can be observed at all points of the tourism process. In recent years,by the support of these factors the Aegean coastal zone gained a new appearance and undertook different functions of usage. Touristic establishments (hotels, motels and campings), holiday villages and summer resorts have rapidly increased on the coasts since 1970 and visible vitality has been experienced on the regional economy. Tourism facilities have gained importance in the Turkish Aegean coastal cities such as Edremit,Ayvalık, Dikili, Foça, Çeşme, Kuşadası,Bodrum and Marmaris.

Throughout the coastal zone, the effects of the weather or climatic conditions on human health and tourism activities mostly increase the value of the usage of these places for recreational purposes . This is the most important point that will be emphasized in this declaration. However, in the limited time it is not possible to examine all climatic elements and events affecting the activities of tourism. For this reason, we will only emphasize the elements of temperature, relative humidity, rainfall and winds. According to the rules that are defined by the expert of medical climatology, heat and temerature, relative humiditiy and wind should be within the submitted comfort limits. The periods that one feels relaxed defined as " comfort zone" reflects the combined effects of the climatic conditions. To some experts 17-25° C is accepted as the most favorable temperature values with a slow wind less than 6 m/ sec. and 30-70 % relative humiditiy. On the Aegean coastal zone , according to the suitable comfort limits , effective temperature period is at least 101 days in Marmaris and most 158 days in Kuşadası (Fig.2). As shown on the diagram , this period includes 27,7 or 43.0 % of the year. Comfort zone is a continuous period in Kuşadası, but comes out as two different periods in other places. Besides, from the meteorological observations it has been established that temperatures only during winter fall under 17,0° C which is the lowest comfort limits (Fig.2). But with the effects of the sea the temperature values stay high, generally close to the monthly means.

Sunshine duration and intensity have important role too. According to our calculation have done with Penman and Kılıç's relations , on the coastal zone the heat that is absorbed is more than the heat that is released. In other words, it is determined that energy balance shows positive values everywhere. For this subject, Edremit, İzmir, Kuşadası and Marmaris meteorological stations have been chosen as pilot stations and to make a comparison Table : 1 has been prepared. According to the values of the table, on the coastal zone, during May-October period favorable sunshine duration is more than 7 hours and the net solar energy reaching to the earth's surface is more than 100 cal/cm^2 per day. In this period atmospheric activity decreases so much that it doesn't give way to cloudness and rainfall and the sky is clear throughout the coastal region. Briefly, under the effects of these conditions on the Aegean coastal zone radiation is high in a period lasting 5 or 6 months. It is clear that temperature values and heat conditions can be an advantage for human health and tourism activities. It can be said that the Aegean coastal zone has a favorable position from the point of view of solar climate.

As pointed out, maritime climate cures mostly take an important part in the treatment of mental diseases and heart, skin and respiration problems (Bahadır and Karagülle,1983). Without doubt,sea water and maritime air provide positive effects for human health. If 20-28° C are

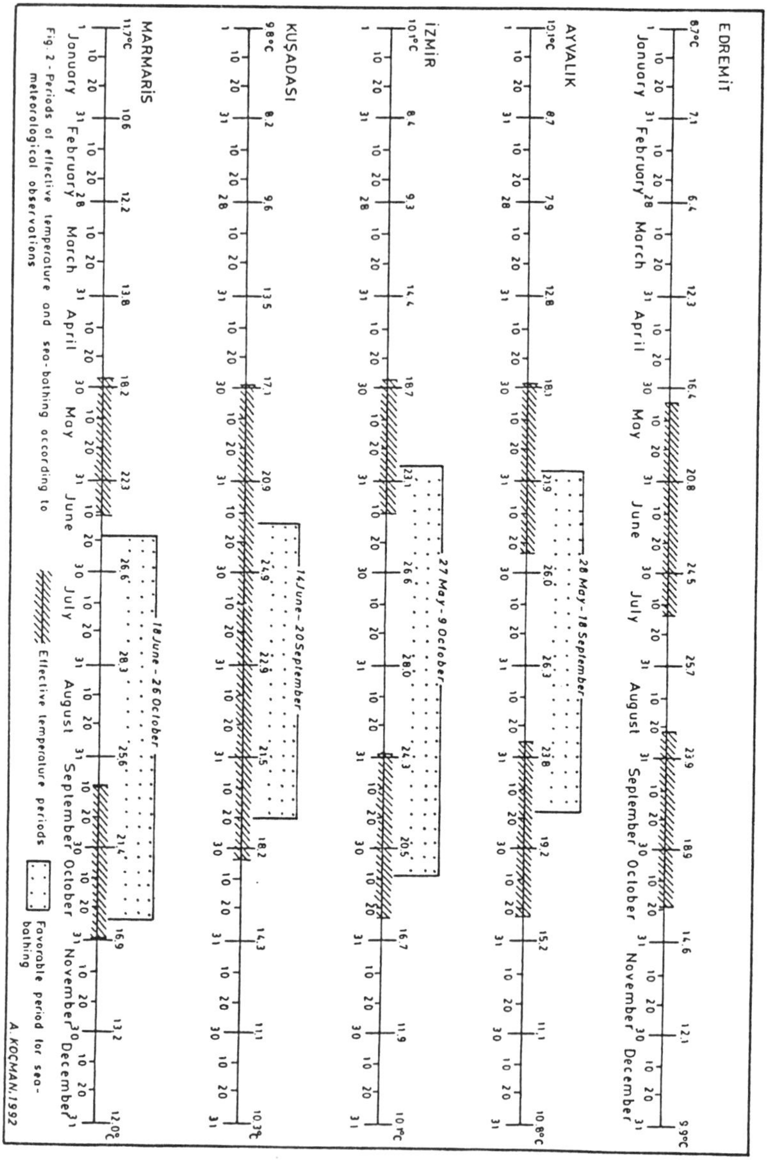

Fig. 2 - Periods of effective temperature and sea-bathing according to meteorological observations

A. KOÇMAN, 1992

taken as the most favorable daily mean temperature and 22-25° C as daily mean temperature of sea water, then, for example, this period will be approximatelly 114 days in Ayvalık, 99 days in Kuşadası and 131 days in Marmaris. If daily mean air temperature between 18-28 C is accepted to be the extreme comfort limits, the emphasized period will be longer such as 174 days in Ayvalık, 203 days in Kuşadası and 226 days in Marmaris (Table 1).

Generally, maximum temperature and humidity proportion in the air have negative effects on the health and activities of the human beings. According to the experts the relative humidity within the considered comfort limits must be 30-70 %. On the Aegean coastal zone throughout the period of the favorable comfort limits daily mean relative humidity proportions are always less than 70 %, but not under 30 %. So, all activities relating to the recreation of the coastal area are realized due to the relative humidity proportion and temperature values become suitable.

On the other hand, five or six months dry period lasts on the coastal zone. Here, the unrainy period can start from the second half of May and can last until the end of October. During the period the number of average rainy days are too little and rainfall lasts too short, not more than a few minutes. During the six-months period between May and October the number of average rainy days are 15 days in Edremit, 13 days in Kuşadası and 14 days in Marmaris. But the period from November to the end of April is cool and rainy. Due to the rainfalls in this period there is no problem of drought throughout the region. As a whole, in the Aegean Sea basin the air masses coming from different direction and related frontal system determine the rainfall, temperature and wind regimes in the year.

According to the medical climatology experts, where the wind speed is less than 6 m/sec. can be accepted favorable (Ülker, 1978). As it is known, on the coastal areas the winds create a feeling of coolness with the help of evaporation. Slow winds and wave motions together cause to renovation of waters and cleaning of the maritime air. Without doubt, these changes provide positive conditions for the human health. In fact, on every part of the Aegean coastal zone the direction of the daily winds and wind speed gain reasonable values. Throughout the year and particularly during the tourism season the speed and frequency of the wind from all directions do not exceed the comfort limits (Fig.3).

Meteorological stations	EFFECTIVE TEMPERATURE VALUES 17.0 – 25.0 °C				Air temperatures for sea bathing (20–28°C)			Water temperatures for sea bathing		
	1. Period Starting and ending dates	2. Period Starting and ending dates	Total number of days	Annual Prop. (%)	Starting and ending dates	Number of days and prop. (%)	Starting and ending dates (22–28°C)	Number of days and prop. (%)	Starting and ending dates (18–28°C)	Number of days and prop. (%)
EDREMIT	5 May – 13 July	22 Aug. – 18 Oct.	128	35.0	22 May – 28 Sept	130; 35.6	—	—	—	—
AYVALIK	29 Apr. – 23 June	26 Aug. – 22 Oct.	114	31.2	18 May – 28 Sept	134; 36.7	28 May – 18 Sept.	114; 31.2	2 May – 22 Oct.	174; 47.7
DIKILI	30 Apr. – 12 July	21 Aug. – 18 Oct.	133	36.4	20 May – 28 Sept	132; 36.2	12 June – 8 Sept.	89; 24.4	16 May – 31 Oct.	169; 46.3
IZMIR	28 apr. – 10 June	29 Aug. – 30 oct.	107	29.3	17 May – 5 Oct.	142; 38.9	27 May – 9 Oct.	136; 37.3	4 May – 9 Nov.	190; 52.0
KUŞADASI	30 April	4 October	158	43.3	23 May – 24 Sept	125; 34.2	14 June – 20 Sept.	99; 27.1	3 May – 21 Nov.	203; 55.6
BODRUM	23 Apr. – 11 June	7 Sept. – 15 Nov.	120	32.9	9 May – 13 Oct.	158; 43.3	27 June – 9 Oct.	105; 28.8	1 May – 9 Dec.	223; 61.0
MARMARIS	25 Apr. – 12 June	9 Sept. – 30 Oct.	101	27.7	17 May – 18 Oct.	155; 42.5	18 June – 26 Oct.	131; 36.0	29 Apr. – 10 Dec.	226; 61.9

Table: 1 – Effective temperature values, air and water temperatures for sea bathing, starting and ending dates annual proportion and mean number of days due to the comfort limits.

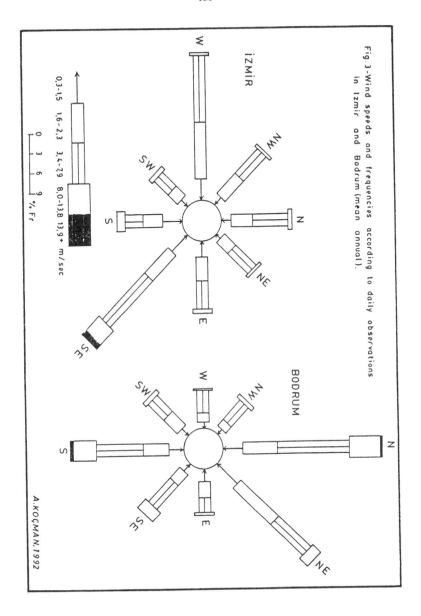

Fig. 3 - Wind speeds and frequencies according to daily observations in Izmir and Bodrum (mean annual).

Conclusion

The Turkish Aegean coastal zone has an important part in recreation and tourism. These activities have improved rapidly. The effects of environmental factors can be observed at all points of the tourism activities. Natural attractions, amenable climate and unique landscape features are the basic resources for tourism. Particularly, climatic conditions are very favorable for the development of tourism in the coastal zone. According to our studies most of the tourists coming to the region prefer the climatic features and attractive environment ranging from the sunshine, sea and sand to historic sites. In fact, the sea and climatic conditions can be effective for the medical tourism and tourism as a whole. As we all know, despite the development in medicine today, resting and climate cures are widely used for treatment.Treatment of some pains and physical strength can be obtained in a healty climate. For this reason, on the Turkish Aegean coastal zone climatic rehabalitation stations can be founded as a social service.

On the other hand, the climatic features of the mountains and highlands rising nearby the coastal zone (such as highland of Kozak, mount of Yamanlar, mount of Manisa and mount of Bozdağ and its highlands) may be the favorable places for medical and mountain tourism. In addition to production ways of some agricultural crops (oranges, olives, grapes, figs and so on) peculiar to Aegean climate can be advertised to attract the tourists' attention. So, it is thought that all these mentioned activities will extend the period of usage of the coastal zone.

However, the growth of tourism activities lead to modifications of the environment.Usage and settlement should be held in a level that will not damage the ecological balance of the natural environment. We should keep in mind that the positive results obtained from tourism activities depend on modern planning mentality.

REFERENCES

BAHADIR,A.S. and M.Z. KARAGÜLLE, 1983. Biyometeorolojik açıdan iklim türleri ve fizyolojik-patolojik-terapötik etkileri.Tıbbi Biyometeoroloji Semineri, 22-23 Sept. 1983, **Bildiriler**, p.81-87, Devlet Meteoroloji İşleri Genel Müdürlüğü,Ankara.

BESANCENOT, J. P. 1990. **Climat et tourisme.** Collection Geographie, Masson, Paris.

EROL,O. 1988. Turkey, in **Artificial Structures and Shorelines.**Ed.by H.J. Walker ,p.241-252, Kluwer Academic Publisehers.

Erol,O. 1989. Türkiye'de kıyıların doğal niteliği, kıyının ve kıyı varlıklarının korunmasına ilişkin "Kıyı Kanunu" uygulamaları konusuna jeomorfolojik yaklaşım. **İstanbul Üniv. Deniz Bilimleri ve Coğrafya Enstitüsü Bülteni,** 6,p.25-46,İstanbul.

HOBBS, J.E. 1980. **Applied Climatology** (Studies in physical geography). Butterworths,London,England.

İLERİ, A. 1983. İklim ve günlük hava olaylarının insan sağlığı ve davranışları arasındaki ilişkiler. Tıbbi Biyometeoroloji Semineri,22-23 Sept.1983, **Bildiriler,** p.141-158, Devlet Meteoroloji İşleri Genel Müdürlüğü ,Ankara.

KAYAN,İ. 1988. Datça yarımadasında "Eski Knidos" yerleşmesini etkileyen doğal çevre özellikleri. Ankara Üniv. **Dil ve Tarih Coğrafya Fakültesi Dergisi,** 11, p.51-70, Ankara.

KILIÇ,A. and A.ÖZTÜRK, 1980. Güneş **Enerjisi.** Kipaş Dağıtımcılık,İstanbul.

OLALI,H. 1968. **Ege Bölgesi Turizmi ve Turizmin Mevsimlik Karakteri.** İzmir İktisadi ve Ticari İlimler Akademisi Yayınları,58,İzmir.

SUNGUR,K.A. 1980. Türkiye'de insan yaşamı açısından uygun olan ve olmayan ısı değerlerinin aylık dağılışı ile ilgili bir deneme. İstanbul Üniv. **Coğrafya Enstitüsü Dergisi,** 23, p.27-36.

ÜLKER,İ. 1978. **Kaplıca ,Deniz ve İklim Kürlerinin Temel Yöntemleri** (Rapor). Turizm ve Tanıtma Bakanlığı,Ankara.

ÜLKER, İ. 1988. **Türkiye'de Sağlık Turizmi ve Kaplıca Planlaması.** Kültür ve Turizm Bakanlığı,Çağdaş Kültür Eserleri Dizisi,1006/129, Ankara.

PROPOSALS AND PROGRAM FOR THE ORGANIZATION OF THE NATURAL PARK ON THE COAST OF SOUTHWEST SARDINIA (IGLESIAS-ARBUS)

P. Castelli, F. Di Gregorio
Università di Cagliari

Summary

Sardinia's regional system of parks, reserves and natural monuments (Regional Law no. 31/89) includes the natural park of Monte Linas, the natural reserves of Capo Pecora, Piscinas/Monte Arcuentu, the San Giovanni Wetlands and the dunes of Pistis, an area of great environmental interest. The recent presentation (1991) of the pilot project for the Monte Linas Natural Park poses the scientific problem of verifying the social goal of bringing the protection of other natural treasures within the framework of the management of the Natural Park and involving the local populations in the cultural planning and the induced productive activities. The method starts from a recognition of the natural (geological, geomorphological, vegetational, faunistic) and cultural (archaeological, architectural and historico-traditional) values of human and patrimonial resources as well as the uses to which the Park is destined. It then goes on to a hypothesis for the singling out of the perimeters and importance of areas (even detached) to be included in the management; and then to the internal zoning into areas of strict protection for ecological and scientific purposes (integral reserve) in which all kinds of productive activity are to be prohibited, areas of general reserve, including those parts of the territory which, although transformed, are the result of a correct equilibrium between man and nature, zones of protection in which traditional uses are to be allowed or encouraged, even through the presence of visitors, and zones of development essentially destined to social life; the organization of roads of access and for visiting places of interest and the setting up of support facilities for park employees and logistic services for the different categories of visitors. Even a rough estimate of the additional costs allows an evaluation of the feasibility of the proposal.

The geoenvironmental characters considered which are decisive in defining the degrees and types of ecosystem protection can be classified as: rocky coast landscapes, beach-outlet-dune systems, wooded hill and mountain or Mediterranean bush landscapes, agricultural or mining landscapes. The main kinds of protection proposed concern:
- protection of the landscapes and natural settings of the steep coasts (from buildings overlooking the sea);
- protection of dunes and sandy beaches (from too much vehicular and pedestrian traffic) through proper planning of access roads;
- the safeguarding and organization of the naturalistic fruition of the wetlands, which are rich in rare or interesting birdlife;
- protection of geological monuments and the paleontological patrimony.

The method and proposed solutions appear to be the most effective answer on the one hand to the desire of the Administration and the local population to create settlements and favour initiatives capable of producing income and jobs, and on the other hand of favouring the conservation of environments and ecosystems which are being drastically reduced in the Mediterranean area owing to increasing pressure for the building of new settlements and creating tourist resorts along the coasts.

1 - Anthropic settlements and frameworks

1.1. The area dealt with in the present work is situated along the southwestern coast of Sardinia (Fig. 1), from Capo della Frasca to Funtanamare, with an area of about 100 km^2. It is an area of low population density whose main settlements, Iglesias and Guspini, have about 15 000 inhabitants. The other towns: Domusnovas, Vallermosa, Villacidro, San Gavino, Arbus, Fluminimaggiore and Buggerru, have populations between 1000 and 9000 inhabitants.

Inhabited areas are situated for the most part in the interior, along the edges of the territory examined, with the exception of Fluminimaggiore and Buggerru (Fig. 2). The latter is the only town of any importance located on the coast, as are the small agglomerates of Nebida and Masua and the tourist developments of Pistis, Torre dei Corsari, Porto Palmas, Marina di Arbus and Portixeddu, still of modest dimensions and of a seasonal nature.

1.2. The guest capacity of tourist resorts admitted by urban plans now in force is over one hundred thousand people, of which more than eighty thousand on the Arbus coast.

1.3. The area is served by State Highway 126 running in a N-S direction, flanked by Highways 130, 196 and 293 and further served by four provincial roads connecting settlements and tourist resorts along the coast.

2. Characteristics of the landscape and naturalistic values

2.1. There are many elements of naturalistic interest, starting from the geological and geomorphological aspects and including the flora and fauna (CASTELLI & FERRARA, 1989).

2.2. From the geological point of view, there are evident outcroppings of terrains of all geological eras, starting from the Cambrian formations which, owing to the completeness of their successions and the presence of rare

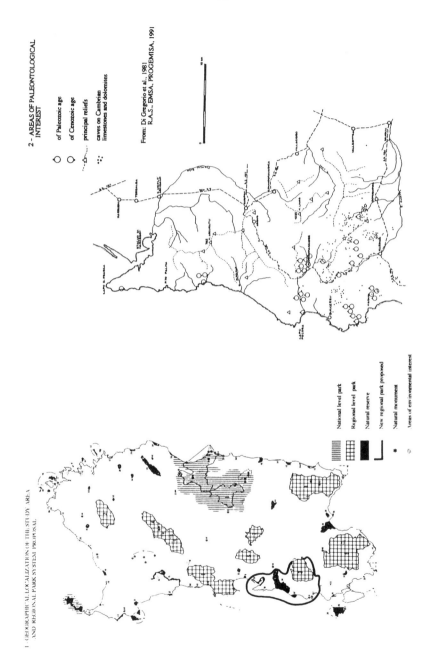

fossiliferous associations (DI GREGORIO et al., 1989), among the most interesting in Europe. Just as interesting are the Ordovician successions owing to the presence of rich associations of brachyopods, trilobites, conularides, etc., and the Silurian successions, especially due to the presence of black limestone lenses bearing cephalopods (Orthoceras) of spectacular dimensions, and rare graptolites (e.g. Monograptus). Still from the geological and paleonthological viewpoints, the tertiary successions in the Arbus area (Funtanazza) are particularly interesting owing to the state of conservation, the wealth of individuals and the variety of species present (molluscs, corals, echynides, algae, etc.)(Fig. 2).

No less interesting are the structural aspects. In this area, as many as three orogenetic phases (Caledonian, Hercynian and Alpine) have been recognized. There are clear signs of these impressed on the landscape, with polyphase folds, overthrusts and the many minerogenetic manifestations, especially those connected with the Hercynian cycle. Those connected with lead-zinc formations are well known and have attracted man's attention starting from prehistoric times and continuing up through the Punic, Roman, Pisan, Spanish and Piedmontese dominations, which have left their marks, many of which worth protecting, on the landscape (DI GREGORIO, 1985).

Also of great interest are the tertiary volcanic manifestations of Arcuentu, especially because of the variety of lithological types and morphological features impressed on the landscape by differential erosion.

Reliefs, prevalently arenaceous, schistose, granitic or carbonatic, rarely reach heights above 1000 meters (Monte Linas, 1236 m) but present constantly different, and sometimes unexpected, morphologies, with deep valleys, upland plains, sharp or rounded crests and rocky cornices, towers, canyons and many karst caves in the Cambrian carbonatic formations (Fig. 3). Worthy of mention in this context are the Cambrian arenaceous and calcareo-dolomitic banks, folded or contorted and sometimes sub-vertical, to be seen in the cliff faces along the coast from Masua to Buggerru, together with the unusual coastal caves (e.g. Canal Grande, Grotta Azzurra, etc.), the mine drifts and *decauville* tracks appearing in the cliff face. Also to be mentioned is the picturesque Porto Flavia, set in a wave-cut cliff, which was used to load the lead-zinc ore mined in the mountain directly onto the waiting ships.

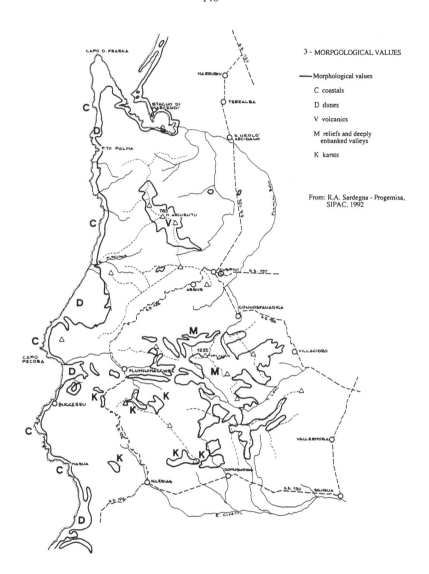

Further north, the coast, sometimes granitic (Capo Pecora) sometimes schistose (Costa Verde, Torre di Flumentargiu) and sometimes basaltic (the Capo della Frasca plateau), is a series of rocky promontories and small bays framing stretches of white sand which are still only slightly exposed to anthropic pressure.

Still from the geomorphological viewpoint, the extreme variety of coastal landscapes is to be underscored, with rocky coasts, wave-cut cliffs up to 300 m in height (e.g. Masua), rias (e.g. Cala Domestica), beaches and free dunes (e.g. Piscinas, Torre dei Corsari) or dunes attached by vegetation l(Buggerru, Scivu, Funtanamare) which are among the most beautiful in Mediterranean Europe. Between Nebida and Masua it is possible to observe one of the oldest coastlines in Europe, with beach conglomerates of the Ordovician lying in clear angular discordance on Cambrian phyllite schists. Also to be seen are islets (Pan di Zucchero) and limestone or calcareo-dolomitic rocks emerging from the sea (Agusteri and Morto), their white or yellowish colours contrasting sharply with the blue of the sea and the red-violet of the conglomerate caused by the ferrous oxides in circulating waters.

2.3. In the most remote areas (e.g. Marganai, Monte Linas, etc.) are present vegetable associations and endemic species of the greatest interest (Fig. 4). Some of these (Marganai, the Dunes of Buggerru, Porto Palmas) have been included in the list of vegetational biotopes of great interest to be protected published by the Italian Botanical Society (1971).

Worthy of special mention are the spontaneous groupings of coastal sclerophyllous with Juniperus macrocarpa, J. phoenicea, Quercus coccifera, Phyllirea angustofolia, P. latifolia, Pistacia lentiscus, etc. found on the Buggerru dunes (Mossa et. al., 1988), the vegetation with Cakile maritima, Agropyron junceum, Ammophila arenaria, Silene corsica, etc. of the stupendous dunes of Piscinas and, in the mountainous interior (Chiappini et al., 1988), the rare white helichrysum (Helichrysum montelinasanum) together with numerous exclusively Sardinian (e.g. Armeria sulcitana, Silene colorata, Festuca morisiana, etc., or Sardinian-Corsican endemic species (Silene nodulosa, Genista corsica, Evax rotundata, etc.) which in late spring explode, creating immense blankets of flowers. In the more impervious valleys, interest is aroused by residual strips of ancient Mediterranean forest with Quercus ilex, Phyllirea angustifolia, P.

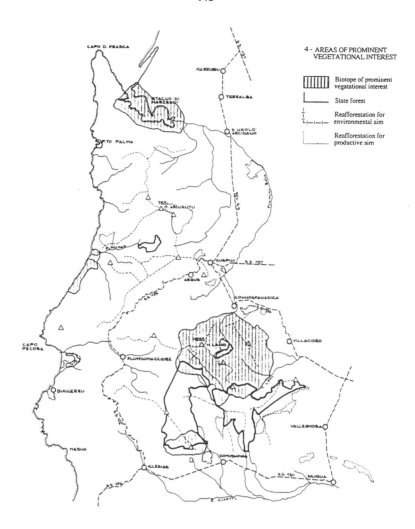

Latifolia, Arbutus unedo, etc., and the presence of relic nuclei of Ilex aquifolium and Taxus baccata, bearing witness to paleoclimatic events.

2.4. The fauna is also of interest (Fig. 5), with the presence of the Sardinian deer (Cervus elaphus corsicanus), the Sardinian cave newt (Hydromantes genei) and many other endemic or exclusive species.

Also worthy of note is the presence along the coast of the peregrine falcon (Falco peregrinus brokei) and the Falco eleonorae, the wild cat (Felix libica), marten (Martes martes) and wild boar (Sus scrofa) in the interior. In the past, the sea turtle (Caretta caretta) was also reported in the past. But certainly richer and more interesting are the aquatic bird species in the Marceddì, San Giovanni and Corru s'Ittiri wetlands where over 10 000 aquatic birds have been counted (Schenck, 1976), among which Phoenicopterus ruber, Pediceps cristatus, Porphirio porphirio, Recurvirostra avocetta, etc., which gained for this area recognition as being of international interest within the framework of the Ramsar Convention.

3. Archaeology, architecture and histrorical monuments

3.1. The mountanous territory under consideration has conserved important traces of Neolithic settlements in the form of karst caves and menhirs, "tombe di giganti" and "domus de janas", not to mention the remains of villages (CHIAPPINI et al., 1988; MOSSA et al., 1988).

3.2. The mining district was an area of intense traffic in the Phoenician-Punic and Roman and Pisan periods (DI GREGORIO, 1985) as was the trading centre of Santa Maria Neapolis, situated at the natural port of the Marceddì wetland along the principal NE-SW artery.

3.3. There are appreciable remains of religious and social (granaries) architecture, starting from the Middle Ages, but especially from the 18th century onwards; of even greater interest is the industrial archaeology of the area, with mines, processing plants and settlements dating from the 19th and 20th centuries, which are gradually being shut down. Nonetheless, their recovery for possible use in the collective interest remains a concrete possibility.

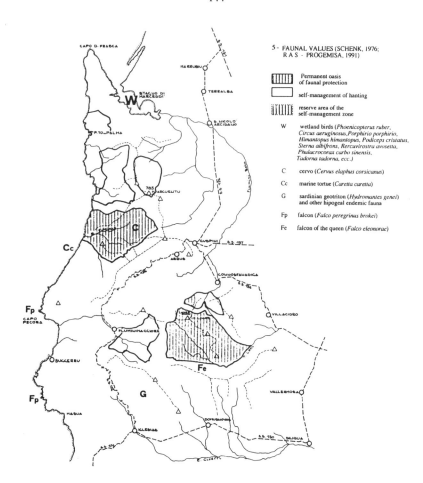

The principal places of scientific and spectacular interest indicated in the literature and found in the field, together with the areas of natural beauty as classified by law, are shown in Fig. 6.

4. Proposals for protection

The Italian Botanical Society was the first to propose (1971) the protection of three natural biotopes in the area, but the first organic proposal for the protection of nature was made by CASSOLA & TASSI (1973) within the framework of the System of Parks and Natural Reserves in Sardinia (Fig. 7). They proposed the creation of the Monte Arcuentu and Foce del Rio Piscinas Natural Park and four natural reserves. Conversely, in 1975, the Central Regional Planning Board proposed the creation of a series of natural reserves, omitting the regional park (Fig. 8), on the basis of the LACAVA team (1975) study.

Despite these proposals, the protection of nature in Sardinia has come up against strong opposition and only in 1989, with Regional Law n° 31 of June 7, 1989, containing regulations for the setting up and management of parks, was this issue reproposed in the strongest terms. In the area considered in the present work, the regional law provides for the realization of the Monte Linas-Marganai-Oridda-Montimannu Regional Park, as well as the natural reserves of Capo Pecora, Coast of Nebida, Dunes of Piscinas and Monte Arcuentu and the natural monuments of Pan di Zucchero with the emerging rocks of Agusteri and Morto and the dunes of Pistis (Fig. 9).

Of these protected areas, only the regional park is now in the planning stage. Now awaiting approval, the project calls for detailed zoning, with the gradual introduction of protective measures depending upon the naturalistic importance of the territory and a series of proposals for steps to ensure the correct tourist and cultural use of the area (Fig. 10).

5. Planning at present

Territorial planning in the area considered is articulated as follows:
a. the territorial landscape plan: the Sardinian Regional Administration/Aru & Chanoux Projects (1990) covering the entire area of the proposed park, which is now in the process of standardization, is awaiting adoption by the Regional Executive, after which it must be published and approved by the

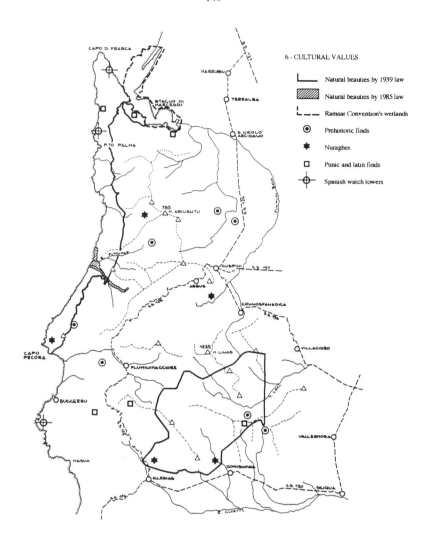

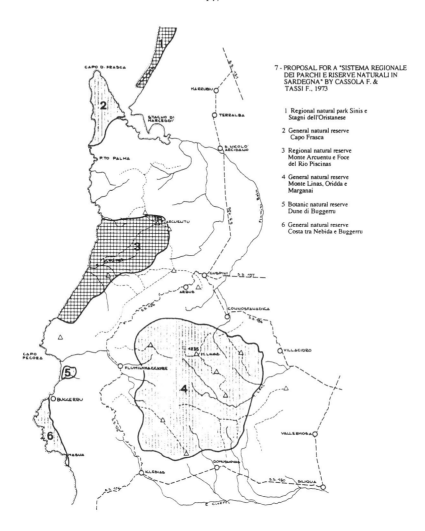

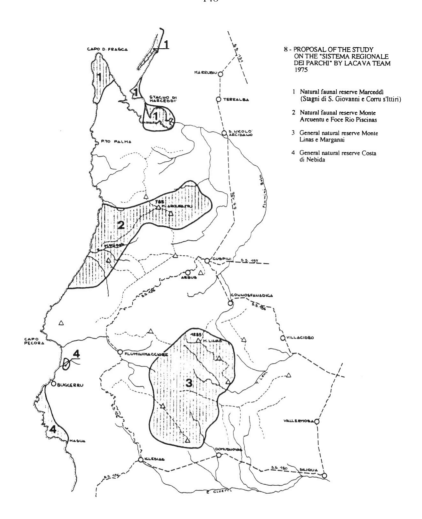

8 - PROPOSAL OF THE STUDY
ON THE "SISTEMA REGIONALE
DEI PARCHI" BY LACAVA TEAM
1975

1 Natural faunal reserve Marceddì
  (Stagni di S. Giovanni e Corru s'Ittiri)

2 Natural faunal reserve Monte
  Arcuentu e Foce Rio Piscinas

3 General natural reserve Monte
  Linas e Marganai

4 General natural reserve Costa
  di Nebida

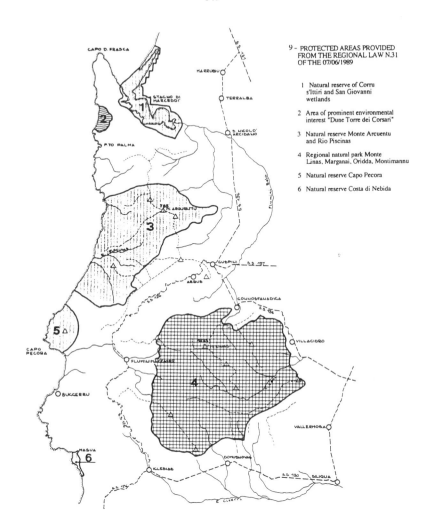

9 - PROTECTED AREAS PROVIDED FROM THE REGIONAL LAW N.31 OF THE 07/06/1989

1. Natural reserve of Corru s'Ittiri and San Giovanni wetlands
2. Area of prominent environmental interest "Dune Torre dei Corsari"
3. Natural reserve Monte Arcuentu and Rio Piscinas
4. Regional natural park Monte Linas, Marganai, Oridda, Montimannu
5. Natural reserve Capo Pecora
6. Natural reserve Costa di Nebida

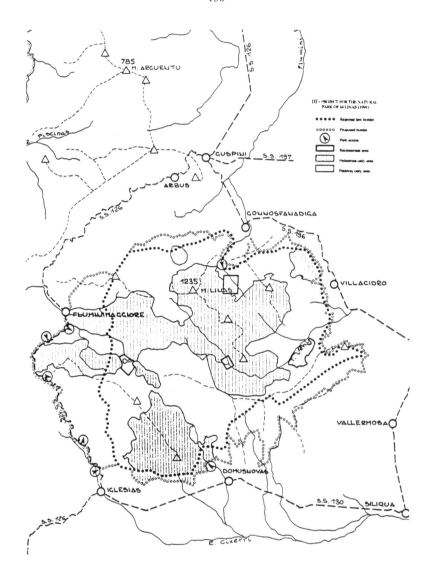

Regional Council. The plan indicates areas in which environmental resources are to be protected;

b. the plan for the Monte Linas natural park: the Sardinian Regional Administration/Lacava-Raspanti Project (1991), covering the southeastern interior area of the proposed park. It has been adopted by the Regional Executive and is now being published to be presented for final approval. It indicates points of access, the internal road system, areas for social and collective fruition, areas where only pedestrian traffic is to be allowed and where this is to be allowed only along marked trails;

c. the mosaic of municipal town plans, all now in force and subjected to repeated variations and adjustments to the new provisions: these plans indicate settlements and areas to be protected. For agricultural areas they refer to the low level regulations included in the regional directives;

d. studies of territorial management for tourist resorts (for the most part along the coasts);

e. municipal or private plans for settlements and main infrastructures.

6. Experimentation of the method proposed for the park plan

6.0. The method for delimiting and organizing the park is based on the proposal previously advanced at the Eurocoast '90 meeting in Marseilles (CAMARDA, CASTELLI & FERRARA, 1990): "identification and location of natural and cultural wealth and its classification by scientific, spectacular and choral interest; location of access and support infrastructures and evaluation of the possibility of their integration into the park organization".

6.1. The delimitation includes the entire Monte Linas-Monte Arcuentu massif, from the Campidano Plain and the Cixerri River plain down to the coast from Masua to Marceddì. It thus includes not only the expanded Monte Linas Park territory but also the five natural reserve areas recognized by law or proposed by experts.

6.2. A first problem to be faced arises from the fact that the territory falls into two different provinces (Cagliari and Oristano) with the inevitable influence that this will have on planning procedures, which are unlikely to proceed in a coordinated way at that level.

6.3. A second problem arises from the fact that State Highway n° 126 (Iglesias - SS 131) separates the Monte Linas Massif from reliefs along the coast. At the same time, however, it supplies an extended base for penetration into both parts of the area; animal crossings will have to be adequately planned, set up and protected.

6.4. The inland perimeter has been drawn following geomorphological and altitude criteria, respecting present, but even more so potential, crop raising. The sea perimeter also includes a part of the sea including bottom areas with geological formations, biotic associations and archaeological remains.

6.5. the "integral reserve" areas have been singled out on the basis of scientific values and specific features to be protected as indicated in the literature and by special studies on specific topics: their perimeters are to be drawn as a function of these recognized features. In this proposal only the most evident values are indicated in draft form, with detailed analyses to be performed at a subsequent stage.

6.6. Areas and monuments of spectacular interest are recognized and delineated contextually; on this basis access trails and points of observation are planned: as a function of the latter, visitors' centres, logistic and cultural facilities are to be provided.

6.7. A series of possibilities is offered by the presence of numerous abandoned or about-to-be-abandoned historic mining settlements as well as many developing seasonal tourist resorts (recently built or with work still in progress).

6.8. Logistic facilities are to be based on restructured mine buildings and the full use of seasonal tourist resorts; cultural facilities (Fig. 11) are located so as to take full advantage of already existing facilities and local vocations, following the principle of having a capillary presence in the various local communities as well as some strong, centralized units.

7. The Southwest Coast Natural Park

7.1. Our proposal aims to reintroduce the overall planning of the entire territory, while the regional law provides for fragmentary planning and protection. By means of an overall analysis of territorial values, the goal is to arrive at land evaluation leading to the definition of future land uses on the basis of specific features of the single parts, without losing sight of the local social, cultural and economic context.

The new structure of the enlarged park will obviously be variously articulated by means of zoning, thus allowing the gradual introduction of protection. Therefore, the most delicate and sensitive areas, which have already been singled out to become natural reserves, will be verified and will

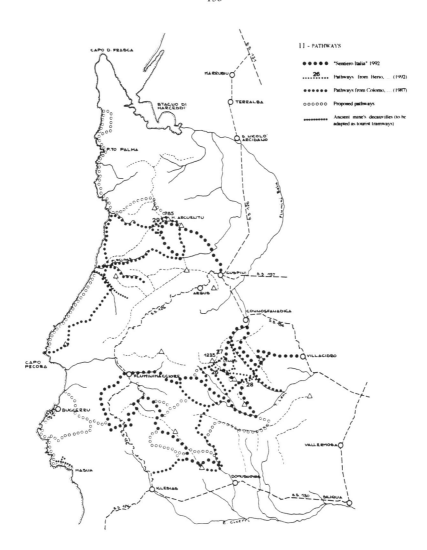

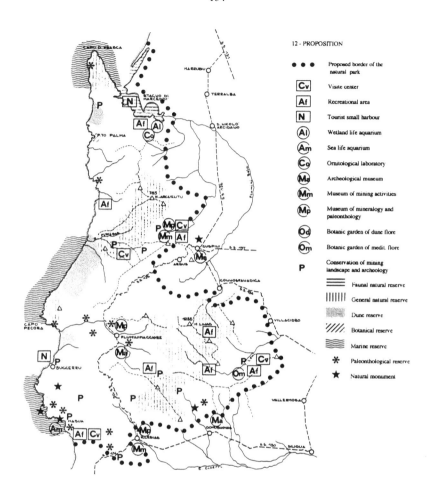

receive proper attention. But it is in the management phase that the greatest advantages will be gained in terms of efficiency and functionality.

As concerns the organization, the pivotal interventions proposed are those indicated in Figs 11, 12). Through them, the intention is to enhance the natural wealth available, but also to take full advantage of the artifacts left by man's labour, which are a part of the traditions of the local populations and impressed on the collective memory.

REFERENCES

BERIO A., CORBELLINI G. & CORTIS R. (1992) - *Itinerari sulle montagne della Sardegna*. Arti grafiche Tamari, Bologna, 26-29.

CAMARDA I., CASTELLI P. & FERRARA C. (1990) - *Hypothese pour l'institution d'un parc naturel sur la côte de Arbus (W Sardaigne), dans le cadre du systeme regional des parcs*. Eurocoast, Marseille, 390-394.

CHIAPPINI M., ANGIOLINO C., FOIS G., MANFREDI P. (1988) - *Monte Linas* - In Biotopi di Sardegna a cura di Camarda I. e Cossu A., Delfino, Sassari, 205-237.

CLUB ALPINO ITALIANO (1991) - *Sentiero Italia*. C.A.I., Milano.

CASSOLA F. & TASSI F. (1973) - *Proposta per un sistema di Parchi e Riserve naturali in Sardegna*. Boll. Soc. Sarda Sc. Nat., VII, XIII, Gallizzi, Sassari, 1-83.

CASTELLI P. & FERRARA C. (1989) - *Antropizzazione ed equilibri naturali: il caso della Costa Verde di Arbus, Sardegna Occidentale*. Atti Congr. di Geoingegneria, Suolosottosuolo, Torino, 27-30 Settembre, 35-45.

COLOMO S. & TICCA F. (1987) - *Sardegna da salvare - Un sistema di parchi e riserve naturali per le grandi distese selvagge della nostra isola*, vol II, Ed. Arch. Fotogr. Sardo, Nuoro, area 124,131,132,135,136.

DI GREGORIO F. (1985) - *Le coltivazioni minerarie dai nuragici ai tempi nostri*. In Sardegna: l'uomo e le montagne, Banco di Sardegna, Sassari, 269-278

DI GREGORIO F., CANNAS E., SPANO C. (1989) - *Il patrimonio paleontologico della Provincia di Cagliari: importanza e tutela*. Atti Int. Conference "Tourism and environment in post- industrial society, T.C.I., Milano.

SCHENK H. (1976) - *Analisi della situazione faunistica in Sardegna. Uccelli e mammiferi*. In S.O.S. Fauna - Animali in pericolo. Camerino, 465-556.

DI GREGORIO F. & GENTILESCHI M.I. (1984) - *Il problema dei parchi naturali nella realtà attuale della Sardegna*. Mem. Soc. Geogr. It., Vol. XXXIII, Pacini, Pisa.

MOSSA L., FOGU M.C., CONGIA P. (1988) - *Complesso dunale di Buggerru Portixeddu*. In Biotipi di Sardegna, a cura di Camarda I. e Cossu A., Delfino, Sassari, 103-122.

PRATESI F. & TASSI F.(1984) - *Guida alla natura della Sardegna*. Mondadori, Milano, 219-235.

REGIONE AUTONOMA DELLA SARDEGNA (c.r.p.) (1975) - *Il sistema dei parchi della Sardegna - parco del Limbara*, a cura del Gruppo LACAVA.

REGIONE AUTONOMA DELLA SARDEGNA (1989) - *Legge regionale 7 giugno 1989, n. 31, Norme per l'istituzione e la gestione dei parchi, delle riserve e dei monumenti naturali, nonchè delle aree di particolare rilevanza naturalistica ed ambientale*. In B.U.R. 16.6.1989, n. 22.

REGIONE AUTONOMA DELLA SARDEGNA - Progemisa (1991) - *Sistema informativo per la pianificazione dell'attività di cava in Sardegna*, Abi2ue, Milano.

SOCIETA' BOTANICA ITALIANA (1971) - *Censimento dei biotopi di rilevante interesse vegetazionale meritevoli di conservazione in Italia*, a cura del Gr. Lav. Cons. Natura, Camerino.

MORPHO-CLIMATIC EVOLUTION OF THE AVEIRO REGION LITTORAL (NW OF PORTUGAL) DURING TERTIARY AND QUATERNARY

F. ROCHA and C. GOMES
Departamento de Geociências, Universidade de Aveiro, 3800 Aveiro, Portugal

ABSTRACT

Clay minerals and the accompanying non clay minerals were utilized for lithostratigraphical discrimination and for paleosurfaces definition. In the "Ria de Aveiro" region (NW Portugal), in particular, clay minerals assemblages reflect paleogeographic and paleoclimatic conditions of formation and deposit, and their nature and crystalochemical features allowed the zonation of Tertiary and Quaternary sediments of the Aveiro sedimentary basin as well as the definition of deposit environment conditioning.

INTRODUCTION

During the last years we have been engaged in studying the sub-surface geology of the Ria de Aveiro region. The area under study represents about 55 km^2, corresponds essencially to the estuary of the river Vouga and it is still an active sedimentary basin. It belongs to the northermost part of the Lusitanian basin. At present the Ria de Aveiro corresponds to a barrier lagoon that communicates with the Atlantic Ocean by an artificial channel. All the information on the sub-surface geology yielded so far is based on mineralogical and sedimentological studies carried out on the cuttings derived from more than twenty deep boreholes used for water prospection and production.

In the Ria de Aveiro region, it has been so far accepted the existence of an important stratigraphical gap between the end of the Maastrichtian and the Plio-Pleistocene. Paleogene and Neogene deposits, up to 550 meters thick, are reported in the continental shelf off the region (Boillot et al., 1972, 1978; Mougenot, 1989). However, in the Ria de Aveiro region just the so-called "Residual, or Silicificated Boulders" are by some authors admitted to be of Tertiary age (Teixeira & Zbyszewski, 1976; Barbosa, 1981; Corrochano & Pena dos Reis, 1986; Marques da Silva, 1990). There are records of Pliocene (Plio-Pleistocene) expressed by raised ancient beachs and fluvial terraces (Soares de Carvalho, 1953; Teixeira &

Zbyszewski, 1976). Lauverjat (1982) refers the possible existence of important marine and continental Pliocene deposits.

MATERIALS AND METHODS

Mineralogical studies were based particularly on X-ray diffraction (XRD) determinations, using the standard techniques. These were carried out both in the less than 38 μm and 2 μm fractions, previously separated. Also, transmission electron micrographs (TEM) of the less than 2 μm fractions were carried out. Physical properties, such as: granulometry, colour, organic matter content, pH and flocculation-defloculation state were determined as well.

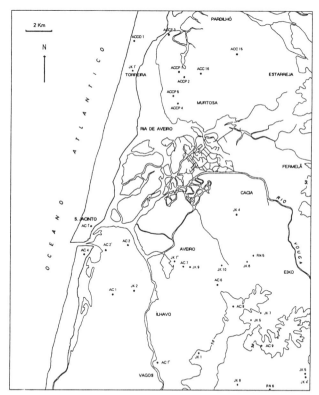

Fig. 1 - Boreholes location.

Clay minerals and the accompanying non clay minerals were utilized for lithostratigraphical discrimination and for paleosurfaces definition. In particular, clay minerals assemblages reflect paleogeographic and paleoclimatic conditions of formation and deposit. Therefore, they are an interesting tool for reconstructing past landscapes and climates. As a matter of fact, in the case of the Ria de Aveiro region, their nature and crystallochemical features allowed the zonation of Tertiary and Quaternary sediments as well as the definition of deposit environment conditioning. In a first stage, we have gathered detailed information from six boreholes: JK1, JK2, JK4, AC2, AC2', and AC6 (see fig. 1). However, for the present work information from more than twenty boreholes has been considered as well.

ANALYTICAL RESULTS

Mineralogical and sedimentological data did show a sharp transition between late Cretaceous sediments (Argilas de Aveiro formation) and the post-Cretaceous sediments. That transition could be expressed in terms of properties such as: compaction, colour, organic content, pH and mineralogical compositior. Indeed, post-Cretaceous sediments are comparatively more loose, dark, organic and acid. In all boreholes, the post-Cretaceous sediments provide water suspensions unstable, that is, showing high flocculation rates. On contrary, water suspensions of the late Cretaceous sediments were quite stable. Besides late Cretaceous and post-Cretaceous sediments could be distinguished in terms of mineral composition, and particularly in their clay mineral suites.

Late Cretaceous sediments (Argilas de Aveiro) are characterized by the following clay mineral assemblage: illite (medium crystallinity), kaolinite (low to medium cristallinity), Ca-smectite (rare), and illite-smectite irregular interstratification. Dolomite is well represented in late Cretaceous sediments, formed mainly by shally grey, green or brown clays. On the contrary, this carbonate is rare in post-Cretaceous sediments. Moreover, the feldspars in the late Cretaceous are richer in K, whereas in the post-Cretaceous sediments they are richer in Na and Ca.

Using the mineralogical data available, Rocha & Gomes (1991) established in post-Cretaceous sediments at least four major sedimentary units, particularly characterized by their clay mineral suites. The composite log of the Tertiary and Quaternary sediments of the Aveiro sedimentary basin is shown at Fig. 2.

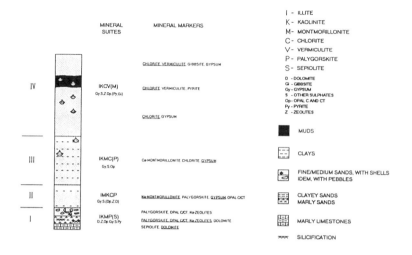

Fig. 2 - Composite log of the Tertiary and Quaternary sediments of the Aveiro sedimentary basin.

A lower unit of the post-Cretaceous sediments (Paleogene?) is characterized by the following mineral assemblage: halite; anhydrite, gypsum and other sulfates (melanterite, natrojarosite, alunite, thenardite and mirabilite); pyrite; iron hydroxides (lepidocrocite and goethite); Na-zeolites (clinoptilolite, phyllipsite and analcite); opal C and CT; dolomite and a clay mineral suite (Na-smectite, Mg-smectite, illite, sepiolite, palygorskite, rare kaolinite, rare chlorite and illite-chlorite irregular interstratification). However, there are variations in the mineral assemblage, and lithologically the unit consists of three distinctive zones (Fig. 2): a lower zone rich in dolostone, a intermediate zone rich in marl and fine to medium sandstone and a upper zone rich in coarse sandstone including gravel and pebbles. Silicification is more developed in the intermediate zone of fine to medium sandstones. Palygorskite is more abundant whenever silicification is more developed whereas sepiolite occurs in the carbonate rich lower zone. Diatom frustules (observed in TEM) exist in the top sediments of this unit.

The second unit (Neogene?) is more uniform in lithologies (fine to medium argillaceous sands and in places clay rich passages). It has a mineral composition similar to the lower unit, less saline, characterized by higher content of Na-smectite and lower content of palygorskite and opal C and CT; dolomite becomes more rare.

The third unit of post-Cretaceous sediments (Pleistocene?) consists of fine and micaceous argillaceous grey sand showing in some zones abundant shells of *Cardium edule* (L.). It is characterized by the following mineral composition: illite (less ferromagnesian than the illite from the two lower units), kaolinite, Ca- and Na-smectite, illite-vermiculite and illite-smectite irregular interstratifications. Carbonates are very rare, as well as opal C and CT.

The upper unit (Holocene?) comprises a lower zone of fine micaceous grey sand with abundant shells of *Cardium edule* (L.), an intermediate zone rich in silty black muds including in places abundant shells and a upper zone of fine to medium brown and grey sands where shells are rare or absent. It is characterized by the following mineral composition: illite (similar to that of the third unit, but with crystallochemistry more diversified), kaolinite, chlorite (this is the unit where this mineral is more commom), vermiculite (absent or very rare in the other units), illite-vermiculite and illite-chlorite irregular interstratifications. The sapropel clay rich/mud contains diatoms, pyrite and marcassite. The top surficial sediments contain small amounts of gibbsite, an unstable land-derived mineral, naturally more abundant landward. This mineral comes from some continental soils (cambisoils) which occur in places surrounding the Aveiro sedimentary basin and where gibbsite is associated with kaolinite, illite and vermiculite-Al hydroxide intergrades.

DISCUSSION OF THE RESULTS

It is considered that late Cretaceous sediments (Argilas de Aveiro) were deposited in a littoral, probably estuarine, environment with low hydrodinamics.

The sediments of the lower unit (Paleogene?) of the post-Cretaceous sediments appear to be mainly derived from a perimarine sedimentation, with confined lagunar character; its pre-evaporitic and evaporitic facies points out towards a dry and hot semi-arid climate. Chamley (1989) reports for Paleogene occurences the presence of fibrous clay minerals, generally associated to Mg-smectite (in vertical and/or lateral sequences), in lagunar environments with carbonates, where sulfates and frequent silicifications can be found also. The Paleogene(?) of the Aveiro basin has similar characteristics. There are variations in the mineral assemblage, pointing out towards alternating less and more saline environments. Anyway, the saline character diminishes and vanishes gradually towards the top of the unit. In fact, in the lower zone, sepiolite indicates a basic lagunar environment, with shallow waters and pH-8 (Galan & Ferrero, 1982). Sepiolite formation is favoured by the high Si/Mg ratio, situation wich facilitates as well the formation of dolomite (Doval et al., 1985; Garcia Romero et al., 1990). This situation is not favourable for opal formation. The variable concentration ratio of sepiolite/Mg-smectite means salinity fluctuations. Mg-smectite increases whenever pH gets close to 9. In the intermediate zone, sepiolite is not present and palygorskite takes its place. This means a climate less arid, with longer wet periods (Chamley, 1989), favouring the action of the chemical weathering in the continental surrounding rocks and as a consequence the higher transportation into the basin of terrigenous phyllosilicates. The deposition environment became brackish and the equilibrium between solutions and unstable phyllosilicates such as smectites favoured the neoformation of palygorskite (Galan & Ferrero, 1982; Galan & Castillo, 1984). Silicification is more developed in the intermediate zone of fine to medium sandstones. It looks to be a ground-water type silicification with a diagenesis contribution. This assumption can be also supported by the abundance of zeolites in the same zone. Silica results either from the dissolution of diatom frustules existing in the top sediments or from the dissolution of clay minerals such as Mg-smectite in alkaline solutions taking place on the basal sediments rich in sepiolite and palygorskite. According to Leguey et al. (1989) the SiO_2/MgO ratio increase to values higher than those found in sepiolite and palygorskite would favour silicification. Velde (1985) reports that in recent non oceanic sediments with zeolites, opal is usually present, and accentuates that alkaline zeolites have less silica than corresponding feldspars. Therefore, feldspar-zeolite transformation would make silica available for the formation of opal. Leguey et al. (1989) stress the important role of water table fluctuation, for the alternating enrichment in fibrous clay minerals and Mg-smectite and for silicification development. A geochemical barrier of oxydising type (Meyer, 1987) and the input of carbonated waters due to water table fluctuation would destabilize the Mg-smectite structure, liberating Si and Mg, and favour the formation of opal, dolomite and sepiolite. Formation of sepiolite would depend of drainage and salinity (Darragi & Tardy, 1987). In periods of dry climate with strong evaporation, pH would increase favouring silica dissolution at pore level, whereas in periods of wet climate, pH would decrease favouring silica precipitation (Meyer, 1987). The lithologie of the upper zone, more detrital and coarse, expresses the existence of periods of high pluviosity and of torrencial type.

The lithological and mineralogical composition of the sediments of the second unit (Neogene?) of the post-Cretaceous point out towards a littoral environment of shallow waters with alternating episodes characterized by higher salt water or fresh water influence. Smectite and illite increase means the transition to climatic conditions significantly more wet (Chamley, 1975; Chamley & Robert, 1980). In particular, the relative increase in smectite, signifies the dominance of warm climate with wet and dry contrasting seasons, that is, from sub-arid to sub-tropical (Galan, 1986).

The terrigenous provenance of the sediments of the third unit becomes more and more notorious. In this unit, illite structure becomes more open and in places less dominant, being compensated by kaolinite and smectite, fact that means the prevailing of hydrolizing conditions wich continue through the lower zone of the fourth unit.

At the fourth unit, the higher contents in kaolinite, illite-vermiculite and illite with a structure more open found in some layers (mainly in the lower zone) correspond to sedimentation conditions where chemical weathering became relatively relevant and in relation with episodes of warm and wet climate. Alternating climate periods, cooler and warmer, are expressed by the following assemblages: illite+chlorite and open illite+kaolinite+illite-vermiculite+vermiculite. The sapropel clay rich mud (with diatoms, pyrite and marcassite) corresponds to depositional conditions with low detritical supply in an anoxic basin, due to the existence of a surficial low-salinity layer. Gibbsite, weathering product in an initial stage of granular igneous and metamorphic rocks, reveals temperate and wet climate (Macias Vasquez, 1980).

The sediments of the two upper units (attributed to Plio-Pleistocene and Holocene) are mainly land-derived and deposited in swamp and estuarine brackish waters, particularly the top unit. In places, the environment has been and still is anoxic, since the sediments are noticeably rich in organic matter and contain sulfides (Berner, 1981).

CONCLUSIONS

The research carried out upon sediments of the Aveiro sedimentary basin did show that in comparison with Cretaceous and ante-Cretaceous sediments clay mineral assemblages in post-Cretaceous sediments become more and more complex, since physical weathering overpass, in general terms, the chemical weathering. Besides, terrigenous input from the river hydrographic basins becomes more and more important in so far as the communication of the estuary with the sea turns out more difficult. Terrigenous materials carried by the several water-courses coming into the Aveiro sedimentary basin have different composition in accordance with the composition of soils existent in their respective hydrographic basins.

Clay minerals assemblages that have been determined allowed the discrimination of sedimentary units both in Tertiary and Quaternary sediments. Also, they provided information on the environmental conditions prevailing at the time of their deposition.

REFERENCES

BARBOSA, B. (1981) - Carta Geológica de Portugal, 1/50000. Notícia explicativa da Folha 16-C, Vagos. Serv. Geol. Portugal.

BERNER, A. (1981) - A new geochemical classification of sedimentary environments. Journ. of Sed. Petr., 51(2).

BOILLOT, G. et al. (1972) - Geologie du Plateau Continental Portugais au Nord du Cap Carvoeiro. La série stratigraphique. C. R. Acad. Sc. Paris, 274.

BOILLOT, G. et al. (1978) - Carta Geológica da plataforma continental. Escala 1/1000000. Serv. Geol. Portugal.

CHAMLEY, H. (1975) - Sedimentation argileuse au Mer Tyrrhénienne au Plio-Pléistocène d'aprés l'étude du forage JOIDES, 132. Bull. Groupe Franç. Argiles, 27.

CHAMLEY, H. (1989) - Clay sedimentology. Springer-Verlag.

CHAMLEY, H. & ROBERT, C. (1980) - Sédimentation argileuse au Tertiaire supérieur dans le domaine méditerranéen. Géol. Méditerr., 7.

CORROCHANO, A. & PENA DOS REIS, R. P. B. (1986) - Analogias y diferencias en la evolución sedimentaria de las cuencas del Duero, Occidental Portuguesa y Lousã (Península Ibérica). Stud. Geol. Salmant., XXII.

DARRAGI, F. & TARDY, Y. (1987) - Authigenic trioctahedral smectites controlling pH, alkalinity, silica and magnesium concentrations in alkaline lakes. Chem. Geol., 63.

DOVAL, M. et al. (1985) - Mineralogia de las facies evaporíticas de la cuenca del Tajo. Trab. Geol., 15.

GALAN, E. (1986) - Las arcillas como indicadores paleoambientales. Bol. Soc. Esp. Miner., 9.

GALAN, E. & CASTILLO, A. (1984) - Sepiolite-Palygorskite in spanish tertiary basins: genetical patterns in continental environments. In: Singer, A. & Galan, E. (ed.) - Palygorskite-sepiolite. Occurrences, genesis and uses. Developments in sedimentology, 37.

GALAN, E. & FERRERO, A. (1982) - Palygorskite-sepiolite clays of Lebrija, Southern Spain. Clays and Clay Minerals, 30.

GARCIA ROMERO, E. et al. (1990) - Caracterización mineralógica y estratigráfica de las formaciones neógenas del borde sur de la cuenca del Tajo (Comarca de Sagra). Bol. Geol. Min., 101.

LAUVERJAT, J. (1982) - Le Crétacé supérieur dans le Nord du Bassin Occidental Portugais. Thése, Univ. P. et M. Curie, Paris VI.

LEGUEY, S. et al. (1989) - Paleosuelos de sepiolita en el Neogeno de la cuenca de Madrid. Est. Geol., 45.

MACIAS VASQUEZ, F. et al. (1980) - Origen y distribución de la gibbsita en Galicia. An. Edaf. y Agrobiol., 39.

MARQUES DA SILVA, M. A. (1990) - Hidrogeologia del Sistema Multiacuifero Cretacico del Bajo Vouga - Aveiro (Portugal). Tese, Univ. Barcelona.

MEYER, R. (1987) - Paléoaltérites et paléosols. B.R.G.M., Man. Méth., 13.

MOUGENOT, D. (1989) - Geologia da Margem Portuguesa. Doc. Técnicos, 32, Inst. Hidr., Lisboa.

ROCHA, F. J. F. T. & GOMES, C. S. F. (1991) - Mineralogical and sedimentological data of the post-Cretaceous sediments of the Ria de Aveiro region. Geolis, in press.

SOARES DE CARVALHO, G. (1953) - Les sédiments pliocènes et la morphologie de la région entre Vouga et Mondego (Portugal). <u>Mem. Not.</u>, 34, Mus. Lab. Min. Geol. Univ. Coimbra.

TEIXEIRA, C. & ZBYSZEWSKI, G. (1976) - Carta Geológica de Portugal, 1/50000. Notícia explicativa da Folha 16-A, Aveiro. <u>Serv. Geol. Portugal</u>.

VELDE, B. (1985) - Clay minerals. A physical-chemical explanation of their occurrence. <u>Developments in sedimentology</u>, 40.

RECENT EVOLUTION OF ÓBIDOS AND ALBUFEIRA COASTAL LAGOONS [1]

FREITAS, M. ([2]); ANDRADE, C. ([2]) & JONES, F. ([2])

ABSTRACT

The Óbidos and Albufeira lagoons are small coastal lagoons located on the Portuguese west coast. At present, these lagoons are isolated from the ocean by sandy barriers and connection with open sea waters is made by artificial opening of temporary inlets.

Recent evolution patterns were investigated using written documents, maps, aerial photographs and periodic field observations; physiographical changes and average sedimentation rates are suggested. Albufeira's tidal inlet morphologic evolution is described and an evolutionary model is proposed.

1. Introduction

The coastal lagoons of Óbidos and Albufeira are located in the portuguese western coast, some 80 Km north and 20 Km south of Lisbon, respectively (Fig. 1). Both structures are partly closed lagoons (**Nichols and Allen** 1981) and coincide with type I of Lankford classification (*in:* **Carruesco** *et al.* 1980).

The flooded area of Óbidos lagoon is about 6 Km^2 and it's maximum length, width and depth are 4.5 Km, 1800m and 4.5m. The correspondent watershed extends through 440 Km^2. The flooded area of Albufeira lagoon is only about 1.3 Km^2. Its maximum length, width and depth are 3.5 Km, 625m and 15m. The corresponding watershed averages approximately 106 Km^2.

Average temperature and precipitation at both sites is about 15°C and 600-650 mm respectively. Precipitation is almost totally concentrated in winter.

---

[1] - This paper is a contribution of Project "Climate Change, Sea-Level Rise and Associated Impacts in Europe", contract EPOC - CT 90 - 0015.

[2] - Geological Department, Faculty of Sciences of the University of Lisbon and "Instituto Nacional de Investigação Científica (INIC)", Campo Grande Edif. C2, 5º Piso, 1700 Lisboa, Portugal.

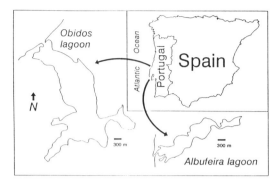

Fig. 1 - Location map of the study area.

Tides are semidiurnal and average tidal range is 2m (high-mesotidal coast according to **Hayes** 1979). When inlets are operating, tidal influence extends to the whole lagoon showing strong dominance of flood currents (**Freitas** 1990).

Wave regime is dominated by NW swell with modal period of 9 - 11 s and significant breaking height of 1 - 2 m. Predominant wave approach induces net littoral drift directed towards south.

Both lagoons are separated from the ocean by continuous sandy barriers. In order to allow water recycling, every year an inlet is opened by artificial means, during Easter. The inlet evolves naturally and finally closes up completely, during summer, mostly by transverse wave-induced sand shifting.

The strong silting of Óbidos lagoon requires periodic dredging of navigation channels used by small fishing boats. A minimal water depth is also attained which is essential for survival of the lagoonal fauna. At Albufeira lagoon the sedimentation rate is not so high and there are no records of dredging.

They represent two sensitive geological systems related to recent evolution of portuguese western coast which have a considerable importance from an economical, ecological and recreational point of view.

A compilation of charts and written documents allowed a reconstruction of the recent evolution of both lagoons, since the XIV century until now. Older charts are interpreted qualitatively, but since the XIX century more accurate maps were produced, allowing the reduction to a common scale and some quantitative investigation. Aerial photographs (vertical and oblique) were also used. Recent cartographic and photographic documents were processed using AUTOCAD Software. On a shorter time scale, seasonal evolution of the Albufeira lagoon inlet was studied, using periodic field observations taken in 1991 and 1992, aerial photographs and a few available topohydrographic surveys.

2. Óbidos Lagoon

Old charts and written documents (Fig. 2) suggest that before the XIX century the lagoon extended further inland, for more than 6 km, until Óbidos village. Ancient maps systematically show this lagoon as an open estuary but there is very strong evidences that the barrier was already formed and persisted since (if not before) the XV century (**Trindade** 1985).

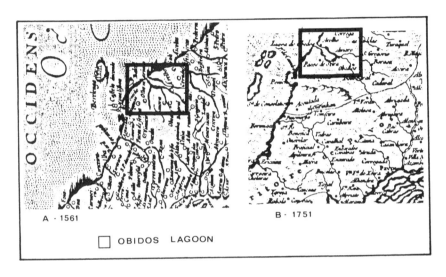

Fig. 2 - Óbidos Lagoon representations on the 1561's and 1751's maps.

A few documents of that century inclusively describe the need to reopen the inlet because of damage produced by flooding of farm land. Between 1867 and 1947 chart comparison shows that the most striking evolutionary pattern of this lagoon is its progressive reduction in size. This change is essentially achieved by intense sedimentation at tributaries mouths, leading to a severe shortening of the lagoonal length, a simplification of its peripheral plan shape and to some general shrinkage of the flooded area (Fig. 3).

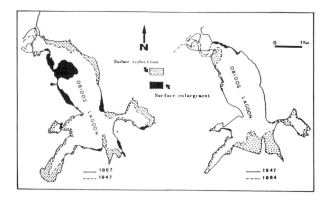

Fig. 3 - Óbidos Lagoon reduction / enlargement surface areas between 1867 and 1964.

Between 1947 and 1964, the same pattern persists: the estuary front of the main eastern tributary ("Real" river) steadily silted up and prograded towards west. In the inner margin of the barrier a few changes are noticeable, suggesting that large volumes of sand entering the lagoon as washovers or tidal sediments were trapped and reshaped in two major flood shoals, leading to the loss of efficiency of the main channel that progressively narrowed and meandered intensively.

A quantitative analysis of this evolution pattern, using hydrographyc surveys dated from 1917 and 1980 (Fig. 4) clearly shows that sedimentation is widespread across the whole lagoonal surface (approximately 3 km^2) and that total erosional surface occupies less then 0.6 km^2. The latter is commonly associated with lateral translation of thalweg lines (e. g. main channel shifts northwards and "Bom Sucesso" creek migrates westwards) but even in this cases laterally accreted surfaces exceed erosioned areas. Total sedimentation corresponds to an input of some 2.7×10^6 m^3 (4.3×10^4 m^3 / year) while erosion totalizes 2.5×10^5 m^3 (4×10^3 m^3 /year). Fig. 5 shows the distribution of sedimentation and erosion rates within the lagoonal area. As expected, inner areas associated to tributary estuaries show higher rates of sedimentation, locally exceeding 4.5 cm/year. Central and deeper region of the lagoon has moderate silting rates. Global average sedimentation rate is about 7 mm/year and average erosion rate is about 0.6 mm/year. Unlike the eastern margin, a general depth reduction is apparent, suggesting that widespread decantation of fine grained materials leading to bottom elevation predominates over lateral (marginal) accretion.

Fluvial input seems to be the major source of lagoonal sediments. An estimate of the fluvial solid discharge ratio was made using empirical methods proposed by **Fournier** (1960), **Langbein** and **Schumm** (1958) and **Teixeira and Romariz** (in press) for watershed erosion. Climatic data considered precipitation averages estimated from 30 years (1952-1982) records and weighted by the Thiessen polygon method. Specific and total erosion rates obtained by these methods were converted in calculated sedimentation rates assuming that:

- no sediment leakage occurs outwards of the system through the inlet;

- there is no significant change in sediment bulk specific weight during transport, sedimentation and burial processes;

- specific weight of sediments is $1 Ton/m^3$, a figure compatible with the predominant (> 95%) muddy nature of surficial sediments (**Freitas** 1989);

- total solid discharge is spread uniformly within the flooded lagoonal surface.

Table I.a. summarizes results obtained from the adequate specific equation proposed by Fournier (1960). As expected, the estimated figure is very high, a result that agrees with the general experience of researchers that used Fournier's method, free of any corrections, in the Portuguese territory. An adaptation of the climatic index proposed by **Teixeira and Romariz** (in press) is included in the second set of figures of the same table.

The method of Langbein and Schumm (table I.b.) was parametrized using three different inputs to calculate "effective precipitation":

a) difference between average precipitation and evapotranspiration;

b) average runoff data obtained from the "Atlas do Ambiente de Portugal" (1975);

c) runoff obtained through Langbein's (1949 *in:* Komar 1976) correlation curve (mean annual precipitation versus mean annual runoff) weighted as a function of average temperature.

Finally, results obtained with the empirical equations proposed by **Teixeira and Romariz** (in press) are summarized in table I.c.

Table I

	INPUT	Sediment yield m³/year	Sedimentation rate mm/year
A Fournier (1960)	Specific equation (SP)	807 840	134,6
	SP with correction parameter (Teixeira & Romariz, in press)	210 232	35
B Langbein and Schumm (1958)	Runoff ("Atlas do Ambiente de Portugal")	134 640	18
	Runoff correction curve (Langbein 1949 in Komar 1976)	134 640	18
	EP = P - EVP (1)	222 640	29,6
C	**Teixeira and Romariz** (in press)	262 240	43,7

Table II

	INPUT	Sediment yield m³/year	Sedimentation rate mm/year
A Fournier (1960)	Specific equation (SP)	269 770	210
	SP with correction parameter (Teixeira & Romariz, in press)	24 168	19
B Langbein and Schumm (1958)	Runoff ("Atlas do Ambiente de Portugal")	32 436	25
	Runoff correction curve (Langbein 1949 in Komar 1976)	46 216	36
	EP = P - EVP (1)	23 850	18
C	**Teixeira and Romariz** (in press)	33 908	26

(1) EP = Effective precipitation
P = Mean annual precipitation
EVP = Evapotranspiration

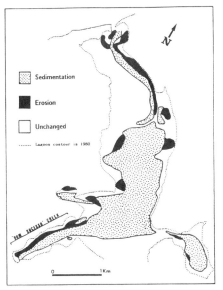 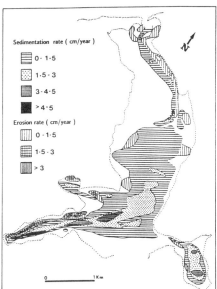

Fig. 4 - Óbidos Lagoon sedimentation / erosion areas between 1917 and 1980.

Fig. 5 - Óbidos Lagoon sedimentation / erosion rates between 1917 and 1980.

There is a very good agreement between observed and predicted average sedimentation rates, though the latter are slightly overestimated. This result seems to confirm:

- an almost perfect trapping capacity of the lagoon for fluvial inputs;

- the absolute predominance of terrestrial (fluvial) inputs as sediment source of the lagoon;

- errors resulting from cartographic comparison are small.

3. Albufeira Lagoon

Analysis of available maps and sketches made since the XV century, suggests that this lagoon evolved alternating an "open estuary" pattern with another of a "closed" or "fully barred estuary" (Figs. 6 and 7). However, contemporary written documents indicate that the sandy barrier was effectively more persistent than suggested by drawings and charts and that it was regularly artificially opened since the XV century (**Castelo-Branco** 1957). In fact, polinic analysis made on peats from Estacada (inner sector of Albufeira lagoon) suggest that at least from 5000 BP this lagoon was already isolated from the sea (**Queiróz** 1989).

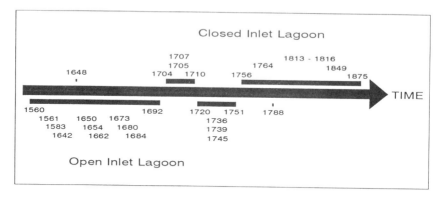

Fig. 6 - Time diagram of available Albufeira lagoon maps arranged by their opened or closed inlet lagoon representation.

Surveys of 1849, 1920 and 1959 (Fig. 8) show persistent shrinkage of the lagoonal margins, specially affecting main eastern tributary mouths and considerable sand retention in shoals located northeast of the inlet mouth. Aerial photographs confirm these trends and in particular the 1986's which clearly shows that small fan-shaped sand bodies did actually accrete and coalesce in the vicinity of active (or formerly active) tidal inlets, as incipient small flood tidal deltas. Later, these sand bodies may suffer remobilization by current activity and mainly by local wind - generated wavelets, very common during autumn, winter and spring, when the lagoon constitutes a fully closed system. Washovers also contributed to the vertical and landward growth of the barrier in the last 30 years. There is no significant return of sand to the nearshore through the tidal channel, indicated by clear dominance of flood currents, a characteristic that was confirmed by field inspection of surface bedforms of intertidal flats.

Albufeira western end is an efficient sand trap for marine sediments, whose intensity has been apparently enhanced in the last 30-50 years by the regular reopening of the inlet (Fig. 9), expressed by an enlargement of the marine sand body thickness towards the inner side of the coastal barrier.

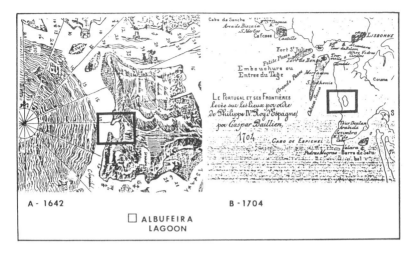

Fig. 7 - Albufeira lagoon coastal region representation on the 1642's and 1704's maps. A - "open estuary" representation; B - "fully barred estuary" representation.

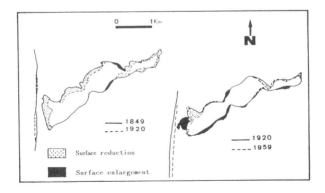

Fig. 8 - Albufeira lagoon reduction / enlargement surface areas between 1849 and 1959.

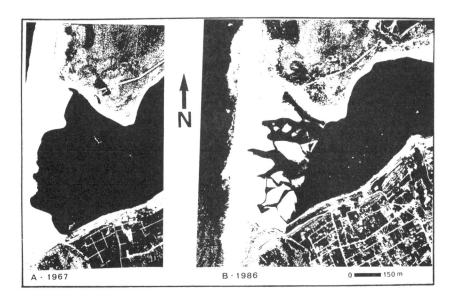

Fig. 9 - Aerial photographs of the Albufeira lagoon barrier system in 1967 (A) and 1986 (B).

Unlike Óbidos, Albufeira lagoon was not surveyed systematically and no series of hydrographic surveys were available to allow an objective estimation of the sedimentation rate. Therefore, an approximate figure was predicted by the same procedure used at Óbidos lagoon.

Figures obtained for Albufeira lagoon range between 18 mm/year and 210 mm/year (Table II). Using the relationship between watershed erosion and sediment delivery ratio obtained at Óbidos we estimate a sedimentation rate of 3-11 mm/year at Albufeira lagoon.

3.1. Seasonal evolution of the barrier-inlet area at Albufeira lagoon

Albufeira coastal barrier is made of clean, coarse, well sorted, quartz sand. Its maximum length and width are approximately 1200m and 210m respectively

(Fig. 10). The northern end contains a small, incipient and hummocky dune field, strongly affected by human activity. The barrier attains there its maximum height - about 10m above H.Z.

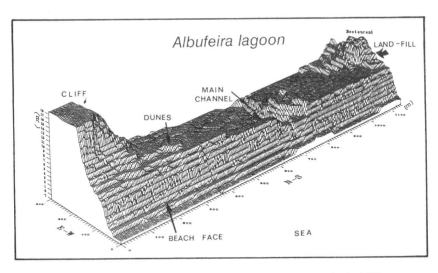

Fig. 10 - 3-D topographic diagram of Albufeira lagoon barrier in 1986.

The inlet is reopened every year (in Easter) by artificial means allowing outflow of brackish water ponded in the lagoon during winter; for a short period of time a limited tidal water exchange occurs.

The inlet is not protected by rigid structures and its efficiency decreases rapidly until it silts up naturally and closes. Its life span varies between 2-3 weeks and 5 to 6 months. Natural reopening of the barrier by overwash is possible in winter, associated to violent storms and extensive flooding of the lagoon but was rarely observed since the beginning of this century. Actually, the regular reopening of the inlet keeps the lagoonal water surface relatively low, inhibiting the development and maintenance of a tidal inlet after an intensive overwash event.

Because of its location, a short distance to the south of a null drift point, littoral drift is not a very important process of sediment transfer both at the barrier and at the inlet gorge. However, slight but persistent oblique approach of dominant swell induces slow downdrift (southwards) migration of the tidal channel that may persist during its lifetime, so contributing to the complete reshaping of segments of the southern half of the barrier.

Several aerial photographs and detailed topohydrographic surveys show very different patterns of inlet morphology, apparently disconnected, but field observations suggested that they could be linked in an evolutionary sequence. In fact, field work conducted during 1991 and 1992 confirmed this hypothesis and first results are present as a series of instantaneous portraits, representing typical evolutionary moments of a continuous sequence that evolves very rapidly (Fig. 11):

Stage 1 typifies artificial reopening. Ponded water is violently flushed out through a narrow opening, ejecting a large volume of sediments to the near shore, mostly eroded from the sandy barrier as a result of the natural enlargement of the artificial channel section. A "v" shaped cross section develops and may persist for a few days. Thalweg line is very deep (more than 2 m below LWM) at the channel gorge and both margins dip very steeply, as a consequence of fast regressive erosion of the sandy enbankments, promoted by tidal currents. Stage 1 is typically short lived and may evolve through two different paths (Fig. 12). A possible path is a direct shortcut to stage 6, probably favored by a combination of a decreasing tidal range, with a change in the morphodinamic stage of the beach towards a more reflective profile. Under these circumstances, a low intertidal beach ridge develops very rapidly at expense of landward migration of swash bars, blocking the outer end of the channel (stage 6.A.). The newly formed bar accretes vertically and shifts upwards as a beach berm, throughout the neap-spring tidal cycle, literally plugging the channel rebuilding and healing the barrier (stage 6.B.). This sequence of events was observed during two weeks in April 91, compelling to a second reopening of the inlet.

A second possible and more common path (stage 2) is triggered by slow downdrift of the main ebb channel, until it leans against the south embankment. A strongly asymmetrical cross section develops and an intertidal point bar-like sand bank starts to grow on the north part of the inlet gorge as a consequence of incipient meandering (Fig. 11 - stage 2A). The bank surface is essentially reactivated by tidal currents and show strong evidence of flood dominance.

On the outer region of the inlet, a swash ramp develops steadily and crescentic swash bars accrete and weld, mostly on the updrift half of the platform. Wave induced sand migration towards the inlet nourishes the external rim of the point bar, allowing the bank to extend outwards and align itself with the nearby beaches (Fig. 11 - stage 2.B.).

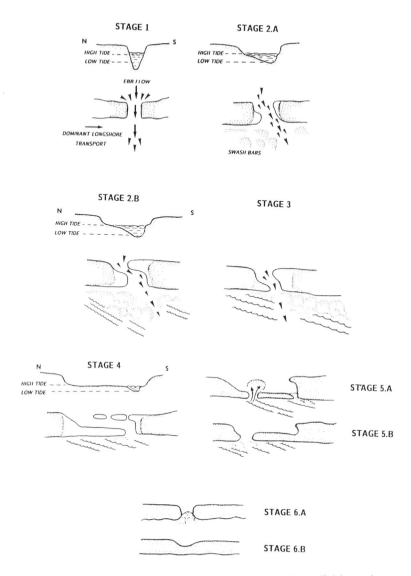

Fig. 11 - Albufeira lagoon inlet evolution stages after its artificial opening.

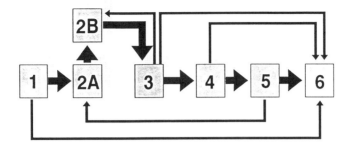

Fig. 12 - Albufeira inlet morphodynamic evolution sketch showing possible connections between several successive stages. Wide arrows indicate 1992's observed sequence.

Meanwhile, linear erosion at the southern margin favours further meandering and extension of the main channel, leading eventually to the development of a second (inner) point bar, producing an overlapping plan shape of the inner inlet area.

Fig. 13 shows two overlapping longshore sections of the barrier drawn after topohydrographic surveys executed in 1975 and 1986. The 1975 situation shows a cross section of a drifting inlet - stage 2.B. On the other hand, the 1986 situation is an almost pure stage 1 example: there is no noticeable drifting of the channel, whose bottom is silting up very rapidly (compare channel depth and cross sectional area with 1975 situation); actually, the inlet survived for just one month in 1986.

Stages 3, 4 and 5 (Fig. 11) may alternate or one of them may predominate for several months, until the inlet starts to silt up irreversibly.

Stage 3 represents a pure ebb delta (though wave dominated) breaching process, similar to Fitzgerald's (1988) description, allowing sand to be released to the downdrift coast. It was observed in 1992, after two and a half months of inlet activity and continuous sand accumulation on the updrift side of the inlet . A huge intertidal, complex swash bar developed, clearly offsetting the surf area until some threshold of sediment retention was reached. After breaching, relocation and clockwise rotation of the ebb-jet, the channel extension was significantly reduced and a continuous, linear swash bar developed on the downdrift coast. This stage may predominate if channel migration towards south is stopped or persists at a very low rate.

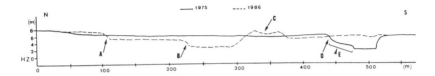

Fig. 13 - Longshore topographic profiles of Albufeira lagoon barrier in 1975 and 1986. A - Ancient channel opening scar, dated 1975 - 1980; B - 1986 channel opening scar; C - Dredge spoil; D - 1975 channel opening scar; E - 1975 inlet drift distance; HZ - Hydrographic zero.

Rapid migration of the channel inevitably leads to stage 4 configuration. This morphodynamic stage is characterized by the enlargement of the inlet width, specially noticeable at high tide, as a result of bank recession on the southern margin of the inlet. Observed channel migration rates range between 0.9 m/day and 3 m/day (Fig. 14). It is closely followed by the northern point bar longshore extension (at approximately the same rate) and vertical accretion up to 4m above H.Z. (Fig. 15). Current activity on the surface of the point bar looses influence to swash activity and eventually the bar is exclusively dominated by overwash, changing first into a low, intertidal ridge and afterwards into a high beach berm. Sand is shifted by transverse movements from the swash bars to the growing intertidal sand bar, consuming the large extension of shoals that previously existed. This change is clearly noticeable in the cross section of the bar that shifts from a convex, symmetrical surface to a high asymmetrical ocean facing sand ridge, carrying typical overwash structures on the lee surface. The second, inner point bar referred in stage 2.B. is cutoff by ebb tidal currents, exploiting gaps developed during flood tide as spillovers and washover channels.

Stage 4 is intrinsically fragile because of very strong channel extension and cross sectional area enlargement. It was observed in 1992 during approximately two weeks but it is inevitably completed by stage 5. This is, once again, an inlet relocation process, involving barrier breaching - stage 5.A. - and discontinuous updrift jumping of the tidal channel - stage 5.B. - induced by overwash during high tide (see **Fitzgerald** 1988 for a detailed explanation of the model).

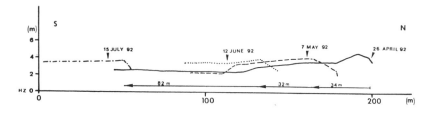

Fig. 14 - Albufeira southern inlet margin retreat between April 26 and July 15 (1992). HZ - Hydrographic zero.

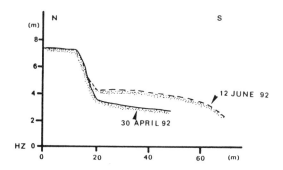

Fig. 15 - Vertical and lateral accretion of the point bar like bank at the gorge of Albufeira tidal inlet, between April 30 and June 12 (1992). HZ - Hydrographic zero.

From stage 5.B, several evolutionary processes are available: shifting to stages 6 (described above) or 2/3 (Fig. 12), are most probable.

During the evolutionary process, channel migration and lateral erosion are very important processes. Both processes induce channel extension or enlargement of the channel diameter, implying a reduction of the self flushing ability associated to ebb currents. This means that every time the cycle is short-circuited and a lower order stage "reboots", hydraulic efficiency of the inlet is reduced. Actually, field measurements suggest that the long term evolution of the inlet is characterized by reduction of the inlet cross sectional area between stages, or even within the same stage of the cycle. In this case, the reduction of cross sectional area is achieved mostly by depth reduction and not by lateral constraint of the channel width.

Shifting of wave obliquity was observed in the summer of 1992, when the inlet was in a stage 2/3 situation. No reversal of the channel drifting direction was observed and the main consequence was a rapid vertical accretion of the point bar like bank and a slight shortening and reorientation of its free end.

The sequence described in this paper corresponds to the logical chain of events according to field observations taken in 1991 and 1992 but it may not represent the unique path or ever the most probable sequence. More field surveys are necessary in order to allow further refinement of this model.

4. Summary and conclusions

The coastal lagoons of Obidos and Albufeira are located in a mixed energy, high mesotidal coast and developed in close association with drowned small scale estuaries. They are separated from the sea by persistent linear sand barriers. Exceptionally violent sea may cut ephemeral inlets in the barriers; however, this process is exceptional in the last decades, due to the regular reopening of the inlets, achieved by artificial means. The life span of these wave dominated structures is small and the barrier heals up naturally after a period of activity ranging from 2-3 weeks to 5-6 months, mostly at expense of cross shore sand movement.

Historical evidence, cartographic and photographic analysis and comparison were used to assess the main recent morphological changes at these lagoons. Both structures are very efficient sediment traps and show evidence of intensive silting.

The outer margin of the lagoons captures sand injected from the nearshore through overwash and flood tidal currents, which are clearly dominant in the inlet sand budget. Sand is later (closed inlet situation) redistributed by intense activity of wind generated wavelets leading to shoal reshaping. These processes extend their activity to the lagoonal beaches, where depth is small and strong refraction and shoaling effects are visible during 2/3 of the year. Thus, the barrier is accreting landwards, but there is no evidence of long-term enlargement of its width. We suggest that the whole barrier may eventually rollover, thrusting washovers and flood delta segments, through quick, step-like, discontinuous jumps, allowing the conservation of the barrier width in a secular time scale. Episodes of null roll over would last longer and must allow coastal retreat of the adjacent cliffy coastline and readjustment of the longshore sediment budget.

The inner side of the lagoons is predominantly affected by sediment input through river discharge. Formerly drowned river estuaries silted up very quickly in the last 3-4 centuries, leading to considerable shortening of the lagoonal length. Shrinkage of the whole lagoonal margin is also observable, to some extent, and ascribed to wave remobilization of river borne sediments. During the last 50 years, and despite very high sedimentation rates occurring near main tributary mouths, the shrinkage effect lost its intensity, namely because of increasing bottom slopes towards the central part of the lagoon. The central area of both lagoons (permanently flooded) acts as a decantation pool for fine grained sediments and silt up mostly by slow bottom raising.

Evolution of tidal inlet morphology was studied at Albufeira. The apparently random morphological variation observed in photographs and maps may be linked in a coherent series of morphodynamic stages, within which a strong tendency for rapid silting/closure is predominant but several back jumps are possible. Some patterns described by several authors (see **Fitzgerald** 1988, for example) for the long term evolution of large tidal inlets were recognized in this site, but operating in a much more rapid time scale (weeks, months) and involving volumes of sediments that differ from the referred examples in several orders of magnitude.

Portuguese coastal lagoon's evolution can be summarized in the following sequence:

1. During glacial low sea level stand occurs an intense valley cutting along main fluvial streams;

2. During subsequent interglacial high sea level stand occurs estuary drowning and intense silting accompanied by sand barrier building and development of the back barrier lagoonal systems.

3. Present day situation is characterized by regular and frequent inlet artificial opening which accelerated sedimentation rates:

   i) they are flood dominated systems: *i.e.* each reopening of the inlet favours the input of marine sand which increases the inner flood shoals volume;

ii) annual reopening of the inlet allows the lagoon water level to drop to a lower position. On a longer time scale this means that all local base levels are (on the average) kept lower than their natural level intensifying the sediment delivery at all river mouths and increasing the silting rate of the inner region of the lagoon.

Bibliographic References

CARRUESCO, C; BIDET, J.C.; KLINGEBIEL, A. (1980) - L'approche géologique des environnements lagunaires. Centre International pour la Formation et les Échanges Géologiques. Paris, 109 p.

CASTELO-BRANCO, F. (1957) - Alguns aspectos da evolução do litoral português. Boletim da Sociedade de Geografia de Lisboa, 75ª série (nº 7-9). Lisboa.

FOURNIER, F. (1960) - Climat et erosion. Presses Universitaires de France, 198 p.

FITZGERALD, D.M. (1988) - Shoreline erosional-depositional processes associated with tidal inlets. Lectures Notes on Coastal and Estuarine Studies, 29. Hydrodynamics and sediment dynamics of tidal inlets. Springer-Verlag, pp. 186-225.

FREITAS, M.C. (1989) - Natureza dos sedimentos do fundo da Lagoa de Óbidos. Geolis, vol. III, fasc. 1 e 2, pp. 144-153.

FREITAS, M.C. (1990) - Lagoa de Albufeira. Estudo preliminar. Geolis, vol. IV, fasc. 1 e 2, pp. 153-172.

HAYES (1979) - Barrier Island Morphology as a Function of Tidal and Wave Regime. *in:* Leatherman (Ed.) Barrier Islands. Academic Press, pp. 1 - 29.

KOMAR, P.D. (1976) - Beach Processes and Sedimentation. Prentice-Hall, Inc., 411 p.

LANGBEIN, W.B. and SCHUMM, S.A. (1958) - Yield of sediment in relation to mean annual precipitation. Trans. Amer. Geoph. Union, vol. 39, nº 6, pp. 1076-1084.

NICHOLS, M. and ALLEN, G. (1981) - Sedimentary processes in coastal lagoons. Coastal Lagoon Research, Present and Future. Unesco Technical Papers in Marine Science, 33, pp. 27 - 80.

QUEIRÓZ, P. F. (1989) - A preliminary paleoecological study at Estacada (Lagoa de Albufeira). Revista de Biologia, 14, pp. 3-16.

TEIXEIRA, S. and ROMARIZ, C. (in press) - Estimativa da erosão hídrica em Portugal continental a partir da precipitação. Método quantitativo para estimar a erosão hídrica em Portugal continental.

TRINDADE, J. (1985) - Memórias de Óbidos. Imprensa Nacional-Casa da Moeda / Câmara Municipal de Óbidos, 289 p.

Cartographic documentation list

- "Carta Militar de Portugal". Scale 1: 25.000
    338 - Óbidos (1964)
    326 - Caldas da Raínha (1964)
    453 - Fernão Ferro [Sesimbra] (1959)

- "Carta Corográfica de Portugal". Scale 1: 50.000
    26-D Caldas da Rainha (1947)
    38-B Setúbal (1920)

- "Filipe Folque" map of Portugal, nº 19. Scale 1: 100.000 (1967).

- "Plano hidrográfico desde o Cabo da Roca até Cezimbra. Secção Hidrográfica da Direcção geral dos Trabalhos Geodésicos do Reino". Scale 1: 50.000 (1849).

- "Plano Hidrográfico da Lagoa de Óbidos". Scale 1: 10.000 (1917).

- "Lagoa de Óbidos. Instituto Hidrográfico de Portugal". Scale 1: 10.000 (1980).

- "Lagoa de Albufeira. Direcção Geral dos Serviços de Urbanização". Scale 1: 1.000 (1975).

- "Lagoa de Albufeira. M.O.P.T.C. - Direcção Geral de Portos". Scale 1: 1.000 (1986).

- "Atlas do Ambiente de Portugal. Direcção Geral dos Recursos Naturais". Scale 1: 1.000.000 (1975).

- Maps and sketches dated from XVI-XIX centuries.

DUNE EROSION AND SHORELINE RETREAT BETWEEN AVEIRO AND CAPE MONDEGO (PORTUGAL). PREDICTION OF FUTURE EVOLUTION[1].

Óscar Ferreira
Instituto Hidrográfico/JNICT

J.M.Alveirinho Dias
Instituto Hidrográfico

ABSTRACT

Aerial photografs from different dates were analyzed with the aim of determining the evolutionary trend of shoreline position in the Aveiro-Cape Mondego coastal stretch. From this analysis it was concluded that the shoreline retreat has been increasing in the last decades, which is mainly due to human actions. Mean rates of shoreline retreat reached a maximum value of -3.9 m/year in the Vagueira-Praia do Areão sector, at the 1980/1990 period.

Future shoreline retreat rates for the studied zone were predicted by linear regression analysis. Using the predicted shoreline retreat rates and knowing the width of the foredunes it was possible to forecast the places where destruction of foredunes will occur along the study area, in the next 30 years.

Nowadays in 12% of the Aveiro-Cape Mondego coastal stretch dunes do not exist anymore. In the year 2020 the complete destruction of the foredunes in 42% of this area is expected.

Taking into account these predictions, some major problems (floods, salt water contamination, roads destruction, etc.) are likely to occur in the near future.

1. INTRODUCTION

The coastal stretch between Aveiro and Cape Mondego (FIGURE 1) is a 50 km long open shore of continuous sandy beach. Foredunes exist along the entire shore except near some villages where they have been destroyed by the man. The height of the dune crest range from 7-8 m above mean sea-level (in the northern part) to 12-13 m (in the southern part).

---

[1] Contribution A25 of DISEPLA project.

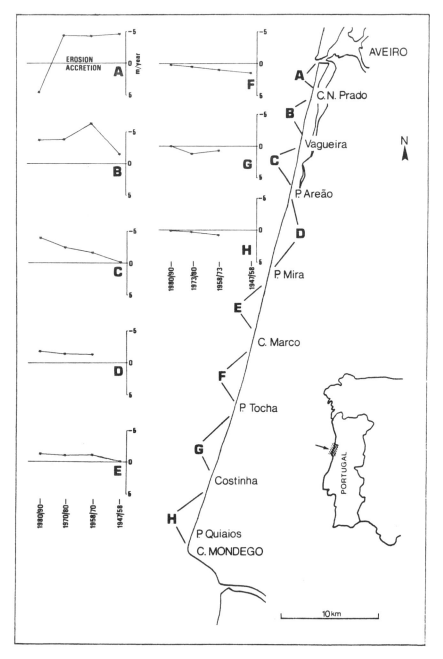

FIGURE 1 - Location of the study area (with sectors specification) and the evolution of the mean shoreline retreat rates for each sector.

The Aveiro-Cape Mondego coastal stretch is the southern part of the natural coastal area Espinho-Cape Mondego, which was formed in the last thousand years. According to GIRÃO (1941) MARTINS (1946) and ABECASIS (1955), in the 10th century the whole of this area was a gulf and only a small sandy spit existed just south of Espinho. Due to a strong sediment supply from the rivers of Northern Portugal (mainly the Douro river) and to a strong net littoral drift from north to south, this sandy spit rapidly grew southward. By the middle of the 18th century the head of the spit reached the sand accumulated against Cape Mondego and the formation of the continuous 100 km long sandy shore was completed. At present this natural sandy shore is artificially divided in two stretches (Espinho-Aveiro and Aveiro-Cape Mondego) by the long jetties of Aveiro harbour. These jetties define an artificial inlet connecting the Aveiro lagoon (the remnant of the ancient gulf) to the sea.

The evolution of all this coast shows a strong regressive trend till the turn of the 19th century. By that time the coastal trend started to be transgressive probably due to both natural and man-made factors, being this transgressive behaviour intensified during the 20th century. In reality, climate parameters changed slightly, leading to somewhat different runoff characteristics (TULLOT, 1986; LOPEZ-VERA, 1986), namely in the flood frequency and intensity, and to the start of the relative sea-level rise. These modifications are probably related to the end of the Little Ice Age (DIAS & TABORDA, 1992). At the same time, the number and magnitude of man interventions in the catchment-basins increased greatly (dam construction, sand and gravel exploitation, dredging, margin stabilization, etc.). For instance, along the Douro river 64 dams were built, having a total water retention of 8165×10^6 m^3 (MONOGRAFIAS HIDROLÓGICAS DOS PRINCIPAIS CURSOS DE ÁGUA DE PORTUGAL, 1986). Due to the mentioned factors, the shoreline began to retreat quickly in the northern part of the Espinho-Cape Mondego coast and as a consequence the old village of Espinho was damaged and partially destroyed by the sea at the beginning of this century. Nevertheless, in the study area (Aveiro-Cape Mondego) the shoreline retreat started only after the stabilization of the artificial inlet of Aveiro with two jetties, in 1949/1950 (ABECASIS et al., 1970; OLIVEIRA et al., 1982). Northward of the jetties a $0,7 \times 10^6$ m^3/year accumulation was reached between 1950 and 1978 (OLIVEIRA et al., 1982), which represents 70% of the net littoral drift (estimated to be about 10^6 m^3/year). Nowadays, with the increase in the jetties length it is assumed that the bypass of sediments downdrift of the jetties it is not significant when compared with the net littoral drift.

This paper focuses on the shoreline changes and evolutionary trends of the coastal stretch between Aveiro and Cape Mondego, from 1947 to 1990, and tries to predict the future behaviour of this coast based on those trends.

2. METHODS

In order to determine the shoreline changes along the studied coastal stretch, aerial photographs from 1947, 1958, 1970, 1973, 1980 and 1990 were analyzed. The beach/dune interface, that is, the bottom of the cliff cut in the dunes was chosen as the shoreline indicator. This geomorphological feature seems to be a much more accurate indicator of a pluriannual shoreline retreat than morphological features on the foreshore or the beach/sea boundary or even the wet/dry beach line.

In a coastal area where the tidal range can reach almost 4 m, the beach/sea and wet/dry beach interfaces are subject to strong daily variations. Changes in the order of 40-50 m in the position of the low and high tide marks are frequent in the study area. That could mean errors of 4 to 5 m/year over a 10 years period of analysis. Also swash can be responsible for 15-20 m changes in the shoreline position, which could represent errors of 1,5 to 2 m/year for the same period of analysis. The pointed errors are too big for a detailed analysis because quite often they have the same order of magnitude of the expected annual shoreline retreat rates. The fact that the tidal stage is not of easy determination, mainly in the older photographs where time is not recorded, is an additional difficulty in applying those indicators.

The beach/dune boundary is a much more reliable indicator when the period of analysis is pluriannual. The identification of this feature is easier and is not subjected to daily or seasonal changes. However, the beach/dune interface can only be used (with minor errors) when dunes are frequently attacked and eroded by the sea. In most of the study area marine erosion of dune cliffs often occurs, mainly during storms but also during spring high tides, that makes the beach/dune interface a reliable and accurate indicator of the pluriannual retreat in the position of the shoreline. Punctual errors committed using this method were estimated to be always smaller than 10 m (1 m/year in a 10 years period) and generally less than 5 m (0,5 m/year in a 10 years period). In some places where the dunes do not exist anymore or where there is no dune erosion (e.g., updrift of groins) it was used the interface between wet and dry beach observed on the aerial photographs. In those cases the results are likely to be much less accurate.

The study area was divided in 8 sectors, which were determined by the position of reliable reference points on the aerial photographs:

A) Barra-Costa Nova do Prado (3.8 km long)
B) Costa Nova do Prado-Vagueira (4.8 km)
C) Vagueira-Praia do Areão (4.6 km)

D) Praia do Areão-Praia de Mira (8.0 km)
E) Praia de Mira-Canto do Marco (7.3 km)
F) Canto do Marco-Palheiros da Tocha (6.7 km)
G) Palheiros da Tocha-Costinha (7.7 km)
H) Costinha-Cabo Mondego (7.1 km)

Measurements were made spaced 275 m along the coast, for the periods 1947/1958, 1958/1970 (or 1958/1973), 1970/1980 (or 1973/1980) and 1980/1990. However, in some sectors it was not possible to determine the shoreline evolution between 1947 and 1958 due to the bad quality of the 1947 aerial photographs.

3. SHORELINE EVOLUTION BETWEEN 1947 AND 1990

From the analysis of TABLE I and FIGURE 1, where mean shoreline retreat rates are expressed for each sector and for different periods, it is obvious that a strong shoreline retreat has affected the sectors south of the Aveiro Harbour.

In the 1947/1958 period only the northern sector revealed high shoreline retreat rates, as a direct consequence of the construction of the jetties in 1949/1950. However, the actual retreat rates caused by those structures were likely to be higher than the expressed ones because the basis of this study are the aerial photographs of 1947, collected before the construction of the jetties.

Since the 1947/1958 period the erosion started to move southwards, reaching maximum values in the sector Costa Nova do Prado-Vagueira (1958/1970) and in the sector Vagueira-Praia do Areão (1980/1990).

TABLE I
Mean shoreline retreat rates (m/year)[2]

	1947/1958	1958/1970	1970/1980	1980/1990
Barra-C.N.Prado	-4.6	-4.3	-4.4	+4.5
C.N.Prado-Vagueira	-1.5	-6.3	-3.8	-3.7
Vagueira-P.Areão	-0.1	-1.6	-2.4	-3.9
P.Areão-P.Mira	---	-1.3	-1.4	-1.8
P.Mira-C.Marco	-0.1	-1.1	-1.0	-1.2
		1958/1973	1973/1980	
C.Marco-P.Tocha	+1.5	+1.0	+0.5	+0.2
P.Tocha-Costinha	---	+0.7	+1.2	0.0
Costinha-C.Mondego	---	+0.8	+0.3	+0.1

[2] By convention accretion is positive and erosion is negative.

In the 1958/1970 period only the 21,5 km southward of Canto do Marco were still in accretion. It is also possible to see a small decrease of shoreline retreat rates in Barra-Costa Nova do Prado sector and a strong increase of erosion in Costa Nova do Prado-Vagueira, which was a consequence of the first defence structures (groin field and rip-rap seawall) built in front of Costa Nova do Prado. Since then the evolution of these two littoral sectors has been completely determined by human interventions. In the following years the groin field was repaired several times and new groins were built, the rip-rap seawall was often damaged, reconstructed and extended further south, and at least one artificial beach nourishment was carried out. At present the shoreline configuration of the northern 9 km of the study area is determined by those structures. The foredunes have generally been destroyed and when they exist overwashes and blow-outs are also present.

For the 1970/1980 and 1980/1990 periods it is clear that a decrease in the shoreline retreat rates occurred in the 2 northward sectors (FIGURE 1A and 1B), which was due to 3 major factors:
- non-existence of beach in front of seawalls (shoreline retreat rate is zero because there is no beach or dunes to erode)
- small accumulations northward of groins
- beach nourishment southward of the jetties

In the coastal sectors between Vagueira and Canto do Marco shoreline retreat rates kept on increasing from 1970 to 1990. The highest mean rate of retreat observed during the last decade was about -3.9 m/year, in the sector Vagueira-Praia do Areão, with punctual retreat rates reaching a maximum of -8 m/year, southwards of Vagueira groin. Between Vagueira and Canto do Marco foredunes were high and well developed. However, a continuous cliff still exists, marking the contact with the beach. This cliff can reach more than 10 m in height and in some places is already located landward of the old foredune crest, which means that the dune was eroded in more than half of its width. Overwashes and blow-outs only exists immediately downdrift of Vagueira groin.

South of Canto do Marco a global dune recession has not yet occurred. Nevertheless, the dune accretion has been decreasing in the last decades and the aerial photographs of 1990 show that in some foredunes a cliff has already started to develop. This was confirmed by field observations where is possible to see the first scarps in the foredunes.

4. TRENDS FOR FUTURE SHORELINE EVOLUTION

The general trend for future shoreline evolution is the expected strong increase in shoreline retreat rates and dune recession, due to a lack of sediment supply, which is caused by the non-existence of important sedimentary sources (at the moment the main source are the dunes themselves) and by the retention of the littoral drift northward of the Aveiro jetties. This increase is generalized to all the study area except for the northern 9 km of the coastline where the shoreline is stabilized through groin fields and rip-rap seawalls.

The analysis of the shoreline retreat rates along the study area for the last 4 decades lead also to the conclusion that the area of maximum shoreline retreat is moving from north to south. Indeed, when the shoreline retreat began the sand supply due to dune erosion in the northern sectors was enough to overcome all the sedimentary deficiency caused by the jetties action. Consequently, the southern dunes were not strongly eroded. However, with the stabilization by hard defences and with the destruction of the dunes in the northern sectors, southern dunes started to retreat and the eroded sand entered into littoral budget.

As a global conclusion from these two major trends it is possible to expect an overall increase in shoreline retreat rates between Vagueira and Cape Mondego and the beginning of dune recession south of Canto do Marco.

5. PREDICTION OF FUTURE RATES OF SHORELINE RETREAT AND FOREDUNE COLLAPSE

Assuming that the shoreline behaviour will not change significantly in the near future and assuming that the evolution of retreat rates can be described, for each sector, by a linear equation fitted to the existing data, it is possible to predict shoreline evolution for the next decades.

Computed linear equations showed good agreement with data for all the sectors, except for Barra-Costa Nova do Prado, Costa Nova do Prado-Vagueira and Palheiros da Tocha-Costinha. In the first two sectors human interventions are so important that the only possible prediction is the maintenance, expansion and enhancement of the almost continuous hard stabilization, with annual repairs and improvements of such structures. In Palheiros da Tocha-Costinha the linear behaviour of shoreline evolution is not evident. This is due to the abnormal value of the mean rate of shoreline position change for the 1973/1980 period, in this sector. However, because there is no reason to this sector to have a diferent behaviour than the adjacent sectors (which is confirmed by the rates in the 1958/1973 and 1980/1990 periods) a linear equation fitted to the data was used and the future retreat rates were predicted. The abnormal value for the 1973/1980 period can be a result of the small period of analysis (7 years only) which increases the posiibility of errors.

FIGURE 2 presents the obtained equations using linear regression analysis applied to data and the respective correlation coefficient (R).TABLE II expresses the mean retreat rates predicted in each sector for the periods 1990/2000, 1990/2010 and 1990/2020.

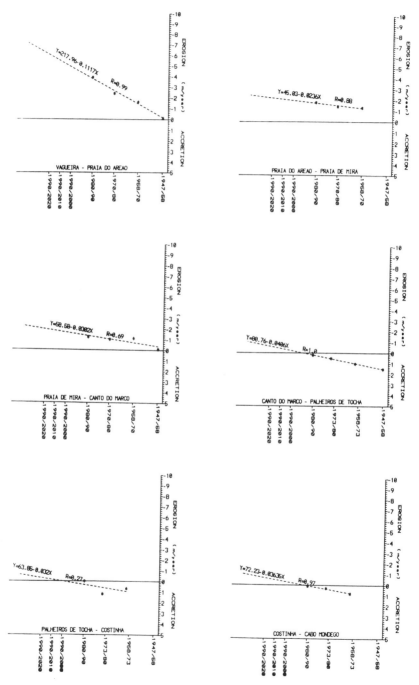

FIGURE 2 - Prediction of mean shoreline retreat rates using linear regression analysis.

The 1990/2000, 1990/2010 and 1990/2020 predictions assume a mean annual shoreline retreat equal to the retreat rate predicted respectively for the year 1995, 2000 and 2005 (mid-point of the studied periods).

TABLE II
Predicted mean shoreline retreat rates (m/year)

	1990/2000	1990/2010	1990/2020
Barra-C.N.Prado	---	---	---
C.N.Prado-Vagueira	---	---	---
Vagueira-P.Areão	-4.9	-5.4	-6.0
P.Areão-P.Mira	-2.1	-2.2	-2.3
P.Mira-C.Marco	-1.7	-1.8	-2.0
C.Marco-P.Tocha	-0.2	-0.4	-0.6
P.Tocha-Costinha	0.0	-0.2	-0.3
Costinha-C.Mondego	-0.3	-0.5	-0.7

Using the values of TABLE II and knowing the foredune width (determined by the 1990 aerial photograph) it was possible to predict the shoreline evolution and the foredune collapse along the coast, for the next decades.

FIGURE 3 shows the places where it is expected to have a complete destruction of the foredunes, for each sector in the next 30 years. TABLE III expresses the evolution of the percentage of shore that it will not have foredunes, relatively to the total length of each coastal sector.

TABLE III
Percentage of each sector without foredunes (%)

	1990	2000	2010	2020
Barra-C.N.Prado[3]	58	58-100	100	100
C.N.Prado-Vagueira[3]	23	23-100	100	100
Vagueira-P.Areão	13	28	100	100
P.Areão-P.Mira	13	14	20	68
P.Mira-C.Marco	2	2	2	7
C.Marco-P.Tocha	8	8	8	8
P.Tocha-Costinha	2	2	2	2
Costinha-C.Mondego	2	2	2	2
Total area	12	14-25	33	42

---

[3] Predictions were not possible to be made. It is assumed that dune recovery will not occur and the present trend for dune destruction and coastal stabilization will continue.

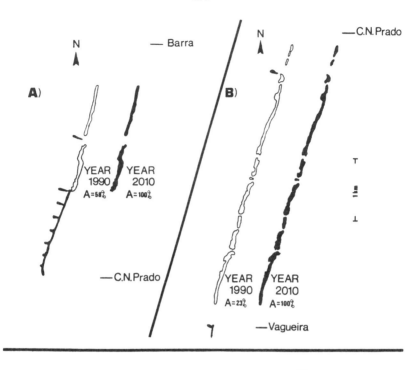
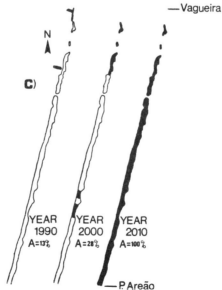

FIGURE 3 - Prediction of complete dune erosion (areas in black) for the next 30 years. "A" means the percentage of dunes completely destroyed for the specified year. The black areas at the year 1990 represent coastal defence structures.

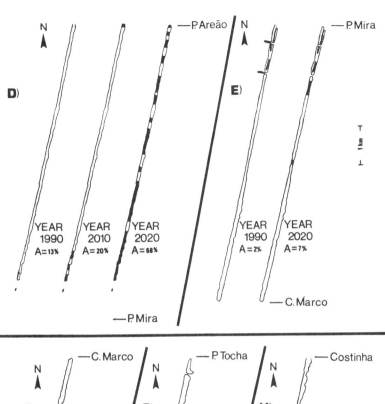
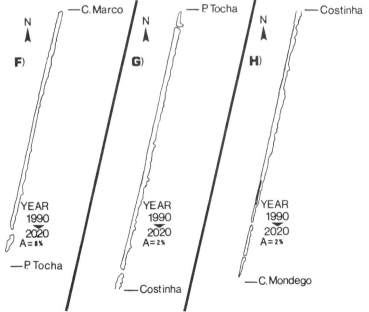

FIGURE 3 (continuation)

Nowadays the northern sectors have strongest dune destruction, due to the stabilization of the coast and to the high erosion to which this area is subject since the construction of the harbour jetties. In the remaining parts of the study area, the foredunes are interrupted only by outfalls or by small villages. The total amount of dunes already destroyed reaches 12% of the study area.

It is predicted that there will be an increase in foredune destruction in the future. The global 12% of shoreline without foredunes in 1990 will increase to a value around 42% in the year 2020. Some sectors will stay completely without foredunes, their only natural protection against the sea. Considering that in the northern part shoreline retreat rates are higher and the foredunes are narrow, it is natural the prediction of an higher percentage of foredunes destruction in the northern sectors and the forecast of total collapse for some of them.

Taking into account that: a) shoreline retreat rates and dune collapse will increase in the next decades; b) landward of the northern foredunes there is a lagoon; c) land elevation between the lagoon and the foredunes is generally less than 3 to 4 m above mean sea-level; d) the distance between the lagoon and the foredunes ranges from 200 to 900 m and e) overwashes will probably continue at higher rates, it is easy to understand that some major problems can occur in the near future. Those problems include the strong possibility of highly destructive floods, the damaging and collapse of buildings and roads, the contamination by sea water of some good agricultural soils and the salting of aquifers.

There is also a probability that a new inlet will form, 900 m south of Vagueira, where the western part of the lagoon is only 200 m landward from the beach. In 30 years (year 2020) the shoreline position is supposed to be about 180 m landward, according to the above predictions. So a new inlet could form in that position in less than 35 years. However, the recent construction (in 1991) of one groin south of Vagueira, just in the place where this new inlet could form diminishes this possibility. Nevertheless, the effects created downdrift by this groin are not currently known, but there is a strong probability that the erosion will increase greatly in the next years.

The shoreline retreat rates and foredune collapse predictions made in this paper do not take into account future hard-structures that almost surely will be built. The alternative scenario for the predicted evolution is an armoured, hard-stabilized shoreline that will replace the natural beach of today (as it already has happened in the northern sectors).

6. CONCLUSIONS

The shoreline between Aveiro and Cape Mondego shows an overall trend of retreat, mainly due to sediment deficiency. This retreat started at the beginning of the second half of the 20th century in the northern part of the study area, following the construction of the jetties in the Aveiro harbour. Since then shoreline retreat has been increasing and moving southwards. Assuming a linear behaviour for the rate of shoreline retreat evolution it is possible to predict the future retreat rates and the recession of the dunes along the study area.

The higher predicted mean value for the rates of shoreline retreat, in the 1990/2020 period, is about -6.0 m/year for the littoral sector Vagueira-Praia do Areão (180 m shoreline retreat in 30 years). For the same period the smaller predicted mean value is about -0.3 m/year for Tocha-Costinha (9 m shoreline retreat in 30 years).

Taking the predicted values for each sector as the basis for the computation of future foredune collapse it was possible to estimate that by the year 2020 almost 42% of the total study area will be without foredunes. This means that the area involved will not have any natural protection against the sea and inland destructions are expected.

Since the beginning of erosion in the study area the "solution" chosen to avoid destruction of properties or inland floods has been hard stabilization using groins and seawalls. However, the seawalls have been responsible for the destruction of the adjacent beaches and the groins for the increase of erosion downdrift. As it is possible to see comparing the mean values of shoreline retreat for different periods, hard structures have not stopped the shoreline erosion and the dune recession.

The only man made solution that achieved good results was the artificial nourishment southward of Aveiro harbour. The maintenance of an annual downdrift bypassing, using the sediments accumulated against the northern jetty, could avoid such a strong increase in shoreline retreat and dune destruction because it will allow the continuity of the littoral drift. Dune reconstruction in some places could also avoid serious problems. Otherwise, with the continuation of the present situation a major part of the study area will be transformed into a field of groins and seawalls.

7. ACKNOWLEDGMENTS

The authors would like to thank Dr. Paolo Ciavola for the good comments and sugestions to the manuscript.

8. REFERENCES

ABECASIS, C. (1955) - The history of a tidal lagoon and its improvement (the case of Aveiro, Portugal). Proceedings of the Fifth Conference on Coastal Engineering, p.329-363.

ABECASIS, F., CASTANHO, J. & MATIAS, M. (1970) - Coastal regime. Carriage material by swell and currents. Model studies and in situ observations. Influence of port structures. Coastal defence works. Memória do LNEC, nº362, 40p.

DIAS, J. & TABORDA, R. (1992) - Tidal gauge in deducing secular trends of relative sea level and crustal movements in Portugal. Journal of Coastal Research (in press).

GIRÃO, C. (1941) - Formas Litorais. in Geografia de Portugal, cap.V, 47p.

LOPEZ-VERA, F. (1986) - Quaternary Climate in Western Mediterranean, ed. F. Lopez-Vera, Univ. Auton. Madrid.

MARTINS, F. (1946) - A Configuração do Litoral Português no Último Quartel do Século XIV. Biblos, vol. XXII, tomo 1, p.163-197.

MONOGRAFIAS HIDROLÓGICAS DOS PRINCIPAIS CURSOS DE ÁGUA DE PORTUGAL (1986) - Direcção-Geral dos Recursos e Aproveitamentos Hidráulicos. Ministério do Plano e da Administração do Território, 569p

OLIVEIRA, I. , VALLE, A. & MIRANDA, F. (1982) - Littoral problems in the Portuguese West Coast. Coastal Enginerring Proceedings, vol. III, p.1951-1969.

TULLOT, I. (1986) - Cambios Climáticos en la Peninsula Ibérica durante el ultimo milénio, con especial referencia a la "Pequeña Edad Glacial". Quaternary Climate in Western Mediterranean, ed. F. Lopez-Vera, Univ. Auton. Madrid., p.273-298.

PRESENT DYNAMICS IN THE ALTO MINHO COAST
(NW PORTUGAL)

Alves, A. M. C.

Ciências da Terra - Universidade do Minho
4700 BRAGA PORTUGAL

ABSTRACT:

The coastline of the Minho region, situated between the Minho River that serves as Portugal's Northern border with Galicia (Spain) and Neiva River in the Southern, is characterized by the presence of estuaries, gravelly beaches and strait, but sometimes relatively extense, sandy beaches interrupted by outcrops of paleozoic schists alternating with hercynian granitoids. These are surrounded by a longshore dune ridge of weak expression and preferentially on the Southern board of the rivers, frequently forming spits. The region is geomorphologically distinct from the rest of the Iberian Peninsula's Occidental coast which to the North of the studied area is dominated by the deep galician "rias" while South of the Douro River is characterized by the presence of extense sandy body and lagoons being the " Ria of Aveiro" the principal example.

The region's present dynamics is, in a general sense and like great part of the Portuguese Occidental coast, controlled by the atmospheric circulation associated with the Azores' Anticyclone, particularly in what refers to the winds that condition the offshore swell from the Northwest with some sporadic occasions, although of great energy, from West and Southwest. The result is a longshore current responsible for the sediment transport from the North to the South. The presence of natural or manmade coastal structures, of which the Caminha Insua in the Minho River Mouth and the North breakwater protecting the Viana do Castelo harbour are examples, provoke important modifications in the waves motion, with the consequent effects on sedimentary dynamics. They can locally induce an inversion of the currents' direction and thereby an inversion of littoral drift of which results the formation of spits and the silting up of the estuaries' mouth.

The present or recent evolution is clearly trangressive, identifiable by the intense erosion of the Quaternary deposits and backshore dunes, and the loss of energy by the rivers manifested by the intense estuary silting up and eventually the entry of sediments from the littoral into the rivers.

I - Introduction

The Alto Minho region corresponds to the Viana do Castelo district, on Portugal Northwest Atlantic coast, limited to the North by the Minho River, that serves as Portugal border with Galicia (Spain), and in the South by Neiva River (fig. 1). This region settles entirely in the Iberian Central Zone of the Hesperian Massif, mainly characterized by hercynian synorogenic granitoids which intruded in Precambrian and Paleozoic metasediments.

It's shore extends up to 35 Km in straight line, in a direction more or less North-South. The present coast is low, with gently sloping rocky coasts and small (between1 and 2 m height) cliffs cut on Quaternary sediments. The backshore is mostly a longshore dune ridge, more or less fixed by vegetation, that in some places extend 3 Km inland and are commonly ocupied by a pine-wood. The adjacent continental shelf dips about 0.05 % and has 35 to 40 Km width (INSTITUTO HIDROGRÁFICO 1984); close to the shoreline there are outcrops of Paleozoic rocks that underlie Mesozoic limestones and sandylimestones in offshore (PEREIRA et al. 1989);

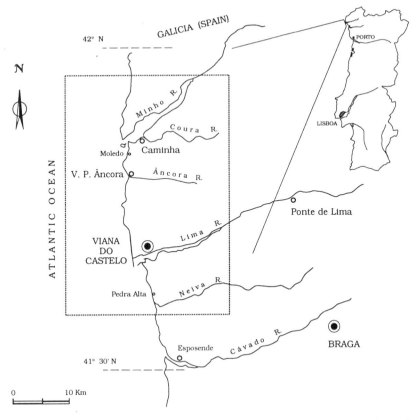

Fig. 1 - Alto Minho region corresponds to the district of Viana do Castelo, in the Northwest of Portugal. Its litoral is confine between Minho River at North side, which serves as border with Galicia (Spain) and Neiva River, the South limit.

sometimes above this assemblage there are recent unconsolidated sediments, gravels to silty sands (MAGALHÃES et al. 1991); the neotectonic in the continental shelf, is manly conditioned by the elongation of the Tomar-Porto Fault (CABRAL 1986).

Studies from INSTITUTO HIDROGRÁFICO (1989) refers that portuguese offshore currents are mainly produced by the North Atlantic central water circulation, integrated in the Azores Anticyclone. Currents in the continental shelf are clearly more frequent from N to S and they depend on the wind regime, showing also daily variations by tide effects. The wave motion presents different situations along the year, being the most common swell (about 80 %) from NW, which theoretically, due to the oblique surf, develops a longshore current, from North to South; a consequence of this fact is the sediment longshore transport, in the same direction, proved by sediments trapped on the northern side of many groins found along the coast line. Regional wind regime have a similar behaviour to the wave motion: prevailing from Northwest, but with higher speed from West and Southwest; winds blowing from East haven't a strong meaning here, probably due to topography protection, nevertheless they are significant in other portuguese places.

Shore zones always have offered many good living conditions, a better climate and easier nourishment, strong reasons promoting fixation of people as it is proved by the testimonies left there along the History.

II - Wave motion and drifting

The wave motion at the portuguese West coast, presents different situations along the year, namely in what refers to differences between winter and summer, either what refers to direction or waves height and period (INMG1982):

- NW - waves with directions in this quadrant they occur about 80%/year, with principal height values about 1 to 1.5 m, 8 s period, in summer, and 2 to 3 m height, 9 s period, in winter; waves mainly have two origins: **sea**, produced by regional wind from N or NW, either due to atmospheric circulation or land/sea thermic differential, and **swell** from NW, generated at high latitudes in the North Atlantic, in this case with higher height and wave period;

- W - from this quadrant becames principally swell, due to the descending of the Polar Front to lower latitudes; this situation occurs mainly in winter, usually with great height (no rarely more than 8 m), and period sometimes reachging 16 s;

- SW - corresponds mainly to sea, due to the approach and passing of frontal systems and/or Low-Pressure Center stationary at SW of Iberian Peninsula; commonly these waves have period of 9 to10 s and 3 to 4 m height.

Specifically for storm situations, waves height higher than 3 m, there was computed values (PONTES & CALADO 1987), of direction and period at portuguese West coast (Tab. I). This table show 3 modal classes for direction. The most important are NW directions (320 to 340 °) to which corresponds the largest period range (7 to 20 s); W directions (280 to 310 °), the second most important, have periods about 8 to 12 s; and to shorter periods, 8 to 10 s, corresponds the SW directions, 230 to 260 °.

Tabel I

DIRECTION (°)	PERIOD T(s)													Total
	7	8	9	10	11	12	13	14	15	16	17	18	19	20	
360	.2	.8	.4	1.0	-	-	.2	-	-	-	-	-	-	-	2.5
350	-	.4	-	-	.2	1.3	.4	2.5	.5	-	-	-	-	-	5.2
340	.2	.5	.4	.2	1.5	2.2	2.6	2.8	.4	1.7	.4	-	-	-	12.9
330	-	.9	-	2.9	.5	1.7	4.8	2.1	1.0	1.0	1.3	-	-	.4	16.7
320	-	.9	-	1.0	2.1	.8	1.4	2.2	2.3	.2	1.0	.4	.4	-	12.8
310	-	-	1.3	1.0	.8	2.9	1.0	1.3	.2	1.0	-	-	-	-	9.6
300	-	1.6	1.0	3.1	1.7	1.8	1.5	1.3	.4	.4	-	-	-	-	12.8
290	.2	.7	3.3	1.0	.5	.5	.2	2.1	-	1.0	-	-	-	-	9.4
280	-	.4	-	1.7	.7	-	-	1.0	.4	1.0	-	-	-	1.0	6.2
270	.2	.5	1.0	-	-	-	.2	-	-	-	-	-	-	-	1.8
260	-	.5	-	.2	.4	1.0	-	-	-	-	-	-	-	-	2.1
250	-	.4	2.3	-	-	-	-	-	-	-	-	-	-	-	2.7
240	-	2.3	-	-	-	-	-	-	-	-	-	-	-	-	2.3
230	-	1.9	-	.2	-	-	-	-	-	-	-	-	-	-	2.1
220	-	.4	-	-	-	-	-	-	-	-	-	-	-	-	0.4
210	-	-	-	-	-	-	-	-	-	-	-	-	-	-	0.0
200	-	-	-	-	-	-	-	-	-	-	-	-	-	-	0.0
190	-	.5	-	-	-	-	-	-	-	-	-	-	-	-	0.5
Total	.7	12.8	9.7	12.2	8.5	12.3	12.3	15.3	5.2	6.2	2.7	0.4	0.4	1.4	100

Tab. I - Computed data of wave motion in storm situations, 3 to 6 m height, at offshore of portuguese West coast. Values are in % of occurrence.

Two types of diagrams were created, based on available swelling values in storm conditions, and bottom morphology, for selected representative events: approaching wave refraction diagrams, considering monochromatic waves with straight initial crests, which permit to represent the transition from deep to shallow water; and local wave refraction diagrams showing local wave motion, with some limitations naturally.

The approaching wave refraction diagrams, as it was expected, show that waves tend to become parallel to the shore line (fig. 2), maintaining however some obliquity downdrift.

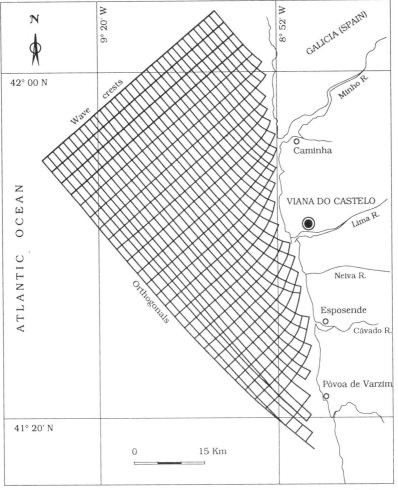

Fig. 2 - Approach refraction diagram, at the offshore NW portuguese coast, for the most common swell, direction (320°) and period (13 s).

These diagrams and wave motion informations permit to infer, excepting local points as Minho and Lima River Mouths, the existence of a net drift from North to South at Alto Minho coast, situation that was already mentioned in other works (CARVALHO & BARCELÓ 1966, CASTANHO et al. 1981, ALVES 1988).

Local refraction diagrams from the coast adjacent to Minho River Mouth and Lima River Mouth show important modifications on swell direction caused by shallow waters, but mainly by shore geomorphological structures.

A breakwater (about 2 Km long) protecting Viana do Castelo harbour, at northern Lima River Mouth, reinforces refraction effects produced by shallow water, also giving rise to wave diffraction (fig. 3), particularly observed with Northwest swell. That creates a longshore current from South to North, beginning at Anha River Mouth, responsible for sediment transport silting up Lima River Estuary and the channel access to harbour. As a way to stop this silting up, another coastal defense structure was constructed, a southern groin about 700 m long, that partially retains those sediments, but does not avoid the permanent dredging necessary to maintain safety conditions of navegability.

At Minho River Mouth, the Caminha's Insua, produces an effect on the swell as a detached breakwater; it reinforces refraction effects and induces wave diffraction (fig. 4), always similar whatever the waves direction. This produces two convergent longshore currents, one N-S from Minho River Mouth, and another S-N from Moledo beach, creating a point-beach (Ponta Ruiva), on shore in front of the Insua, and silting up the Minho River South Channel (named "barra portuguesa").

III - Recent shoreline evolution

With the exception of punctual situations we can say that nowadays, the Alto Minho coast presents, in general an erosion process. Although the coastal morphology can't be considered cliffty, due to the existence of a low coastal platform, the coast presents a morphology typical of eroded regions independently of it's litological composition:

- rocky backshore areas present typical corroded forms, without any covering sediments;

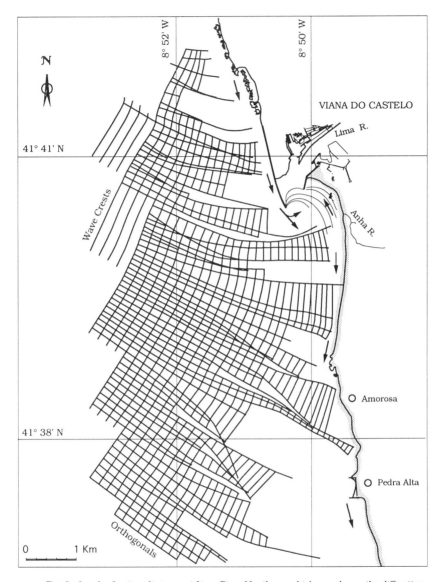

Fig. 3 - Local refraction diagram, at Lima River Mouth, on which was drawn the diffraction (tiny lines), produced by the North breakwater. Offshore origin of waves 320°, period 13 s. Arrows represent the longshore transport.

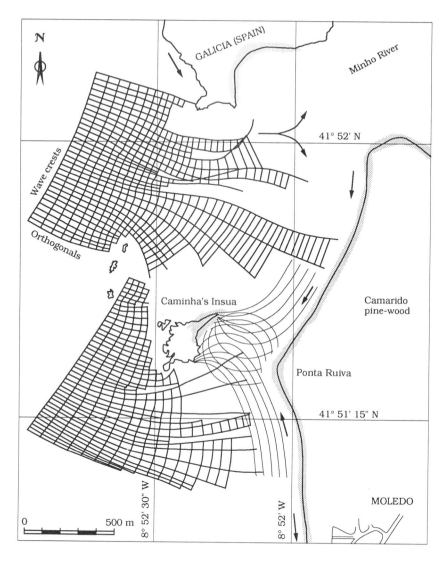

Fig. 4 - Local refraction diagram, at Minho River Mouth, on which was drawn the diffraction (tiny lines), produced by the Insua of Caminha that play a role of a detached breakwater. Waves direction at offshore is 320° and its period 13 s. Arrows represent the longshore transport.

- on quaternary backshore areas there is always a small cliff, about 1-2 m height, that has a slow retreat;
- dunar backshore areas present stronger erosion, which is evident by the continuous and intense cliff retreat, because of the undercutting by swash during storms.

There is few information available, so the evolution of the Alto Minho coast, during a long period, can't be made with enough accuracy. Nevertheless, in some areas there was some information, maps and aereal photographs of different epochs, that allow to analyse the evolution during the last 120 years. These areas are namely South of the Minho River Mouth and the Lima River Mouth.

III.1 - Southern Minho River Mouth Area

Studies comparing maps, using the level line 0 as a reference, allow to conclude that since 1885 to 1983, the coast line advanced about 70 to 100 m, towards the sea, between the River Mouth and Ponta Ruiva, and retreated about 30 to 50 m, inland, between Ponta Ruiva and Moledo beach (fig. 5). This shows a mean advance ratio of about 0.7 - 1.0 m/year, at the northern side of Ponta Ruiva, and a retreat of about 0.3 - 0.5 m/year, at the southern side.

The accretion verified on the northern side, proved by the gradual development of the foredunes area, obeys to an evolution scheme of morphogenetic systems with great complexity of variables. This is a system where many processes play a role at the same time and way:

- estuarine processes, namely the nourishment in sediments to the longshore drift;
- spits origin and evolution processes - in West portuguese coast spits "roots" in the southern board of rivers and grow up to North (ALVES 1989 b);
- processes developped in the presence of morphologic structures that function as detached breakwaters, in case, the Caminha's Insua;
- eolian processes - the persistence of winds blowing from Northwest, acting perpendicularly to the shoreline, push the sediments left by the swash to the backshore, forming foredunes;

The evolution on the southern side of this area, where an important erosion since the last 100 years is noticed, must be integrated in the general erosive situation of this portuguese coastal segment.

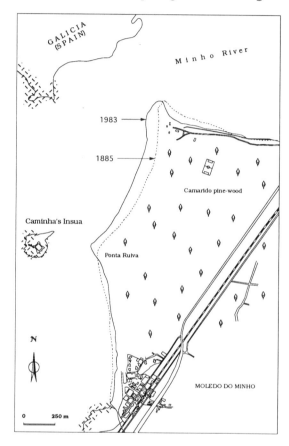

Fig. 5 - Evolution of shore line at Minho River Mouth, since 1885 to 1983, showing the shore grow up, from Ponta Ruiva point beach to North and the thinning from that point to South.

III.2 - Southern Lima River Mouth Area

The evolution study of this area, mainly nearby the Lima estuary, must be done considering two different epochs: before the building of breakwaters to protect the Viana do Castelo harbour, 1865 to 1980 period; and after that, which means the events created by manmade structures.

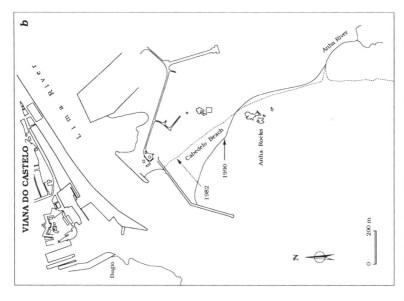
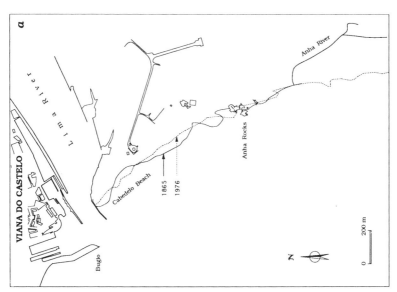

Fig. 6 - Evolution of shoreline at Lima River Mouth since 1865 to 1977 before the last defense structures (a), and after (b), 1982 to 1990 period.

Since 1865 to 1976 there isn't a well defined direction of evolution, advance or retreat, although when comparing maps of shorter periods it is verified the alternate migration of level line 0 of reference; however, the coast line morphology was modified, it became apparently straighter with time (fig. 6 a).

After the construction of the last defense structures, during 1979/80, until now, litoral morphology evolution was complete dependence of them. The great accretion which occured at updrift of the southern groin, constrained the movement of level line 0 about 350 m towards the sea (fig. 6 b).

Meanwhile, from the Anha River Mouth, inversion point of longshore currents, toward South, erosional effects were if not started at least reinforced, specially at Pedra Alta beach, creating a cliff on longshore dune (fig 7); here, some coastal protecting measures only increased erosive processes (ALVES 1989 a), in such a way that since 1981 until 1991 the cliff backshore retreat reaches in some places about 100 m.

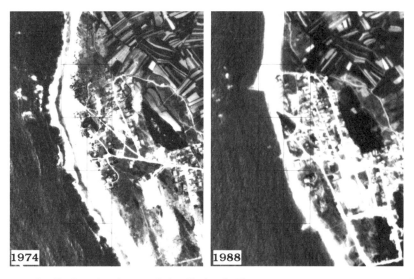

Fig. 7 - Recent evolution at Pedra Alta beach. The erosion occuring since 1981, it's well define by the retreat of the beach contact to vegetation. In the Northern side the erosion was stopped due to the construction in 1984 of a groin that trapped the sediment transport (white largest spot in right photo), howhever increasing the process towards South. Scale - 1: 10 000. Aerophotographs are from Instituto Geográfico e Cadastral.

IV - Conclusions

Present sedimentary dynamics on the portuguese West coast results from litoral processes related with wave motion, mainly Northwest swell. This orientation creates a longshore current from North to South, giving rise to sediment transport in the same direction, that are trapped in northern side of the coastal groins. Occasionally inversion occurs on this litoral drift due to natural or manmade geomorphological structures.

Regional rivers nowadays have a small contribution, generally of fine material, on sediment supplies of solid charge to littoral drift; this is one of the principal causes of beach "thinning" and the consequent dismantling of Quaternary deposits and backshore dunes.

Anthropic action on sedimentary dynamics in the Alto Minho region is an indubitable fact, and it can reach a great importance in recent evolution of this segment of portuguese coast.

Biboigraphy

ABECASSIS F., CASTANHO J. e MATIAS M. F. (1970) - Coastal regime. Carriage of material by swell and currents. Model studyes in situ observations. Influence of port structures. Coastal defense works. Breakwathers. *Memórias do L.N.E.C.* 362, Lisboa.

ALVES A. M. C. (1987) - Ocorrência de Seixos de Silex nas Praias do NW de Portugal. *Boletim Cultural* 4, Centro de Estudos Regionais, Viana do Castelo. 157-164

ALVES A. M. C. (1989 a) - O litoral e a defesa do património. *Estudos Regionais*, 5. Viana do Castelo. 69-81

ALVES A. M. C. (1989 b) - O porquê da inversão das restingas na costa ocidental portuguesa (sua relação com a erosão provocada por efeito de estruturas portuárias). Actas do 4º (SILUSB) Simpósio Luso-Brasileiro de Hidráulica e Recursos Hídricos. 303-312

CABRAL J. (1986) - A neotectónica de Portugal Continental - estado actual dos conhecimentos. *Maleo*, 2, 14, 3-5.

CARVALHO P. e BARCELÓ J. (1966) - Agitação marítima na costa Oeste de Portugal Metropolitano. *Memórias do L.N.E.C.* 290, Lisboa.

COVAS J.M.A. (1987) - Diagramas de Refracção para as Zonas Marítimas Adjacentes a Caminha e a Viana do Castelo. Relatório particular, 6/88, LNEC, Lisboa.

MAGALHÃES F., RODRIGUES A., DIAS J. A. (1991) - Potencialidades em inertes na plataforma continental norte portuguesa. *Geonovas* especial 2, Lisboa. 155-166

INSTITUTO HIDROGRÁFICO (1984) - Carta Hidrográfica, do rio Minho a Espinho, nº 1, na escala 1: 150 000.

INSTITUTO HIDROGRÁFICO (1989) - Roteiro da Costa de Portugal, 1ª parte (costa ocidental entre o rio Minho e o cabo Carvoeiro), Lisboa.

INSTITUTO NACIONAL DE METEOROLOGIA E GEOFÍSICA (1982) - Anuário do INMG. Lisboa.

PEREIRA E., RIBEIRO A., CARVALHO G., MONTEIRO H. (1989) - Carta Geológica de Portugal, folha 1 na escala 1/200 000. Serviços Geológicos de Portugal.

PONTES M.T e CALADO F.J. (1987) - Direcções e Períodos de Tempestades na Costa Ocidental de Portugal. Relatório particular. DER 23/87, LNETI, Lisboa.

DETERMINATION OF THE EVOLUTIONARY CONDITION OF COASTAL CLIFFS ON THE BASIS OF GEOLOGICAL AND GEOMORPHOLOGICAL PARAMETERS

V. Rivas and A. Cendrero.

DCITTYM (Earth Sciences), Fac. Science, Univ. of Cantabria. 39005 Santander, Spain.

ABSTRACT

A method is described for the characterisation of cliffs in terms of erosive rates, based on the correlation between observed erosion and erosional micromorphology, on the one hand, and a series of geomorphological parameters (bedrock, structure, height of the cliff, slope, orientation, gradient of the adjacent coastal platform) using regresion and principal component analysis.

These analysis, applied to the north coast of Spain, show that 4 types of cliffs can be distinguished in terms of erosion, corresponding to rates of approximately: <1mm/year; mm/year; cm/year; dm-m/year. The factors that appear as most important for determining erosion in the cliffs of this region are bedrock and the gradient of the adjacent platform, with the following regresion equation:

Erosion grade= 1.62+0.164 bedrock-0.079 platform gradient.

Correlation coefficient: 0.6151

The principal component analysis shows that 100% of the observed variance can be explained by three axes: First axis, form variables (height and slope), second axis, material variables (bedrock, structure, erosive condition), third axis, process variables (orientation, gradient of the platform).

Although the method used needs refining, it offers good posibilities for the prediction of erosive rates in cliffs, on the basis of parameters which are easy to determine.

Finally, observations carried out in this region, indicate that there has been an intensification of erosion rates in recent years, which does not seem to be related either with an increase in storm frecuency or intensity or with human intervention.

INTRODUCTION

The aim of this work is to define the possible relationships between certain geological and geomorphological features and the rate of erosion in coastal cliffs in the eastern part of the Bay of Biscay (Figure 1), as well as to analyse possible recent changes in the rate of erosion in sectors with intensive retreat.

The total lenght of the coastline studied is 331.2 km, out of which 259.6 km correspond to cliffs, most of them quite stable, as they are formed by compact rocks, mainly massive limestones, but some showing clear indications of retreat, perceptible even within a period of a few years.

METHODOLOGY

Normally, in cliffs with small rates of erosion, as is often the case in the Bay of Biscay, it is not possible to make direct measurements of the rate of retreat, because changes within a period of years, decades or even centuries cannot be established with certainty. Therefore, an indirect approach was followed. First, a series of geomorphological indicators of the erosive condition of cliffs were indentified. These were: presence of human structures affected by erosion and degree of destruction; frequency of recent landslide scars; presence of eroded soil profiles at the edge of the cliff; presence of "hanging valleys" and their height above sea level; altitude of the unvegetated part at the base of the cliff; volume and type of erosion debris at the foot of the cliff, and width of the presently active abrasion platform.

The systematic observation of those indicators along the coast enabled the definition of four categories with respect to erosive condition. They were:

1.- Stable: compact rocks, such as massive limestones or quartzites; smooth convex rupture of gradient at the cliff´s edge, without sharp edges; total absence of eroded soil profiles; total absence of "hanging valleys"; vegetation developed right to the foot of the cliff.

2.- Slightly erosive: fairly compact rocks, such as sandstones, bedded limestones or marls; accumulation of blocks and boulders at the foot of the cliff, but without scars of falls or slides; absence of eroded soil profiles; very scarce "hanging valleys", at low elevation over present sea level and corresponding to small streams; vegetation developed till practically the base of the cliff; orientation sheltered from prevailing wave direction.

3.- Moderately erosive: presence of non-compact formations, such as flysch; bedding planes roughly coincident with cliff slope; sharp slope break at the top of the cliff; eroded soil profiles; visible, relatively well preserved scars of surficial slides; some "hanging valleys"; very scarce vegetation over non-consolidated formations; present abrasion platform well developed.

4.- Intensely erosive: unconsolidated formations, such as clays or large slope deposits; very sharp slope break at the top of the cliff; systematic presence of eroded soil profiles; abundant, recent slide scars; systematic presence of "hanging valleys" at high elevations over present sea level; no vegetation over unconsolidated formations; ver well developed abrasion platform; human structures affected.

Erosion rates for these qualitative categories have been tentatively estimated, by comparison with observations in other regions (Thorarinsson, 1966; Horikawa and Sunamura, 1977; Cendrero and Díaz de Terán, 1985; Bird and Rosemberg, 1986; Clayton, 1989) in the order of 1: < 1mm/year; 2: mm/year; 3: cm/year; 4: dm-m/year. In the case of category 4, erosion rates have been determined by means of direct observations and/or analysis of historical maps and air photographs.

In principle, the erosive condition must reflect the combination of two groups of factors: (a) factors related to the intensity of the erosive agent and (b) factors depending on the resistance of the material. There is a third group of geometrical factors (c) that should be a consequence of the intensity of erosion.

Probably the main factor with respect to the activity of the erosive agent (a) is the frequency and intensity of storms in the area, but as the coast considered is fairly straight and covers a limited lenght of littoral, with uniform "ocean climate", this factor was considered as a constant for the area of study. The other two factors which determine the incidence of wave energy on the coast are the orientation with respect to prevailing storms -which come from the northwest- and the gradient of the continental shelf inmediately adjacent to the shore.

The factors depending on the resistance of the materials (b) include: rock compacity, spacing of structural discontinuities and their position with respect to the topographic slope.

Finally, geometrical factors (c) which are a consequence of erosion but at the same time also have an influence on erosive rates are: height of the cliff and gradient.

The next step was the identification of "homogeneous cliff sectors"; that is, stretches of cliff in which all the variables (erosive condition, process, material, geometrical) defined above are homogeneous, within a certain range of variation. This requires the definition of the "types" to be considered for each variable, be it the presence of a certain material or a given interval of values. Table 1 lists the types considered. Figure 1 shows a part of the coast in which "homogeneous cliff sectors" have been represented.

In order to establish the possible relationships between the erosive condition of cliffs and the variables indicated above, multiple regression and principal component analyses were carried out for the "homogeneous cliff sectors" identified, using, respectively, the programmes 2R and 4M of the BMDP package (Regents of University of California, 1983). Erosive condition was taken as the dependent variable in the multiple regression analysis. In the principal components analysis all variables were considered together, in order to determine the minimum number of variables which could describe the system.

RESULTS AND DISCUSSION

The best correlation between erosive condition and the independent variables considered, obtained by multiple regression analysis, was expressed by the equation:

$$y = 1.62 + 0.146\, x_1 - 0.079\, x_2$$

where: y = erosive condition; x_1 = bedrock; x_2 = gradient of the platform adjacent to the cliff.

The correlation coefficient was 0.615, not too high, but with a level of significance $P > 0.001$, which is quite good for correlations with field data. According to these results, 37.84% of the "erosive condition" of cliffs can be explained as a result of the two variables indicated. It appears that, although these variables do help to characterise the degree of cliff erosion, their predictive value is still limited.

The results obtained could be explained in two ways. One posibility is that there are other variables which could have an influence on differential cliff erosion, but this does not appear to be very likely. The second possibility is that the variables have not been adequately considered. The multiple regression analysis used should, strictly speaking, be applied to continuous variables. In this case, although all the variables considered (with the exception of bedrock) are continuous, they have been treated as discreet variables, because of the need to characterise "homogeneous cliff sectors" by means of certain intervals (or "types") of each variable. It is very likely that better results would be obtained if instead of using such "sectors", measurements are carried out at specific "points", so that the different variables can be expressed quantitatively in the corresponding magnitudes. Finally, other, non-linear correlations could exist between erosion and the independent variables considered.

The principal component analysis shows that the present geomorphological or evolutionary condition of cliffs can be completely described and represented by three

LITHOLOGY	km	%	STRUCTURE	km	%	ORIENTATION	km	%
Massive limestones	67,075	24,74	Subhorizontal	25,450	9,39	N	170,500	62,89
Calcarenites	16,525	6,09	Paral./sea/<30º	21,350	7,87	NW	40,125	14,80
Marly limestones	30,950	11,42	Paral./sea/30-60º	24,575	9,06	W	13,000	4,79
Marly flysch	20,975	7,74	Paral./sea/>60º	8,850	3,26	SW	1,750	0,64
Sandstones	20,750	7,65	Paral./land/<30º	5,625	2,07	S	0,000	0,00
Siltstones	17,125	6,32	Paral./land/30-60º	8,375	3,09	SE	1,200	0,44
Sandy flysch	26,750	9,87	Paral./land/>60º	14,950	5,51	E	18,725	6,91
Luutic flysch	23,425	8,64	Perpendicular >60º	3,450	1,27	NE	25,800	9,52
Marls	25,425	9,38	Perp./S/<30º	12,625	4,66			
Claystones	4,825	1,78	Perp./S/30-60º	6,980	2,57	PLATFORM GRADIENT	km	%
Diabases	2,200	0,81	Perp./N/<30º	1,475	0,54			
Quaternary deposits	15,075	5,56	Perp./N/30-60º	2,400	0,88			
			Perp./E/<30º	0,250	0,09	< 1,5 %	36,875	13,60
GRADIENT	km	%	Perp./E/30-60º	2,925	1,08	1,5-3 %	31,300	11,54
			Perp./W/<30º	8,000	2,95	3-5 %	126,950	46,83
0-10 %	25,100	9,26	Perp./W/30-60º	23,320	8,60	5-8 %	37,575	13,86
10-30 %	36,100	13,32	Obl./E/>60º	1,250	0,46	8-10 %	13,575	5,01
30-50 %	74,925	27,64	Obl./E/sea/<30º	2,750	1,01	> 10 %	24,825	9,16
50-100 %	104,075	38,39	Obl./E/sea/30-60º	27,075	9,99			
> 100 %	30,900	11,40	Obl./E/land/<30º	6,100	2,25	HEIGHT	km	%
			Obli./E/land/30-60º	9,050	3,34			
EROSIVE CONDIT.	km	%	Obl./W/>60º	0,000	0,00	0- 10 m	34,550	12,74
			Obl./W/sea/<30º	5,975	2,20	10 - 25 m	38,100	14,05
Stable	85,625	31,58	Obl./W/sea/30-60º	4,250	1,57	25-50 m	97,225	35,86
Slightly erosive	98,375	36,29	Obl./W/land/<30º	5,600	2,06	50-100 m	56,375	20,79
Moderately erosive	77,650	28,64	Obl./W/land/30-60º	1,950	0,72	>100 m	44,850	16,54
Intensely erosive	9,450	3,48	Massive	36,500	13,46			

Table 1.- "Types" identified for the different variables, indicating the lenght of cliff corresponding to each one in the coastal stretch studied. The position of structural planes with respect to the cliff (parallel, perpendicular, oblique) is indicated, as well as the direction and magnitude of dip.

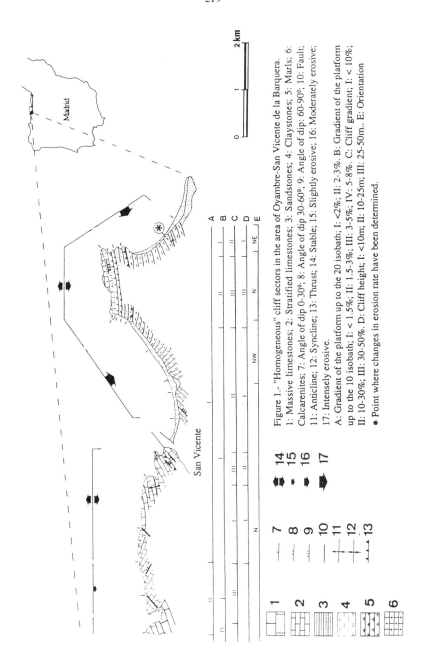

Figure 1.- "Homogeneous" cliff sectors in the area of Oyambre-San Vicente de la Barquera. 1: Massive limestones; 2: Stratified limestones; 3: Sandstones; 4: Claystones; 5: Marls; 6: Calcarenites; 7: Angle of dip 0-30º; 8: Angle of dip 30-60º; 9: Angle of dip: 60-90º; 10: Fault; 11: Anticline; 12: Syncline; 13: Thrust; 14: Stable; 15: Slightly erosive; 16: Moderately erosive; 17: Intensely erosive.

A: Gradient of the platform up to the 20 isobath; I: <2%; II: 2-3%. B: Gradient of the platform up to the 10 isobath; I: <1.5%; II: 1.5-3%; III: 3-5%; IV: 5-8%. C: Cliff gradient; I: <10%; II: 10-30%; III: 30-50%. D: Cliff height; I: <10m; II: 10-25m; III: 25-50m. E: Orientation

* Point where changes in erosion rate have been determined.

Table 2.- Number of days, per year, with wind velocities higher than 50km/h. Data from the "Centro Meteorológico del Cantábrico", Santander.

YEARS	No. DAYS	YEARS	No. DAYS
1.956	65	1.974	101
1.957	(15)	1.975	96
1.958	120	1.976	92
1.959	109	1.977	101
1.960	148	1.978	106
1.961	(53)	1.979	96
1.962	(38)	1.980	62
1.963	99	1.981	84
1.964	(64)	1.982	93
1.965	(68)	1.983	80
1.966	41	1.984	102
1.967	(65)	1.985	74
1.968	110	1.986	45
1.969	79	1.987	67
1.970	73	1.988	61
1.971	71	1.989	40
1.972	71	1.990	51
1.973	60		

( ) Year with incomplete data

axis grouping the variables selected. The first axis (Factor 1) corresponds to the variables "height" and "gradient" of the cliff, or geometrical variables; the second axis (Factor 2) includes "bedrock", "structure" and "erosive condition", or material variables; the third axis (Factor 3) groups "orientation" and "gradient of the platform", or process variables. The equations of the three axis that can describe any sector of the cliffs along this littoral are:

Axis 1 : 0.925 gradient + 0.908 height + 1.831
Axis 2 : 0.814 bedrock + 0.787 erosive condition - 0.487 structure + 0.621
Axis 3 : 0.811 orientation - 0.725 platform gradient + 1.226

Therefore, it seems that the variables selected do provide the means to describe and characterise the evolutionary condition of coastal cliffs. However, the method described must be refined in order to obtain results with greater predictive value.

It is obvious that cliff sectors with the most intense erosion (category 4) are the ones where determination of erosive rates, and their possible changes in recent times, is of particular interest. The area where this has been best documented corresponds to the clayey cliffs by the beach of Oyambre (point * in figure 1), in which buildings provide a good reference point, whose position with respect to the cliff´s edge can be determined on old maps and air photographs. The analysis of such documents has shown that between 1946 and 1970 the cliff retreated less than 5 m; between 1970 and 1987 a retreat of approximately 15 m took place. Direct field determinations after 1987 have shown retreats greater than 2m/year.

Data about storm frequency and intensity show no significant variation for the same period (table 2). Similarly, there are no causes related to human activities along this rather undeveloped part of the littoral, nor in the adjacent stream basins, which could explain the observed changes.

It is tempting to explain the observed increase in erosion rates as the result of a possible, recent elevation of sea level, but the evidence so far available is by no means conclusive. In this context, it is interesting to point out that in the last 5-10 years, all dune fronts along this coast show clear signs of erosion. More observations of this kind, in different points along the coast would be necessary in order to assess that possibility.

CONCLUSIONS

The characterisation of coastal cliffs using a series of descriptive observable parameters related to process, material and geometry, and their correlation using multiple regression and principal component analyses, provides the means for a preliminary diagnosis of the erosive condition of the coastline. It also helps to determine the factors with the greatest influence on erosion, which seem to be, for the littoral studied, lithology and gradient of the coastal platform. However, the results obtained are only partly satisfactory and a more refined analysis is needed, using continous, not discreet variables.

In sectors where cliff erosion is intense, an increase of erosion rates has been observed since around 1970. This increase does not seem to be related to changes in storm intensity or frequency nor to human activities. To what extent this increased erosion reflects a possible rise in sea level is an open question.

REFERENCES

BIRD, E.C.F. and ROSENBERG, N.J. (1.987): Coastal cliff management: an example from Black Rock Point, Melbourne, Australia. Journal of Shoreline Management, 3: 39-51.

CENDRERO, A. y DIAZ DE TERAN, J.R. (1.985): Caracterización cuantitativa de los procesos de erosión en las series volcánicas de la isla de La Gomera (Canarias). Actas I Reunión del Cuaternario Ibérico, 1: 531-543.

CLAYTON, K.M. (1.989): Sediment input from Norfolk cliffs, Eastern England. A century of coast protection and its effect. Journal of Coastal Research, 5 (3): 433-442.

HORIKAWA, K. and SUNAMURA, T. (1.977): A study of coastal cliffs by using aerial photographs. In: Air photography and coastal problems, ed. M.T. El Ashry. Benchmark Papers in Geology, 38. Dowden, Hutchinson and Ross, Stroudsburg, Penn: 178-183.

REGENTS OF UNIVERSITY OF CALIFORNIA (1983): BMDP Statistical Software. University of California, Los Angeles.

THORARINSSON, S. (1.966): Surtsey; the new island in the North Atlantic. Viking Press. N.York, 47 p.

THE TURBULENT RESUSPENSION OF COHESIVE INTERTIDAL MUDS:
some new concepts and ideas

By Kevin Black

School of Ocean Sciences, University College of North Wales, Menai Bridge, Anglesey, GWYNEDD

Abstract

The direct, *in situ* evaluation of the behaviour of cohesive marine and estuarine sediments under tidal flows has become of increasing relevance today, because it provides an inherently more accurate and realistic description of the sediment hydraulic properties. Laboratory flume studies, despite their usefulness, are unfortunately unable to recreate either the natural environment or *in situ* properties of these sediments.

Experiments using a purpose built, field-portable recirculating seawater flume (which can be used directly on undisturbed sediments), have yielded new insights hitherto unavailable into the way in which flocculated intertidal muds respond to periods of increasing bed shear stress, similar to natural tidal boundary layer flows. Results presented from within the intertidal zone of a Welsh estuary are compared to laboratory experiments and used to demonstrate many of the features associated with the erosional phases of these sediments.

Erosion of sediments through time under periods of increasing bed stress is a combination of time-dependant linear and non-linear processes and may be classified in terms of the Type I/Type II patterns of Mehta and Partheniades (1982). Patterns of erosion are contingent upon the microstructural, textural and microbiological properties of the floc aggregates, which contribute to micro-scale gradients of cohesion within the surface 1-2 mm. Although sediment transport is primarily via suspension, some concommitent bedload transport is observed.

Specific manipulative field experiments demonstrate the emerging influence of biological processes on sediment resistance to erosion; forced chemical removal of the biological and organic phases within the surficial floc aggregate network reduces the threshold condition by a single order of magnitude, and increases initial entrainment rates by between 160-300%. Evidence suggests the role of benthic diatoms in particular may be important in this respect.

1. INTRODUCTION

Naturally formed cohesive coastal and estuarine sediments – those containing generally >10% by weight of electrochemically charged clay mineral particles – are a complex and heterogeneous milieux in which a multitude of physical, chemical and biological reactions occur (Hayes, 1964). These sediments may be considered fundamentally different to coarser, predominantly sand-sized sediments with respect to their behaviour under environmental fluid stresses. As cohesive sediments or "muds" are net integrators of the many contemporaneous processes which are responsible for their formation, their transport characteristics, in particular the way in which they are entrained by tidal flows, cannot be easily parameterised through a simple combination of physical indices like grain size and density.

Erosion of these substrates is a complex and at present poorly understood interplay of such variables as nearbed hydrodynamic conditions during bed formation, bulk density, water content, microstructure and the activities of living micro- and macro-organisms (Paterson and Daborn, 1991). Furthermore, erodibility is observed to vary with time (Paterson, 1989; Paterson et al., 1990) and in the natural environment the distribution of those properties contributing to sediment cohesion are prone to vary with space (McCall and Fisher, 1980; Amos et al., 1988). This outstanding natural complexity means that it is virtually impossible to recreate the natural environment and in situ properties of these sediments in laboratory flumes or other laboratory apparatus, and consequently realistic information regarding the erosional behaviour of these sediments can only usefully be obtained within a framework revolving around field experimentation (West et al., 1990). In any case, serious doubts surround those studies which remove these sediments, however carefully, advocating a "nominally undisturbed" approach. Cohesive sediments are notoriously non-Newtonian materials and deform easily and quite often irreversibly when sampled (Young and Southard, 1978).

This contribution presents the results of some recent experimental work conducted directly on undisturbed inter-tidal muds using a specially developed field-flume. The data are used to demonstrate the physical processes which achieve resuspension and to outline some important qualitative and quantitative aspects of the manner in which natural muds respond to fluid stresses; the in situ experiments reported are compared and contrasted to previous similar work in the laboratory. Specific manipulative experiments are then described which demonstrate the emerging influence of micro-biological and organic processes on the stability of intertidal muds.

2. METHODOLOGY

In situ techniques

In order to examine in a controlled manner the resuspension of cohesive intertidal muds, a mobile recirculating seawater flume (MORF) was designed and built (Figure 1). This device and its calibration are described more fully in Black and Cramp (submitted). The flume is a small portable mechanical device able to be emplaced directly on the mud surface; ambient seawater is circulated within a 10 cm wide channel at a pre-determined speed over an area of exposed mud (0.025 m^2). The flume is instrumented with a turbidity sensor to monitor the response of the bed and water through time, and observation ports enable a detailed visual record of the erosion process to be gathered. Floc erosion is assessed in terms of the time-evolution of suspended particulate matter in the flume water column. Estimates of the boundary shear stress over the mud bed were obtained from spatial measurements of the vertical velocity structure in the channel using a laser doppler anemomenter.

Flow within the flume was able to be varied manually and thus the operator was able to "drive" the experiment. Note the flume is suitable for subaerial deployment only.

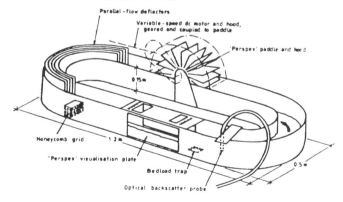

Figure 1. The recirculating field flume (MORF). Motor not shown.

3. RESULTS

Figure 2 shows the type of data collected with the flume. It consists of time-series of the mean concentration of suspended sediment ($\overline{SSC}$) in the flume water. The imposed bed stress regime, consisting of periods of constant flow velocity *above* the threshold for significant entrainment and equal in time, is shown separately (Figure 2b). For these experiments, Δt=10 minutes and $\Delta\tau_0$=0.11 Nm^{-2}.

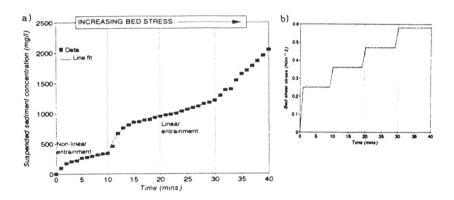

Figure 2. a) SSC-t data collected over 10 minute constant stress intervals in the MORF. The bed stress regime is illustrated in b).

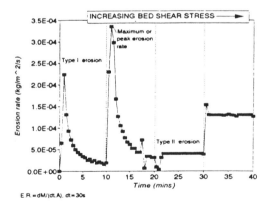

Figure 3. Erosion rate (Δ)-time data abstracted from the time series in Figure 2(a).

There is a positive and immediate response of the mud to the imposed bed stress, and erosion is evident at the lowest τ_0. Broadly, two different types of erosion are evident. The first pattern of erosion, which appears to be more characteristic of low imposed stresses (or in other words floc aggregates at or very close to the sediment-water interface), is where the variation of $\overline{SSC}$ is non-linear with time for a constant bed stress. Erosion is therefore a continuously decreasing function with time after an initial high peak. The time variability of the erosion rate ε, computed as $\Delta\overline{SSC}/(A.\Delta t)$ where $\Delta t = 30$ seconds and A is the area of the exposed mud, and corresponding to the time-series data is shown in Figure 3. Non-linear entrainment is, in fact, represented by an exponential decay in ε from a peak rate (ε_p). The erosion rate for these sediments is $O(10^{-4}\ kgm^{-2}s^{-1})$. Although ε_p at $\tau_0 = 0.36\ Nm^{-2}$ is greater than the corresponding value at $\tau_0 = 0.25\ Nm^{-2}$, more generally it is observed to be independant of the absolute magnitude of τ_0.

Higher bed stresses are characterised by linear or quasi-linear patterns of erosion. The erosion rate, defined by the slope of the line on the $\overline{SSC}$-t plots (Figure 2a), is therefore also invariant through time (Figure 3). Order of magnitude differences exist between the two types of erosion where rates of linear entrainment are usually much less $(O[10^{-5}-10^{-6}\ kgm^{-2}s^{-1})$; they are, however, theoretically continuous within the timescales involved.

The total quantity of mud transported via suspension during each erosion experiment is often considerable. Table I summarises some statistics relating to a number of deployments of the MORF in the Carew Estuary, S. Wales. Some approximate estimates of concommitent bedload transport were determined using a flush-mounted bedload trap in the flume structure, however bedload transport of mud aggregates by rolling and saltation is limited in comparison to suspension transport.

Table I. Computed total suspended- and bed-load sediment transport for selected Carew Estuary muds.

EXPERIMENT TYPE	TOTAL SSC AT $\tau_0=0.58$ Nm^{-2}, $(M_s)/g$	TOTAL BEDLOAD MASS $(M_b)/g$	M_s/M_b
Series I			
Control bed	14.82	0.48	30.88
Freshly deposited mud	32.08	0.70	42.83
Series II - biocide treatments			
+ Formalin	32.57	0.31	105.1
+ $CuSO_4$	21.20	0.16	132.5
+ H_2O_2	17.57	0.24	73.2
+ $HgCl_2$	15.18	0.57	26.6

Biogenic Stabilisation

Commensurate with each field deployment of the MORF was the measurement of a suite of physical and biochemical bed properties integrated over the top millimetre of the sediment-water interface. The bed was sampled using plastic 50 ml syringe barrels, 28 mm in diameter and truncated to facilitate insertion into the bed. Specific sedimentological indices were examined in relation to the sediment hydraulic indices such as τ_{cr} and ε.

Figure 4 shows the relationship between the threshold condition for Carew Estuary muds and the 'soluble' or liquid phase carbohydrate content (LPC) of the surficial floc aggregates. LPC content was determined using the phenol-sulphuric assay of Dubois et al., (1956) and is expressed in terms of glucose equivalents per gram dry sediment(GE/g); LPC's are notionally thought to represent the products of microbial excretion rather than particulate polysaccharide-rich debris found often in muddy intertidal zones (Grant et al., 1986). The positive linear relationship suggests a causal relationship and ca. 60% of the observed varation in τ_{cr} can be accounted for by changes in the LPC concentration within the mud.

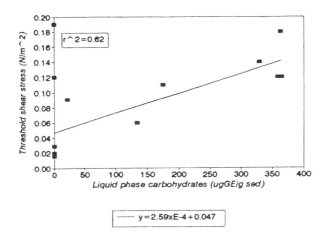

Figure 4. Dependance of the threshold shear stress (τ_{cr}) on the sediment soluble carbohydrate content ($n=12$).

Although no statistically important relationship was discernable between τ_{cr} and the sediment chlorophyll a content, it is interesting to note the strong relationship ($r^2=0.948$) between chlorophyll a and the sediment LPC content (Figure 5).

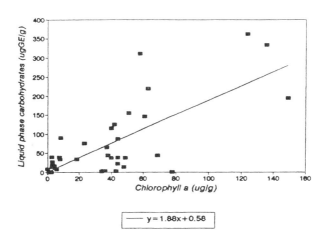

Figure 5. Relationship between sediment chlorophyll a and soluble carbohydrate content ($n=35$).

Specific experiments designed to examine the extent to which microbial processes govern fine sediment properties and dynamics were conducted. These involved deliberate and selective chemical inhibition of various living and non-living organic components of the sediment.

Areas of sediment 2m×3m were marked and then fine-sprayed with two litres with one of four chemicals (strength 1 Molar). Flume measurements were made after single tidal inundations to restore the interstitial electrochemistry and natural sediment microtopography. Table II summarises the results of some experiments from the Carew Estuary.

Table II. Initial (time-mean) erosion rates and threshold shear stresses of Carew Estuary muds.

EXPERIMENT TYPE	ε_5 $(kgm^{-2}s^{-1})^a \times 10^{-4}$ at $\tau_0 = 0.25$ Nm^{-2}	τ_{cr} (Nm^{-2}) $(\pm 1\sigma)^b$
Series I		
Control bed	0.48	0.12±0.03
Freshly deposited mud	0.81	0.18±0.04
Series II – biocide treatments		
+ Formalin	1.46	0.12±0.03
+ $CuSO_4$	1.34	0.016±4E^{-3}
+ H_2O_2	0.79	0.091±0.02
+ $HgCl_2$	0.93	0.029±9E^{-3}

[a] Computed as the arithmetic mean erosion rate over the first five minutes of each stress increment.
[b] Standard deviation of τ_{cr} evaluated in terms of the accuracy of calibration within the field flume.

With the exception of the application of formalin, all treatments appear to decrease the erosion threshold by a single order of magnitude and increase the initial entrainment rate for $\tau_0 = 0.25$ Nm^{-2}.

4. DISCUSSION

Physical Processes

Despite their limited applicability to muds in their natural environment, some of the concepts and ideas derived from previous manipulative laboratory experiments may be useful in understanding how these muds behave under fluid stressing. The linear and non-linear patterns of erosion observed in the ε-t series are qualitatively similar to the Type I/Type II classification of Mehta et al., (1982) and Mehta and Partheniades (1982) and have been reported by a number of laboratory (e.g. Thorn and Parsons, 1980; Kusuda et al., 1984; Kuijper et al., 1989). and other field (e.g. Amos et al., 1992a,b) researchers. The prevalence of Type I (non-linear) entrainment at low super-critical stresses and continuous or Type II erosion at high(er) τ_o is consistent with a strength stratified interface underlain by a more homogeneous fabric.

Type II erosion has been modelled in laboratory flumes using thoroughly remoulded and manually placed beds in which isotropic physico-mechanical gradients are totally erased; it is simply aggregate-by-aggregate detachment through time by powerful bed stresses of deeper buried floc layers which have been over-consolidated to some relatively uniform, low porosity mass (Krone, 1976). However, controversy surrounds the causal explanation of Type I erosion in cohesive sediment research. Variously, the decay in ε has been attributed to simultaneous macroscopic floc deposition (e.g. Peirce et al., 1970; Kang and Lick, 1982), development of resistive bed roughness patterns through time (Karcz and Shanmugam, 1974) or rheopexic orientation of surficial floc networks (Kusuda et al., 1982). Perhaps the most popular and plausible mechanism thus far advocated suggests Type I erosion to reflect a gradual increase in inter-floc strength with depth wherein erosion proceeds at a diminishing rate to a datum where the sediment shear resistance τ_R is equal to the imposed bed stress (Mehta et al., 1982; Mehta and Partheniades, 1982).

Unfortunately, specific hypotheses utilised by Mehta and his colleagues in attempting to explain Type I erosion are coupled to macroscopic, *centimetre* scale depthwise variability in properties that control cohesion, like density and water content. Depths of sediment erosion in the MORF and other benthic flume studies (*e.g.* Güst and Morris, 1989; Amos *et al.*, 1992a,b) and under natural tidal conditions (*e.g.* Rhoads *et al.*, 1978; Halka *et al.*, 1991) are at most a few *millimetres*. Consequently, these concepts cannot be applied to this new data. Evidently some process(es) operative on a very micro scale must be important in controlling floc erosion and contributing to micro-gradients of cohesion in the surficial floc aggregate networks.

Partheniades (1990) noted recently that although gradients in the bulk physical indices due to mechanical overburden, for instance, are unlikely over the surface 1 *mm*, it is possible that decreases in erodibility of the type observed may be attributable to higher individual floc densities and diameters which would preferentially settle at the bottom of a depositional basal layer from a tidal flow (Stow and Bowen, 1980; Kranck, 1984). Higher density flocs are associated with higher inter-bond strengths (Krone, 1976) and are therefore more able to withstand the maximum velocity shear at the top of the viscous sub-layer. Provided they are nominally cohesive they will be bonded to the bed before, and more strongly than, smaller and lighter flocs. This may prove to be a more attractive explanation for the observed Type I erosion of these muds, particularly since hydrodynamic sorting of natural estuarine particulates has recently been observed in controlled laboratory experiments (*e.g.* Muschenheim, 1989; Ockenden, 1991) and in the field (Daborn and others, 1991), and very thin (<0.1-0.2 *mm*) discrete laminations have been recorded for natural tidally deposited muds (*e.g.* Lee *et al.*, 1987; Kirby, 1990). Certainly this theory could be easily tested under controlled laboratory conditions.

Biological Processes

The modification of sediment properties by marine organisms is not a particularly new concept (Rhoads and Boyer, 1982) however relatively few studies have explored the structure and function of benthic communities in terms of their influence on the *hydraulic* properties of fine-grained muds. Indeed, the role of micro-organisms as modulators of sediment stability more generally has to date received only limited attention (Probert, 1984).

Data reported here suggest a potentially very important control by microbiota on the stability of intertidal muds. Figure 4 reveals a trend in sediment behaviour in terms of its carbohydrate content; this relationship is not unequivocal, but natural muds like these often display considerable patchiness in many sediment properties and it becomes difficult to derive high correlation coefficients in such circumstances (Thorn, 1981). Black (1991), in fact, demonstrated varying degrees of inter-property covariation for both biochemical and physical sedimentary indices in the Carew Estuary, which reduces the coefficient of determination (r^2) in such mono-parameter comparisons.

Increased concentrations of soluble carbohydrate in marine sediments have been linked to increases in sediment stability through their propensity to form thread-like and sheet-like interconnections between mineral grains (*e.g.* Paterson *et al.*, 1986; Decho, 1990). These polymers have also been noted as net aggregagators of fine sediment particles (Bailey *et al.*, 1973; Holland *et al.*, 1974).

Manzenrieder (1983) advocated the use of a 'biological stabilisation factor' S_b to express biogenic retardation of floc resuspension; S_b is the ratio of the critical friction velocity of biologically active sediment to its sterile equivalent. Paterson and Daborn (1991) recently cited $35 < S_b < 800\%$ for a variety of studies on cohesive and non-cohesive sediments. A corresponding range for the data reported here attributable to independant variation in sediment LPC content is 200-300%.

It is interesting to note that LPC content but not chlorophyll *a* relates to increases in τ_{cr}. This is consistent with recent other studies which have demonstrated direct relationships with measures of microbial exudate rather than microbial biomass (Boyer, 1980; Grant et al., 1986; Dade et al., 1990). Although bacteria are known to secrete adhesive mucus which may similarly bind together mineral grains (and may therefore be a component of the LPC content as measured here), nevertheless the tripartite relationship [diatom-exudate-sediment property] is clear from the chl a/LPC data in Figure 4 and has been confirmed using SEM images of the sediment in this and other studies (e.g. Gouleau, 1977; Paterson, 1986). A number of investigators have reported coincident areas of high chlorophyll a content and sediment accumulation rate (Coles, 1974; Mayer et al., 1985), further suggestive of diatomic stabilisation.

The biocide experiments afford a useful insight into the influence of organic processes on sediment stability. De Boer (1981) initially used metabolic poisons to inhibit microbial activity on an intertidal mega-ripple shoal and observed significant increases in net sediment transport. The data presented here are consistent with his observations and interpretation. Application of lethal quantities of chemical poisons causes dramatic increases in the time-mean erosion rate at $\tau_0 = 0.25 \ Nm^{-2}$ - $160 < S_b < 300\%$. - and order of magnitude decreases in the threshold condition. Analysis of the size spectra of treated muds revealed commensurate de-seggregation of the floc aggregate networks and consequently a decrease in net sediment cohesion. Very similar results have been reported by Daborn and others (1991) for Bay of Fundy muds, which changed virtually from cohesive silts to a primarily non-cohesive silt-flat under chemical treatments. The benthic biology at this site, rather than classic electrochemical interactions, was almost exclusively responsible for the cohesive nature of the substrate (Paterson, pers. com.).

Formalin (40% formaldehyde in water) is an antisceptic and biological preservative rather than an oxidant or metabolic inhibitor and therefore interstitial and bonded organic material should theoretically retain its structural integrity. Consequently, τ_{cr} exhibits virtually no change relative to the control bed (Table II).

This closely echoes the results and observations in the laboratory of McCall and Fisher (1980) and emphasises the role organic matter may play in aggregate formation and bed stability.

5. CONCLUSIONS

The physical processes characterising the response of cohesive intertidal muds to fluid shear are complex and heavily contingent on the (excess) bed shear stress. The data examined demonstrate qualitative similarities to previous and contemporary laboratory experiments, however they are useful in that they represent accurate quantitative information and are therefore inherently more realistic as estimates of the sediment hydraulic paramaters.

Erosion through time under periods of increasing excess bed stress is observed to be a combination of linear (Type II erosion) and non-linear (Type I erosion) processes wherein specific patterns of erosion may be tentatively related to micro-gradients of cohesion within the sediment fabric. At present these gradients are inferred rather than measured.

Assessment of those primary variables governing the resistance to erosion of natural muddy sediments is complicated. However, it is becoming increasingly apparent that benthic diatoms and possibly also bacteria may be responsible for modifying sediment stability by increasing the interparticle bond strengths over and above any electrochemical interaction. Note, however, that the envelope of potential biological binding effects is finite and biogenic stabilistation is ultimately offset by severe physical disturbances such as storms. Consequently stabilisation may be omni-present but vary seasonally, for instance. Furthermore, within natural fine-grained sediments, the situation is generally more complex than perceived here and factors which may also govern net sediment stability in the field include benthos interaction (e.g. invertebrate herbivory), solubilisation or consumption of biogenic mucus or subaerial effects (e.g. insolation, precipitation).

The studies reported here represent part of a wider trans-disciplinary effort designed specifically to further understand the dynamics of cohesive sediment erosion. In many ways the observations are probably the tip of a great 'sedimentological iceberg', and further field and laboratory experimentation within a suitable framework are most sorely needed if we are to progress in our understanding of the behaviour of these sediments.

6. REFERENCES

AMOS, C.L., VAN WAGONER, N.A., and DABORN, G.R., 1988 The influence of subaerial exposure on the bulk properties of fine-grained intertidal sediment from the Minas Basin, Bay of Fundy. *Estuar. Coastal Mar. Sci.* 27:1-13.

AMOS, C.L., GRANT, J., DABORN, G.R., and BLACK, K.S., 1992a Sea Carousel – a benthic, annular flume. *Estuarine Coastal Shelf Sci.* 34:557-577.

AMOS, C.L., CHRISTAN, H.A., GRANT, J., and PATERSON, D.M., 1992b A comparison of *in situ* methods to measure mudflat erodibility. *Proc. 2nd. International Conference on Modelling Coastal, Estuarine and River Waters*, Bradford, England, September (in press).

BAILEY, B., MAZURAK, A.P., and ROSOWSKI, J.R., 1976 Aggregation of soil particles by algae. *J. Phycol.* 9, pp.99-101.

BLACK, K.S., 1991 The Erosion Characteristics of Cohesive Estuarine Sediments: some *in situ* experiments and observations. Unpublished PhD. Thesis, Univ. Wales, 313pp.

BLACK, K.S., and CRAMP, A., 1992 A device to examine the *in situ* response of cohesive intertidal muds to fluid shear. *Marine Geology* (submitted).

DADE, W.B., DAVIS, J.D., NICHOLS, P.D., NOWELL, A.R.M., THISTLE, D., TREXLER, M.B., and WHITE, D.C., 1990 Effects of bacterial exopolymer adhesion on the entrainment of sand. *Geomicrobiol. J.* 8:1-16.

De BOER, P.L., 1981 Mechanical effects of of micro-organisms on intertidal bedform migration. *Sedimentology* 28:129-132.

DECHO, A.W., 1990 Microbial exopolymer secretion in the oceans: their rôle in the food webs and marine processes. *Oceanogr. Mar. Biol. Ann. Rev.* 28:73-153.

DuBOIS, M., GILLES, K.A., HAMILTON, J.K., REBERS, P.A., and SMITH, F., 1956 Colorimetric method for determination of sugars and related substances. *Analytical Chemistry* 28:350-356.

DUCK, R.W., AND McMANUS, J., 1991 Cohesive sediments in Scottish freshwater lochs and reservoirs. *Geomarine Lett.* 11:127-131.

GOULEAU, D., 1976 Le role des diatomees benthiques dans l'engraissementrapide des vassieresAtlantiques decouvrantes. *C. R. hebd. Seanc. Acad., Paris,* **D283**:21-23.

GRANT, J., and GÜST, G. 1987 Prediction of coastal sediment stability from photopigment content of mats of purple sulphur bacteria. *Nature* 330:244-246.

GRANT, J., BATHMANN, U.V., and MILLS, E. L. 1986 The interaction between benthic diatom films and sediment transport. *Estuar. Coastal Mar. Sci.* 23:225-238.

HAYES, F.R., 1964 The mud-water interface. *Ann. Rev. Oceanogr. Mar. Biol.* 2:121-145.

HALKA, J., PANAGEOTOU, W., and SANFORD, L.P., 1991 Consolidation and erosion of deposited cohesive sediments in Northern Chesapeake Bay, USA. *Geomarine Letters* 11:174-178.

HOLLAND, A.F., INGMARK, R.B., and DEAN, J.M., 1974 Quantitative evidence concerning the stabilisation of sediment by benthic diatoms. *Marine Biol.,* 27:191-196.

KANG, S.W., and LICK, W., 1982 Net entrainment and deposition of fine-grained sediments in freshwater. (Cited in Lick, 1982).

KARCZ, I., and SHANMUGAM, G., 1974 Decrease in scour rate of fresh deposited muds. *J Hydraulics Division,* Proc. ASCE 100(Hy 11):1735-1738.

KRANCK, K., 1984 The role of flocculation in the filtering of particulate matter in estuaries. *In: The Estuary as a Filter* (Kennedy, V.S., Ed.):159-175.

KRONE, R.B., 1976 Engineering interest in the benthic boundary layer. *In: The Benthic Boundary Layer* (McCave, I.N., Ed.), Plenum Press, New York:143-156.

KUIJPER, C., CORNELISSE, J.M., AND WINTERWERP, J.C. 1989 Research on the erosive properties of cohesive sediments. *J. Geophys. Res.* 94(C10):14,341-14,350.

KUSUDA, T., UMITA, T., and AWAYA, Y., 1982 Erosional process of fine cohesive sediments. *Memoirs of the Faculty of Engineering,* Kyushu University, **42**(4):317-333.

KUSUDA, T., UMITA, T., KOGA, K., FUTAWATARI, T., and AWAYA, Y., 1984 Erosional process of cohesive sediments. *Wat. Sci. Tech.* 17:891-901.

LEE, D., LICK, W., and KANG, S.W., 1981 The entrainment and deposition of fine-grained sediments in Lake Erie. *J. Great Lakes Res.*

7(3):224-233.

LEE, H.J., CHOUGH, S.K., JEANG, K.S., and HAN, S.T., 1987 Geotechnical properties from sediment cores from the SE Yellow Sea: effects of depositional processes. *Marine Geotech.* 7:37-50.

LICK, W., 1982 Entrainment, deposition and transport of fine-grained sediments in lakes. *Hydrobiologia* 91:*31-40*.

MANZENRIEDER, 1983 Die biologische Verfestigungen von Wattflachen aus der sicht des Ingenieurs, Miteilungen des Leichtweiss-Instituts fur Wasserbau, TU Braunschweig, Heft 79, S.135-193.

MAYER, L.M., RAHAIM, P.T., GUERIN, W., MACKO, S., WATLING, L., and ANDERSON, F.E., 1985 Biological and granulometric controls on sedimentary organic matter of an intertidal mudflat. *Estuar. Coastal Mar. Sci.* 20:491-503.

McCALL, P.L., and FISHER, J.B., 1980 Effects of tubuficid oligochaetes on physical and chemical properties of Lake Erie sediments. *In: Aquatic Biology* (Brinkhurst, R.O., and Cook, D.G., Eds.), Plenum Press, N.Y.:253-317.

MEHTA, A.J., and PARTHENIADES, E. 1982 Resuspension of deposited cohesive sediment beds. *18th. Conference Coastal Engineering*:1569-1588.

MEHTA, A.J., PARCHURE, T.M., DIXIT, J.G. and ARIATHURAI, R. 1982 Resuspension potential of deposited cohesive sediment beds. *In: Estuarine Comparisons* (V.S. Kennedy, Ed.), Academic Press:591-609.

MUSCHENHEIM, D.K., 1987 The dynamics of near-bed seston flux and suspension-feeding benthos. *J. Mar. Res.* 45:473-496.

OCKENDEN, M.C., 1991 Field measurments of cohesive sediment processes, 1990-1992. MAST-I Section G6-M *Coastal Morphodynamics* Book of Extended Abstracts, Report 4.5.

PARTHENIADES, E., 1990 The effect of bed shear stresses on the deposition and strength of deposited cohesive cohesive muds. *In:* Microstructure of Fine-Grained Sediments: from mud to shale (Bennet *et al.*, Eds.). Frontiers in Sedimentary Geology, Springer-Verlag:175-183.

PATERSON, D.M., 1986 The migratory behaviour of diatom assemblages in a laboratory tidal micro-ecosystem examined by low-temperature scanning electron microscopy. *Diatom Res.* 1:227-239.

PATERSON, D.M., 1989 Short-term changes in the erodibility of intertidal cohesive sediments related to the migratory behaviour of epipelic diatoms. *Limnol. Oceanogr.* 34:233-234.

PATERSON, D.M., and DABORN, G.R., 1991 Sediment stabilisation by biological action: significance for coastal engineering. *In:*

(Peregrine, D.H., and Lovelace, J.H.) *Developments in Coastal Engineering*, Univ. Bristol Press:111-119.

PATERSON, D.M., CRAWFORD, R.M., and LITTLE, C., 1986 The structure of benthic diatom assemblages: a preliminary account of the use and evaluation of low-temperature scanning electron microscopy. *J. Exp. Mar. Biol. Ecol.* 96:279-289.

PATERSON, D.M., CRAWFORD, R.M., and LITTLE, C., 1990 Sub-aerial exposure and changes in the stability of intertidal estuarine sediments. *Estuar. Coastal Mar. Sci.* 30:541-556.

PEIRCE, T.J., JARMAN, R.T., and de TURVILLE, C.M., 1970 An experimental study of silt scouring. *Proc. Instn. Civil Eng.* 45:231-243.

PIERCE, J.W., 1990 Microstructure of suspensates: from stream to shelf. *In:* (Bennet et al., Eds.) *Microstructure of fine-grained sediments: from mud to shale.* Frontiers in Sedimentary Geology, Springer-Verlag:139-145.

PROBERT, P.K., 1984 Disturbance, sediment stability and trophic structure of soft-bottom communities. *J. Marine Res.* 42:893-921.

RHOADS, D.C., and BOYER, L.F., 1982 The effects of marine benthos on physical properties of sediments: a successional perspective. *In: Animal-Sediment Relations: the biogenic alteration of sediments* (Tevesz, M.J.S., and McCall, P.L., Eds.), Plenum Press, N.Y.:3-52.

RHOADS, D.C., YINGST, J.Y., and ULLMAN, W., 1978 Seafloor stability in central Long Island Sound: Part 1. Temporal changes in erodibility of fine-grained sediment. *In: Estuarine Interactions* (Wiley, M.L., Ed.), Academic Press:221-244.

STOW, D.A.V., and BOWEN, A.J., 1980 A physical model for the transport and sorting of fine-grained sediment by turbidity currents. *Sedimentology* 27:31-46.

THORNE, C.R., 1981 Field measurements of rates of bank erosion and bank material strength. *Erosion and Sediment Transport Measurement* (Proc. Florence Symp.), IAHS Publ. No.133:503-512.

THORN, M.C.F., and PARSONS, J.G. 1980 Erosion of cohesive sediments in estuaries: an engineering guide. *Proc. 3rd. Int Symp. on Dredging Technology,* Paper F1, BHRA, Bordeaux:349-358.

WEST, J.R., ODUYEMI, K.O.K., and BALE, A.J., 1990 The field measurement of sediment transport parameters in estuaries. *Estuarine Coastal Shelf Sci.* 30:167-183.

YOUNG, R.A., and SOUTHARD, J.B. 1978 Erosion of fine-grained marine sediments: seafloor and laboratory experiments. *Geol. Soc. Am. Bull.* 89:663-672.

PERSISTENT MARINE DEBRIS ALONG THE GLAMORGAN HERITAGE COAST, UK; A MANAGEMENT PROBLEM

S L SIMMONS & A T WILLIAMS

Coastal Research Unit, Science and Chemical Engineering Dept, University of Glamorgan, Pontypridd, Mid Glam. Wales, UK.

ABSTRACT

The Glamorgan Heritage Coast is located in a high energy wave environment in South Wales, UK. Eight beaches - Gileston, Col-huw, Tresilian, Nash, Temple Bay, Dunraven, Ogmore-by-Sea and Merthyr Mawr were investigated for persistent marine debris. Debris was collected in three 5m wide transects down each beach from high to low tide locations in summer and winter. At Tresilian, debris was collected from the entire beach in one summer and two winter periods, 1989/90. Debris composition was similar at the above sites and compared favourably with results found by other investigators ie, plastics made up approximately 75% and 25% ; paper 6% and 54%, and metals 20% and 21% of the total debris by number and weight respectively. Apart from metal containers which were found in greater quantities on sand beaches, beach substrate (pebble or sand) was shown to have little influence on abundance and composition of debris. This is in contrast to findings of other workers. Negligible seasonal variation in debris was found, disproving views that beach debris is solely a result of visitor discards. Adjacent random belt transects studied at Tresilian brought into question the applicability of random belt transects on certain beaches due to non-uniform debris deposition. Seventeen days after all debris had been cleared from Tresilian beach, it was replaced by an even bigger volume. This suggests that beach clearance is not a sound long term management option as it does not tackle the debris source.

INTRODUCTION

Impacts resulting from the growing popularity of disposable products over the last decade are becoming more apparent in the general environment. With some items conceivably persisting for hundreds of years, concern has increased regarding the type and amount of debris which subsequently is deposited on shorelines and beaches.

Marine debris consists of a wide assortment of plastics, metal, glass and paper with persistent plastic debris being especially problematical due to its longevity (Pruter, 1987). In many cases, due to their advantageous properties, plastics have been substituted for traditional alternatives: polyesters and nylons instead of cotton, wool and linen; plastic and polystyrene instead of glass and paper, together with numerous other applications.

The importance of Persistent Marine Debris (PMD) may be questioned as impacts appear to be viewed mainly from an aesthetic point on beaches. Research however has indicated that the problem is far more widespread, having adverse effects on animals and man. Laist (1987) stated that five groups of wildlife existed which were particularly susceptible to death or injury from discarded plastic: marine and terrestrial mammals, birds, sea turtles, fish and crustaceans. Man may also be at risk; for example from containers of chemicals. In 1990, 6 canisters of potassium cyanide and 18 mainly of xylol and resorcinol were found on Sussex beaches between Brighton and Newhaven, along with medical wastes. Onions (pers comm) has found medical waste at Merthyr Mawr beach (Fig. 1). Additionally, munitions are not an uncommon occurrence in some areas (Dixon & Dixon, 1979). Injuries to beach users, especially bathers, have been recorded particularly foot lacerations from encounters with broken glass and nails protruding from driftwood (Dixon, 1987).

There seems only one coastal management viewpoint for marine debris, and that is to collect and deliver it to the local council refuse collection service. Management views epitomise the Dickensian Micawber attitude of "it will go away, or, we will find something to counter it in the near future". The problem is huge and growing each year. Extremely little work has been done on marine sources, transport patterns and sinks and virtually nothing on its riverine counterparts. It is blithely assumed that most marine debris is from visitor discards or is washed in from the sea, but what proportions of beach debris are of riverine, marine and visitor origin? This is a fundamental question for any effective management plan, yet answers to this question are not known for virtually any coastal area in the world.

LOCATION

The Glamorgan Heritage Coast extends for 22km along the northern shore of the Bristol Channel (Fig. 1). The philosophy of such coasts is to leave the land in private ownership and conserve these natural areas for future generations (Williams, in press). Some 33% of England and Wales have been designated Heritage Coasts. Zoning policies differ from area to area, but in the Glamorgan Heritage Coast only remote and intense zones exist. Remote zoning is adjacent to the intensely populated ie "honeypot" locations which have refreshments, toilets etc for visitors. There are four in the area; Col-huw, Dunraven,, Nash and Ogmore-by-Sea (Fig. 1). With some 1.5 million people living just forty minutes drive from the Glamorgan Heritage Coast (Williams and Sothern, 1986) beach cleanliness is imperative for beach areas to maximise their tourist potential.

Fig. 1   Location Map - The Glamorgan Heritage Coast.

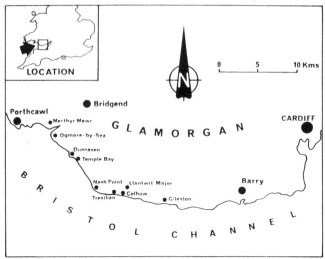

The area is quasi-estuarine with an open sea fetch on a 240 degree bearing. The tidal range is of the order of 6m and the strong prevailing south westerly winds drive powerful waves against the predominantly Lias Limestone vertical sea cliffs which reach up to 80m in height (eg Whitmore Stairs). As a result of the marine regime, longshore and subsequent debris movement is from west to east. The area therefore lies in an aggressive morphogenic region which is undergoing erosion of some 6-8 cm per year (Williams et al, 1991). Only three sand beaches exist backed by a pebble ridge - Dunraven, Merthyr Mawr (which merges into a 250Ha dune system), Ogmore, the latter two being a continuum separated by the Ogmore river. All other (Gileston, Col-Huw, Tresilian, Nash, Temple Bay) beaches are rock and shingle and are typical pocket beaches.

METHODOLOGIES

The eight beaches mentioned above and shown in Fig. 1, were sampled for debris deposited at both low and high tide marks in summer and winter surveys in 1989/90. Analyses were carried out on samples to assess composition and, where possible, origin of the debris.

For each of the beaches studied, a fixed back-shore point was chosen which could be easily identified on subsequent visits. A random number table was used to determine a distance, in paces, of the transect from this point. Using a 30m tape the transect was marked out at right-angles to the high tide mark, encompassing all wind-blown litter, and stretching 30m down the beach from this point. A second 30m tape was placed 5m along the beach, parallel to the first, clearly marking the study area. All litter within the transect was collected and analyzed. Three transects were set up on each of the eight studied beaches. The exact position of the transect was noted so the same area could be sampled via the second sample taken in the winter.

This method was applied at all sites with the exception of Tresilian Cove where the entire beach was sampled. Obviously, a large amount of data was obtained, but analysis was limited in that numbers were recorded only for each class of debris type. Tresilian Cove was sampled once in the summer and on two occasions in the winter.

The collected debris was sorted into various categories using a checklist approach (Dixon and Cooke, 1977). Composition and relative abundance of material was considered first, with item classes recorded by both number and dry weight. For containers, contents were determined from labels, bottle design and odour. Country of manufacture of containers or contents was used to determine the range of geographical origins. These data were obtained from labelling, overprinting and embossed instructions. Date of manufacture was also noted if printed or embossed on the debris.

Statistical analyses of results were undertaken to determine if differences between summer and winter periods for metals, plastics, paper and glass were statistically significant at the 0.05/0.01 level. The number of items in the different classes for each of the transects on every beach were input into a data file using the SPSSx computer statistical package (Norusis, 1985).

RESULTS AND DISCUSSION

(i) TRANSECT STUDIES: SEVEN SITES

(a) SEASONAL VARIATION IN DEBRIS

Numbers of container and non-container items found in summer and winter samples at all beaches except Tresilian, were compared to ascertain differences in debris quantity found over the entire length of the Glamorgan Heritage Coast at different times of the year. The average number of containers and non containers for each beach, were summed to give total values per transect (Table 1). Differences were apparent in debris composition and quantities. In the container category, plastic, metal and paper receptacles were all more abundant in the summer than winter (Table 1). This may be due to accumulation in a less hostile marine condition, or increased input of debris from beach visitors, but t-tests (prob. 0.01) of container material types showed only metal and paper containers were significantly different. In contrast, non-container results show plastics in particular to be more abundant in the winter. Such increased amounts of non-container material are likely to be found due to debris fragmentation in harsher conditions prevalent in winter. Although there appear to be inequalities in numbers of debris items at different times of the year. t-test analysis of non-container results indicated that none of the material types in this category were found in significantly different numbers at the times surveyed.

Table 1. Seasonal variations in average number of container/non-container items per transect in the Glamorgan Heritage Coast.

	SUMMER				
	PLASTIC	METAL	GLASS	PAPER	MISC.
CONTAINERS	7.8	2.5	0.1	1.0	0.2
NON-CONTAINERS	18.9	0.4	---	0.1	0.8
	WINTER				
	PLASTIC	METAL	GLASS	PAPER	MISC.
CONTAINERS	7.4	0.8	0.2	0.0	0.1
NON-CONTAINERS	24.3	0.5	---	0.5	1.2

Indications were that very little alteration overall in beach debris quantities were found during different seasons. Table 1 shows the preponderance of plastics in this period compared to other items. Although as stated above, metal and paper containers varied with season, their contribution to the whole was minimal. A common misconception by the public is that most beach debris is solely a result of careless visitor discards in the area, but this is not completely true for all debris types ie plastics (Table 1). Other possible sources are riverine and marine.

(b) INFLUENCE OF BEACH SUBSTRATA ON DEBRIS

The diverse nature of beach types along the 22km stretch of the Glamorgan Heritage Coast allowed for a comparison to be made as to the effect of beach substrata on the quantity and composition of shoreline debris. Dixon and Cooke (1977) carried out a similar investigation, comparing a shingle and sand beach. Their results indicated that sandy beaches, because of their shallow gradient and broad reach zone, retained debris for longer periods and accumulation would be expected to occur on such beach types.

The numbers of container and non-container items from Merthyr Mawr (sand) and Gileston (pebble) were compared and results contrasted with those found by Dixon and Cooke (1977). These beaches were chosen as both areas attract very few beach-users, probably due to inaccessibility (parking necessitates a long walk to the beaches - 10 to 20 minutes!) and thus direct visitor input of debris to the areas is minimal. The choice of sites minimized the likelihood of the sand beach having a greater input of debris due to increased popularity.

With the exception of plastics - the largest container category - all other container material types were found in greater abundance on the sand rather than pebble beach (Table 2). The former had a higher number of glass and paper containers, probably due to the more yielding beach substrata; these do not however constitute a large proportion of the total debris.

Table 2. Influence of beach substrate on average numbers of container/non-container items per transect

	SAND	PEBBLE	SAND	PEBBLE
	CONTAINERS		NON-CONTAINERS	
PLASTIC	9.0	12.0	18.9	24.2
METAL	3.3	1.0	0.4	0.5
GLASS	0.3	0.0	0.0	0.0
PAPER	1.3	0.3	0.1	0.6

Bagnold's dispersive stress phenomenon may account for the elevated numbers of plastic containers on the pebble beach (Williams and Davies, 1989). With the perpetual motion of beach sediment, larger objects, such as plastic containers are usually pushed to the surface. Additionally, the larger pore spaces allow infiltration and burial of smaller size items.

Plastic durability may also partly explain the elevated numbers of containers found on the pebble beach, the very nature of which mitigates against debris persistence. In the case of Gileston pebble beach, sediment is moving at some 400m per year in an easterly direction. This movement could result in fragmentation of containers, especially those of fragile composition. Such container fragmentation possibly resulted in the elevated numbers of non-container materials found on the pebble beach (Table 2).

Paired t-tests showed that no significant differences (at the 0.05 level) were present in any of the non-container classes. Of the container classes the only one found with a significantly different distribution on the two beach types was metal receptacles, whose abundance was greater on the pebble beach.

Overall, results appear to support a divergence from Dixon and Cooke's (1977) theories on the influence of different beach substrata on abundance and composition of beach debris. In order for more definite conclusions to be drawn, further extensive surveys would have to be executed over a long time span and in many diverse areas.

(ii) INTERVAL SURVEY: ONE SITE

A pilot interval survey was undertaken at Tresilian Cove in order to investigate the re-accumulation of beach debris. This survey involved high manpower usage and as such was only undertaken once. All debris was removed from the beach on November 15, 1989 and again 17 days later on December 2, 1989 (Table 3). Containers were only categorized into general plastics and metals as other material types were found only in insignificant quantities.

Table 3. Seventeen Day Interval Survey of Tresilian Cove.

	15 NOV	2 DEC	15 NOV	2 DEC
	CONTAINERS		NON-CONTAINERS	
PLASTIC	282	117	106	206
POLYSTYRENE	0	0	10	150
METAL	18	26	7	9
GLASS	1	0	0	0
NET/ROPE	0	0	43	47

After the 17 day period, the number of plastic containers found was approximately half the number recorded on the first collection (282 vs 117 items). Contrastingly, the number of metal containers found were elevated in the second collection, showing no effect of the beach cleaning (18 vs 26 items). Unexpectedly, a vast increase in the number of non-container materials was seen in the December collection. Large increases were found in the number of general plastics and polystyrene items. Polystyrene was designated as an individual category due to its irregular and greatly fluctuating appearance which could lead to misleading results if included in the general plastics grouping. Even disregarding the polystyrene category as a localized input, non-container items were still more abundant 17 days after beach clearance. These results bring into question the effectiveness of small scale beach clearing programs as a means of solving or even effectively alleviating the problem of PMD.

It appears that beach debris is part of a mobile phase in which items are deposited and recovered by tidal ebb and flow and the influence of wind and currents. This continual movement means the removal of debris from one beach may prevent it being washed away onto another area of the shoreline but that it will be replaced quickly by debris from elsewhere. This pilot scheme has indicated the necessity for replicates of the above survey, a fact of which the authors are well aware. Results highlight the enormity of the problem and the need for a more preventative approach if a long-term management solution is to be found.

ANALYSIS OF SAMPLING METHODS

Random belt transect methods have been adopted in the field of marine debris surveillance with very little justification as to their suitability for such investigations (Dixon and Dixon, 1981). A pilot study was set up at Tresilian to look at data variations between transects, and possible implications of using results from only three transects to represent an entire beach. Tresilian was split into fourteen adjacent 5m transects, stretching from the back shore to high tide marks, labelled A to N in an west to east direction. Debris items were sorted into one of eleven categories and numbers in each transect recorded (Fig. 2). This initial survey indicated the kind of variations possible along a beach and brings into question the use of this method on beaches with non-uniform debris deposition. Further results will be collected to enable future statistical analysis.

Classification of quantities of container and non-container materials was carried out by both number and weight. This procedure is contrary to methods applied by previous workers, in which only one type of measurement has been used (Horsman, 1985; Dixon and Cooke, 1977; Winston, 1982). Dual measurements were undertaken as it was felt that using only numbers or weights could lead to extraneous results. Examination of the data proved this to be the case.

Summer results for quantities by number and weight of different container materials at Temple Bay have been used as an example. By number, plastics constitute 74% of the debris. This is reduced to mere 24% if calculated by weight. Metal remains fairly constant by either measurement ie number, 20%; weight, 21%. Paper showed a large discrepancy, accounting for only 6% of total debris by number but 54% by weight. Similar non-conformities for all other sites were found which emphasizes the need for dual measurements to be undertaken.

The importance of dual measuring is even greater in the quantification of non-container materials. With many objects in the form of fragments, and items such as fishing line, it seems inadequate and misrepresentative to quantify them by number alone. It is felt that for future investigations,

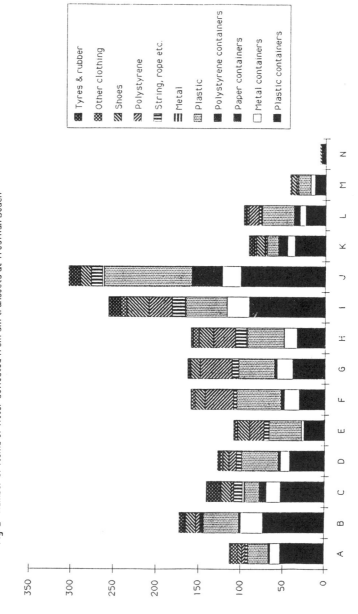

Fig. 2 Number of items of litter collected from 5m transects at Tresilian Beach

measurement by number and weight is advisable if results are to be truly representative.

CONCLUSIONS

The investigation of Persistent Marine Debris along the Glamorgan Heritage Coast not only highlighted the scope of the problem in this area, but also the complexity. The study brought to light both contradictions and confirmation of previous studies. Analysis of seasonal variation in the quantities and composition of marine debris were not significantly different between summer and winter periods, apart from metals and paper. This contribution was minor compared to others notably plastics. Therefore it must be concluded that visitor discards of beach debris are only a part contributor to the problem and that the marine and riverine inputs should perhaps be given more emphasis in any future research.

Results of the 17 day interval survey carried out at Tresilian Cove, confirmed the need to view the situation from a broader perspective if solutions are to be found which will alleviate the problem. The effectiveness of random belt transects as a uniform method of beach debris surveillance was also questioned. Whilst oceans are used as dumping grounds and the rivers as waste disposal systems, a mass of floating garbage will proliferate and plague coastlines indefinitely.

A comparison of the effect of beach substrate on debris showed that no significant differences were apparent in the quantities or composition of debris on differing beach types, apart from metal containers which contradicted Dixon and Cooke's (1977) findings. Predominant debris classes found within the study area were also divergent from those discovered by Dixon and Cooke (1977) in their investigation of Sandwich Bay, Kent. It became evident from the study that in order to obtain truly representative results, dual analyses, both of number and weight, need to be carried out. Persistent Marine Debris is a problem not only along the Glamorgan Heritage Coast but in most other coastal areas. A tightening of legislation concerning dumping of debris and improved public awareness would greatly improve the situation; but what is really needed is for money to be made available for fundamental research into the problem, as without facts, effective management cannot occur.

ACKNOWLEDGEMENTS

The authors would like to thank Richard Date\James Mendelssohn and the environmental group at Atlantic College, St Donats, for manpower provided in beach clean-ups. Thanks also to Jim Short (technician) for assistance in the field. We would also like to thank Trevor Dixon for his guidance throughout this research.

REFERENCES

Dixon, T. R. (1987). Operational Discharges from Ships and Platforms. (Garbage, Packaged Dangerous or Harmful Goods and Pyrotechnics.) North Sea Forum Report. pp. 57-59.

Dixon, T. R. & Cooke, A. J. (1977). Discarded containers on a Kent Beach. Mar. Poll. Bull. Vol. 8. No 5. pp. 105-109.

Dixon, T. R. & Dixon, T. J. (1979). Munitions in British Coastal Waters. Mar. Poll. Bull. Vol. 10. pp. 352-357.

Dixon, T. R. & Dixon, T. J. (1981). Marine Litter Surveillance. Mar. Poll. Bull. Vol. 9. No. 9. pp. 289-295.

Horsman, P. V. (1985). Garbage Kills. BBC Wildlife. Aug. pp. 391-396.

Laist, D. W. (1987). Overview of Biological Effects of Lost and Discarded Plastic Debris in the Marine Environment. Mar. Poll. Bull. Vol. 18. No 6B. pp. 319-326.

Norusis, M. J. (1985). Advanced Statistical Guide -SPSSx. McGraw-Hill Book Company. pp. 505.

Pruter, A. T. (1987). Sources, Quantities and Distribution of Persistent Marine Plastics in the Marine Environment. Mar. Poll. Bull. Vol 18 (6B). pp. 305-310.

Williams, A. T. (in press). The Quiet Conservators.

Williams, A. T. & Davies, P. (1989). A Hard Rock Sediment Budget for the Inner Bristol Channel. In (ed) S Y Wang, "Sediment Transport Modelling", Amer. Soc. Civ. Eng., New York, pp. 474-479.

Williams, A. T., Morgan, N. R. & Davies, P. (1991). Recession of the Littoral Zone Cliffs in the Bristol Channel, UK. In (ed) O T Magoon "Coastal Zone '91". Amer. Soc. Civ. Eng., New York, pp. 2394-2408.

Williams, A. T. & Sothern, E. T. (1986). Recreational Pressure Along the Glamorgan Heritage Coast, Wales, UK. Shore and Beach. Vol 54(1). pp. 30-37.

Winston, J. E. (1982). Drift Plastic - An Expanding Niche for a Marine Invertebrate. Mar. Poll. Bull. Vol 13. No 10. pp. 384-351.

BEACH AESTHETIC VALUES: THE SOUTH WEST PENINSULA, UK

A.T.WILLIAMS[1]; S.P.LEATHERMAN[2] and S.L.SIMMONS[1]

1. Coastal Research Unit, Department of Science and Chemical Engineering, University of Glamorgan, Pontypridd, Wales, UK.
2. Centre for Global Change, University of Maryland, College Park, Maryland, USA.

ABSTRACT

An innovative check list was devised to rate beaches on their aesthetic characters. Physical, biological and human usage parameters were selected and a five point scoring scale devised so that assessment was objective. Summation of the scores for each sub-heading mentioned above and the obtaining of a grand total, enabled percentage values to be given for any investigated beach. One hundred and eighty two such beaches were analysed for the South West Peninsula, UK. The highest rating value obtained was 86%, the lowest 55.6%. The median value was 73.2%. Ratings tended to follow the pattern set out by European Blue Flag beaches but there were several discrepancies. Differences were found between beaches located on the north and south Peninsula shorelines.

INTRODUCTION

Literature on landscape evaluation is vast, but when the landscape is coastal, this literature volume shrinks appreciably. When the topic is beach aesthetics there is a dearth of work and no one has attempted to rate beaches on a semi- quantitative scale. The Oxford Dictionary defines aesthetic as " concerned with, capable of, appreciation of the beautiful". Perhaps beautiful beaches only exist in the eye of the beholder as the concept of beauty is intangible; perhaps fleeting and incapable of being quantified (Williams and Lavalle, 1990). However, the massive growth in tourism has meant that beach resorts have many parameters that can be classified on an objective basis, and this paper is an innovative and pilot attempt at classifying 182 such beaches located in a prime tourist area in England - the South West Peninsula of the United Kingdom (Fig.1). Tourism is both the greatest and fastest growing asset and problem for the area. At the height of the tourist season - the first week in August - over a quarter of a million people are to be found in each of the counties of Devon and Cornwall, 90% within 2 miles of the coast (Devon Coastal Conservation Study,1988). Much of the increased sewerage due to this population is discharged untreated into tidal and coastal waters (Rivers Information,1991). Tourism brings in over $1000 million per year, producing 50,000 full time jobs and some 25,000 seasonal ones.

There is currently a move towards cleaning up the beaches of Britain in accordance with European legislation - EC Directive 76/160 - on improving the quality of bathing waters, and some $49 billion will be spent by the Water Authorities mainly on sewage treatment works in order to attain this objective. It is envisaged that by 1998 all dumping of sewage sludge at sea will have stopped. In the UK, there exists (1991) some 453 designated bathing waters, of which 76% meet European quality standards (in 1988 it was 67%). These beaches are monitored by the National Rivers Authority (NRA) 20 times during the bathing season for sewage bacteria, and twice for viruses. The NRA is responsible for safeguarding and improving water quality in the UK and their main powers relating to bathing water came from the 1989 Water Act. Bathing water is defined as 'all running or still fresh water or parts thereof and sea water in which bathing is explicitly authorized by the competent authorities of each member state or, bathing is not prohibited and is traditionally practised by a large number of bathers'; (76/160/EC). Originally a very narrow definition of bathing waters was used eg, 500 people in the water at the same time, so that only 27 such beaches were designated in the UK, of which 11 were in the South West Peninsula. In 1985, bathing waters were re designated

according to criteria relating to popularity with 134 being located in
this region. All are tested by the NRA for 19 physical, chemical, and
biological parameters, monitoring points on beaches being where the
highest densities of bathers occur. In 1990, 15 beaches in the south
west region failed to satisfy mandatory bacteriological standards of
the EC Directive.

The top accolade for European beaches is the Blue Flag award and to
date 29 British beaches have qualified for this award, 14 of them are
located in the south west peninsula, and 11 were studied in obtaining
data for this paper. This award is operated by the Foundation for
Environmental Education. Strict criteria of cleanliness and management
is the essence of obtaining Blue Flag status. Water bacteria levels
must be within permitted limits, sea and beach must be free of sewage,
oil and industrial pollution, there must be a lifeguard presence,
plenty of litter bins, signboards giving graphs of monitored data. In
the European context,the numbers of northern Mediterranean Blue
Flagged beaches are: Spain 86; France 36; Yugoslavia, 0; Greece, 6;
Italy 17.

Environmentalists the world over are concerned at the current state
of beaches .This paper is a pilot attempt to classify beaches based on
many factors. In the past, researchers have concentrated on but one
aspect of the problem eg. erosion, pollution. The aesthetic approach
postulated, is an innovative attempt to rank beaches in many different
areas in accordance with a strict objective criteria set encompassing
all main elements, and could be used in any beach assessment.

PHYSICAL LOCATION

The South west peninsula of England has long been isolated from the
rest of the country by virtue of the Somerset Levels, Blackdown Hills
and Exmoor. The area covers some 8.5 thousand square miles and has a
very small urban population. The greater part of the region is covered
by Palaeozoic rocks, mainly Devonian and Coal Measures.
Armorican earth movements forced these rocks into roughly east west
folds. Pre Cambrian rocks are exposed especially in southern coastal
Devon and Cornwall, whilst batholithic granites can be seen in the

centre of the region eg. Dartmoor, Bodmin Moor. The geology of the region is complex and coastal exposures are frequently contorted when seen. The coastline varies from high rocky promentories separating secluded shingle rich pocket beach areas, especially in the north , to wide sweeps of sand eg Exmouth , Sidmouth in the south. It has over 700 km of open coast and 300 km of estuaries. The geology determines beach composition with wide sand beaches mainly to be found in the south east part of the region , deriving from breakdown of the Keuper marls and sandstones, Permian sandstones and Upper Greensand sedimentary rocks. Because of the marine influence, 365 growing days per year is possible in coastal areas, and the southern coastline receives over 1650 annual hours of sunshine. The region is therefore much sought after by tourists and retiree's. The region has 134 out of the 407 'Eurobeaches' in England and Wales.

Large contrasts exist between the north and south coastlines. The north Devon coast has many areas of sand dunes and shallow estuaries, whilst the south coast has numerous drowned valleys with wide bays fringed with beaches and cliffs. In the case of Cornwall, the north coast is subject to the full force of Atlantic gales. No real harbours exist, but high cliff scenery interspersed with small coves is the norm. It is a relatively treeless area. The western part has some wide sand bays composed of soft shales surrounded by hard igneous rock headlands, whilst the southern sector resembles Devon.

METHODOLOGY

A checklist procedure was devised encompassing physical, biological and human interest parameters. In this pilot study, no weighting has been carried out and little feedback of beach user perception on this matter has been obtained. This is a topic for further research. Each parameter shown in Table 1, has associated with it a five scale index. The observer places a tick in the corresponding box according to site evaluation. When the observer has worked through the table, box summation was carried out (the value of one being given to the left hand box, five to the right hand box etc). Sub- totals were also obtained for values representing the physical, biological and human interest categories. Grand totals were turned into percentages and

Table 1. BEACH RATING SCALE QUESTIONNAIRE

	PHYSICAL FACTORS	CATEGORIES				
		1	2	3	4	5
1.	Beach width at low tide	narrow, <10m	10-30m	30-60m	60-100m	>100m, wide
2.	Beach material	cobbles	sand/cobbles	coarse sand	- - -	fine sand
3.	Beach condition or variation	erosional	- - -	stable	- -	depositional
4.	Sand softness	hard	- - -	- - -	- - -	soft
5.	Water temperature	cold/hot	- - -	- - -	- - -	warm (70°-80°F)
6.	Air temperature (midday)	< 60°F >100°F	- - -	- - -	- - -	80°-90°F
7.	Number of sunny days	few	- - -	- - -	- - -	many
8.	Amount of rain	large	- - -	- - -	- - -	little
9.	Wind speeds	high	- - -	- - -	- - -	low
10.	Size of breaking waves	high/dangerous	- - -	- - -	- - -	low/safe
11.	Number of waves/width of breaker zone	None	1-2	3-4	5	6+
12.	Beach slope (underwater)	steeply sloping bottom	- - -	- - -	- - -	gently sloping bottom
13.	Longshore current	strong	- - -	- - -	- - -	weak
14.	Rip currents present	often	- - -	- - -	- - -	never
15.	Color of sand	grey	black	brown	light tan	white/pink
16.	Tidal range	large (>4 meters)	3-4m	2-3m	1-2m	small (<1 meter)
17.	Beach shape	straight	- - -	- - -	- - -	pocket

| 18. | Bathing area bottom conditions | ☐ ☐ ☐ ☐ ☐ rocky,—————————→fine cobbles, mud   sand |

BIOLOGICAL FACTORS

19.	Turbidity	☐ ☐ ☐ ☐ ☐ turbid—————————→clear
20.	Water color	☐ ☐ ☐ ☐ ☐ grey——————————→aquablue
21.	Floating/suspended human material (sewerage, scum)	☐ ☐ ☐ ☐ ☐ plentiful—————————→none
22.	Algae in Water Amount	☐ ☐ ☐ ☐ ☐ infested—————————→absent
23.	Red Tide	☐ ☐ ☐ ☐ ☐ common——————————→none
24.	Smell (e.g., seaweed rotting fish)	☐ ☐ ☐ ☐ ☐ bad odors————→fresh salty air
25.	Wildlife (e.g., shore birds)	☐ ☐ ☐ ☐ ☐ none——————————→plentiful
26.	Pests (biting flies, ticks, mosquitos)	☐ ☐ ☐ ☐ ☐ common————————→no problem
27.	Presence of sewerage/ runoff outfall lines on/across the beach	☐ ☐ ☐ ☐ ☐ several—————————→none
28.	Seaweed/jellyfish on the beach	☐ ☐ ☐ ☐ ☐ many——————————→none

HUMAN USE AND IMPACTS

29.	Trash and litter (paper, plastics, nets, ropes, planks)	☐ ☐ ☐ ☐ ☐ common—————————→rare
30.	Oil and tar balls	☐ ☐ ☐ ☐ ☐ common—————————→none
31.	Glass and rubble	☐ ☐ ☐ ☐ ☐ common—————————→rare
32.	Views and Vistas Local scene	☐ ☐ ☐ ☐ ☐ obstructed————→unobstructed
33.	Views and Vistas Far vista	☐ ☐ ☐ ☐ ☐ confined————→unconfined
34.	Buildings/Urbanism	☐ ☐ ☐ ☐ ☐ Overdeveloped———→pristine/wild

35.	Access	Limited ☐ ☐ ☐ ☐ ☐ → Good
36.	Misfits (nuclear power station; offshore dumping)	Present ☐ ☐ ☐ ☐ ☐ → none
37.	Vegetation (nearby) Trees	None ☐ ☐ ☐ ☐ ☐ → many
38.	Well-kept grounds/ promenades or natural environment	No ☐ ☐ ☐ ☐ ☐ → Yes
39.	Amenities (showers, chairs, bars, etc.)	None ☐ ☐ ☐ ☐ ☐ → Some
40.	Lifeguards	None ☐ ☐ ☐ ☐ ☐ → present
41.	Safety record (deaths)	Some ☐ ☐ ☐ ☐ ☐ → None
42.	Domestic animals (e.g., dogs)	Many ☐ ☐ ☐ ☐ ☐ → None
43.	Noise (cars, nearby highways, trains)	Much ☐ ☐ ☐ ☐ ☐ → Little
44.	Noise (e.g., crowds, radios)	Much ☐ ☐ ☐ ☐ ☐ → Little
45.	Presence of seawalls, riprap, concrete/ rubble	Large amount ☐ ☐ ☐ ☐ ☐ → None
46.	Intensity of beach use	overcrowded ☐ ☐ ☐ ☐ ☐ → ample open space
47.	Off-road vehicles	common ☐ ☐ ☐ ☐ ☐ → None
48.	Floatables in water (garbage, toilet paper)	common ☐ ☐ ☐ ☐ ☐ → None
49.	Public Safety (e.g., pickpockets, crime)	common ☐ ☐ ☐ ☐ ☐ → rare
50.	Competition for free use of beach (e.g., fishermen, boaters, waterskiers)	many ☐ ☐ ☐ ☐ ☐ → few

the beaches ranked; the higher the percentage the better was the aesthetic environment. One hundred and eighty two beaches were investigated in this manner, eighty six in Devon and ninety four in Cornwall (Table 1). Statistical testing of the significance of the results, involved testing for normality via the Kolmogorov - Smirnov Goodness of Fit. If a normal distribution was found - it was in all cases except one, Table 2 - testing was done via a t test. Testing was via the Mann-Whitney U and Wilcoxon Rank Sum W test in the one case that was non - normal . (Table 2).

RESULTS AND DISCUSSION

The joint top beach rankings were Coverack and Camel Bay in south and north Cornwall respectively; and the lowest ranked beach was Jennycliff Bay in south Devon (Fig.1). Median percentages for Devon, north and south respectively, were 72.4 and 70.4; for Cornwall, north and south repectively, 80.4 and 72.4. Similarly, median percentages for all Devon were 69.2; all Cornwall 78.4; all north shore sites 78.4; all south shore sites 71.6. The overall median value for 182 sites was 73.2% typified by Portholland beach in south Cornwall.

From an aesthetic viewpoint the north Cornwall beaches stand out from all others. If one selects the median value of 73%; 33 out of the 34 beaches investigated in north Cornwall exceeded this value against 35 out of 60 in south Cornwall. In the case of Devon, the northern shore had 12 out of 22, and the south 18 out of a total of 66. Table 2 indicates that all parameters, differ for the areas in question - a reflection of both the physical and cultural environment.

Examination of results obtained for the county of Devon, show that only in human usage factors was a significant statistical difference obtained between north and south shores (Table 2). This was expected given that the south Devon shoreline is called the 'English Riviera', with large pleasure resorts eg. Torquay, Paignton acting as fulcrums for entertainment. Nothing on this scale exists to the north. Much of this area has National Park and Heritage Coast status and is wild and uninhabited except for small hamlets eg Coombe Martin.

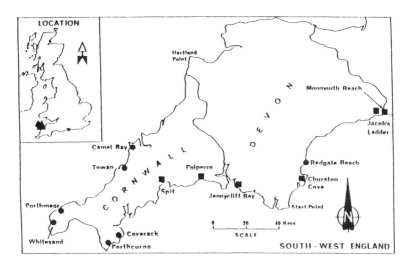

Fig. 1 Location. Circles represent top high ranked beaches; squares bottom low ranked beaches.

When the whole of the county is tested against Cornwall, differences are apparent in all except the human usage parameter (Table 2). The larger grouping makes no difference to physical and biological parameters but the small cove villages and larger resorts of both counties blend into one another when analysed in the cultural.

Cornwall has to date only 3 Blue Flag beaches, one in the north (Porthmeir) and two in the south. The only Blue Flagged Cornish beach investigated in research for this paper was Porthmeir beach which rated a value of 85% - a very high figure when the top rating value was 86%. North Devon has no Blue Flag beaches, but fifteen occur in south Devon. Ten of these beaches were rated in the present study. Taking the median cut off value of 73 %, two beaches failed to reach this value, Jacob's Ladder (59.6%), and Babbacombe (68.4%), Sidmouth (73.6%) just reaches it. These beaches ranked 64th, 42nd, and 22nd

TABLE 2. Statistical testing of results. Two tailed Probability values.

		Site 1 vs. Site 2.	Kolmogorov-Smirnov Goodness of Fit.	
			Site 1	Site 2
DEVON	%	.17	.61	.87
North (Site 1)	P	.57	.76	.24
vs.	B	.23	.78	.83
South (Site 2)	H	.01'	.65	.64
			N = 22	66
CORNWALL	%	.00'	.40	.72
North (Site 1)	P	.01'	.29	.35
vs.	B	.00'	.50	.41
South (Site 2)	H	.00'	.90	.87
			N = 34	60
ALL DEVON (Site 1)	%	.00'	.97	.29
vs.	P	.00'	.24	.29
ALL	B	.00'	.81	.33
CORNWALL (Site 2)	H	.45	.98	.72
			N = 88	94
ALL DEVON & CORNWALL	%	.00'	.14	.58
North Shore (Site 1)	P	.05'	.08	.07
vs.	B	.02'	.61	.05*
ALL DEVON & CORNWALL	H	.00'	.83	.73
South Shore (Site 2)			N = 56	126

* Non normal. Testing was done via the Mann-Whitney U & Wilcoxon Rank Sum W test.
' Statistically significant difference between the two groups
N Number of beaches in sample.
% Percentage; P Physical; B Biological; H Human Parameters (Table 1).

out of 66 for the county as a whole. Redgate, Meadfoot and Dawlish all scored highly (84%, 78.8% and 78.4%; ranking 1st, 4th and 5th respectively in the county .The remaining four Blue Flag beaches, Budleigh Salterton, Exmouth, Teignmouth and Paignton had aesthetic values ranging from 76.4% to 75.6% There seems to be a trend for these beaches to have a high aesthetic value but Jacob's Ladder was an anomaly.

Blue Flag beaches are highly sought after by developed resorts, but a new award has been introduced by the Foundation for Environmental Education for non developed resorts, termed the Golden Starfish Award. This could suggest a two pronged checklist for beach ratings, one for developed and one for undeveloped beaches. A less rigorous classification of good beaches is that of the Good Beach Guide (1991). Of their top beaches – termed four star beaches – 15 UK beaches qualified on high water quality out of a sample of over 450 bathing beaches. Five of them are located in the study area, and four were analysed by the check list approach. In South Cornwall, Kennack Sands and Porthoustock scored 78.4% and 74.8% respectively; in north Cornwall, Constantine Bay scored 76.8%; and in south Devon, Bigbury scored 77.6% None of these beaches rated in the top six in their respective counties according to the check list showed in Table 1.

Conservationalists and environmentalists are demanding new legislation regarding Britain's beaches, and their arguments usually embrace pollution. For example, in Great Britain, some 250 pipelines empty millions of gallons of raw sewage into the sea daily according to the Good Beach Guide (1991). This guide rates beaches mainly on pollution grounds and on an annual basis. For example, in 1989, Porthminster beach in north Cornwall did not qualify, but was successfull in 1990. Its rating on the checklist parameters (Table 1), was 80.8% ie. a median value for this stretch of coastline. Biological characteristics however are but one facet of an aesthetic assessment, and we maintain that it is more important to quantify the whole beach area rather than simply one aspect ie. as carried out by the Good Beach Guide (1991). This is bourne out by comments made by the Water Services Association – which represents Water Companies. They plan to spend some $49 billion on improvements by the year 2000AD and argue that if water quality was the sole criteria, some 318 beaches should have been Blue Flagged in 1990. The fact that there were only 29 reflected poor beach management by local authorities ie. sewage discharge improvements were not being matched by cleaner beaches (South Wales Echo, 20 May, 1991).

CONCLUSIONS

Assessment of the aesthetic character of a beach as shown by parameters in Table 1, can give a semi- quantitative ranking to beaches in a district, county, country or continent. Results from this pilot survey of 182 beaches in England indicated that the fundamental parameters needed for aesthetic assessment are the ones of physical, biological and human usage. The highest ranked beach had a rating of 86%; the lowest 55.6%. Further refinement is needed on these parameters but the ranking scheme produced is claimed or thought to be a sound reflection of conditions experienced at the various beaches. It is hoped that further testing, possibly on a developed / undeveloped resort basis will be carried out, for example on the coastal beaches of the USA, Wales, as results obtained will be of benefit to the tourist industry as well as the local community.

ACKNOWLEDCEMENTS

We are very grateful to Ms Jane Driver, National Rivers Authority, Exeter branch, who allowed us access to a team of coastal pollution monitors. In 1990 the team filled in the checklist at each South West Peninsula beach monitored by the NRA.

REFERENCES

Devon Coastal Conservation Study, , 1978. Devon Conservation Forum, Exeter, 136pp.
EC. Directive on Bathing Water Quality 76/160.
Rivers Information, NRA, 1991.
South Wales Echo, 20 May, p.2, 1991.
The Good Beach Guide, 1991. Marine Conservation Society.UK.
Williams, A,T. and Lavalle, C, 1990. Coastal landscape evaluation and photography. Journal of Coastal Research, 6 (1),p 1011 -1020.

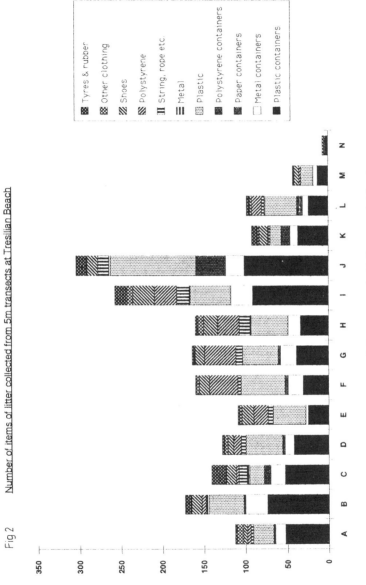

Fig 2 Number of items of litter collected from 5m transects at Tresilian Beach

AN APPROACH FOR THE REHABILITATION OF COASTAL WETLANDS AND ESTUARINE AREAS IN NORTHERN SPAIN

Francés, E.; Rivas, V. and Cendrero, A.
Div.Earth Sciences, DCITTYM. Fac. Sciences. Univ. of Cantabria. 39005 Santander, Spain.

ABSTRACT

During the last two centuries a large part of intertidal and wetland zones in the north coast of Spain have been subject to reclamation through enclosures, draining and filling, with the consequent environmental degradation and loss in biological productivity. Recent initiatives have been launched to reverse this trend and plans are being prepared for the rehabilitation of some of those degraded areas.

This work presents two examples of rehabilitation plans which have been designed using a methodology based on the identification, mapping and assessment of integrated homogeneous units. Both intertidal and permanently emerged zones were mapped and a series of units with similar characteristics with respect to landforms, materials, processes, biological assemblages and human features were represented. These integrated units were the basis for description, diagnosis and definition of management actions.

A series of environmental indicators, reflecting the potential for conservation and recovery of the units, as well as the technical feasibility of restoration actions, were defined and used to obtain a classification of such units according to their capability for rehabilitation. These indicators were also used to establish the most advisable kind of action in each unit, in order to restore the area. The final product obtained was a map showing the recommended types of use and the rehabilitation actions to be undertaken.

INTRODUCTION

Estuaries and coastal marshes are highly dynamic environments with ecosystems of high productivity and biodiversity, especially in the temperate regions (Odum, 1972; Ketchum, 1972; Margalef, 1974, Vélez, 1979; Miracle, 1981; Beanlands, 1983; Guilcher *et al.*, 1985; Canteras, 1987; Perez and Canteras, 1987); furthermore, because of the bigeographic situation of Cantabrian estuaries, at the southwestern edge of Europe, they represent a border zone for several European taxons, such as Galium arenarium or Medicago maritima, both species in danger of extinction within the Iberian Peninsula (Aseginolaza *et al.*, 1985). On the other hand, these environments are extremely fragile because of their limited area and easy accessibility, which result in an intense human use.

Despite these characteristics, up to recent times marshes and coastal wetlands were considered improductive areas and their reclamation was encouraged by most governments (Moss, 1980; Resh and Balling, 1983; Azurmendi, 1985; Cencini *et al.*, 1988; Ramón, 1988; Spilotro and Roccanova, 1990). Cantabrian estuaries are no exception and have been subject to reclamation since, at least, the sixteenth century and especially in the last two centuries (Rivas and Cendrero, 1991).

Nowadays, however, the coastal fringe is recognized by the population and national and regional administrations as one of the major conservation areas. This has brought about a change in attitude, and some degraded or partilly reclaimed estuarine and wetland areas have started to be restored in order to improve environmental conditions and to establish management plans which make conservation compatible with human activities.

This paper presents examples of two restoration plans in the Basque Country: the Iñurritza estuary, in Zarauz, and the Txingudi Bay in the Bidasoa estuary, both areas with a high degree of degradation and with a great social demand for recuperation.

METHODOLOGY AND RESULTS

The main objetive of the work carried out, in both cases, was to preserve presently existing valuable environments, to recover areas susceptible of re-incorporation to natural processes, and to improve the quality of the environment in adjacent urban areas.

In Zarauz, a diagnosis of the environmental quality and the restoration potential of the small estuary was made, followed by a zoning proposal for a management plan. In the Bidasoa estuary these initial stages were followed by the desing of specific technical projects to be undertaken, for the restoration of the area and its recovery for public use.

A similar method was followed in both cases. The first step was the identification, characterization and mapping of a series of homogeneous units, defined on the basis of a variety of biotic and abiotic features, such as surficial materials, landforms, processes, biological assemblages, soils, land uses, etc. (Cendrero and Díaz de Terán, 1987). Figure 1 shows an example of the kind of map obtained in the Bidasoa estuary.

The second step was of the evaluation of units taking into account their environmental quality and their potential for recovery. A diagnosis of the present situation was made. This included an assessment of environmental quality, present use, identification of impacts and conflicts, and evaluation of possible future uses.

The specific indicators used for assessment in the Zarauz area were:
-Proximity to climax, defined by the degree of evolution and/or alteration of the biotope and of the biological communities.
-Presence of certain bioindicators, such as the abundance of rare or fragile taxons or the condition of authocthonous plant communities.
-Degree of persistence of natural estuarine dynamics.
-Reversibility potential, defined by the technical posibility to restore natural conditions.
-Absence of land-use conflicts, taking into account land tenure and present land use.

All units were assessed using those indicators (table 1) and then classified in a rank of five terms, reflecting potential for preservation and restoration. Figure 2 shows the results of the application of the indicators described in the Zarauz area. The diagnosis described was the basis for the design of the management plan. The general criteria followed were: a) priority should be given to the preservation of units with active estuarine dynamics or in which such dynamics can be restored easily; b) "natural" or preservation areas should be surrounded, wherever possible, by a buffer zone of parks, to shelter them from the influence of adjacent urban areas and to improve the environment of the latter. The results of the application of these criteria is shown in figure 3.

In the Bidasoa estuary a similar procedure was followed, both for the identification and mapping of units and for their diagnosis. Proposals for zoning and management were made on the basis of the following criteria: a) priority sould be given to the protection of (scarce) units with well preserved ecosystem structure and natural dynamics; b) impacts from consolidated, irreversible types of land-use should be reduced, for instance by landscape protection measures; c) natural conditions should be recovered wherever technically feasible and ecologically advisable; d) the recreational and educational potential of the area should be developed. The distribution of activities and actions proposed is shown in figure 4.

In both areas, specific actions recommended for improving the condition of units to be preserved included: removal of fill material in estuarine areas where natural circulation could be restored; breaking-up of artificial levees, to increase the area of tidal influence; erradication of certain exotic, invader species, such as Baccharis halimifolia; reforestation of fringe areas with authochtonous species; strict control of activities in perservation areas; monitoring the evolution of plant and animal communities, after implantation of the recommended measures.

Areas less valuable from the ecological point of view, or where recovery of natural conditions is more difficult, should be devoted to public parks. These areas should fullfill the demand for "green space" in neighbouring urban areas, and shelter protected zones from their influence. The design of parks contemplates the creation of an irregular topography, establishment of abundant tree barriers of authocthonous species, utilisation of existing channels and water bodies, creation of a minimum of artificial structures, etc. Footpaths, observation and panoramic viewpoints should be established throughout the parks, in order to facilitate educational activities.

Finally, in areas where existing uses are consolidated, no specific actions were proposed, apart from the normal measures to reduce pollution impacts, be it from agriculture or from urban-industrial areas.

An example of the kind of technical proposals made for the restoration of one of the channels in Txingudi is shown in figure 5.

CONCLUSIONS

The approach presented here provides a simple and straight-forward method for the assessment of coastal areas where restoration is contemplated and for the design of zoning and management plans.

Initial descriptive maps (morphodynamic units or homogeneous environmental units) can be made on th basis of selected, relevant parameters, both biotic and abiotic. In a second, diagnosis step, the units represented in the former maps can be assessed, using a series of indicators and establishing ranks according to different combinations of such indicators. In a third step certain criteria are defined to derive prescriptive maps form diagnosis maps, with specific proposals for land use or for management actions. These can, in turn, be developed through the design of specific projects.

REFERENCES

ASEGINOLAZA, C., GOMEZ, D., LIZAUR, X., MONTSERRAT, G., MORANTE, G., SALAVERRIA, M.R., URIBE-ECHEVARRIA, P.M. y ABJANDRE, J.A. (1985): Catálogo florístico de Alava, Vizcaya y Guipúzcoa. Servicio Central de Publicaciones del Gobierno Vasco. Vitoria, 1149 p.

AZURMENDI PEREZ, L. (1985): Molinos de mar. Colegio Oficial de Arquitectos de Cantabria. S.A.U.R. Santander, 71 p.

BEANLANDS, G.E. (1983): Land use pressures on coastal estuaries in Atlantic Canada. Coastal Zone Mgt. Journal, 11 (1-2): 117-132.

CANTERAS, J.C.(1987): Condiciones ecológicas en los estuarios cantábricos. Seminario Internacional Sobre Problemas de Uso del Territorio y Manejo de Zonas Litorales. Council of Europe, Bilbao, 10p.

CENCINI, C., MARCHI, M., TORRESANI, S. and VARANI, L. (1988): The impact of tourism on Italian Deltaic coastlands: four case studies. Ocean and Shoreline Mgt., 11 (4-5): 353-374.

CENDRERO, A. and DIAZ de TERAN, J.R. (1987). The environmental map system of the University of Cantabria, Spain. In: Mineral resources extraction, environmental protection and land-use planning in the industrial and developing countries, eds. P. Arndt and G. Lüttig, E. Schweizerbartsche Verlagsbuchhandlung, Stuttgart: 149-181.

GUILCHER, A., PONCET, F., HALLEGOUET, B. y LE DEMEZET, M. (1985): Breton Coastal Wetlands: Reclamation, Fate, Management. Journal of Shoreline Mgt., 1: 51-75.

KETCHUM, B.H (ed.) (1972) The water's edge. Cultural problems of the coastal zone. MIT Press, Cambridge, Mass. 393 p.

MARGALEFF, R. (1974): Ecología. Omega. Barcelona, 951 p.

MIRACLE, M.R. (1981): Análisis ecológico del estado actual de degradación de los ecosistemas de agua dulce y salobre del Mediterráneo español. Coloquio Hispano-Francés sobre Espacios Litorales. M.A.P.A., Madrid, pp. 213-223.

MOSS, D. (1980): Historic changes in terminology for wetlands. Coastal Zone Mgt. Journal, 8 (3): 215-225.

ODUM, P.E. (1972): Ecología. Interamericana. México, 639 p.

PEREZ GARCIA, M.L. y CANTERAS, J.C. (1987): Aplicación del método de ^{14}C para la medida de la producción primaria en un estuario de Cantabria. Actas, XIII Reunión de Estudios Regionales. Santander: 147-156.

LLAMAS, R. (1988): Conflicts between wetland conservation and groundwater exploitation: two case histories in Spain. Environmental Geol. and Water Sci., 11 (3): 241-251.

RESH, V.H. and BALLING, S.S. (1983): Tidal circulation alteration for salt marsh mosquito control. Environmental Mgt., 7 (1): 79-84.

RIVAS, V. and CENDRERO, A. (1991): Use of natural and artificial accretion on the north coast of Spain; historical trends and assessment of some environmental and economic consquences. J. of Coastal Res., 7 (2): 491-507.

SPILOTRO, G. and ROCCANOVA, C. (1990): Sea level changes and ancient mapping of the Taranto area. Proc. 6th International I.A.E.G. Congress, Balkema, Rotterdam: 235-241.

VELEZ SOTO, F. (1979) Impactos sobre zonas húmedas naturales. Ministerio de Agricultura. Servicio de Publicaciones Agrarias. Monografías, 20. 29 p.

Table 1.- Diagnosis matrix

Units \ Indicators	Environmental quality			Potential for recovery		Sum
	Proximity to climax	Bioindicators	Natural dynamics	Reversibility	Land-use conflicts	
1.- Sandy beach	4	2	4	4	4	18
2.- Pebble beach	5	2	5	5	5	22
3.- Dune front	4	2	4	5	3	18
4.- Vegetated foredunes	4	5	4	5	4	22
5.- Altered dunes	3	4	2	4	2	15
6.- Vegetated backdunes	4	5	4	5	5	23
7.- Sandy intertidal area	5	2	5	5	5	22
8.- Silty intertidal area	4	5	5	5	5	24
9.- Isolated intertidal area	3	3	3	3	5	17
10.- Filled and cultivated former intertidal area	2	2	1	2	1	8
11.- Filled former intertidal area	2	3	1	2	2	10
12.- Built-up former intertidal area	1	1	1	1	1	5
13.- Partly built-up dunes	1	1	1	1	1	5
14.- Vegetated cliff	4	4	5	5	5	23
15.- Excavation	1	1	1	1	1	5
16.- Slope with campsite	2	1	2	2	1	8

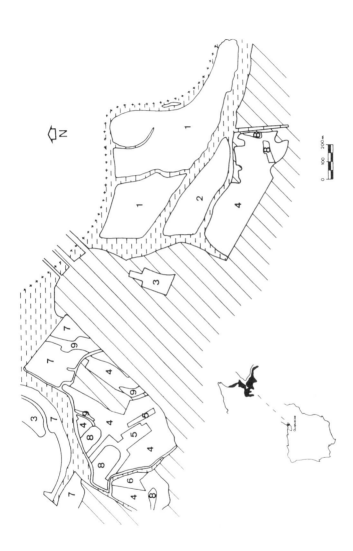

Figure 1.- Map of homogeneous units in Txingudi Bay, Bidasoa estuary. 1: Intertidal silty banks covered with halophytes; 2.- Drained silty banks with orchards; 3: Coastal wetlands with halophytes and reed; 4: Cultivated alluvial floodplain; 5: Floodplain with prairies; 6: Humid floodplain areas, with Juncus ssp. and Scirpus ssp.; 7: Supratidal sandy areas with vegetation; 8: Heterogeneous rubble; 9: Intertidal silty areas with Phragmites australis. Horizontal dashes: Water; Oblique lines: Existing built-up areas; Crosses: French-Spanish border.

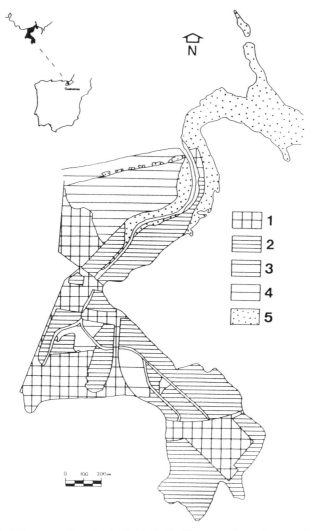

Figure 2.- Potential for preservation and restoration in the Iñurritza estuary, Zarauz. 5: very well-preserved untis, with the highest quality (values 21-25 in Table 1); 4: fairly well-preserved units (values 16-20) in which restoration actions are advisable and require practically no physical intervention; 3: moderately preserved untis (values 11-15) which can be recovered through technical actions, feasible under existing socio-economic conditions; 2: very altered units (values 6-10) which requiere complex technical actions for restoration; 1: Completely degraded units (value 5) which cannot be restored, due to physical or to legal constraints.

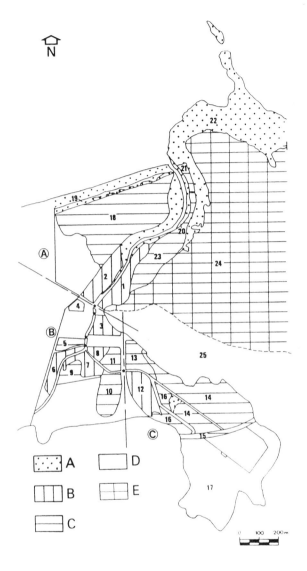

Figure 3.- Proposal for use in the Iñurritza estuary, Zarauz. A: preservation; B: restoration of natural conditions; C: creation of public parks; D: low-density urbanisation; E: landscape protection. Serial numbers for the identification of units are also shown.

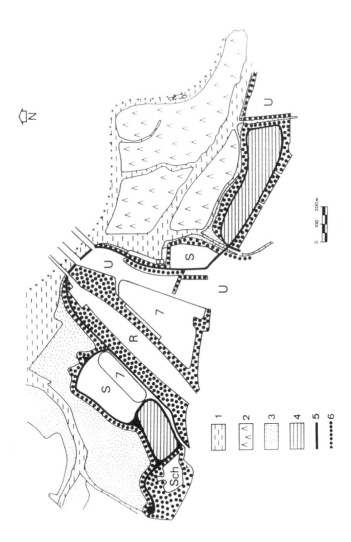

Figure 4.- Zoning and management actions proposed for the Txingudi Bay area, Bidasoa estuary. 1: Water; 2: Preservation and restoration of the natural systems; 3: Landscape rehabilitation of shore areas; 4: Urban parks; 5: Nature trails; 6: Tree barriers; 7: Parking; 8: Marina. Existing uses are indicated as follows: U: Urban; S: Sports grounds; R: Railway station; Sch: School

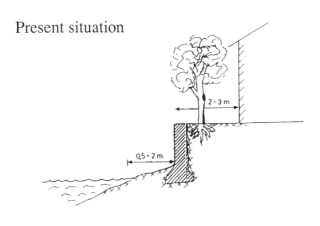

Present situation

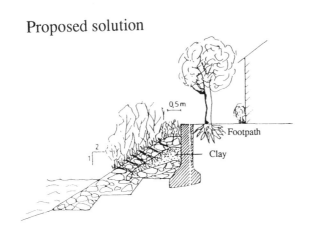

Proposed solution

Figure 5.- Technical proposal made for the restoration of channel margins in Txingudi Bay.

Planning and Design for the Leisure Craft Within a Traditional Fishing Harbour.

António Trigo Teixeira
Teresa Gamito

Hidrotécnica Portuguesa, Consulting Engineers
Apartado 5058 - 1702 Lisboa Codex - Portugal

ABSTRACT

The debate over the advantages and the disadvantages of the presence of the leisure and fishing boats within the same harbour has been going on for some time. It seems that two requirements should be met before the planning and the design work proceed towards a successful planning solution. First of all, the harbour must be large enough to accommodate the two fleets keeping them apart in separate mooring basins. Secondly, there must be enough grounds surrounding the basin where to create all the facilities for the leisure craft and for the fishing fleet.

The authors present in this paper the planning and design of a new mooring basin that will be created within a traditional fishing harbour in the north of Portugal. Within this harbour it is planned to create two hundred and sixty berths for the leisure craft ranging from 8m to 18m in length. Draft will range from 1.8m to 3.0m. The grounds surrounding the leisure basin constitute also a feature that is at least in importance to the water area containing the pontoons for berthing of the boats. Special attention was paid to the location of the port facilities and buildings for the following activities: administration, bathhouses and the boat storage sheds. In addition, space was allocated for the access road, walkways and parking areas for boats and automobiles. An open service yard, open dry storage for boats, fuelling activities and boat handling equipment was also considered in the design.

1. INTRODUCTION

The construction of the first facilities in the Póvoa de Varzim harbour dates back to the end of the last century.The harbour is located in the portuguese west coast near the Póvoa the Varzim village, in a densely populated area, not far from the Oporto city - Figure 1. The harbour is sheltered by two breakwaters which close a small bay in the Póvoa de Varzim coast where the harbour is situated.

Figure 1 - Map of Europe showing the location of the Póvoa de Varzim harbour on the portuguese coast.

Over the last couple of years the harbour has suffered major redevelopment with the construction of a new fishery on the northern area. The fishery handling equipment was renovated, an auction hall was built and the mooring capacity for fishing vessels was greatly increased. All the investment in recent years has been made to support the development of the traditional fishery activity.

In previous studies and in the harbour master plan [1], the possibility of building, within this harbour, some facilities for the leisure craft were considered. The idea is not new and has its roots on the potential of the harbour and the village to attract visitors. With the decision of constructing facilities for the leisure craft one will be making the area more attractive for the yacthmen. In addition, in the village there are a nautical club, the "Clube Naval Povoense" running a sailing school and being engaged in many other nautical activities. The small premises the club used to have in the harbour were completely deactivated due to the expansion of the fishery.

The central question that remained to be answered was how to accommodate within the same harbour the leisure and the fishing craft [2]. Considering the grounds available it was clear that the southern part of the harbour, still unaffected by the fishery, would be the one with the greatest potential to develop for the leisure craft. A decision was taken by the Municipality and the Port Authorities to go ahead with the planning for leisure in this area. The study was commissioned to HP - Hidrotécnica Portuguesa, Consulting Engineers in Lisbon.

The southern part of the port meets the two following requirements: keeps the fishing and leisure craft in separate mooring basins and has enough grounds inland to create the port facilities.

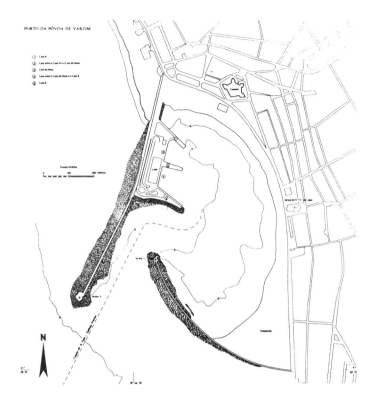

Figure 2 - General view of the Póvoa de Varzim harbour.

2. LAYOUT STUDIES.

The first phase of the planning programme was to study different layouts for the leisure area. Four layouts were under consideration and compared with each other according to the following criteria: berthing capacity, water area, inland area, sheltering conditions and cost. At this point of the planning it was essential to define an aim for the berthing capacity in the harbour. The composition of the leisure fleet and the dimensions of the berths are presented as a function of the class of the boat, Table I. This table was made after a statistical analysis was performed on the dimensions of a sample containing more than eight hundred boats.

The sample has mainly sailing boats and was taken from different sources. Boats registered on portuguese ports and commercially available boats. Due to the available area it was decided to keep the port capacity below the figure of 300 berths. The layouts studied are shown on Figure 3.

Table I - Classification of the leisure fleet.

Class	Boats Characteristic dimensions			Fender	Berth Dimensions		
	lenght (m)	width (m)	draught (m)	widths (m)	lenght (m)	width (m)	Water depth (m) (Datum)
I	6.0						
II	8.0	2.7	1.5	0.30	8.0	3.00	-1.60
III	10.0	3.3	1.8	0.30	10.0	3.60	-1.90
IV	12.0	3.8	2.3	0.30	12.0	4.10	-2.40
V	15.0	4.4	2.5	0.60	15.0	5.00	-2.60
VI	18.0	5.4	3.0	0.60	18.0	6.00	-3.00

The mooring capacity decreases from solution D to solution A. Because of the small natural depths it is necessary to dredge the mooring basin. The bottom is mainly sand over hard rock. The solution D is clearly the more favourable as it offers the largest grounds inland, 6.2ha, and is also the one which can accommodate more boats. In favour of this solution are the smaller volumes of material to be dredge in order to open the basin.

In what concerns the sheltering conditions all the solutions are almost equal. The exception is solution D that has two extra pontoons closer two the harbour entrance, and for that reason, might be under more sever wave conditions. All the measurements for the material to be dredge were made using a digital representation of the terrain.

3. GEOLOGICAL SURVEY.

According to what was mentioned previously there were indications from an old geological survey that the bottom of the basin was of very hard rock. Since the existing natural depths will not allow the navigation of boats having a draft greater than 1.0m it was necessary to carry out some dredging to open the mooring basin. To calculate more precisely the volume of dredge material to remove a geological survey was carried out in the area, from April to June 1991, based on a rectangular grid with a spacing of (30.0m X 20.0m).

On Figure 4 a plan of the drilling holes is shown along some typical cross sections. The dark area on these sections show the amount of the hard rock to be removed. This geological survey enabled the authors to modify slightly the layout of the solution in order to reduce dredging costs.

4. PLANNING AND DESIGN

The planning and design of the leisure area was made in order to provide good functionality for the port and to reduce to a minimum the interference with the movements of the fishing vessels. On the other hand, access roads were carefully located to ensure good links with the village. The access to the port will be made from the existing marginal road. From this point to the left the area is allocated for the civic centre. The centre will include a nautical club, swimming pool, restaurant, cafes, supermarket and parking. The architecture is done for the nautical club and restaurants to be overviewing the port.

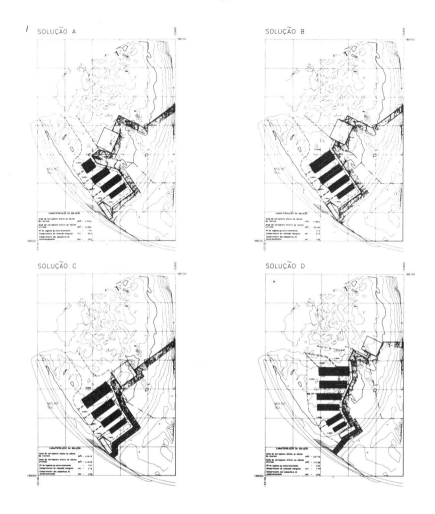

Figure 3 - Póvoa de Varzim. Layout studies.
        a) 119 berths b) 145 berths c) 151 berths
        d) 228 berths.

Close to the mooring basin will be located the port facilities directly related with the yachtsmen activity. The facilities are as follows: boat handling and storage equipment; toilets, showers and maintenance office; reception and enquiries, administration, finance office, meteorology and customs. Figure 5 depicts schematically a diagram showing the location of the facilities.

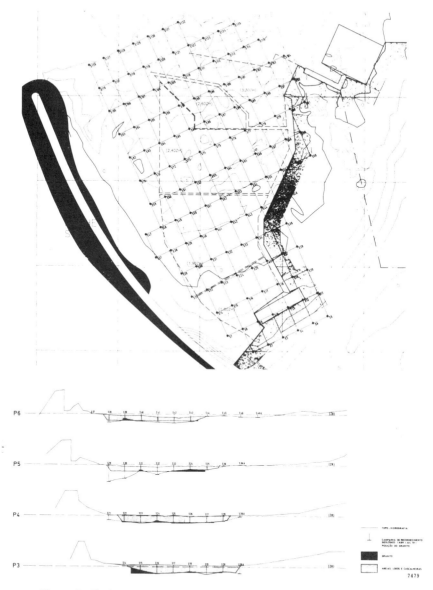

Figure 4 - Geological survey. Plan and typical cross section.

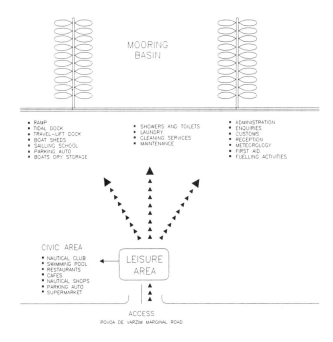

Figure 5 - Scheme showing the access road to the leisure area of the harbour and the location of the port facilities.

5. PORT FACILITIES

The definitive port layout with an "L" shaped mooring basin occupies an water area of some 4.5 ha. Berthing will be made along six floating pontoons, at different drafts, allowing the mooring of two hundred and sixty boats within a range from 8 to 18 m in length.
Another floating pontoon near the entrance of the leisure craft area and of the administration and reception services will be the Reception Quay for the visiting yachts as well as a pontoon for the fuelling activities.

The transference land-water of the boats will be possible through a ramp for the smaller boats or using the travel-lift dock. A 32 tonne hoist and boat mover will allow the lifting and the moving to and from the dry storage of the boats that are less than 5.10 m wide.
A tidal dock having a platform at 1.2 m above the lowest low water level will provide for cleansing, painting and small repairs.
These three facilities are located in the sheltered area of the harbour. Since the draft available is 1.60m, the access for the larger boats will be somehow conditioned.

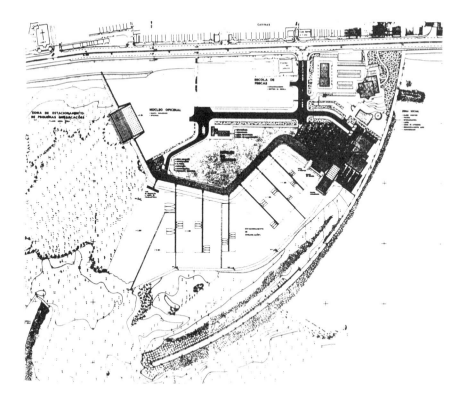

Figure 6 - Mooring Basin.

6. CONCLUSIONS.

The goal was to create some port facilities for the leisure craft within a traditional fishing harbour. The Póvoa de Varzim Harbour. Special attention was paid to the planning and design in order to reduce to a minimum the interferance with the fishery, which is located in the northern part of the harbour.

The mooring basin is based in six rows of floating pontoons, with a capacity for two hundred and fifty boats. On the grounds inland there will be some port facilities grouped into four areas: the civic area having restaurants, cafes, nautical club a swimming pool and shops; the boat handling area, including a ramp, dry storage, boat sheds, and a sailling school; the service area which includes, showers, toilets and a laundry; and finaly the administration area, including customs, finance office, reception, enquiries and meteo.

Figure 7 - General view of the Póvoa de Varzim Harbour.
Fishery (existing) and Leisure area (in project).

7. REFERENCES

[1] - Hidrotécnica Portuguesa. Porto da Póvoa de Varzim. Revisão do Plano Geral. Relatório Final. Julho de 1979.

[2] - Avantages et Inconvenients de la Presence Simultanee dans un port de Plaisance, des Navires de Peches et de Plaisance. AIPCN. Commision Internationale pour la navigation de Sport et de Plaisance. Question B2. 1971.

ENVIRONMENTAL DEVELOPMENT AMONG THE TENMILE CREEK COASTAL DUNES IN OREGON, USA

Olavi Heikkinen
Department of Geography, University of Oulu, SF-90570 Oulu, Finland

Abstract

The article describes and explains aeolian forms and processes and their relations to vegetational and hydrological conditions which have markedly changed since about 1930, mainly due to the introduction of the grass *Ammophila arenaria*. The dates and other information obtained by thermoluminiscence, radiocarbon and dendrochronological methods and from observations available in the literature help to provide quite a good understanding of environmental changes in the Tenmile Creek dune field in Oregon.

Introduction

Very spectacular coastal dunes are found along the Pacific coast from southern Washington through Oregon to northern California (Figs. 1 and 2). These Pacific Northwest coastal dunes have been studied comprehensively by Cooper (1958,1967). Kumler (1969) and Wiedemann et al. (1969), for instance, have described the plants and plant successions on them, and Wiedemann (1984, 1990) gives a recent review of a wide variety of physical and biotic aspects. Cooper (1958), Lund (1973), Hunter et al. (1983) and Wiedemann (1984) have discussed the Tenmile Creek dune area in greater detail.

The development of coastal dunes in the Pacific Northwest is favoured by certain climatic and geomorphological factors. The dunes are situated in the zone of westerly winds, and a cell of subtropical high pressure lies quite deep in the south in winter, with the Aleutian Low forming in the polar front over the Pacific, whereas a center of high air pressure lies over the cold continent. Under these pressure conditions the Oregon coast is subject to strong, moist south-westerly winds (Fig. 1). In summer, when the subtropical high pressure centre and the polar front have moved northwards and the air pressure is lower over the warmed continent, dry, cool winds blow towards the coastal dunes from the northwest (e.g. Muller & Oberlander 1984:88). These semiglobal wind systems, plus more local land-sea breezes in summer, keep the sand in motion, and mainly dictate the aeolian forms and processes in the area.

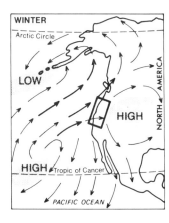
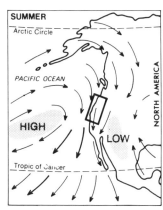

Figure 1. Air pressure and wind conditions prevailing in the Pacific Northwest of North America.

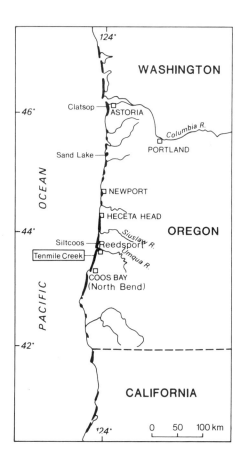

Figure 2. Coastal dunes (black) in the Pacific Northwest. The name of the area studied here, Tenmile Creek, is highlighted.

Old wave-cut terraces formed by ancient sea erosion serve as a base for the coastal dunes. Older dunes accumulated on marine terraces during the Pleistocene epoch and are now represented by yellow-brown, weathered and badly eroded forms which lie well above the present sea level and extend further inland in places than the recent "Holocene or Flandrian dunes" do (Cooper 1967:7; Lund 1973; Wiedemann 1984:8). There has been no space for dunes to develop in the streches where steep cliffs or high rocky promontories front upon the ocean and constitute the coastline.

Thus, strong winds from the open sea and the existence of gently sloping terraces fronting upon quite a straight shoreline have enabled the development of prominent coastal dune fields in the Pacific Northwest. The most impressive dune region is located in Oregon between Heceta Head and Coos Bay. This is 86 kilometres long and 2 to 4 kilometres wide and dissected only by the Siuslaw, Siltcoos, Umpqua, Tenmile and Coos rivers. The present study area is a part of this dune field (Figs. 2 and 5).

Visible human influence on the coastal dunes began only in the mid-1800s, when the first farms and towns were established. Since then the vegetation has changed rapidly (Cooper 1958; Lund 1973; Beaulieu & Hughes 1975; Wiedemann et al. 1969; Hunter et al. 1983; Wiedemann 1984, 1990).

At the beginning, human intervention caused erosion of the sand dunes, which led to the planting of vegetation for stabilization purposes. The most notorious plant species used for this purpose was *Ammophila arenaria*, the European beachgrass, which was introduced to California in 1896 and had already spread or been planted all along the Oregon coast by the early 1900s (Wiedemann et al. 1969; Wiedemann 1984). *Ammophila arenaria* is now naturalized and occurs wild almost everywhere behind the high tide line (Franklin & Dyrness 1973:291).

Ammophila arenaria has occupied the habitats of numerous native plants and is also principally responsible for the development of a mighty foredune ridge immediately beyond the high tide line. Such high foredunes were fairly uncommon prior to 1930 (Wiedemann et al. 1969). The large foredune prevents sand from moving further inland and impedes percolation of the inland groundwater into the ocean. This changed situation has resulted in expanding deflation and forestation behind the foredune.

Aims and methods

The purpose is (1) to characterize local climatic features, aeolian forms and resulting vegetational conditions, (2) to obtain more accurate information about the age of the

"Pleistocene dune sand", (3) to trace the speed of aeolian processes and other environmental changes, and (4) to predict the future development in the Tenmile Creek area.

Two samples of dune material were taken for thermoluminescence (TL) dating, one from the grey "Holocene dune sand" and the other from the yellow-brown "Pleistocene dune sand" (Fig. 5). The TL dates give estimated ages for the accumulation of the sand deposits. The TL determinations were carried out at the Nordic laboratory in Roskilde, Denmark. The TL method is discussed by Mejdahl (1988).

Five ^{14}C samples were collected from snags or partly rotten exposed stumps for dating at the Radicarbon Dating Laboratory, University of Helsinki, in order to explain past aeolian processes. Tree-ring samples were extracted with an increment borer from almost 100 recently established conifers to date the forestation process which is still continuing in the area.

Elevations of numerous dune-field points and the levels of 12 interdune pools were measured with an altimeter to obtain a general view of the topography and hydrological features of the area. A geomorphological map with some vegetational information (Fig. 5) is based on fieldwork and aerial photographs (1:24 000) from the year 1982. Further fieldwork was carried out in the summers of 1984, 1986 and 1989.

Climatic characteristics

The area has a maritime climate. Within the Köppen system of climatic classification it falls into type Csb (Dierke Weltatlas 1974:176), i.e. a Mediterranean climate with a dry, cool subtropical summer. The winter winds from the SW are often so stormy that, although they are usually accompanied by rain, they transport more sand than do the relatively dry summer winds blowing approximately from the NW (Hunter et al. 1983).

The average annual temperature at Reedsport (18 m a.s.l.) to the north (Fig. 2) in the period 1951-80 was 11.2 °C, the corresponding temperature in August, the warmest month of the year, being 16.1 °C and that in January, the coldest month, 6.4 °C (Fig. 3). The figures show that the temperature regime is by no means extreme.

Precipitation shows considerable seasonal fluctuations, with the highest rates in winter and the lowest in summer (Fig. 3). Annual mean precipitation at Reedsport in 1950-80 was 1940 mm. The wettest month was January, with 337 mm, and the driest July, with 12 mm. Only 4.5 % of the yearly rain falls in the three driest months (June, July, August) compared with 48.4 % in the three wettest months (November, December, January).

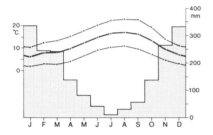

Figure 3. Monthly mean temperatures, means of maximum and minimum temperatures and total monthly precipitation at Reedsport, alt. 18 m, for the period 1951-80.

The wind rose diagram (Fig. 4) gives wind directions at Newport (see Fig. 2) in sectors of 15° and wind speeds in three classes. The diagram shows that the strongest winds blow from the S-SW. Even though most of these winter storms bring heavy rain, they are able to deflate and transport sand (Hunter 1980), the quantity of which would, of course, be larger if the sand were dry. In addition to winter winds, only the N-NW summer winds are sufficiently intensive to move and drift sand in any great amounts. The forms of the Tenmile Creek dune field are governed by quite a regular wind regime, and their appearance changes markedly from season to season.

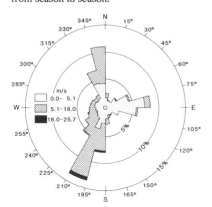

Figure 4. Wind rose for Newport, Oregon, based on data from June 1, 1973 to May 31, 1974. Redrawn after Hunter et al. 1983.

Aeolian forms and vegetation

The Tenmile Creek dune field is an active mosaic, a result of the interplay of sand, wind, water, plants - and more recently also of Man. The terminology for coastal dune forms applied here (Fig. 5) follows closely the usage of Cooper (1958), Hunter et al. (1983) and Wiedemann (1984).

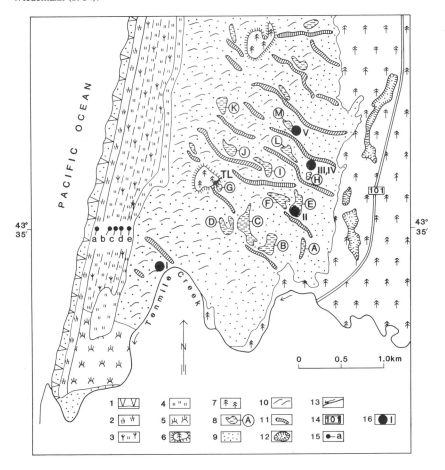

Figure 5. Geomorphological interpretation of the Tenmile Creek dune area, based on fieldwork and air photographs (1:24 000) from 1982. 1 - beach (littoral zone), 2 - grass-covered foredune, 3 - forested deflation plain, 4 - treeless deflation plain, 5 - grass-covered sand hummocks, 6 - forest remnant (TL refers to thermoluminescence samples), 7 - old forest on stabilized dunes, 8 - swale (dune pond), 9 - open sand surface, 10 - transverse dunes, 11 - oblique dune, 12 - blowout, 13 - creek, 14 - road number, 15 - site for sampling of recently established trees, 16 - radiocarbon - dated stump.

The beach or littoral zone, which extends from the extreme low water level up to the extreme high water level is about 100 metres wide in the Tenmile Creek area. A grass-bound foredune ridge rises up steeply to about 10 metres just landward of the beach (Fig. 6). This massive, continuous foredune has accumulated during the present century due to the introduction of *Ammophila arenaria*, which binds sand effectively and is highly tolerant of burial in sand. Nowadays the foredune is so high that the wind can no longer carry sand over it from the beach to the interior and there are some wind-battered clusters of lodgepole pine, *Pinus contorta*, growing on its crest.

On the leeside and beyond the recent foredune, especially in the south (Fig. 5), one can see rows of grass-covered sand hummocks and driftwood logs between them. The sand hummocks represent a zone of parallel smaller foredunes and dune mounds that existed on the shore before the development of the one huge foredune. In those days the storm waves were able to hurl driftwood far inland.

The deflation plain lies east of the foredune and is attributable to its development. The wind, which cannot carry sand over the foredune, erodes the sand surface close to the groundwater level behind it, whereupon the eroded moist sand surface quickly becomes covered by vegetation. The whole process is a new phenomenon, for no deflation plain or continuous foredune is present in photographs from 1939. The deflation plain expanded eastwards by about 5.7 m a year in the period 1954-75 (Hunter et al. 1983).

Ammophila arenaria acts as a pioneer species in occupying the deflated sand plain, and once the sand surface is stabilized it slowly dies out and gives way to herbaceous communities, which in turn are followed by certain shrub communities and tree species (Wiedemann 1984). The zone which stabilized first houses thickets of *Pinus contorta* and sitka spruce, *Picea sitchensis*. The wind-tolerant *Pinus contorta* is commonly the pioneer tree species at such sites.

Figure 6. Foredune cut through by a bulldozer. The person standing in front of the foredune gives an idea of its dimensions. Siltcoos, 1984. All photographs by the author.

East of the deflation plain is the reducing open sand surface with its various dune forms. The transverse dunes are 1 to 6 m high and some hundred metres long, with a gentle windward slope and a very steep lee slope (about 33°). They are particularly prominent in summer because most of them are accumulated by summer winds blowing from the N-NW. During the winter they are flattened and partly obliterated by the strong S-SW winds.

The oblique dunes are relatively permanent ridges about one kilometre long and some 20 to 50 m high. It is thought nowadays (Hunter et al. 1983) that the winter winds give them their basic form and orientation and move them towards the N-NE, their average speed between the years 1939 and 1975 having been approx. 4.4 m a year. The dry N-NW summer winds then modify the upper parts of the oblique dunes.

Figure 7. This precipitation dune at Tenmile Creek is currently moving eastwards at a rate of more than 1 m a year, burying the trees deep in sand 1986.

Swales or dune ponds are wet troughs between large dune ridges. They are commonly filled with standing water, and in some cases with ephemeral ponds.

The precipitation or retention ridges found on the eastern edge of the dune area are asymmetric dune ridges with a steep lee side 10 to 20 m high invading the forested land (Fig. 7). The "precipitation" of sand takes place as the winds carrying it lose their power at the edge of the forest. According to Cooper (1958:115) these ridges migrate eastwards at a rate of no more than 1.6 m a year. There are no distinct parabolic dunes in the Tenmile Creek area.

Most of the stabilized dune forms between the open dune field and highway no. 101 are forested (Fig. 5). There are three large blowouts in the forested dune zone.

There are two "forest remnants" in the area, mainly representing old "Pleistocene" dunes and supporting an old growth of forest. There is an exposure on the SE side of the more southerly forest remnant. Two thermoluminescence samples were taken here, one from recent grey sand deposits and the other from old weathered deposit (Fig. 8).

Figure 8. Ephemeral dune pond (G in Fig. 5) in the foreground. In the background one can see an oblique dune on the left and the southerly forest remnant on the right. The thermoluminescence samples (Tenmile 1 and 3) were taken 1 m into the face of the section in 1986.

Ages and processes

The following thermoluminescence dates were obtained from the sand samples (Fig. 8):

Tenmile 1   1060 +/- 100 years (R-863203)
Tenmile 3   18 300 +/- 1500 years (R-863204)

As expected, the younger, grey dune material had accumulated during the Holocene, as shown by the date for the Tenmile 1 sample, and although there are many possible sources of error attached to thermoluminescence dating, the yellow-brown sand forming the core of the forest remnant, represented by the Tenmile 3 sample, was deposited towards the end of the Pleistocene. Thus the earlier broad notions of the age of the ancient dune formation are seen to be in principle correct. It should also be mentioned that Wiedemann (1990) obtained a thermoluminescence date of 11 200 years for a sample taken from the surface of the yellow-brown sand at Sand Lake (see Fig. 2).

Samles for ^{14}C dating were taken from five tree stumps that had been exposed from beneath the sand in depressions between large oblique dunes in order to define the rate of movement of the dunes (Fig. 5). The youngest tree rings were not represented in the samples, as the outermost layers of the wood on the stumps concerned had rotted away.

Table 1. Conventional radiocarbon dates for five exposed stumps located as shown in Fig. 5.

Stump	Height (m a.s.l.)	^{14}C age
I	8	390 +/-100 BP (Hel-2049)
II	20	180 +/-100 BP (Hel-2050)
III	33	90 +/-100 BP (Hel-2051)
IV	29	170 +/-100 BP (Hel-2052)
V	27	570 +/-130 BP (Hel-2053)

The resulting radiocarbon dates are relatively recent, and after conversion to calendar years by reference to a set of calibration curves (e.g. Stuiver & Kra 1986), taking account of the standard deviations, enable the following conclusions to be reached. The material from stumps II, III and IV is very recent and could certainly not date from earlier than the mid-17th century, while stump I must have grown around 1420-1640 (Fig. 9) and stump V around 1280-1420.

Figure 9. Exposed stumps in the southern part of the Tenmile Creek dune field. Sample I (see Fig. 5 and Table 1) for radiocarbon dating was taken here. 1986.

The trees concerned must have once grown in inter-dune depressions and have been buried beneath the oblique dunes as these moved northeastwards. As the movement of the dunes then continued they will have eventually been exposed again from beneath the rear edges. Since the oblique dunes have been moving at a rate of 4 m a year in recent times and the interval between them is less than half a kilometre, one could conclude that it would have taken just over 100 years for the next inter-dune depression to reach the same site. This means that stumps II, III and IV could have been buried beneath a dune only once but stumps I and V could have been covered more than once, although it may well be too presumptuous to even put forward such calculations.

As mentioned above, the crest of the foredune and the deflation surface behind it, like the inter-dune depressions, have rapidly become forested in recent decades, and dendrochronological cores were taken from the largest trees at various sites in order to date this process. The oldest identified tree may then be taken as giving a minimum age for the onset of forestation at each site.

As may be seen in Tables 2 and 3, forestation beyond the actual forest remnants evidently began only after 1950, the pioneer tree species on the deflation plain being *Pinus contorta*, which is capable of withstanding wind and a poor soil, followed relatively soon by *Picea sitchensis* (Table 2). The swales in the dune area are invaded only by *Pinus contorta*, which gained root first of all in the dampest, most low-lying of them, to be found in the southern part of the area. There are many swales, however, which remain treeless for the moment (Table 3).

Table 2. Years of establishment of the oldest trees on the foredune and deflation plain (sites a-e in Fig. 5).

Site	Year	Tree species
a	1952	*Pinus contorta*
b	1961 (1965)	*Pinus contorta (Picea sitchensis)*
c	1964 (1964)	*Pinus contorta (Picea sitchensis)*
d	1968 (1968)	*Pinus contorta (Picea sitchensis)*
e	1976 (1973)	*Pinus contorta (Picea sitchensis)*

Table 3. Years of establishment of the oldest trees and water levels in 12 swales, as measured on 27th July 1984 (swales A-M in Fig. 5).

Swale	Water level (m)	Establishment year	Tree species
A	13	1966	*Pinus contorta*
B	14	1950	*Pinus contorta*
C	15	1967	*Pinus contorta*
D	14	1947	*Pinus contorta*
E	18	1954	*Pinus contorta*
F	18	1957	*Pinus contorta*
G	20	-	-
H	20	-	-
I	21	-	-
J	21	-	-
K	21	-	-
L	22	-	-
M	27	-	-

The developement in the future

Without future human intervention, both the forested coastal deflation plain and the eastern edge of the open sand field will migrate to the east. As the expansion rate of the deflation plain is much faster than that of the eastern border, the active dune field at Tenmile Creek will diminish in size. As the trees on the deflation plain and in the swales grow to their full height, they will serve as a very effective wind barrier, encouraging the establishment of vegetation in downwind areas (Hunter et al. 1983).

The "sand starvation" caused by the huge foredune may prevent the development of new oblique dunes (Wiedemann 1984:27), while the present oblique dunes will probably advance to the N-NE in their present order, ultimately to merge with the precipitation dunes. This would also be one step towards the disappearance of the open sand field at Tenmile Creek. It is also possible that the natural elements, particularly the winds, may erode and break down the "mature" foredune before long. This would again enhance the supply of sand from the beach and start new aeolian activity.

References

Beaulieu, J.D. & P.W. Hughes (1975). Environmental geology of western Coos and Douglas Counties, Oregon. *Oregon Department of Geology and Mineral Industries Bulletin* 87. 148 pp.

Cooper, W.S. (1958). Coastal sand dunes of Oregon and Washington. *Geological Society of America. Memoir* 72. 169 pp.

Cooper, W.S. (1967). Coastal sand dunes of California. *Geological Society of America. Memoir* 104. 131 pp.

Dierke Weltatlas (1974). Westermann Schulbuchverlag GmbH, Braun-schweiz.

Franklin, J.F. & C.T. Dyrness (1973). *Natural vegetation of Oregon and Washington*. USDA Forest Service General Technical Report PNW-8, Portland, Oregon. 417 pp.

Hunter, R.E. (1980). Quasi-planar adhesion stratification - An eolian structure formed in wet sand. *Journal of Sedimentary Petrology* 50: 263-266.

Hunter, R.E., B.M. Richmond & T.R. Alpha (1983). Storm-controlled oblique dunes of the Oregon coast. *Geological Society of America Bulletin* 94:1450-1465.

Kumler, M.L. (1969). Plant succession on the sand dunes of the Oregon coast. *Ecology* 50:695-704.

Lund, E.H. (1973). Oregon coastal dunes between Coos Bay and Sea Lion Point. *Ore Bin* 35:73-92.

Mejdahl, V. (1988). *The Nordic laboratory for thermoluminescence dating*. Riso National Laboratory, DK-4000 Roskilde, Denmark. 15 p.

Muller, R.A. & T.M. Oberlander (1984). *Physical geography today*. Third edition. Random House, New York. 591 pp.

Stuiver, M. & R. Kra (ed.)(1986). Calibration issue. *Radiocarbon* 28, 2B:805-1030.

Wiedemann, A.M., L.J. Dennis & F.H. Smith (1969). *Plants of the Oregon coastal dunes*. Oregon State University Bookstores Inc., Corvallis, Oregon. 117 pp.

Wiedemann, A.M. (1984). *The ecology of Pacific Northwest coastal sand dunes: a community profile*. U.S. Fish and Wildlife Service. FWS/OBS-84/04. 130 pp.

Wiedemann, A.M. (1990). The coastal parabola dune system at Sand Lake, Tillamook County, Oregon, U.S.A. In: *Proceedings Canadian Symposium on Coastal Sand Dunes* 1990, 171-194.

FLUIDIZATION AND BEACHFACE DEWATERING: RECENT PROGRESS AND CASE HISTORIES

PARKS, J. President,
DYNEQS Ltd., PO Box 2043, Tampa FL 33601 USA

Abstract

Fluidization is a process for maintaining navigability of marina and small harbor entrances and of sand bypassing from a permanent buried system of perforated pipes and pumps.

Beachface dewatering is the opposite process, of pumping water out of a shore-parallel buried drain pipe, to enhance beach accretion and stability.

Improvements in means for burying pipe in saturated non-cohesive sands have made both of these new technology methods economically attractive with advantageous cost-benefit ratios.

Fluidization works by pumping excess water into the sediment producing a fluid mixture of sand and water that can be pumped for appreciable distances, leaving a channel behind. Beachface dewatering works by removing water from the tidally-elevated vadose saturated zone beneath the beach swash zone, allowing part of each wave runup to soak into the beach, thus reducing the volume and velocity of backswash and leaving additional sand on the active beachface.

Much of beach erosion occurs during the few days per year of significant storm activity. "Standard" beachface dewatering offers some protection against storm erosion by providing a wider base beach. However, storm surges allow waves to attack at the foot of the dunes (if present), and create a temporary "perched" water table on the upper beach. Another drain-pipe can be placed well above normal ground water level at the dune foot. With connection to the main drain and to the normal pump system by gravity, activating the upper drainpipe during storm surges retards storm beach erosion.

Fluidized sand bypassing can mitigate beach erosion caused by improved inlets. Beachface dewatering may retain a wider nourished beach for many years. A wider and higher beach may protect against sea cliff erosion.

INTRODUCTION

Shore erosion control is facing a dilemma (defined as a choice between equally unsatisfactory alternatives). Orrin Pilkey says we have three choices: (1) do nothing and let the shoreline retreat; (2) armor the coastline with "hard" structures such as seawalls, breakwaters, groins and jetties; or (3) nourish -- and renourish -- and renourish forever. Many states have essentially outlawed Option 2 except for dire emergencies. Option 3 is quite expensive, even if the sand is readily available. And Option 1 is unthinkable except in undeveloped areas.

There are new technologies coming along that offer some hope for alleviating this situation. Two "soft" solutions will be described in this paper: both are continuous "active" processes, rather than "passive" structures or applications of new sand every few years (Weisman et al, 1982; Parks, 1989).

DESCRIPTION OF NEW METHODS

Fluidized sand bypassing is a method for bypassing sand around an inlet, using fluidizers to feed a fixed sand-transfer pump (Figure 1). The fluidized sand above the pipe will flow by gravity downslope to the "cone of depression" around the mid-length jet-eductor pump (Parks, 1989; 1991).

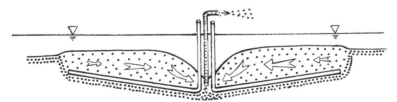

Figure 1 - Longitudinal section of fluidization

The fluidizer pipes can be placed within the navigable channel, using the channel as a "sand trap" (Figure 2). While fluidized, in a 50:50 sand:water slurry condition, the sand can be pumped some distance to a downdrift eroded beach. This system serves a dual purpose: maintaining navigable depths in the channel; and nourishing the downdrift beach (Parks, 1990).

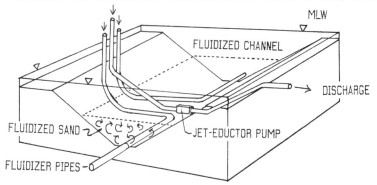

Figure 2 - Perspective sketch of fluidized sand bypassing

Beachface dewatering is the generic term for actively pumping down the local sub-beach watertable, to provide space for percolation of part of each wave uprush, so that the volume and velocity of backswash is reduced, thus enhancing accretion and retention of sand on the beach (**Figure 3**). In effect, the normal summertime accretionary process is maintained year round, and storm erosion effects are reduced.

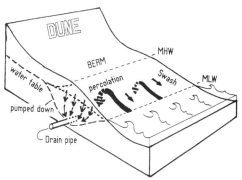

Figure 3 - Perspective view of beach

For long stretches of beach, 300 m modules of drains may be separated by longer untreated zones that will benefit from sand eroded from the bulges seaward of the drains (Fig. 4).

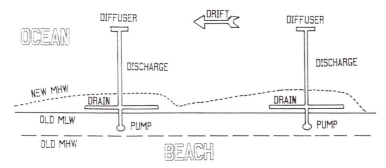

Figure 4 - Map View of beach

OTHER APPLICATIONS OF NEW TECHNOLOGIES

FLUIDIZED SAND BYPASSING

In addition to maintaining navigable depths in a boat channel, fluidization can be used to enhance the bypassing of sand around an inlet, using fluidized trenches as sand traps updrift of the updrift jetty. Sand generally accumulates against the updrift jetty in large volume, thus starving downdrift beaches. When sand begins to "spill around" the outer end of the updrift jetty into the channel, there is often political pressure to extend the jetty, which can only result in trapping more sand. A better solution is to move the sand to the downdrift side, but most fixed bypass pumps have been inadequate, and mobile pumps (Indian River Inlet, Delaware) or multiple pumps (Nerang River, Australia) have been quite expensive solutions. Using fluidizers to "feed" a fixed sand transfer pump seems a feasible method, provided costs can be brought down from that of the Experimental Sand Bypass System at Oceanside, California, under the Los Angeles District of the U.S. Army Corps of Engineers.

HOLDING A BEACH NOURISHMENT IN PLACE FOR MANY YEARS

The American prototype of beachface dewatering at Sailfish Point, near Stuart on the East Coast of Florida, has been in operation since September 1988. It has been successful in stabilizing the beach (damping out summer-winter beach width oscillations) by comparison with untreated updrift and downdrift control zones, without detectable adverse downdrift effects (Terchunian, 1990). The updrift control zone has continued the historical rate of retreat (1.5 m per year), while the treated zone showed some slight advance.

Most artificially nourished beaches require re-nourishment after a few years. Pilkey (1989) estimates that beach life span (expressed as the time to lose 50 percent or more of the "artificial") is less than 2 years for 40 percent of U.S. East Coast barrier island replenished beaches, and between 2 and 5 years for another 48 percent.

An optimum arrangement might combine artificial beach nourishment with beachface dewatering to provide a wider beach that could last for many years before needing renourishment. One option would be to do the initial nourishment to only about one-half the usually over-designed added width, and then install an improved beachface dewatering system along the seaward margin of the nourished beach with the savings from the narrower nourishment. Annual operating costs for the power for the pumps and pump maintenance or replacement will be a small fraction of the cost of a renourishment, so that over a 30 year period, cumulative costs would be less than half that to be expected from periodic renourishments.

EFFECT OF WIDER BEACH ON CLIFF EROSION

Erosion and retreat of sea cliffs is the result of a combination of several processes; including undercutting of the toe of the cliff by wave-induced abrasion and solution, and groundwater-induced slumping of cliff materials (Palmer, 1973; Leatherman, 1986).

"The levels of storm tidal surges and wave activity are probably the most important factors accounting for coastal land loss on the northern Chesapeake Bay in Maryland. At least three to four times each year, storm waves overtop the beaches that generally lie at the bases of bluffs and high banks" (Pilkey & Zabawa, 1989).

Some workers have suggested that the beach at the foot of the cliff does not materially influence cliff erosion rates, as aerial photos from previous years show that the width of the beach has remained constant during many years of cliff and shoreline retreat. However, it seems reasonable to expect that increasing the high-tide width of a beach and raising the elevation of the berm, such as would be attained by nourishing the beach to a new profile, could effectively retard the undercutting of the cliff (Figure 5).

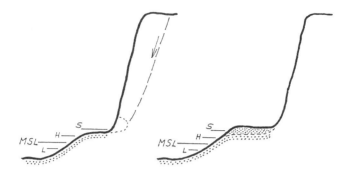

Figure 5 - Cross-sections of sea cliff and beach
(MSL = mean sea level; H = high tide line;
L - low tide line; S = storm surge line)

PROTECTING DUNES AGAINST STORM SURGE EROSION

In many parts of the world, most beach erosion and shoreline retreat occurs during the relatively few days per year of significant storm activity. "Standard" beachface dewatering offers some protection against storm erosion by providing a wider base beach. However, storm surges cause a temporary local rise in sea level, allowing storm waves to attack the "loose" sand at the foot of the dunes (if present). The higher sea level creates a temporary "perched" water table on the upper beach, as the permeability of the upper beach wind-blown sand is not great enough to drain away the large volumes of added storm water.

A possible solution is to emplace another one or more drain-pipes well above normal ground water level at the dune foot (Figure 6). These are connected, via a shut-off valve, to the normal pump system by gravity. Activating the upper drainpipe during storm surges will retard storm beach erosion.

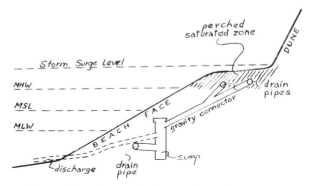

Figure 6 - Storm drain at dune foot

EXAMPLES AND DEMONSTRATIONS

One locale that combines many of the features discussed above will be the site of a demonstration of the improved new technologies beginning in the fall of 1992 (Figure 7).

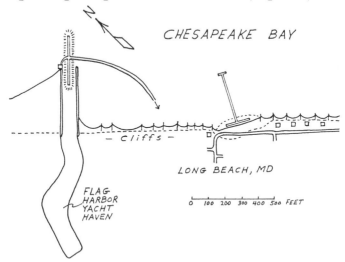

Figure 7 - Map of Flag Harbor Marina and Long Beach, MD

Jetties around an improved harbor entrance (which must be dredged annually) have cut off the normal north-to-south longshore drift causing downdrift beach erosion along a coast with cliffs and bluffs. A beachface dewatering system will be emplaced at the "public" beach, and fluidized sand bypass system will be installed later in the harbor entrance.

CONCLUSIONS

1. Two new technologies, fluidization and sub-beach watertable lowering, have potential for treating two long-standing problems of the coast: shoals in boat channels, and beach erosion.

2. These are active processes, requiring pumping of water to or from perforated pipes, rather than passive structures such as seawalls, groins, jetties and breakwaters, or episodic actions with intervening periods of passive response, such as channel dredging or conventional beach nourishment.

3. Environmentally, these new technologies are much to be preferred over conventional methods, as they cause little turbidity or disruption of marine bottom life.

4. The "bottom line" is that over a reasonable lifetime for a project (20 to 30 years or longer), these new technologies appear to be quite cost effective, and cumulative costs are on the order of one-half of those of conventional methods.

REFERENCES

LEATHERMAN, S.P., 1986, Cliff stability along western Chesapeake Bay, Maryland. Marine Technology Society Journal, 20(2), 28-36.

PALMER, H.D., 1973, Shoreline erosion in Upper Chesapeake Bay: The role of groundwater. Shore & Beach, 41(2), 19-22.

PARKS, Jim, 1991. Implementing Cost-Effective Environmentally-Safe Well-Spaced "Harbor of Refuge" Marinas. Proc. 2nd

Intl. Symp. on Coastal Ocean Space Utilization (COSU II), Long Beach CA, 2-4 April 1991, 9 pp.

PARKS, J.M., 1991b. New "Dredging" Technology for Inlets and Beaches: Move Sand to the Pump. Proc. Coastal Sediments 91, Specialty Conf/WR Div/ASCE, Seattle, WA, June 25-27, 1991, p. 1943-1954.

PARKS, Jim, 1991c. Pumping In and Pumping Out: Case Histories of Fluidized Sand Bypassing for Channels and Beachface Dewatering for Beaches. Proc. 7th Symp. on Coastal & Ocean Mngmnt, ASCE, Long Beach CA July 8-12, 1991, 193-203.

PARKS, Jim, 1991d. Fluidized Sand Bypassing for Inlet Channel Maintenance. Proc. 3rd Intl. Conf. on Coastal & Port Engineering in Developing Countries (COPEDEC III), Mombasa, Kenya, 16-20 September, 1991, 9 pp.

PARKS, J.M. 1990. Fluidized Sand Bypassing: Cost Effective and Environmentally Acceptable Alternative to Dredging of Marina Entrances. Proc. Intl. Marina Inst., Sept 5-7, 1990, Washington, DC, 10 p.

PARKS, J.M. 1989a. Fluidization: Channel Maintenance & Sand Bypassing. Proc. 6th Symp. Coast. & Ocean Mgmnt, ASCE, Coastal Zone 89, Charleston, SC, pp. 1711-1723.

PARKS, J.M., 1989b. Beachface Dewatering: A New Approach to Beach Stabilization. The Compass, v 66(2), 65-72.

PILKEY, O.H., & C. ZABAWA, 1989, Shoreline erosion in the upper Chesapeake Bay (IGC Field Trip T233). In Coastal and Marine Geology of the United States, 28th Int. Geol. Congr, American Geophysical Union, pp. T233:1-13.

TERCHUNIAN, A.V., 1990. Performance of Beachface Dewatering: The STABEACH System at Sailfish Point (Stuart), Florida: Proc 1990 Natl Conf Beach Pres Tech, Beaches: Lessons of Hurricane Hugo, St. Petersburg, FL, 16 Feb 1990, 185-201

VESTERBY, H., 1991. Coastal Drain System: A New Approach to Coastal Restoration, Proc. GEOCOAST 91, 3-6 Sept 1991, Yokohama, Japan, 651-654.

WEISMAN, R.N., COLLINS, A.G., and PARKS, J.M. 1982. Maintaining Tidal Inlets by Fluidization. J. Wtrwy. Port, Coast. and Ocean. Engr., ASCE, 108 (ww4) 526-538.

Chemical and biological characteristics of different water masses in the German Bight.

Dieter Gerdes[1] and Karl-J. Hesse[2]

Alfred -Wegener- Institut für Polar- und Meeresforschung
2850 Bremerhaven, FRG [1]

Forschungs- und Technologiezentrum Westküste
2242 Büsum, FRG[2]

Abstract: Especially in shallow sea areas with rather complicated hydrographic structures like the German Bight, actual distributions of plankton organisms are supposed to be very heterogeneous, because the patterns result from the physical processes to a large extend. This article shows that nevertheless the German Bight´s pelagic system can be subdivided into at least 3 distinct natural regions, being distinguished on the basis of physical, chemical, and biological variables. Data interpretation makes use of the cluster analysis.

1. Introduction: The hydrography of the German Bight is rather complicated due to topography, the tidal regime, and especially to continental runoff discharged by Elbe, Weser, and Ems, being mixed up with water of the open North Sea. As revealed by satellite photographs and shipborne recordings of chlorophyll fluorescence, the space scales of phytoplankton patches in this area often range in the order of 0.01 to 5 km. Various attempts have been made towards a more integral view of the pelagic system in the German Bight, which covers an area of about 100 x 100 km.

With respect to the biological inventory, different water masses have been identified in the last years (Bosselmann, 1984; Gerdes, 1985; Witzel, 1989). The distribution of organisms, however, often was further complicated by the existence of highly dynamic fronts occurring regularily in the German Bight (Krause et al., 1986; Schaumann et al., 1988).

Support by the ´Deutsche Forschungsgemeinschaft´is gratefully acknowledged.
Contribution No. 342 of the Alfred-Wegener-Institute for Polar- and Marine Research.

The present study continues these investigations with the aim to find out, whether natural regions with specific distribution patterns of organisms and chemical substances can be identified inspite of the highly variable oceanographical environment.

2. Material and Methods

The investigations covered different seasonal aspects of the German Bight's pelagic system. They were performed during 6 cruises on board of RV *Victor Hensen*, each comprising 2 transects with 10 stations. The cruises took place in February, March, July, and October 1984 and in February and April 1985.

Fig. 1: Location of sampling stations on 2 transects in the
German Bight

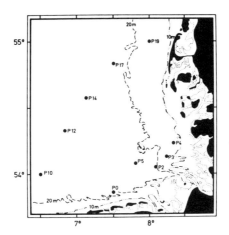

Positions of stations were chosen on the basis of former investigations (Wulff, 1935; Gerdes, 1985), in order to account for the variability of the system.

On each station salinity and water temperature were recorded by means of a CTD probe (ME Meerestechnik - Elektronik GmbH). Discrete water samples from 2 m water depth were taken with a Niskin bottle. Chl $\underline{a}$ concentration was determined for the total amount and the fraction < 20 µm according to Jeffrey & Humphrey (1975). Determination of inorganic nutrients (NH_4^+, NO_3^-, NO_2^-, PO_4^{3-}, Si) was done according to Parsons et al. (1977). For microscopic evaluation of phyto- and protozooplankton, subsamples were preserved with Lugol's iodine and analyzed according to the Utermöhl method (1958).

Zooplankton samples (>250 µm fraction) were taken from 2 m water depth with a Lenz-Vacuum Planktonpumpe (Lenz, 1972) and stored in hexamine buffered formalin (4 %) prior to further examination in the laboratory.

For the purpose of water mass analysis, data sets were processed in a Siemens 7.865 computer using the Clustan 1C package (Wishart, 1978). Stations were clustered into groups of similarity by a numerical classification, using the algomarative method of Ward in combination with the Bray-Curtis measure.

3. RESULTS

3. 1. Seasonal aspects

Chlorophyll $\underline{a}$ and nutrient concentrations exhibited typical seasonal patterns for boreal regions, i.e. nutrient maxima and chlorophyll minima in late winter and autumn and an opposite situation in spring and summer (Fig. 2 a, b).

Throughout the year a significant inverse relationship between salinity and nutrient concentrations stressed the importance of continental runoff as a dominant source of nutrient supply into the German Bight (Tab. 1).

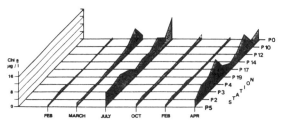
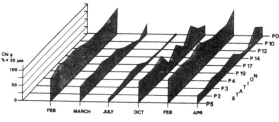
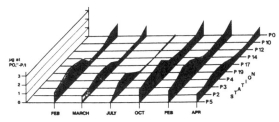
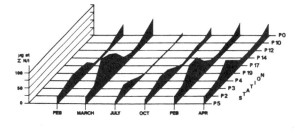
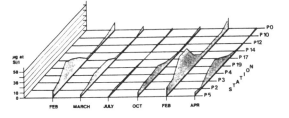

Fig 2: Seasonal variablity of Chl a and inorganic nutrient concentrations at 10 stations in the surface water of the German Bight.

a) total Chl a
b) Chl a-fraction < 20 μm
c) PO_4^{3-}
d) total inorg. N
e) Si

Tab. 1: Linear correlation between nutrient concentrations and salinity; data from 6 cruises in the German Bight during 1984/85.

	1984				1985	
	February	March	July	October	February	April
S°/oo						
to						
N	-xx	-xxx	-xx	-xxx	-xxx	-xxx
PO_4	-xxx	-xxx	-xxx	-xxx	-x	-xx
Si	-xxx	-xxx	-xxx	-xxx	-xxx	-xx

level of significance: -xxx = 0.1 %
- xx = 1 %
- x =   5 %

3.2. Phytoplankton

In late winter, about 80 % of total chl.a was contributed by the nanophytoplankton fraction. Dominant forms were nanoflagellates and small benthic and tychopelagic diatoms, especially *Paralia sulcata, Odontella aurita, Thalassionema nitzschoides, Actinoptychus senarius,* and *Delphineis surirella,* forming local growth-patches in near shore areas already in February. The relatively high abundance values of resuspended bottom forms underlines the importance of shallow sea areas (especially the Wadden Sea) during this season.

With the beginning spring bloom, however, these forms lost their dominant role and bigger diatom species like *Coscinodiscus radiatus , Thalassiosira punctigera, Coscinodiscus wailesii* and *C. concinnus* became more and more dominant, especially in the offshore areas of the German Bight.

The summer situation (July) was marked by a relatively high mean chl.a content (8 µg l^{-1}), the nanophytoplankton fraction constituting only about 14 % of this. There was a marked spatial heterogeneity in species composition and distribution. In the south western parts of the German Bight , a typical bloom of dinoflagellates occurred with *Ceratium fusus* being dominant, but also the toxic *Dinophysis acuminata* was quite abundant (> 20.000 cells l^{-1}) in more offshore areas. Diatoms, however, dominated the phytoplankton in the eastern parts of the German Bight: *Odontella sinensis* and *O.regia* occurred favouredly in the vicinity of the estuaries of Elbe and Weser, while *Rhizosolenia shrubsolei, R. setigera, Chaetoceros socialis,* and *Guinardia flaccida* made up the bulk in the more north eastern parts. Locally, the phototrophic ciliate *Mesodinium rubrum* occurred in high numbers.

The situation in autumn (October) already was very close to winter conditions. About 58 % of the generally low chl. a values were made up by nanophytoplankton fraction. Among the diatoms, small *Paralia sulcata* and *Thalassionema nitzschoides* made up the major part. In the more saline water further offshore some dinoflagellates, among them *Prorocentrum micans, Torodinium robustum* and several *Ceratium* species, still occurred.

3.3. Zooplankton

Comparable seasonal fluctuations as described above were obvious in the distribution of the mesozooplankton, too, implying both, differences in abundance and in composition (Tab. 2).

Low densities (< 1000 Ind. m^{-3}) characterized the late winter situation. Holoplanktonic forms dominated (> 98 % of total abundance), and copepods among them accounted for more than 95 %. Always present but of less importance were the copelates *Oikopleura dioica* FOL and *Fritillaria sp.*, whereas species of other groups occurred only occasionally.

With ongoing season zooplankton abundance increased and meroplanktonic forms, making up in late winter only 1 % of total organisms, gained importance. In April 1985 meroplanktonic larvae accounted for about 33 % of total organism density. Polychaete larvae, most frequently Phyllodocidae, Polynoidae, *Nephtys sp.* and *Harmothoe sp.*, but also different Spionidae made up about 74 % of this pool,

Tab. 2: Composition of the mesozooplankton (< 250 μm) from surface water of the German Bight; investigations of the years 1984/85.

	February	March	July	October	February	April
mean zooplankton density (Ind m^{-3})	560	752	11507	2369	661	1921
% holoplankton	98.7	96.1	68.7	98.0	98.8	67.4
	composition of holoplankton (%)					
Copepoda	98.7	90.3	50.0	94.2	95.2	90.1
other Crustacea	0.5	0.2	0.6	0.6	-	0.1
Medusae	0.3	0.5	14.3	2.4	-	0.6
Ctenophora	-	0.2	0.5	-	-	-
Copelata	0.5	8.3	20.9	1.7	4.8	8.8
Chaetognatha	-	0.4	1.2	1.0	-	-
Rotatoria	-	-	-	-	-	0.2
Turbellaria	-	-	12.5	0.1	-	0.1
Tomopteris helgolandica	-	0.1	-	-	-	-
Unidentified	-	0.2	-	-	-	-
% meroplankton	1.3	3.9	31.3	2.0	1.2	32.6
	composition of meroplankton (%)					
Polychaeta	20.0	30.5	70.6	29.6	16.7	73.8
Gastropoda	20.0	-	1.0	16.3	-	0.2
Bivalvia	-	4.5	0.7	13.5	33.3	0.2
Asteroidea	-	4.5	0.2	-	-	1.2
Ophiuroidea	-	-	1.5	2.8	16.7	-
Echinoidea	-	-	17.8	-	-	-
Phoronida	-	-	2.2	13.3	-	-
Bryozoa	40.0	26.0	0.2	5.5	33.3	-
Crustacea	-	34.6	4.6	19.0	-	21.8
Pisces	-	-	0.4	-	-	2.8
Others	20.0	-	1.0	-	-	-

followed by larvae of crustaceans (nauplii of cirripeds on the nearshore stations P 19, P 5, P 4, P 0, and P 3; additionally 1 decapod larva). Larvae of other taxa were less frequent (compare Tab. 2). The holoplankton, which at that time made up about 67 % of total mesozooplankton, had become more diverse but copepods were still dominating elements of this fraction.

In July, the meroplankton shared about the same proportion of total mesozooplankton abundance as in April. Larvae of polychaetes and echinoids mainly accounted for this high abundance (Tab. 2). Considering the polychaetes, typical and frequently occurring species were *Lanice conchilega* (PALLAS), *Magelona papillicornis* (F. MÜLLER), different species of spionids and phyllodocids, and occasionally *Pectinaria sp.,Harmothoe sp., Sabellaria sp., Poecilochaetus serpens* ALLEN, and different polynoid larvae.

Among the echinoderms, larvae of the heart urchin *Echinocardium cordatum* (PENNANT) were prevailing; ophiopluteus larvae of *Ophiura albida* (FORBES) and *O. texturata* (LAMARCK) were less abundant. A single specimen of *Ophiopholis aculeata* (O.F. MÜLLER) was found at station P 14, on three further stations bipinnaria larvae occurred in low numbers, too.

Among the poorly represented crustacean larvae dominated by far those of cirripeds (about 80 %). The remaining larvae belonged to the decapods *Crangon crangon* L., *Macropipus holsatus* (FABRICIUS), *Carcinus maenas* (L.), and *Callianassa sp.*.

Only about half of the highly abundant summer holoplankton consisted of copepods. Other crustaceans, especially the normally very abundant cladocerans *Podon intermedius* LILLJEBORG and *Evadne nordmanni* LOVEN occurred only in very low numbers (compare Gerdes, 1985). The other 50 % of the holoplankton fraction mainly were made up by copelates, turbellarians and the coelenterate plankton. Typical species of the latter were *Lizzia blondina* FORBES, being present on 90 % of all stations, followed by *Eutima gracilis* (FORBES & GOODSIR), *Obelia sp., Bougainvillia ramosa* (VAN BENEDEN), and sporadically occurring species like *Eucheilota maculata* HARTLAUB, *Sarsia sp., Octorchis gegenbauri* (HAECKEL), and *Rathkea octopunctata* (SARS).

Drastically decreased abundance values for total mesozooplankton characterized the October situation, comprising a high proportion of holoplanktonic forms and a poorly developed meroplankton fraction. Copepods, medusae, and copelates mainly constituted the holoplankton, whereas relatively high proportions of mollusc larvae, actinotrocha larvae of phoronids and cyphonautes larvae seem to be characteristic for the autumn meroplankton, as also shown in an earlier investigation (Gerdes, 1985).

3.4. Local aspects

Cluster analyses of the chemical/biological data sets revealed the existence of at least three regions with specific water mass characteristics throughout all seasons. However, due to our selected sampling strategy no statements can be done about the extension of these water masses and the variability of extension in time.

Fig. 3: Regions with specific water masses, resulting from clusters of chemical/ biological parameters of surface water in the German Bight; 1984/85.

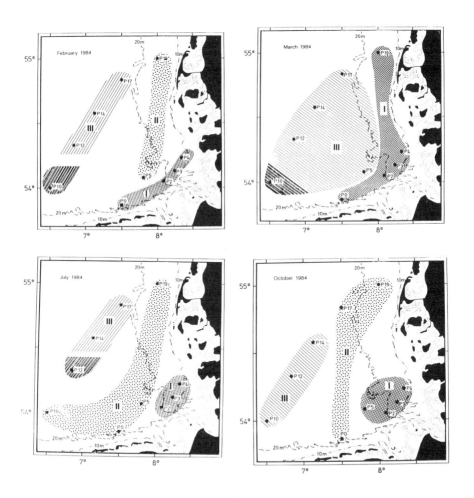

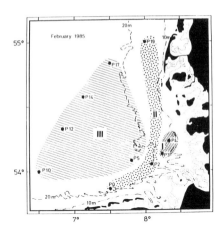 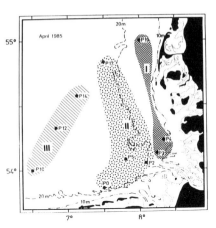

1. *Region I* in the inner German Bight in front of the estuaries of Elbe and Weser is especially represented by stations P 3 and P 4. Other stations in the vicinity, however, like P 2, and, to a certain extend, also P 0, P 5 and sometimes even P 19 in the NE German Bight belong to this water mass.

The water of this region is heavily influenced by the river discharge of Elbe and Weser. This is reflected by lowest salinity and highest nutrient levels and, especially in winter, high Chl. $\underline{a}$ concentrations. Small phytoplankton species as well as benthic and tychopelagic forms are more abundant than elsewhere, the latter indicating the proximity of the wadden sea.

2. *Region II* is a mixture of two different water masses, belonging to water of region I and more saline water influenced by the open North Sea (*Region III*, see below).

This water mass is the most heterogeneous one with
stations P 0 in the SW bight and P 19 in the NE part being
most representative, but occassionally P 5, P 2 and P 17
also fit into this cluster.

According to its intermediate position it is difficult to attribute any characteristic
features to this water mass. Anyhow, nutrient levels and chlorophyll concentrations
are generally lower than in type I water. The composition of phytoplankton and
zooplankton assemblages can be regarded as a mixture of species from both
adjacent water masses.

3. The 'offshore water' of *Region III* is strongly influenced by
   water of the open North Sea. This water was typically found
   at stations P 12 and P 14, but sometimes it also extends to
   stations P 10, P 17 and even to P 5 in the vicinity of the
   ´ Helgoländer Tiefe Rinne´.

Region III is characterized by the lowest nutrient concentrations and, in winter, by
very low chlorophyll levels. Phytoplankton composition differs from the other regions
by a higher proportion of large pelagic diatoms and dinoflagellates. Copepods are
dominating in the zooplankton. Indicator organisms like *Aglantha digitale*, *Sagitta
elegans* and *Ophiopholis aculeata* occassionally facilitate the identification of
this water mass.

Beside these persisting main water masses, smaller water patches with distinct
biological features were occasionally identified by means of the cluster procedure.
This especially holds true for station P 10 in the SW part of the bight. Within the
poorly developed pelagic system of the early year, this station exhibited highest
abundance values and the most diverse composition in the phyto- and zooplankton
communities.

The special status of station P 12 in July 1984 was due to a local red tide of
Noctiluca scintillans and its effects on other components of the pelagic system
(see Schaumann et al., 1988).

4. DISCUSSION

Situated at the SE border of the North Sea, the German Bight is influenced by very different water masses. According to an ICES study (1983), Atlantic water deriving from the English Channel (about 3200 $km^3 y^{-1}$) and water of the central North Sea (about 1600 $km^3 y^{-1}$) are the main sources for marine influx into the German Bight. Additionally, a high amount of freshwater enters the bight via Elbe, Weser, and Ems (yearly average about 35 $km^3 y^{-1}$). Tidal stirring and residual currents lead to a more or less intense mixing of these water masses. Furtheron, different types of fronts occurring regularily in the German Bight (Krause et al., 1986; Budeus, 1989; Hesse et al, 1989 b) further complicate the hydrography, especially in the surface water layer.

In view of this complicate hydrographic structure one would hardly expect any specific long-term distribution pattern of substances or organisms . On the basis of our comprehensive chemical and biological data sets, however, cluster analysis clearly revealed the existence of at least three regions with water masses, having their own specific characteristics, and prevailing throughout all seasons.

Traditionally, hydrographic parameters like water temperature and salinity provided the basis for water mass analysis (Dietrich, 1950; Tomczak & Goedecke, 1962; Lee, 1970). In several studies also chemical parameters have been used (Goedecke, 1936; Kalle, 1956; Martens, 1978).

Water mass differentiation by means of most of their chemical and especially biological inventories is complicated by the fact that both parameter sets behave like non-conservative tracers. It appears, however, that organisms are very sensitive indicators, integrating a wide variety of environmental properties, including physical and chemical conditions and reflecting them in their distribution patterns. This is supported by the fact, that comparison of the resulting chemical/biological clusters with corresponding T/S diagrams shows very good agreement in differentiation (Fig. 4a-f) for the February 1985 and March 1984 studies. In all other cases, however, this didn´t fit so well, since accordance was only found for the two extreme water masses of the offshore and coastal region (*Region III* and *I*). In these cases, both methods separated the stations in the same configuration. Water of Region II, however, combines stations with a considerably high hydrographic variability (compare Fig. 4).This result is consistent with the fact, that this water is not homogeneous, neither from the chemical nor from the biological point of view, thus underlining our difficulties to attribute distinct biological or chemical characteristics to it. Nevertheless, the separation of Region II water is possible by combining the total information of our chemical and biological data.

Fig. 4: T/S diagrams from the surface water layer in the German Bight. Station groups obtained by cluster analysis are marked by different symbols: *Region I* water
Region II water
Region III water

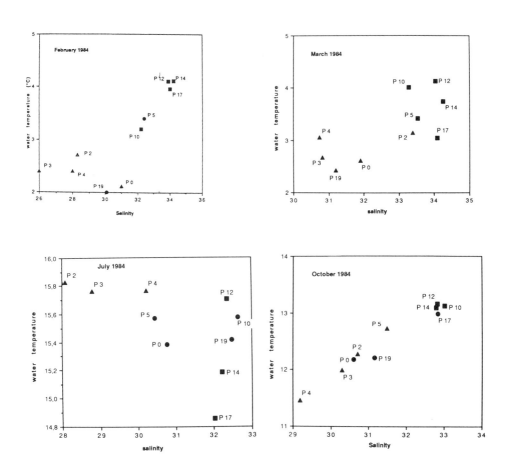

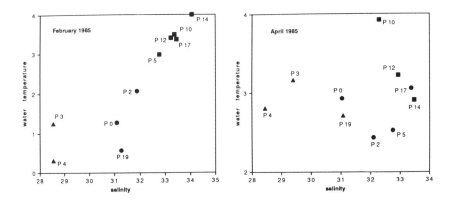

All data used in the present investigation were obtained from the surface water layer. In which way might this have affected the results?

During late autumn, winter and spring, when 5 of our studies were performed, the whole water column in the German Bight - with the exception of the frontal region of the Elbe river plume - is normally well mixed (Budeus, 1989; Hesse et al., 1989 a). During these times of the year, our surface data are suggested to be representative for the whole water column.

In summer, however, stratification also occurs in the deeper parts of the bight. The typical depth of the horizontal boundary layer lies in the range of about 10 to 15 m. Hence, during summer the surface water mass distribution cannot be considered representative for the whole water column. This holds especially true for nutrient and phytoplankton characteristics, whereas for the majority of the zooplankton organisms, which have a higher mobility, these differences may be less pronounced.

Comparison of our results with those of former investigations shows a fairly high degree of constancy in the water mass distribution patterns. This holds true for the studies of Wulff (1935), who differentiated seven water masses on the basis of their phytoplankton associations, and also for those of Goedecke (1936), who worked out very similar distribution patterns based on the lime content of the water masses. More recently, comparable patterns were achieved on the basis of the abundance and composition of zooplankton organisms (Bosselmann, 1984; Gerdes, 1985; Witzel, 1989). It seems, that these observed patterns in the pelagic system even show certain similarities to the benthic macrofauna community patterns in the German Bight, as worked out by Salzwedel et al., 1985.

Despite the high dynamic in the hydrographic regime of the German Bight, a certain constancy seems to exist in the pelagic system, thus leading to specific water mass characteristics with specific distribution patterns. No statements can be done about the extension of the distinct water masses, because the number of stations in our fixed station grid was too small.

Literature cited

Bosselmann, A. (1984): Untersuchungen zur Verteilung von Zooplankton in der zentralen Nordsee und in der Deutschen Bucht unter besonderer Beachtung des Transportes von Meroplankton. Dipl.-Arb., Univ. Bochum

Budeus, G. (1989): Eine tidal-mixing Front in der südlichen Nordsee. Struktur, Auftreten, Entstehungsbedingungen. Diss. Univ. Hamburg

Budeus, G. (1989): Frontal variability in the German Bight. Topics in Marine Biology. Ros, J.D. (ED.). Scient. Mar. 53 (2-3): 175-185

Dietrich, G. (1950): Die natürlichen Regionen der Nord- und Ostsee auf hydrographischer Grundlage. Kieler Meeresforsch. 7: 35-69

Gerdes, D. (1985): Zusammensetzung und Verteilung von Zooplankton sowie Chlorophyll- und Sestongehalte in verschiedenen Wassermassen der Deutschen Bucht in den Jahren 1982/83. Veröff. Inst. Meeresforsch. Bremerh. 20: 119-139

Goedecke, E. (1936): Der Kalkgehalt im Oberflächenwasser der Unterelbe und der Deutschen Bucht. Arch. Dtsch. Seewarte 55 (1): 1-37

Hesse, K.J., Gerdes, D., Schaumann, K. (1989 a): A winter study of plankton distribution across a coastal salinity front in the German Bight. Meeresforsch. 32: 177-191

Hesse, K.J., Liu, Z.L., Schaumann, K. (1989 b): Phytoplankton and fronts in the German Bight. Topics in Marine Biology. Ros, J.D. (ED.). Scient. Mar. 53 (2-3): 187 - 196

ICES (1983): Flushing times of the North Sea. Coop. Res. Rep. 123, Intern. Council Explor. Sea, Kopenhagen

Jeffrey, S. W., Humphrey, G. F. (1975): New spectrophotometric equations for determining chlorophylls a, b, c_1 and c_2 in higher plants, algae and natural phytoplankton. Biochem. Physiol. Pflanz. 167: 191-194

Kalle, K. (1956): Chemisch- hydrographische Untersuchungen in der inneren Deutschen Bucht. Dt. Hydrogr. Z. 9: 55 - 65

Krause, G., Budeus, G., Gerdes, D., Schaumann, K., Hesse, K.J. (1986): Frontal systems in the German Bight and their physical and biological effects. In: Nihoul, J.C.J. (ed.): Marine interfaces ecohydrodynamics. Elsevier Oceanogr. Ser. 42: 119 - 140

Lee, A. J. (1970): The currents and water masses of the North Sea. Oceanogr. Mar. Biol., Annu. Rev. 8: 33 - 71

Lenz, J. (1972): A new type of plankton pump on the vacuum principle. Deep-Sea Res. 19: 453 - 459

Martens, P. (1978): Contribution to the hydrographical structure of the eastern German Bight. Helgoländer wiss. Meeresunters.31: 414 - 424

Parsons, T. R., Takahashi, M., Hargrave, B. (1977): Biological oceanographic processes, 2nd edition. New York: Pergamon Press

Salzwedel, H., Rachor, E., Gerdes, D. (1985): Benthic macrofauna communities in the German Bight. Veröff. Inst. Meeresforsch. Bremerh. 20: 199 - 267

Schaumann, K., Gerdes, D., Hesse, K. J. (1988): Biological characteristics of a Noctiluca scintillans red tide in the German Bight, 1984. Meeresforsch. 32: 77 - 91

Tomczak, G. and Goedecke, E. (1962): Monatskarten der Temperatur der

Nordsee, dargestellt für verschiedene Tiefenhorizonte. Ergänzungsh. Dt. Hydrogr. Z., B, No. 7: 16 pp

Utermöhl, H. (1958): Zur Vervollkommung der quantitativen Phytoplankton Methodik. Mitt. internat. Verein. theor. angew. Limnol. 9: 1 - 38

Wishart, D. (1978): Clustan User Manual (3rd Edition). Program Library Unit, Edinburgh University, 175 pp.

Witzel, B. (1989): Schwermetallkonzentrationen in Copepoden aus verschiedenen Wassermassen der Deutschen Bucht. Z. angew. Zool., Jahrg. 76 (3): 303 - 332

Intertidal zone of Spitsbergen and Franz Josef Land

Jan Marcin Węsławski, Marek Zajączkowski, Tomasz Suryn
Arctic Ecology Group
Institute of Oceanology, Polish Academy of Sciences
Sopot 81-967, Powstańców Warszawy street 55, P.O.Box 68
Poland

Abstract

Zone between high and low water marks have been studied on 242 stations at Spitsbergen and 25 on Franz Josef Land in summer 1988-1991. The aim of the study was to compare "warm" and "cold" coasts at equally high latitudes defining their importance for coastal waters. Seven general coast types were defined from the geomorphological point of view. In Svalbard intertidal biomass range from 1 to 4000 kJ/m2. In Franz Josef Land, intertidal biomass range from 1 to 100 kJ/m2. Intertidal in Svalbard equals in biomass the sublittoral, while on FJL it contributes to no more than 15 % of sublittoral biomass.

Introduction

Arctic coasts have received little attention, when compared to Arctic shelf waters. Few papers have been published on geomorphology and geology of Arctic shores (McLaren 1980), Sempels (1982). The universal features of Arctic littoral biology are known from works by Thorson (1933), Madsen(1936), Steven(1938), Stephenson and Stephenson (1949). In recent years many impact studies related to the oil spill have been performed, especially in the Canadian Arctic (Owens et al 1987). Other threat - the global climatic change may affect Arctic coast with increase volume of freshwater, melted from retreating glaciers and ice cover.

Svalbard littoral fauna was studied by Summerhayes and Elton(1923), recently by Ambrose and Leinaas (1989, 1990), Hansen and Haugen (1990), Węsławski et al (1992).

The region of Franz Josef Land is less known, data on coastal biocenoses are scarce (Golikov and Averincev 1977).

The aim of present study was to compare the coasts of two European Arctic archipelagos and to answer the questions:
How important for the coastal waters are intertidal areas of Arctic islands ?
What kind of effect may have the freshwater on Arctic littoral ?

Materials

Two European Arctic archipelagos were object of this study. Svalbard (with Spitsbergen island) at 76 - 81 deg N and Franz Josef Land at 80 - 82 deg N (fig.1).

Survey on the Svalbard coasts have been made in

1988-1991 in the frame of joint Norwegian-Polish project, sponsored by Norsk Polarinstitutt and Institute of Oceanology Polish Academy of Sciences. About 1300 km of coast line have been studied at southern part of the archipelago (fig.1). Part of the data have been published in Węsławski et al (1992), Węsławski (1992), where detailes on the sampling methods can be found. Franz Josef Land coasts have been visited in 1990 and 1992 during Sov-Nor-Pol Expeditions (Gjertz and Wiig 1992). Data from circa 500 km of coast line were collected, in central and southern part of archipelago (fig.1). Computer system for storing and processing the data from intertidal was prepared on the base of DBase III plus programme application by Swerpel (1988), adjusted and widened by Suryn and Rószkowska(1992).

Results and discussion

coast types

For present, comparative study some generalisation was needed and Table 1 shows seven major coast types. From that number five types were common for both archipelagos. The skjerra coast, typical for West Spitsbergen, have not been observed at Franz Josef Land. The cryolittoral was observed on Franz Josef Land only. The cryolittoral name was given by Golikov and Averincev (1977) who described such coast type from Franz Josef and New Siberian Islands (pers. comm.). It occurs at places with permanent presence of very cold water (temperatures below -1.7 deg C) and perennial ice foot on coast. There, the tidal zone down to some one to three meters depth is made of ice, scattered with stones. The most common coast type on Svalbard are low gravel and stony beaches which made up to 50% of investigated coast line (Table 1). Ice coast of different type was dominating Franz Josef Land (60%).

physical parameters

Temperature of tidal water range in Svalbard from - 1.88 deg C throughout the winter to plus 14 deg at maximum in summer (Węsławski et al 1988, Węsławski et al 1992). In Franz Josef Land it never exceeds - 1 deg C (Swerpel 1992, Averincev 1989).

Salinity ranged from 0 to 35 ppt, typical values were noted between 28 and 30 ppt (fig.2).

Tides are regular, semidiurnal in Spitsbergen, with amplitude up to 1.8m (Swerpel and Siwecki 1975). On Franz Josef tides are irregular, semidiurnal, with 0.5m amplitude (Atlas Arktiki 1980, Swerpel 1992).

Ice is present as ice foot in littoral. At Svalbard ice foot lasts for three to nine months (Węsławski et al 1992). At FJL ice foot lasts all year round, only in some, south-exposed sites it disappears for August-September.

Water masses are of mixed character with the predominance of Atlantic current influence in Svalbard coasts (Swerpel 1985, Węsławski et al 1992). Franz Josef Land lays within Arctic water mass domain, although traces of Atlantic water are also common (Matishov 1992).

littoral biomass

Arctic littoral communities were described by Thorson (1933), Madsen (1933), Stephenson and Stephenson (1949), Golikov and Averincev (1977), Węsławski et al (1992). Similarily to studies mentioned above, five major assemblages, were described from the littoral of investigated coast. The richest in species and biomass was *Balanus-Fucus* assemblage, other were *Gammarus, Onisimus,* Oligochaeta and Cryolittoral assemblage.

Major part of investigated coast line was inhabited with measurable ammount of macroorganisms. The poorest were exposed, smooth rocks, when only thin film of filamentous algae was observed. Calving glaciers cliffs do not have marine benthic organisms connected specifically with them. If fauna occurs there, it must be plankton or stray organisms drifted towards glacier front. At solid coast the most common assemblage was "Oligochaeta" community, typical for low beaches in both archipelagos (Table 1). Rich inhabitation of littoral observed in Svalbard (at maximum over 2000g/m2, Węsławski et al 1992) have not been recorded at Franz Josef Land.

Table 1 Characteristics of Svalbard and Franz Josef Land coasts, data on biomass in kJ/m2 from Węsławski et al 1992.

general coast type	range of biomass	mean % share of coastline	
		SVAL.	FJL
perennial ice foot	50	0	2
glacier cliffs	10	8	59
rocky cliffs	05	12	1
boulders & stony beach	50	16	30
skjerra	900	7	0
low beach	10	50	8
watt & river mouth	15	7	0
mean value of intertidal biomass		78	22
mean value of sublittoral biomass		60	120
maximal tidal amplitude (cm)		180	50

sublittoral biomass

Benthic biomass in Svalbard coastal waters range from 1 to 200 g/m2 (Gorlich et al 1987, at FJL from 1 to 800 g/m2 (Golikov and Averincev 1977, Węsławski and Zajączkowski 1992). Using conversion from g to kJ as described in Węsławski et al 1992, respective mean values of 60 and 120 kJ are presented (Table 1).

relation between littoral and sublittoral

Locally abundant sublittoral macrophytes may create large deposits in tidal zone, especially after autumn storms. Several hundreds meters long and some meters thick kelp deposits are common at Spitsbergen island west coast (Węsławski et al 1988). In FJL such deposits are rare, and not exceed the biomass of 1000g/m2 (pers. obs). Due to the freezing of intertidal in winter time, motile animals are moving to sublittoral contributing somehow to its biomass. The relation between littoral and sublittoral biomass is more balanced in Svalbard, than on FJL. In Svalbard the littoral constitutes up to 110% of sublittoral biomass, where in FJL no more than 15% was noted (Table 1).

fresh water increase scenario

Increase of air temperature may cause increase of freshwater volume released to coastal waters every summer from glaciers and ice. In connection with this process three factors have important influence on coastal zone:
- lowering of salinity
- increase of suspended matter ammount
- water level raise

All littoral organisms observed in the investigated area belong to the euryhaline forms, they withstand sharp salinity changes in tidal cycle (Legeżyńska et al 1984, Węsławski et al 1992). It is unlike, that slight changes in salinity alone, may affect directly tidal organisms of European Arctic.

Melting water in Arctic is usually connected with great ammount of suspended mineral matter. In Svalbard, glaciers discharge contains from 20 to 500g mineral matter per l (Gorlich et al 1987), at FJL values of 5 to 50g were noted (Wiktor and Zajączkowski 1992). Ammount of sedimenting matter seems to be one of the major factors controlling benthic biomass in Arctic waters (Gorlich et al 1987). Absence of rich littoral biocenoses in the inner fjord basins at Svalbard may illustrate the negative effect of muddy water discharge from glaciers.

Water level rise is a neglible fact for present - day littoral biocenoses in Arctic. There are very few places with perennial intertidal vegetation. Majority of intertidal organisms may migrate or inhabit new areas in very short time (Różycki and Gruszczyński 1981).

Acknowledgments

Present study was sponsored with the grant 6020101 from the Committee on Scientific Research of Polish Academy of Sciences.

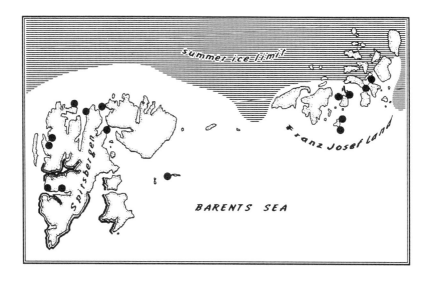

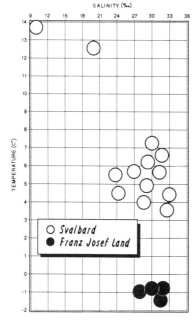

Fig.1 Investigated coastline, data from summer 1988 - 1992. Dot indicates mareographic stations, thick line - continous surrey.

Fig.2 Temperature and salinity values of intertidal waters found in the peak of summer (August) in the investigated coasts.

REFERENCES

Ambrose W G, Leinaas H P (1988) Intertidal soft-bottom communities on the West coast of Spitsbergen - Polar Biology 8: 393 - 395

Ambrose W G, Leinaas H P (1990) Size specific distribution and abundance of amphipods (Gammarus setosus) on an arctic shore. Effects of shorebird predation ? In : Proceedings of 24 th European. Marine Biological Symposium M. Barnes, R. N. Gibson (eds), Aberdeen University Press: 239 - 249

Atlas Arktiki (1980) Izd.Ministerstva Oborony SSSR, Moskva, in Russian, 360pp

Cross W E, Martic CM, Thomson DH (1987) Effects of experimental releases of oil and dispersed oil on Arctic nearshore macrobenthos. II Epibenthos. Arctic 40: 201 - 210

Golikov AN, VG Averincev (1977) Biocenoses of the Franz Josef Land and adjacent waters. Issledov Fauny Moriej 14: 5 - 55; in Russian

Hansen JR, Haugen T (1990) Some observations on intertidal communities on Spitsbergen (79 o N), Norwegian Arctic. Polar Res 00:000- 000

Legezynska E, Moskal W, Weslawski JM , P Legezynski (1984) Influence of environmental conditions on the benthos distribution in shallow water bay, Nottinghambukta, Spitsbergen. Oceanografia 10:157 - 172; in Polish

Madsen H (1936) Investigations on the shore fauna of East Greenland with a survey of the shores of other Arctic regions. Medd Gronl 100, 8: 112pp

Matishov GG 1992 Oceanographic survey on Franz Josef Land in 1991. Apatity, in Russian, 25pp

McLaren P (1980) The coastal morphology and sedimentology of Labrador: a study of shoreline sensivity to a potential oil spill. Geol Surv Canada Pap 79-23: 41pp

Owens EH, Harper JR, Robson W, Boehm PD (1987) Fate and persistence of crude oil stranded on a sheltered beach. Arctic 40: 109 - 123

Rozycki O, Gruszczynski M (1981) The inhabitation of coastal rocks at Hyttevika and Steinvika (West Spitsbergen). Materialy Sympozjum Polarnego PTG, Sosnowiec: 213 - 224; in Polish

Sempels JM (1982) Coastlines of the Eastern Arctic. Arctic 35:170 - 179

Siwecki R, Swerpel S (1979) Oceanographical investigations in Hornsund, 1974 - 1975. Oceanografia 6: 45 - 58; in Polish

Stephenson TA, Stephenson A (1949) The universal features of zonation between tidemarks on rocky coast. Journ Ecol 37: 289 - 305

Summerhayes VS, Elton CS (1923) Contribution to the ecology of Spitsbergen and Bear Island. The Journ Ecol 11: 214 - 287

Suryn T, Rószkowska M (1992) Computer graphic visualisation for studies on coastal ecology. Master of Sciences Thesis, Technical University of Gdańsk

Swerpel S (1984) Temperature and salinity of surface water at coastal measuring point, Isbiornhamna, Spitsbergen. Pol Pol Res 8: 57 - 64

Swerpel S (1992) Hydrometeorological conditions of Tichaia Bay, Franz Josef Land, summer 1991. Meddr Norsk Polarinst 00: 00-00

Thorson G (1933) Investigations on shallow water animal communities in the Franz Josef Fjord (East Greenland) and adjacent waters. Meddr Gronl 100 (2): 70pp

Wiktor J., Zajączkowski M. (1992) Phytoplankton and suspended matter in Tichaia Bay, Franz Josef Land. Meddelelser om Polarinstitutt, 00,00-00

Węsławski JM (1992) Genus Gammarus (Crustacea, Amphipoda) from Svalbard and Franz Josef Land. Distribution and density. Sarsia 77, 00-00

Węsławski JM, Kwaśniewski S., Zajączkowski M., Jezierski J. Moskal W. (1988) Seasonality in an arctic fjord ecosystem. Polar Research 00, 000 000

Węsławski JM, Zajączkowski M (1992) Sediments and benthic fauna from Tichaia Bay, Franz Josef Land. Meddel.om Polarinstitutt 00,00-00

Węsławski JM, Wiktor W., Zajączkowski M., Swerpel S. (1992) Intertidal waters of Svalbard. 1. Macroorganisms biomass and distribution. Polar Biology, 00,00-000

Coastal zone management; tools for initial design

H.G. Wind[1] and E.B. Peerbolte[2]

1. Introduction

In the development of coastal zone protection schemes various stages can be envisaged, starting at the initiative phase via the design phase towards excecution and maintenance. In the initiative phase it is decided which type of solution is feasible, given the values at stake, such as recreation, fishery, navigation, floodprotection etc. In the design phase the selected solution is worked out.

The initial design phase, the first step in the design process, however is hampered by uncertainty. Too often the physical and economic values can only roughly be estimated and the geometry of the site and buildings have to be derived from a map giving positions of a few years ago. If the effects of the uncertainty on the outline of the solution (and hence the losses and benefits) could be made clear, then the coastal zone manager has a basis to decide which additional information he needs to improve the outline of the solution. If the cost of obtaining additional information do not outweigh the expected benefits, he can decide that the initial design phase is finished. In this paper the following question will be addressed: what is the effect of uncertainty in the various types of information on the initial design of a coastal zone measure? As an example a coastal zone replenishment has been taken. It will be clear that the results of this study can easily be extended to other questions and construction types in the intial design phase.

The study can be separated in two parts. First the framework of analysis, as outlined by Peerbolte and Wind (1991), will be applied to a sandy coast with beaches and dunes, where the morphological processes will be related to economic impacts. In the second part of the study the internal consistancy of the framework of analysis is studied; hence the relevance and required accuracy of morphological and economic information within the framework of analysis is investigated.

2. Framework of analysis

In the coastal zone societal benefits can be envisaged for instance from nature, recreation, fishery etc. The present example will concentrate on economic benefits resulting from activities in the dune section. These yearly economic benefits V consist of services, wages etc. or in a period T: V_T. The economic value of for example a hotel is regarded as the economic benefits, discounted over the lifetime of the hotel.

---

1) Civil Engineering and Management, University of Twente, PO Box 217, 7500 AE Enschede, The Netherlands; Scientific advisor Delft Hydraulics.
2) Delft Hydraulics, PO Box 152, 8300 AD Emmeloord, The Netherlands

Due to long term shoreline erosion and storm surges the existance of the hotel and hence of the economic benefits can be threatened. The probability of loss of economic benefits E in a period T can be reduced by measures in the coastal zone at a cost K. The resulting societal cost Z is found by adding the investment cost K and the remaining loss of benefits E:

$$Z = K + E \tag{1}$$

Supposing that both K and E are a function of the co-ordinate X normal to the shore, then the minimum of Z is found from:

$$\frac{dZ}{dx} = 0 \tag{2}$$

The resulting societal benefits V_r are defined as the difference between V_T and Z. From an economic point of view the coastal protection policy becomes attractive if V_r is positive. Further V_r should be larger in the case with coastal protection than without coastal protection. Basically we compare the expected loss of benefits or erosion damage with the investment cost to reduce such damage as is illustrated in Figure 1.

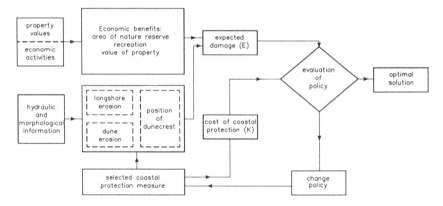

Figure 1 Framework of analysis

An outline of the calculation of the loss of economic benefits E is shown in Figure 2. On top the geometry of the building (or group of buildings) is shown. At time t=0 the building is located at a distance x from the edge of the coastal protection. The annual economic benefits of the building are shown in Figure 2b. If the dune crest recedes, the building may be damaged and hence the economic benefits may be reduced.

The probablity that in a given year the erosion will reach or even exceed the building is shown in Figure 2c. This probability of exceedance follows from the condition that

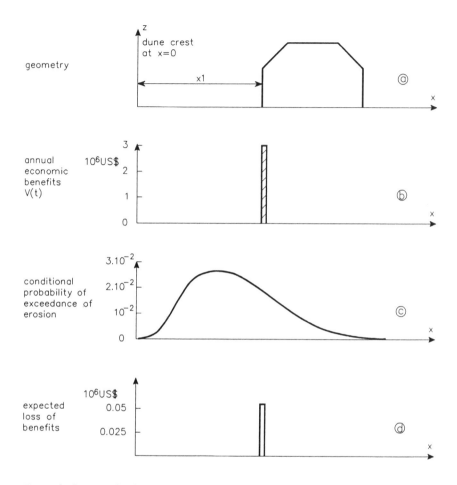

Figure 2 Steps in the determination of damage of property due to dune erosion

if failure due to erosion occurs in a certain year, no failure has occurred in the preceeding period of time, because it is assumed that no reconstruction takes place. This procedure was applied in a benefit-cost analysis of beach nourishment measures at Bloemendaal (Dutch coast), which location was subject to beach erosion (Peerbolte et al, 1989). The economic loss distribution in a certain year t, denoted by dE/dt in Figure 2d, follows from the product of the economic benefits in Figure 2b and the conditional probability of exceedance in Figure 2c. The damage dE/dt in year t can be found from the integration of the effective damage distribution in Figure 2d along the X-coordinate. The damage E in a period T follows from integration dE/dt over the time T. In the following we elaborate this case.

Geometrical representation

The co-ordinate system is fixed to the position of the dune crest at time t=0. The y-axis runs parallel to the shore and the x-axis points land inward. A building is located at time t=0 at a distance x1 from the dune crest. The distance x between the building and the edge of the replenishment can be represented as:

$$x = x_r + x_1 + g(t) \tag{3}$$

where:
x_r = (fictitious) width of the replenishment
x_1 = distance between the building and the edge of the dunerow at time t=0
$g(t)$ = longterm coastline evolution

The width x_r is termed "fictitious" because in practice the volume A will be distributed below the level of the dune crest and hence will extend over much greater width. In the sediment balance this difference is accounted for.

The economic benefits

The annual economic benefits V of the activities related to the hotel such as services and wages at time t=0 is represented by V_0. The increase of the value V with time is expressed by the growth rate of capital γ and is reduced by the long term level of interest r (discount rate). If γ and r are constant then

$$\frac{dV}{dt} = (\gamma - r)V \tag{4}$$

or

$$V = V_0 \, e^{(\gamma - r)t} \tag{5}$$

Frequency of exceedance μ and economic loss

In the present model the economic loss in the dune area is related to the recession of the shoreline. The longterm recession in eq. 3 is represented by the function g(t). The stochastic process of dune erosion is characterized by the longterm averaged annual frequency μ that the dune erosion exceeds a width x: $\mu = \mu(x)$. Further μ is time dependent because of the longterm shoreline evolution, therefore in a period of T years the frequency of exceedance will be:

$$\mu_T = \int_0^T \mu(x) dt \tag{6}$$

The probability of k events in a given year that the erosion exceeds a given distance x follows a binomial distribution. Under weak conditions we may adopt a Poisson distribution and for k=0 we find:

$$\Pr\{\bar{x} < x\} = e^{-\mu(x)} \tag{7}$$

or in a period of T years:

$$P_1 = \Pr\{\bar{x} < x \mid 0 < t < T\} = e^{-\int_0^T \mu(x)dt} \tag{8}$$

Economic loss in any one year is defined as the economic benefits in a certain year multiplied with the probability of erosion in that year. The probability that dune erosion in year T exceeds x can be derived from eq. 7:

$$P_2 = \Pr\{\bar{x} > x \mid t=T\} = 1 - e^{-\mu(x)} \tag{9}$$

Loss of economic benefits result only if in the preceeding period the erosion has not exceeded x. This conditional probability of failure in year T can be expressed as:

$$\Pr\{\bar{x} > x; t=T \mid \bar{x} < x; 0 < t < T\} = P_1 P_2 = \{1-e^{-\mu(x)}\} e^{-\int_0^T \mu(x)dt} \tag{10}$$

The damage in year T in this case can be represented as:

$$\frac{dE\{t\}}{dt} = \int_0^\infty V P_1 P_2 \, dx = \int_0^\infty V_o e^{(\gamma-\tau)T} \{1-e^{-\mu(x)}\} e^{-\int_0^T \mu(x)dt} dx \tag{11}$$

If a policy is selected in which the property is reconstructed after each failure, reconstruction costs have to be added to the term K. In that case the longterm average number of exceedances of x determines the loss of economic benefits:

$$\frac{dE\{t\}}{dt} = \int_0^\infty V_o e^{(\gamma-\tau)T} \{1-e^{-\mu(x)}\} dx \tag{12}$$

3. Analysis of uncertainty in data and models

In this paragraph the framework of analysis will be applied to the initial design of a replenishment. In order to investigate the effect of uncertainty in the economic and morphological data, a sensetivity analysis will be carried out.

Outline of the case study

The geometrical representation of the case study is shown in Figure 2. Parallel along the beach at a distance of 15 m from the dune front some hotels and houses are located. The annual economic benefits are estimated at 10 million US $. In this example a discount rate of 8% and a growth of capital of 10% has been applied.
The cost of the replenishment consists of a fixed cost of US $ 50.000 and a variable cost of 3 US $/m³ sand. The length of the replenishment is 2500 m and the height is 20 m. The total volume A and hence cost of the replenishment varies with linearly the width x_r. In the present example the volume A will be computed corresponding with minimum total cost Z.

The frequency of exceedance curve, shown in Figure 3, has been derived for conditions along the Dutch coast using the DUROSTA model (Steetzel, 1992). In the present example the curve will approximated by the following exponential distribution:

$$e^{-m(x-x_0)} \tag{13}$$

The values for m and x_0 are derived from Figure 3 and have been estimated at respectively .044 and -20.

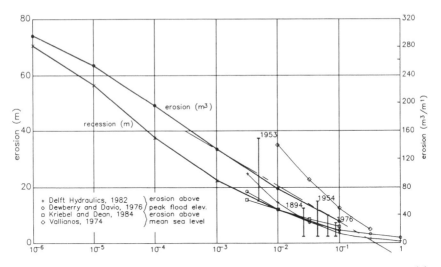

Figure 3 Frequency of exceedance distribution obtained with the DUROSTA model and compared with field data

The exponential distribution is a simplification of the real frequency of exceedance curve and deviates both for small and large values of the frequency of exceedance. However in view of the wide scatter in field data as shown in Figure 3, this approximation seems to be justified. The influence of uncertainties in m and x_0 will be evaluated by means of a sensitivity analysis. Instead of the deterministic approach used in this example, also DUROSTA can be included in a probabilistic approach directly yielding the error bands.

Shoreline erosion and retreat of the replenishment represented by g(t) have been set at 1 m per year. The time horizon of 10 years is chosen.

Optimal volume of the replenishment

The optimal volume A of the replenishment has been obtained by varying x, i.e. the distance between the buiding and crest of the replenishment. The results are shown in Figure 4.

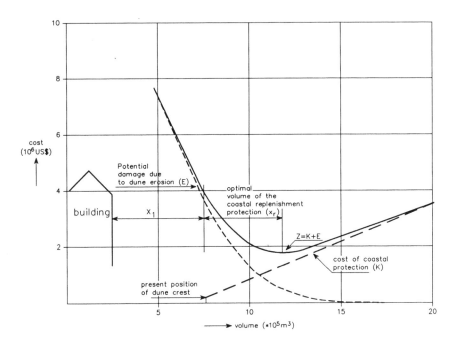

Figure 4 Potential damage and cost of coastal protection as a function of the volume of the replenishment

With increasing value of x or A, the probability that the building will be damaged reduces also. This implies that the economic losses E reduce with increasing value of x. For x larger then 15 m the costs K of the replenishment start to build up. The sum of the cost of the replenishment and the economic losses are represented by Z in Figure 4. The minimum value of Z is achieved for a volume of 430.000 m^3.

Sensitivity analysis of data and models

In order to investigate the effect of a variation in the data and models on the volume of the replenishment, a sensitivity analysis has been carried out. The result of the sensitivity analysis is shown in Figure 5. On the vertical axis the volume of the replenishment is shown relative to the volume for the parameters presented at the beginning of this paragraph (zero condition).

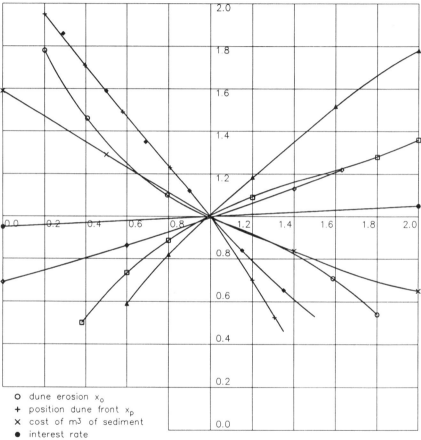

○ dune erosion x_o
+ position dune front x_p
× cost of m³ of sediment
● interest rate
◇ long term shore line recession C
□ value of property
▲ time T
◆ position dune front x_p

Figure 5 Sensitivity analysis of the volume of the replenishment

The value of the parameters on the horizontal axis is expressed relative to the zero condition. The variation in value of the non-dimensional parameters ranges from zero to two, i.e. the longterm erosion rate of the shoreline in the zero condition is 1m/year and is varied between 0 m and 2 m/year. year. It follows from Figure 5 that the effect of this variation on the volume of the replenishment is about 30%. Even larger variations in volume of the replenishment follow from varying the timehorizon and the geometrical position of the building relative to the dunefront. At the other hand, the effect of variation in mobilization costs is negligible, clearly because these costs are small relative to the total costs of the project. The effects of changing the interest rate ($\gamma - r$) is surprisingly small. This can be explained by the fact that both the growth of capital and the discount rate have been taken into account. If the growth of capital is neglected, then the discount rate is a sensitive parameter.

6. Discussion and recommendations

A framework of analysis has been developed for the initial design of coastal zone protection projects. A strong point of the framework is that it integrates a strong body of knowledge in the field of morphology with economic practice.

In the present study the framework of analysis is only applied to a one dimensional beach section. Extension to the two dimensional case, as recommended by Dean (1988), seems rather straightforward and is included in the outline in Figure 1.

A theoretical weakness in the framework of analysis is hidden in the step from the binomial distribution towards the poisson distribution in chapter 2. One of the prerequisites of this step is that the parameters are not dependent on time. However in particular the value of x is time dependent due to the long term shoreline evolution. A first investigation seems to indicate that the prerequisite is weak.

A key element in the framework of analysis is the frequency of exceedance of dune erosion. In this report the DUROSTA model is suggested as a means of obtaining this distribution. However, as can be seen in Figure 3, the spatial distribution of the dune erosion is rather large. This effect on the effective damage has not been taken into account.

In the sensitivity analysis and the derivation of the frequency of exceedance curve, use has been made of a deterministic approach. The formulation of the problem is clearly suitable for a probabilistic approach.

7. Acknowledgements

During this study valuable contributions and comments have been made by H.A. Pennekamp, C.H. Hulsbergen and F.M.J. Hoozemans. H.J. Steetzel provided the results of the dune erosion computations. Discussions with C.F. de Valk on the statistical aspects of this study have been very helpful. Dr. A van der Veen made helpfull comments on the economic framework.

Literature

Steetzel, H.J. Cross-shore transport during stormsurges. Proc. of the Coast. Eng. Conf., pp. 1922-1934, 1990.

Hallermeyer, R.J. Applying large replicas of shore erosion by storms. Coastal Sediments' 87, 1415-1429.

Kriebel, D.L. and Dean, R.G. Beach and dune reponse to severe storms. Proc. of the Coast. Eng. Conf., 1984, pp. 1584-1599.

Vallianos, L. 1974. Beach fill planning - Brunswich County etc. Proc. of the Coast. Eng. Conf., 1974, pp. 1350-1369.

Delft Hydraulics, 1982. Rekenmodel voor de verwachting van duinafslag tijdens stormvloed; Report M1263 part IV (in dutch); Delft, The Netherlands.

Dewberry and Davis, 1986. Description and assessment of coastal dune erosion. Report for Federal emergency management agency, Fairfax, Va.

Dean, R.G. (1988). Realistic economic benefits from beach nourishment. Proc. of the Conf. on Coast. Eng., pp. 1558-1572.

Peerbolte, B. and Wind, H.G., 1992. Policy implications of sea level rise in the Netherlands. Proc. of the Int. Conf. on Climate Impacts on the environment and Society (CIES). Univ. of Tsukuba, Ibaraki, Japan, pp. B-51 B-55.

Peerbolte, E.B. and Pennekamp, H.A., 1989. Kustachteruitgang Bloemendaal. Verslag beleidsanalytisch onderzoek, Delft Hydraulics, H1064 (in Dutch).

MORPHODYNAMICS OF INTER-TIDAL AREAS - THE WADDEN SEA

H.H.S. Noorbergen
National Aerospace Laboratory NLR
Remote Sensing Department
Marknesse, The Netherlands

1. Introduction and study area

Inter-tidal areas, such as the Dutch Wadden Sea, are dynamic areas that are constantly subject to morphological changes. Tidal channels shift and tidal flats are varying in height. The spatial distribution of tidal channels and tidal flats can strongly vary in several years. Until now geomorphological changes in an area like the Dutch Wadden Sea are measured by depth-soundings from vessels. Soundings are made in a pattern with a grid size of 200 metres. A complete depth-sounding for the Dutch Wadden Sea requires a surveying period of about five years. The derived depth-sounding maps at scale 1:10.000 are therefore not a presentation of one moment but the compiled result of soundings during a five-year period. With available satellite images it is possible to determine the geomorphology of the Dutch Wadden Sea at one certain moment. The advantage of this is that geomorphological changes during the acquisition period are excluded thus enabling a better link between the geomorphology and other factors like meteorological conditions. The project presented here aims at demonstrating the usefulness of remote sensing in determining the geomorphology of an inter-tidal area, the area between high and low water, such as the Dutch Wadden Sea. By using more satellite images of different acquisition dates (years) geomorphological changes can be detected.

2. Data collection

2.1 Satellite and acquisition

For this project the chosen satellite imagery is coming from the Landsat MSS satellite. In fact, from 1972 till now, there was a series of satellites; Landsat 1-5. The present build up database of 20 years remote sensing provides a useful archive for multitemporal research.

MSS stands for Multi Spectral Scanning, a technique by which the reflected electromagnetic radiation from the earth's surface is detected in several (in this case 4) spectral bands or certain wave length intervals of the electromagnetic spectrum. Landsat MSS detects the following four spectral wavelengths:

MSS bands		wavelength (μm)	
4	1	0.5 - 0.6	green (visible)
5	2	0.6 - 0.7	red (visible)
6	3	0.7 - 0.8	near infrared
7	4	0.8 - 1.1	near infrared
(Landsat 1, 2, 3)	(Landsat 4, 5)		

Each part of the electromagnetic spectrum contains specific information about objects and phenomena on the earth's surface. For this project the following MSS-bands are of importance:

band 4: penetrates the most in water (in clear water about 15 metres). A tidal channel that is already surrounded by water (incoming tide) can be followed farthest with band 4. Higher sediment concentrations reduce this effect.

band 5: like band 4 but to a lesser degree; for this project although an advantage because it enables to make subtle distinctions between sandbanks and flooded tidal channels.

band 7: land (vegetation) reflects electromagnetic radiation to a large extent while water absorbs all radiation (no reflection). This property enables to make the best discrimination between water and land.

The spatial resolution of Landsat MSS, that is the ability to detect different objects separately, is 56 x 79 metres (79 metre in the flight direction and 56 metre in the scanning direction). These smallest distinguishable picture elements are called pixels. A complete Landsat MSS image is 185 by 185 kilometres.
Landsat is a polar satellite i.e. the satellite describes orbits in north-south direction that cross both poles. The height of the orbit is about 700 kilometres and it costs 100 minutes for the satellite to complete one full orbit around the earth. In the meantime the earth rotates underneath the satellite so that the next pass (or satellite track) will be 2750 kilometres in westward direction. In a twenty-four hours period the

satellite makes about 14.5 orbits and it takes 16 days before the whole
earth is covered. Then the satellite crosses the same strip on earth again.
Because of this orbital pattern, acquisition dates and local times coming
over are fixed and of importance to this project. For the Dutch Wadden Sea
the time of coming over is about 10.45 hours A.M.

Landsat images can be selected by judging 'quick-looks' (preprocessed B/W
prints) and establish cloud cover and location of the area of interest.
Then the images can be purchased on computer compatible tape (CCT) or as
photographic product. For this project the first selection contained some
160 suitable Landsat MSS images.

2.2 Water levels

From the selected cloud free imagery of the Wadden Sea only the images
taken at low tide are of interest to this project, because only then one
can observe as much as possible inter-tidal area. However in a large area
like the Dutch Wadden Sea it never is low tide at the same moment.
Two main factors are of importance:
1. tidal differences in metres
2. differences in tidal phase.

In the Wadden Sea the tidal difference increases from the west (about 1.5
m) to the east (about 3 m). In addition, the tidal differences also
increase from the sea gates between the isles to the main land (north-south
direction).

The tidal phase between Den Helder (in the west) and Delfzijl (in the east)
is about 5 hours. Differences between times of high and low tide also occur
in north-south direction between the sea gates and the main land.
Furthermore there are important phenomena like storage basins of tides and
slack water (neat tide) directly south of the Wadden isles.

The factors mentioned above cause that the spatial distribution of water
levels in the Dutch Wadden Sea is very complicated. In order to select
satellite images from different years and to make good comparisons between
those years, for each year of the interest one has to know exactly how the
tidal conditions were at the acquisition time and how the water levels are
spatially distributed.

Analysis of the "Tide tables for the Netherlands" learns that the largest
area with low tide at one moment is situated between the "Waardgronden"
(south of the isle of Vlieland) and East Groningen (Eemsharbour). On that

moment it is already incoming tide in the west and still falling tide in the east.

From 13 local tidal stations the actual water level measurements could be obtained and plotted in charts (one chart per satellite acquisition date). By continuously refining the selection procedure, eventually three satellite images (acquisition years) could be selected that meet the criteria that are briefly summarized below:
- an overview of the whole Dutch Wadden Sea area
- water levels must be as low as possible through the area
- different years must be comparable in a preferably long time period.

The selected images are:
1. 1975, 5th of May.
   The Dutch Wadden Sea fits in one satellite track
2. 1980, 1st and 2nd of November.
   Because of different location of the satellite tracks, this image had to be composed of two images that overlap. The seam in the created mosaic is chosen at the point where the smallest differences in water levels between both acquisition dates occur.
3. 1987/1988, 11 May 1988 - 5 July 1987 - 2 October 1987.
   This mosaic is composed of satellite images from three different satellite tracks and dates. Seams in the mosaic are chosen in the same way as described for 1980.

3. Satellite image processing

3.1 Geometric correction

All, initially six, MSS images were geometrically corrected to the Netherlands rectangular coordinate system (stereographic projection). The 1980 image was corrected by indicating ground control points on the topographic map sheets 1:50.000 of the Wadden Sea area. After geometric correction of 1980, this image became 'master' image for the correction of the other images (1975, 1987, 1988). The geometric correction included a reduction of the original pixelsize (79 x 56 metres) to an output pixelsize of 50 x 50 metres. The corrected area measures 190 kilometres in west-east direction and 85 kilometres in north-south direction.

3.2 Derived products

In order to establish the geomorphological changes between 1975, 1980 and 1987/1988 several approaches were chosen, of which the two most important will be explained here. As stated before, from the four MSS bands, band 5 and 7 contain most information that is of relevance to this project. These bands were used to detect the differences between years.

1) On both bands a gradient filtering routine was applied that works in two directions. The gradient calculations are done by means of convolution filters, with the following filter coefficients:

in x-direction

$$\begin{matrix} -1 & 0 & +1 \\ -1 & 0 & +1 \\ -1 & 0 & +1 \end{matrix}$$ and

in y-direction

$$\begin{matrix} -1 & -1 & -1 \\ 0 & 0 & 0 \\ +1 & +1 & +1 \end{matrix}$$

The result is two files with digital values in the range 0 - 225. Digital value 128 represents pixels with no sharp changes while in the direction of digital values 0 and 255 the changes become sharper. Both files are stretched on contrast in such a way that the sharp changes get the digital value 255 and the 'non-changes' digital value 0. Thus white **contour lines** are created, one or two pixels wide, that indicate sharp changes, in this case the change between land and water. By summation of the two files the contour lines of x- and y-direction are joined together so that the complete contour image is created. This was done for the three years on bands 5 and 7. Analysis of the results pointed out that the sharpest contour lines are generated from band 7. The tidal channels can be followed longer with band 5, which was also the theoretic expectation, but the image also contains more 'noise' so that land-water differences are less clear.

Any two years can be combined in order to detect the differences in geomorphology. This is done by presenting the most recent year in red and the other year in cyan. These two colours provide a good colour contrast. In this way difference images were made from the years 1975-1980, 1980-1987/1988 and 1975-1987/1988.

2. MSS bands 5 and 7 of the three satellite mosaics were contrast-stretched with emphasis on water features. Therefore land features in the image become saturated and will be presented in white. Clear water becomes almost black while suspended sediment and wet tidal flats will present any greytone between black and white.
One now has two possibilities to combine years, both methods were used:
- Combination of two years by presenting the most recent year in red and the other year in cyan, similar to the combining of contour line images. However, instead of line information, now the area information is combined. For example dry tidal flats (that were white in a single image) are presented in bright red on unique locations for the recent year and in bright cyan on unique locations for the previous year. In other words red indicates the formation of new tidal flats and cyan indicates the creation of new tidal channels. The colour white in the combined image are the stable tidal flats i.e. not changed between the two years. Greytones, sometimes in the direction of red, sometimes in the direction of cyan indicate subtle differences in more or less wet tidal flats of both years.
- Combination of three years by presenting each year in a so-called basic colour (red, green or blue). In this project the year 1987/1988 was presented in red, 1980 in green and 1975 in blue. The colours indicating uniqueness (red, green and blue), the combinations of two basic colours (cyan, magenta and yellow) and various greytones, sometimes in the direction of the six colours mentioned, give an enormous amount of information. It will be obvious that this demands skilled interpretation. The big advantage is that in one image a complete overview is created of a large area for a long period of years.

4. Interpretation

In this part an interpretation of the images is given. This interpretation is done particularly from a remote sensing point of view. For a more complete analysis the reader is referred to the report "Morfodynamica van intergetijde-gebieden" by van der Spek and Noorbergen (ISBN 90-5411-039-2). This brief interpretation also refers to the enclosure (NLR flyer).

- false colour composites

    The composites with band combination 6, 5, 4 (R, G, B) show the inter tidal area in great detail. The darker parts along the dikes of the mainland and in the slack water areas indicate higher moist contents in fine-grained sediment; the areas with lower currents. It can also be seen that the eastern part of the Dutch Wadden Sea contains more fine-grained sediment than the western part. Higher sandbanks can be recognized by the white colours. South of the German Leybucht a strange rectangular pattern can be seen (diking-in?).

- contour line images

    It appears that the contour line images based on band 7 give a very sharp contrast between water and land. Although band 5 penetrates better in the water, this does not give an essential improvement because also less important contrast differences are enhanced, resulting in more 'noise'. A disadvantage of this method is that when the whole image has a lower intensity (for example 1980 which was of November), also contrast differences are harder to detect. Also the small changes (one or two pixels) cannot be detected by this filtering method.

- difference images based on coloured areas

    The combined image with three years, presented in red, green and blue, gives very much information. This can complicate the interpretation. Phenomena that are univocal and continuous, such as the erosion of East-Terschelling, can be detected very well. Also the future land reclaims, north of the main land and displayed in orange to red, can be seen clearly.

    In order to detect the geomorphological development between 1975 and 1987/1988 in more detail, one has to change over to the images in which two years are compared (presented in red and cyan). In this respect the combination 1975-1987/1988 is very well interpretable because of similar water levels.

    The biggest changes occur in the sea gates.

    From west to east (see also Fig. 1):

    - the sandbank "Noorderhaaks" is shifting and rotating clockwise in the direction of the isle of Texel. Texel itself expands in south-

west direction. The tidal flats north of "Kornwerderzand" change strongly.
- the isle of Vlieland develops in southward direction. The tidal channels in "de Waardgronden" (tidal flats) shift eastwards.
- the sandbank on "de Noordwestgronden" has almost disappeared while the sandbanks of "de Jacobsruggen" spread out north. The northwest point of the isle of Terschelling is eroded by about 500 metres.
- the sandbanks directly northwest of the isle of Ameland shift in landward direction. The "Boschgat" (tidal channel) moves its course directly to the outer delta and causes erosion of the east point of the isle of Terschelling with 2000 metres. The "Boschgat" widens out and also changes the drainage pattern in the slackwater area south of Terschelling. The tidal flats between "Boschgat" and "Amelander gat" expand strongly in northward direction. The tidal channels of "Dantziggat" and "Kikkertgat" expand eastwards, causing a shift of the slackwater area in eastward direction.
- the eastpoint of the isle of Ameland accretes with about 1000 metres in eastward direction, also causing a shift of the "Holwerder Balg" (tidal channel) eastwards. On the north west point of the isle of Schiermonnikoog a sandbank develops in east to west direction, followed by an embranchment in southward direction (after 1980).
- the "Eilander Balg", eats of the isle of Schiermonnikoog, shifts eastward causing accretion of East Schiermonnikoog. North of "Simonszand" (tidal flat) new tidal flats develop. Rottumerplaat (isle) expands in northward direction. The tidal channels in the water storage area shift and seem to expand.

5. Conclusions

In general one of the conclusions of this analysis is that the geomorphological changes that can be detected from the satellite imagery (a period of 12 years) are part of cycles that span a period of about 50 years. This can be verified by making use of the older soundings-charts. The satellite images are judged to be very useful and cost effective. Moreover one can now create an overview of a large area with good resolution at **one** moment. The image processing techniques to combine satellite images in order to detect morphological changes are relative simple. The results are easy to interpret. An important requirement when selecting satellite images is to have sufficient pre-knowledge about water levels in the area. Remote sensing techniques for an application like this can be used very well to indicate certain developments (changes in geomorphology) after which one can decide on more detailed studies.

6. Acknowledgement

The Wadden Sea project was carried out by the National Aerospace Laboratory (NLR) under contract to the Netherlands Department of Geology (RGD), with financial support from the Netherlands Remote Sensing Board (BCRS).

7. Literature

- Jensen, J.R. (1986). Introductory digital image processing, a remote sensing perspective

- Rijks Geologische Dienst, Haarlem (1977). Geologisch onderzoek van het Nederlandse Waddengebied

- Spek, A.J.F. van der, Noorbergen, H.H.S. (1992). Morfodynamica van intergetijdegebieden. BCRS rapport 92-03, ISBN 90 5411 039 2

- Ven, I. van der (1988). Het gebruik van remote sensing voor een geomorfologische kartering van de Waddenzee. Stageverslag Rijkswaterstaat-Meetkundige Dienst, Delft

Simulation of morphodynamics of tidal inlets in the Wadden Sea

by

J. van Overeem[1], R.C. Steijn[1] and G.K.F.M. van Banning[1]

Summary - For a sound management of complicated systems such as the Wadden Sea, it is important to have a thorough understanding of the morphodynamic behaviour of the tidal inlets to the Wadden Sea. Tools are available to analyse the governing processes. The COMOR-model, being a numerical model for coastal morphology, showed to be a powerful tool for the analysis of the morphodynamics of these complicated systems. After a brief introduction of some characteristics of tidal inlets, examples of model results are given for a number of tidal inlets to the Wadden Sea.

1. Introduction

The Wadden Sea (Figure 1) is an international inner sea separated from the North Sea by means of a number of barrier islands, which belong to Denmark, Germany and The Netherlands. The Wadden Sea is internationally acknowledged as wetland, such as defined in the framework of the Ramsar Convention. The Wadden Sea, as a wetland, is an important nursery area for juvenile fish and feeding area for birds.

For a sound management of this coastal system, it is of utmost importance to have a thorough understanding of the functions of The Wadden Sea and their interrelations. Management decision related to issues such as coastal protection of the islands, sand mining, closure of inner seas or channels, exploration and exploitation shall only be taken after a sound analysis of the policy to be followed, taking into account the effect of the human interference on the system and its functions (Overeem, 1990). An example of a method to evaluate different policy options for coastal zone nourishment is found in Wind et al. (1992).

In this type of policy analysis, the knowledge concerning the morphodynamic behaviour of the system is an important input. In particular, the morphodynamic behaviour of the tidal inlets to the Wadden Sea is complicated and therefore difficult to predict. This is because of the complex interaction of a number of relevant processes, which play a role both at the outer and inner delta and in the inlet itself. Changes of the hydraulic and morphologic characteristics of the Wadden Sea, caused by natural changes and/or human interference will have clear effects on the behaviour of the inlet and consequently on the adjacent coastlines and inner and outer delta's. The morphology-shaping forces are tide, wind and wave action, coriolis forces and last but not least relative sea level rise.

---

1) DELFT HYDRAULICS, PO Box 152, 8300 AD Emmeloord, The Netherlands

DELFT HYDRAULICS has a powerful numerical model available to analyse the governing processes, which usually take place at different time and length scales, and to predict the morphodynamic behaviour. This model has been used in a number of projects related to tidal inlets to the Wadden Sea, such as the "Friesche Zeegat" between the Dutch Islands of Ameland and Schiermonnikoog and the "Eyerlandse Gat" between the Dutch Islands of Texel and Vlieland (see Figure 1).

In the following, the methodology which has been applied for the prediction of the morphodynamic behaviour of these inlets using numerical modelling systems, will be described and a number of results will be highlighted and discussed.

2. Tidal inlets

A tidal inlet is defined as the (short) waterway connection between the sea and a bay or lagoon, through which mainly tidal currents flow. Tidal inlets consist of three interconnected units, namely:
- the outer delta or ebb tidal delta (seawards of entrance);
- the tidal throat (narrow channel at entrance); and
- the inner delta or flood tidal delta (landwards of entrance).

Flood or ebb-dominance in a tidal inlet occurs when after integration of the currents over the tidal cycle, a residual remains. Flat tidal inlet basins (like the Wadden Sea) favour ebb dominance of the inlet throat. Especially where both waves and tides are important, inlets develop an ebb dominant main (central) channel, with flood dominated secondary channels between the main channel and the barrier islands. Ebb dominated channels bring sediment to the seaward edge of the outer delta, where wave action is generally responsible for the 'keeping together' of the delta. When under storm conditions a breaker zone is created along the periphery of the outer delta this deposited material can be transported backwards to the barrier islands. In Figure 2, the basic principles of sediment fluxes in a tidal inlet are shown. A more extensive description of tidal inlet morphodynamics is given in Steijn (1991).

3. Morphological modelling system COMOR

Introduction
In 1983 DELFT HYDRAULICS started with the development of a system of interconnected computer programs, the so-called COMOR-system, to simulate morphological changes in coastal areas. Nowadays we may say that we have succeeded in making such a compound modelling system.

The design philosophy of COMOR can be summarized as follows:
- one file system including uniform read and write routines for all models;
- one general purpose drawing package for this file system;
- each model handles its own characteristics on its own internal grid.

The output of each model is given to the general file system on one uniform grid.

The COMOR system contains 4 different type of modules, being:
- a module for steady state wave propagation;
- a module for tide-, wind- and wave-driven flow;
- a module for cohesive and non-cohesive sediment transport; and
- a module for morphodynamic developments.

Especially the latter module is still in the stage of development and therefore most of the practical applications still deal with initial bottom changes. An aggregate flow chart of the modelling system is given in Figure 3.

Schematization
The basis for the simulation of coastal morphology is formed by its governing physical processes, being the hydrodynamics of waves and tides. Especially the interaction between the deterministic character of the tidal wave propagation and the statistical character of the wind and waves is of importance. The wind as well as the waves have their effect on the tide (wind and wave driven currents), where the current has its effects on the waves (current refraction).

The art of schematization is to reduce the unlimited amount of information of the real world to a limited set of boundary conditions, each with its own weighing factor, in order to give a correct representation of the mean annual sediment transport. The required detail in place and time should be in balance with the required computer time and costs.

Flows
For the proper simulation of the tide in the Wadden Sea, the flow model is nested in a sequence of larger tidal flow models, starting with a tidal model of the entire Northwest European continental shelf. As a schematization, the neap-spring tidal cycle is schematized in one representative so-called "morphological tide". Usually this representative tide is ten percent above the mean tide.

Waves
The boundary conditions for the wave models are generated, starting with a wave climate study. This wave climate study is mostly based on wave measurements from a buoy or ship observations. This deep water wave climate is transferred to the shore by using one or two dimensional wave propagation models. The nearshore wave climate at the boundary of the area of interest is schematized into a limited number of wave heights, periods and directions combinations, taking into account the effect of the waves on the sediment transports, being the stirring, the cross-shore transportation of sediment particles and the generation of longshore flow. Since steady state wave propagation models are applied, usually a number of 10 to 20 runs is required during one tidal cycle.

Sediment transports
A number of cohesive and non-cohesive sediment transport formulae is available within the COMOR system, varying from simple local sediment transport formulae to three

dimensional time dependent convection diffusion solvers, taking account of the effects of time-lag and space-lag of suspended sediment.

In the following a number of applications of this modelling system for tidal inlets in the Wadden Sea will be dealt with.

4. The tidal inlet "Friesche Zeegat"

Background
One of the subjects under study within the Dutch Coastal Genesis research programme is the interaction between the morphodynamic behaviour of tidal inlets and their adjacent coastal sections (Steijn et al., 1992). The study has been concentrated on one of the tidal inlets to the Wadden Sea the so-called "Friesche Zeegat" (Figure 1). The reason for this is that after the closure of the Lauwerszeepolder in 1969, the tidal storage capacity was almost instantly reduced by some 35 percent. Consequently, significant changes took place in the inlet's morphology. These well-recorded changes make this inlet particularly interesting for numerical simulations, as described below.

Observed changes
The semi-diurnal tide has a mean tidal range of 2.2 metres. The local deep water wave climate can be classified as medium wave energy, with an annual mean significant wave height of some 0.75 m. The sediment characteristics vary over the area. However, the grain sizes are generally rather fine with D_{50} values ranging from 160 - 200 μm.

After the closure of the Lauwerszeepolder in 1969, the basin area reduced from about 210 million m^2 to 120 million m^2. Due to the presence of high tidal flats in the Lauwerszee this caused a somewhat smaller reduction in the tidal storage capacity of the inlet. Nevertheless, such a reduction will result in a reduction of the main channels' cross-sections and in a reduction of the volume of sand stored in the outer delta. The changes in bottom topography in the whole area from 1970 onwards have been recorded by echo soundings every four to five years.

In summary, the most pronounced changes over the period 1970-1987 were:
- the entire basin showed a tendency of sedimentation, whereas the outer delta eroded;
- the periphery of the outer delta has moved landwards;
- a typical L-shaped sand bar, elevated above MSL, gradually has developed in line with the shoreline of Schiermonnikoog;
- the watershed south of Schiermonnikoog has moved about two kilometres eastwards, in this period, thereby increasing the basin area again;
- a distinct sedimentation at longshore bars near the edges of the outer delta and at the tidal channels in the inner delta.

From the observed changes (1970-1982) in sediment volume as a function of the original depth it can be seen that erosion occurs at the outer delta at greater water depths and shoaling of the almost entire inner delta.

The morphodynamic modelling
For the simulation of the morphological evolution of this area, it is necessary to identify the relevant physical processes. Waves and current are the primary driving mechanisms behind sediment motion in a tidal inlet environment. Waves are responsible for the stirring, the cross-shore transportation of sediment particles and the generation of longshore flow. The various physical processes have been simulated by the two-dimensional horizontal model COMOR.

An overall tidal model for the whole Wadden Sea was used to obtain a consistent set of boundary conditions for the detailed flow model of the inlet. The overall model solves the usual depth-averaged shallow water equations on a 500x500 m^2 rectangular computational grid.

For the detailed flow model DELFT HYDRAULICS' model TRISULA has been applied with a non-uniform computational grid to allow for a good representation of the physical processes. The density of computational grid points (some 20,000 in total) is concentrated in those areas where relatively large spatial gradients in flow patterns exist (like in the breaker zones along the barrier islands). The shallow water equations include wave- and wind-induced forcing terms, and wave effects on the bottom friction. Relevant wave parameters inside the grid are obtained from the wave models.

For the simulation of the wave conditions the 2D directionally decoupled parametric shallow water wave model HISWA has been applied. This model solves the action balance equation and accounts for the effects of depth and current refraction, directional energy spreading, energy dissipation due to bottom friction and wave breaking and energy gain due to local wind. The ambient current is derived from the regional tidal model.

Sediment transports and patterns of erosion and sedimentation are computed on the same grid as the detailed flow model, using data from the flow model and the wave model. Different sediment transport formulae can be used, like Bijker, Van Rijn in which time-lag and space-lag effects can be accounted for, or Bailard, which combines wave-asymmetry induced and flow-induced transport components.

Boundary conditions
Hydraulic conditions are either deterministic (like the tides, which can be predicted beforehand) or stochastic (like waves, which depend on meteorological conditions). Besides the wide spectrum of possible hydraulic conditions over a whole year, each of wave field and sediment transport computation assumes steady wave conditions. The art of schematization is to find a limited set of boundary conditions, each with its own weighing factor, which give a correct representation of the mean annual sediment transport. For the time being a state of the art schematization procedure has been applied (Steijn, 1989), which should result in correct annual-averaged sediment transport patterns.

Schematization of the tide has resulted in the selection of a representative tide, with a tidal amplitude slightly above the mean tidal value. The tide has been split up into ten carefully selected stationary conditions, which are further used in the steady state models for waves and sediment transport. The wave climate has been represented by two wave conditions only, one originating from a northerly direction (355° with $H_{s,deep}$ = 1.7 m) and the other from a north-westerly direction (305° with $H_{s,deep}$ = 1.3 m).

Numerical model results
As a full picture of model results is impossible to give on a few pages, we will restrict ourselves to only one hydraulic condition with waves from a north-westerly direction (305°).

The current and sediment transport (using Bijker's formula) patterns are shown in Figures 4 and 5, respectively. Although the ebbing currents keep relatively confined in the main tidal channels, the sediment transport capacity is largest near the edges of the outer delta. Note that the wave-induced longshore flow (and littoral sand drift) along Ameland is larger than the opposite ebb tidal flow.

Integration of the model results over a full tidal cycle gives the residual flow and sediment transport patterns. The latter is given in Figure 6 for northwesterly waves. The residual flow pattern appears to be a combination of a residual through-flow and isolated residual eddies along the periphery of the outer delta. The residual eddies are found at those places where shallow and deeper areas are relatively close.

As a final result of the computations the pattern of annual sedimentation and erosion has been computed by combining the results obtained for the three hydraulic conditions (no waves, northerly waves, northwesterly waves). The combined annual result (based on the Bijker formula) is given in Figure 7.

Interpretation
Although the analysis of the model results still continues, some remarkable features have already been observed:
- Under 'average' conditions, the tidal inlet 'sprinkles' sediment to the foreshore, where it settles between MSL -5 m and MSL -10 m. Computations with extreme wave conditions shows that during storms this deposited material is transported towards the islands along the periphery of the outer delta.
- The shallow areas tend to migrate southwards. Although observed for all transport formulae, this is most pronounced for the Bailard formula, which includes transport due to wave asymmetry.
- The largest dynamics in morphology is found in the western inlet, most probably because this part has to redistribute large quantities of littoral drift input from the Ameland breaker zones. The morphological time scale of the eastern inlet, even after its abrupt disturbance by the closure of the Lauwerszeepolder, is much larger (adjustments take longer period of time).
- The main central tidal channels are ebb-dominant, while the areas in between these channels and the barrier islands are slightly flood dominated. The larger channels in the inner delta (Zoutkamperlaag) seems to be relatively neutral.

The numerical computations deal with transport by sand only. In reality, however, siltation of deep sections of the inner delta tidal channels has been observed. Furthermore, it should be noted that till now only initial patterns of sedimentation and erosion can be calculated.

5. The tidal inlet "Eyerlandsche gat"

The North Sea Coast of the Island of Texel suffers from erosion. The northern coastline of the island is strongly influenced by the morphodynamic processes that play a role in the tidal inlet "Eyerlands Gat". The average coastline retreat of the 5 km long most northern coastal section is about 4 m/yr. Since 1979 about 8 million m^3 of sand has been nourished to mitigate the coastal erosion.

Since the beginning of 1990 the Dutch Ministry of Public Works has decided for the policy to maintain the position of the coastline of 1990. Eroding coastal sections are in principle protected by means of beach nourishment, although other type of structures should also be a possible option (Hillen, 1992).

In order to come up with an optimum solution for the coastal protection of northern part of the Island of Texel a study is performed in which a number of protection schemes are to be considered. For each of the schemes the effectivity for coastal protection is determined together with its impact on the environment. A final decision for the scheme is going to be taken after a thorough analysis of effectivity, impact and cost. Four basic solutions are to be considered, being continuous nourishment, a long dam (upto 1000 m), a groyne-system and a system of offshore breakwaters.

DELFT HYDRAULICS has been commissioned by the responsible Regional Direction of Rijkwaterstaat to build a mathematical morphological model of the tidal inlet and the adjacent coastlines in order to understand the morphodynamics of the area and to assess the impact of the proposed measures on the coastline and the channel and bar system of the tidal inlet (Ribberink et al., 1992). To that aim the modelling system COMOR has been applied in a similar way as described above for the study of the "Friesche Zeegat".

Sediment transport patterns and patters of erosion and sedimentation have been obtained for one characteristic tidal condition ("morphological tide") and for two characteristic wave conditions, namely a typical northern and a western wave condition. Also computations have been performed for a storm condition. The resulting sediment transport patterns for the actual situation and for a situation with a long dam have been given in Figures 8 and 9. The effect of a westerly storm on the sediment transport patterns can be seen in Figure 10.

From the study the following results and conclusion have been obtained:
- the erosion of the northern part of Texel is caused by a type of valve mechanism. The in northern direction increasing flood transport picks-up the sediment of the coastline, which is temporarily deposited in the inlet and the inner delta. The

ebb-currents picks this sediment up and transport it mainly through the central and northern channels to the outer delta and the coastline of the Island of Vlieland;
- part of the sediment that is deposited at the outer delta is transported in an offshore direction under conditions with high waves. So only little sediment returns to the coastline of Texel;
- all protection schemes (except nourishment) stop the erosion, so less sand is picked-up from the northern coastline and as a consequence less sand is provided to the system, which may suffer from a deficit of sand;
- The 1000 m long dam blocks a large volume of the sand and therefore may have the greatest impact on the tidal inlet system; a deep erosion hole is expected at the head of the dam;
- the groyne and offshore breakwater systems also prevent that sand is provided to the morphodynamic system and therefore will also have an impact on the system, although to a lesser extent than the long dam.

Presently, the study is underway and it shows that the COMOR modelling system provides a good tool for the evaluation of the impact of protection schemes on larger morphodynamic systems such as the tidal inlet "Eyerlandse Gat".

6. The German Wadden Sea

Besides the above Dutch tidal inlet projects DELFT HYDRAULICS is presently involved in a number of projects related to tidal inlets to the German Wadden Sea.
The Niedersächsiches Landesamt für Wasserwirtschaft, Forschungsstelle Küste at Norderney commissioned the construction and implementation of a dedicated two dimensional tidal model of the western part of the German Wadden Sea. The aim of this dedicated model, once implemented at the computer of the Landesamt, is to provide a management tool for the definition and verification of development plans for the area.

In order to assess the effect of the construction of a pipeline for gas transportation to the German Coast on the hydrodynamics and the morphology of a number of tidal inlets, numerical models were constructed. Special attention was given to the modelling of extreme design conditions providing the boundary conditions for the pipeline design.

7. Conclusions

For a sound management of complicated systems such as the Wadden Sea, it is important to have a thorough understanding the effect of human interference on the system and its functions. In this respect the morphodynamic behaviour of the tidal inlets to the Wadden Sea plays an important role and should be known and preferably predictable.

Tools are available to analyse the governing processes of such morphodynamic systems. From a number of projects the COMOR-model, being a numerical model for coastal morphology, showed to be a powerful tool for the analysis of the morphodynamics of these complicated systems.

References

Hillen, R., 1992. Dynamic preservation of the coastline in the Netherlands. How does it work. Int. Coastal Congr., Kiel, Germany.
Overeem, J. van, 1990. Coastal zone management and coastal defence in The Netherlands. OAS seminar on erosion, pollution and recovery of the coastal zone and its resources. Buenos Aires, Argentina.
Steijn, R.C., 1989. Schematization of the natural conditions in multi-dimensional numerical models of coastal morphology. DELFT HYDRAULICS report H526-1.
Steijn, R.C. and Louters, T. 1992. Hydrodynamic and morphodynamic modelling of a mesotidal inlet in the Dutch Wadden Sea. Oceanology International 92, Brighton, U.K.
Ribberink J.S. and De Vroeg J.H., 1992. Coastal protection Eyerland. Hydraulic and morphologic effect study. DELFT HYDRAULICS report H1241.
Wind, H.G. and Peerbolte E.B., 1992. Coastal Zone Management; tools for initial design. Int. Coastal Congr., Kiel, Germany.

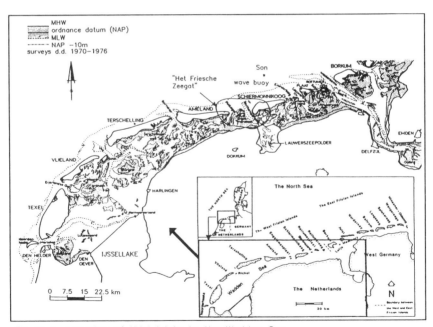

Figure 1 Location of tidal inlets to the Wadden Sea

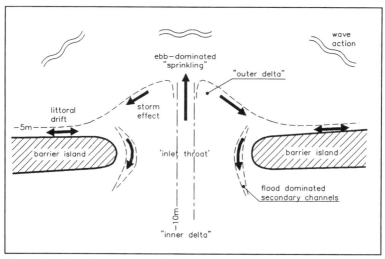

figure 2 Simplified pattern of sediment transports in a tidal inlet

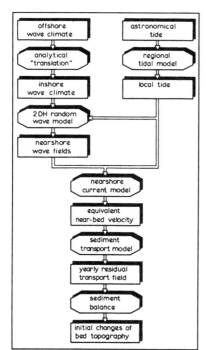

Figure 3 Aggregate flow chart of the numerical model

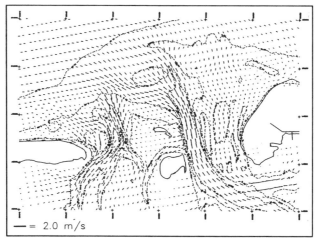

Figure 4 Flow pattern at maximum ebb and NW waves

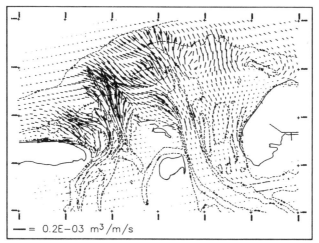

Figure 5 Sediment transport patterns at maximum ebb and NW waves

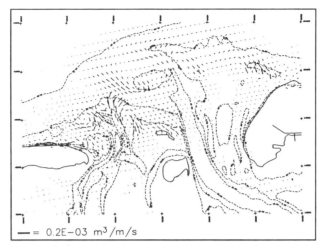

Figure 6 Residual sediment transport pattern for NW waves

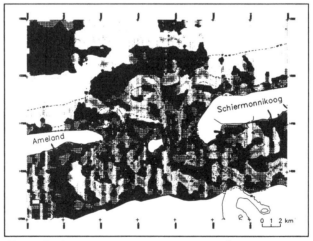

Figure 7 Computed annual sedimentation/erosion pattern

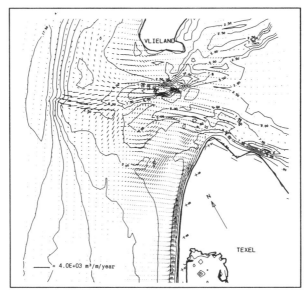

Figure 8 Annual sediment transport patterns

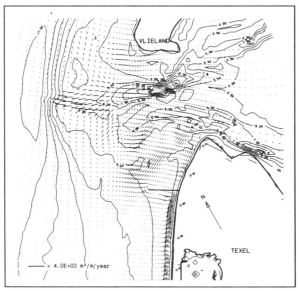

Figure 9 Annual sediment transport patterns for sedimentation with dam

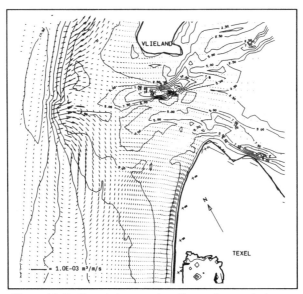

Figure 10 Sediment transport patterns for westerly storm waves

MORPHOLOGICAL STRUCTURES IN GERMAN TIDAL FLAT AREAS

by

Dr.-Ing. Reinhard Dieckmann

Esinger Steinweg 98, 2082 Uetersen

Fed. Republic of Germany

1 INTRODUCTION

The nearshore region of the German Bight is characterised by tidal flats covering an area of 7.500 km^2 (Fig. 1). These flats are a region of extensive alluvial activity. Sedimentation and erosion occur in the long term but also on a daily basis. The

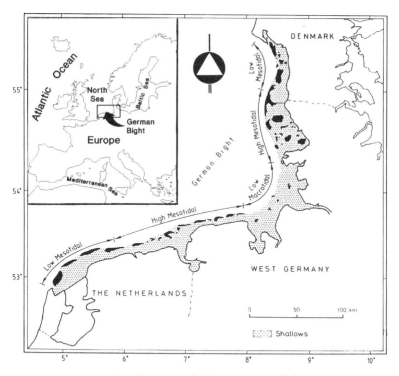

Fig. 1: Map of the German Bight

forces of tidal currents and wave action create a manifold and fascinating pattern of forms in the flat area. Hydraulic forces and the response of the morphology are in a mutual relationship. Special patterns of forms can be found in the entire scale range of morphological investigations, i.e. from small ripple marks in the micro scale to the general morphology of large coastal sections in the macro scale. This paper deals with special morphological structures in the meso scale which can be found in the flats of the German Bight.

2 MORPHOLOGICAL STRUCTURES
2.1 Definitions

The tidal flat area along the German Bight consists of approximately 40 tidal basins. Tidal flooding and drainage of the basins occurs in streams, creeks, and gullies. Looking at these tidal basins as shown in aerial photographs, it can be seen that the morphology of the flats is very complex and dissected by numerous small to large watercourses creating a typical shape pattern. Within this pattern there exist several repeatedly occurring sand bodies called morphological structures. A distinctive feature of these structures is the mutual margin with the watercourses. The shapes of the watercourses are also characteristic. Sand bodies with a special shape are defined as morphological structures when their position and form indicate which forces are responsible for their formation [6]. Basic forms are described in [7] and [8]. In general, there is a distinction between:
- channel-like structures such as ebb and flood channels,
- areal structures such as different kinds of plates and ridges.

The evolution of ebb and flood channels and their importance in coastal morphology is well known [2]. Special conditions pertaining to ebb and flood channels in estuaries are pointed out in [5].

Areal morphological structures are a part of the inventory of tidal flat forms and are visible in aerial photographs as well as in bathymetric charts with detailed contour lines. A suitable set of coastal charts with a scale of 1:25.000 is available for the German North Sea coast.

The following areal morphological structures have been detected to date [1]:

Tongueshaped ebbridges are extended sand bodies separating ebb and flood channels. These structures are connected with larger plates of flat areas at the landward end.
Hook plates can be found at bifurcations of tidal streams and tidal creeks. A hook plate projects into the main channel and acts as a leading structure influencing the flow conditions in the bifurcation area.
Middleground ridges are formed in a very broad channel creating two subchannels. Coriolis force effects are responsible for the different flow conditions during ebb and flood tide in these subchannels.
Saddle ridges are extended sand bodies creating a connection between two different plates. These ridges are lower in the middle part than at either end. The morphology of a saddle ridge is similar to a hyperbolic paraboloid.
Marginal ridges are local elevations at the margins of larger plate areas. Most of the marginal ridges are partly or totally separated from larger tidal flat areas by small ebb or flood channels.

Figure 2 shows as an example of how morphological structures are incorporated into the morphology of a tidal flat area.

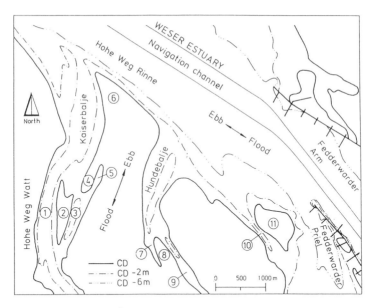

1 Main channel
2 Middleground ridge
3 Side channel
4 Tongueshaped ebbridge
5 Seaward limited ebb branch
6 Hook plate
7 Flood branch
8 Tongueshaped ebbridge
9 Ebb branch (transistional)
10 Split runnel
11 Marginal ridge

Fig. 2: Arrangement of different morphological structures in a tidal flat area beside the main channel of the Weser estuary

2.2 Occurrance of morphological structures

Morphological structures are best visible at mean low water (MLW). Nevertheless, it may be assumed that special morphological structures also occur at higher levels in the tidal flats. These structures are of lesser vertical extension and therefore not easy to detect [3,4].

A systematic catalogue of morphological structures is not available at present. There is also no information on the systematic distribution of the structures within a tidal basin. Structures formed mainly by ebb currents are predominant along the German North Sea coast.

The length of time these morphological structures exist depends on the hydrological boundary conditions; there are no generally valid rules. However, some morphological structures in the Weser estuary have been observed since their formation for a period of more than 30 years ([1], see chapter 3).

Though the German Wadden Sea shows many morphological structures that presumably are also found in many other similar coastal areas, it shows unique structures due to different hydrological, morphological and sedimentological conditions.

2.3   Importance of morphological structures

Morphological structures are always the result of sediment transport processes. The present knowledge on the mechanisms of sediment transport is not sufficient to explain the evolution of these structures. To date, some attempts have been made to explain the causality between the different morphological structures and the hydrological conditions from which these special forms of sand bodies originate.

The morphology of the tidal flats is very dynamic and stable conditions are not attained because the hydrological boundary conditions are always changing in short as well as in long time periods. However, some morphological structures are quite stable and therefore relatively resistant to short time changes. Examples are hook plates, marginal plates and ebb channels. Others, such as tongueshaped ebbridges, flood channels, and middleground ridges are sensitive to changes. If the general evolution of morphological structures is known, it is possible to use these structures as an early indicator for changes in the hydrological system of a tidal basin. This also gives one the opportunity to make better predictions for morphological changes occurring in the vicinity of the observed stuctures.

The course of development including the formation, the mature state and the process of degeneration is known today only for some morphological structures. In order to progress in this area, it is necessary to describe the development of morphological structures in a quantitative manner. Finally, it should be possible to use morphological structures as a tool for monitoring tidal flat areas.

3 EXAMPLE: THE TONGUESHAPED EBBRIDGE EVERSANDSTEERT

One of the most striking morphological structures in the Wadden Sea is the tongueshaped ebbridge. The tongueshaped ebbridge Eversandsteert, situated at the margin of a side channel in the Weser estuary, is an example of a morphological structure whose development we can describe quantitatively. The development of Eversandsteert took place without any artificial influence.

The development of the Eversandsteert from a small round plate to a typical tongueshaped ebbridge is shown in 8 steps in Figure 3. The 8 frames all show the same section of the overview map. The growing of the plate in the ebb direction and to a lesser degree in flood direction is visible. A charac-

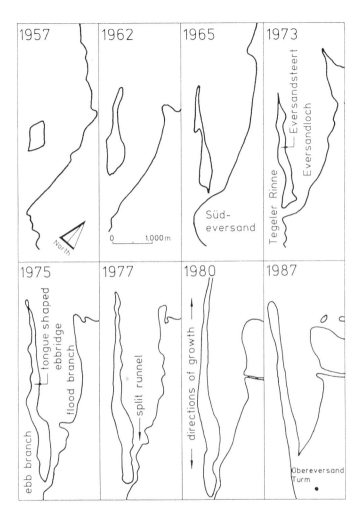

Fig. 3: Evolution of the tongue shaped ebbridge Eversandsteert since 1957

teristic split runnel has been formed between the structure and the main flat area. This split runnel silted up after 1980 as a part of the process of degeneration. The width of the opening of the adjacent flood channel decreased. Reasons for these changes include normal sedimentation processes in the flood channel, the shifting of the structure, and the swaying of the split of in the main channel.

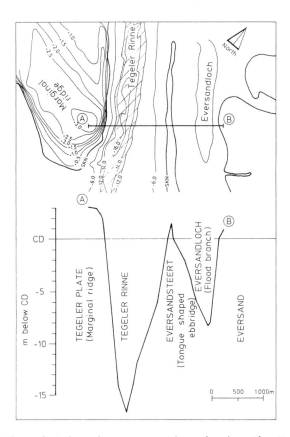

Fig. 4: Plan sketch and cross section showing the Eversandsteert (1980 conditions)

Tongueshaped ebbridges are part of the morphological system of the large main channels (ebb channels) as shown in Fig. 4. The cross section of the main channel decreases during its development. Scouring took place in the main channel parallel but opposite to the ebbridge, thus fulfilling the requirements of the continuity equation. The main channel forms a asymetrical cross section as a result of this process. Figure 4 also shows very clearly that the ebbridge consists of a large sand body. Only a small part is visible at low tide.

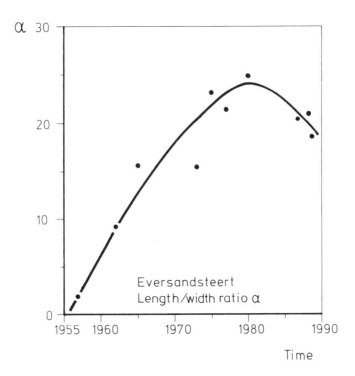

Fig. 5: Development of the length/width ratio α of the Eversandsteert between 1957 and 1989

A very easy way to characterise the development of an ebb ridge in a quantitative manner is to specify the length/width ratio α of the structure. The length/width ratio is given by

$$\alpha = \frac{max.\ length}{mean\ width}\ [-]$$

for every stage of development. Fig. 5 presents the variation of α for the Eversandsteert as a function of time. At the stage of the largest extension in length α was determined to be approximately 25. Since that stage (1980) the Eversandsteert is in a process of degeneration. This indicates that the hydrological boundary conditions in this area have changed significantly. Unfortunately, it is not possible to verify this due to a lack of measurements.

When comparing α values for ebbridges from different locations along the Wadden Sea coast it is evident that the Eversandsteert is a special case (see table 1). Values of α vary between 5.1 and 24.9. To interpret this wide range of values it is necessary to get detailed information on the stages of development of the different tongueshaped ebb ridges and on prospective local particularities. This is a task to be undertaken in the future.

(1)	(2) Length [m]	(3) Area [m^2]	(4) Mean width [m] (3)/(2)	(5) length – width ratio α (2)/(4)
Eversandsteert 1980 Eversandsteert 1989	5.640 3.700	1.276.000 742.980	226 200	24,0 18,5
Hohenhornsand Tidal basin Till	4.400	1.613.200	366	12,0
Tertius-Sand Nord Tertius-Sand Süd Tidal basin Meldorfer Bucht	4.000 7.300	2.668.800 7.244.200	667 993	6,0 7,35
Purrenplate Eider estuary	3.000	1.362.000	454	6,61
Kohlhof Tidal basin Hever	2.600	1.083.600	416	6,2
Heversteert Tidal basin Norderhever	5.600	6.140.800	1.096	5,1
Westerbrandung Tidal basin Norderaue	3.000	675.200	225	13,3
Jungnamensand Vortrapptief	3.200	682.800	213	15,0

Table 1: Some characteristic parameters of different tongueshaped ebbridges

4 REFERENCES

[1] DIECKMANN, R.: Morphologische Strukturen im Weserästuar. Deutsche Gewässerkundliche Mitteilungen, Bd. 33 (1989): 104 - 112.

[2] DIECKMANN, R.: Geomorphologie, Stabilitäts- und Langzeitverhalten von Watteinzugsgebieten der Deutschen Bucht. Mitteilungen des Franzius-Instituts für Wasserbau und Küsteningenieurwesen der Universität Hannover, Heft 60 (1985): 133 - 361.

[3] JACOBSEN, N. K.: Geomorphology of the Wadden Sea: 5. Form elements of the Wadden Sea Area. In: Wolff, W.J. (Ed.): Ecology of the Wadden Sea, Vol. I. A.A. Balkema, Rotterdam (1983).

[4] JAKOBSEN, B.: Vadehavets Morfologi. Folia Geographica Dania, Tom XI No. 1 Kobenhavn (1964): 169 - 176.

[5] ROBINSON, A.H.W.: Ebb - Flood Channel Systems in Sandy Bays and Estuaries. Geography, Vol. 45 (1960): 183 - 199.

[6] RODLOFF, W.: Über Wattwasserläufe. Mitteilungen des Franzius Instituts für Grund- und Wasserbau der TU Hannover, Heft 34 (1970): 1 - 88.

[7] SCHELL, R.: Zur Frage der Sandwanderung in Küstengebiet und im Mündungsgebiet eines Tideflusses. Wasserwirtschaft (1953): 292 - 302.

[8] VEEN, J. van: Eb- en vloedschaar systems in de Nederlandse getijwatern. Tijdschrift van het Koninklijk Nederlandsch Aardrijkskundig Genootschap, Wadden Symposium (1950).

Dynamics of the Eider Channel
An Analysis of the Precision of Profiles

G. Gönnert
Technische Universität Berlin, Institut für Geographie

1. Abstract
2. Introduction and Formulation of the Central Question
3. The area of research: The Eider
4. Presentation of the problem
5. The methodology
5.1 The Profile and the "angle of deviation"
5.2 Interpretation of area dynamics using "Central Profile"
6. Further research
7. Literature

1. Abstract

This project analyses the movement of the Eider meander. In addition to a qualitative study of the channel form in horizontal position and vertical line, the changes in sedimentology must be quantified. Calculation of the MORAN-function with depth - changing seems to be the most exact method to balance an area and to give a statement on the dynamics of a certain area.
To visualize the morphology, other methods are necessary. Profiles are suitable, but until now, there had been no possibility to give a statement on the precision of the calculation of cross-sections and of the interpretation of such profiles. One way of solving this problem is to plot an additional profile perpendicular to the isobath. The difference between the two profiles gives the exactness of the calculation of cross-sections and the interpretation of the original profile in percent.
Furthermore, with this procedure it is possible to visualize the dynamics of the area by drawing a map with serial sections perpendicular to the isobath. Further research will have to show whether the formation of new channels can be predicted.

2. Introduction and the formulation of the Central Question

The project, headed by Prof. Voss, cooperates practically with the Amt für Strom- und Hafenbau Hamburg, especially with Prof. Siefert, and the Amt für Land- und Wasserwirtschaft Heide, Dezernat Gewässerkunde Büsum, especially with Dr. Wieland.
The objective of this project is to analyse the influence and effects of protective constructions with special reference to the barring of the Eider (1967-1973) on the morphodynamics of the channel in the Eider. The priority is to investigate the increase of the meandering on account of the disturbance of the dynamic balance.

3. The area of research

The Eider is located in the northwest of Schleswig-Holstein, south of Eiderstedt and north of Büsum. The area of research encloses the estuary of the Eider from the barring to the North Sea.
The research includes the inside of the estuary, which reaches from the barring to 10 km westwards and from the outside of the estuary west of the 10 km line to 14 km. This limitation is due to the existing maps and data.
The reason for the investigation of both the inside estuary and the outside estuary, is that the influences on the channels are different for each. While the outside is influenced by the independent processes of the North Sea with horizontal and vertical moving sediments and with sediment passages, the inside is influenced from both - the North Sea and the River-Eider.
Alongside the 190 km course of the river are a lot of protective buildings. The morphodynamics of the whole area is heavily unbalanced. One example is the barring of Nordfeld built in 1934/36. One effect of this barring was that the sedimentation rose until the diameter of the channel was reduced to one third. Consequently navigation was endangered.
The barring of the Eider had been built up between 1967 and 1973 to save the coast from storm tides and protect navigation. The tidegate is in operation since 1979, regulating the tidestream and stopping sedimentation.

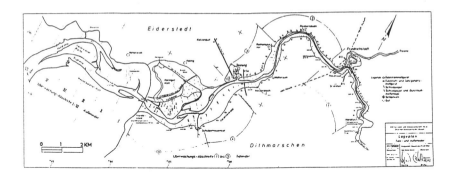

Chart 1: The Eider (in: Amt für Land und Wasserwirtschaft Heide 1986)

4. Presentation of the problem

On account of the protective constructions, the morphodynamics have been heavily unbalanced. The most spectacular episode was the northwards changing of the northern channel. Until the barring had been built, the northern channel was only more or less an insignificant branch of the southern channel. While the channels in 1966 were able to swing back and forth between both coastlines, Hundeknöll and Vollerwiek, the river diversion by ponding 500 m north of the natural river reach stopped the room to move. The consequence was that the channels in the estuary were heavily meandering and sharply changed their course. Under the influence of the Coriolis force the low tide stream shifted into the northern channel, which moved towards the dike with an increasing speed. By 1979 the speed of this motion was 1 meter per day and there was a distance of only 60 meters to the foot of the dike.

Chart 2: Aerial photograph of the research area (in: Amt für Land- und Wasserwirtschaft Heide 1986)

To protect the dike and the coast a sand dam had been built, which divided the
north channel into two arms. A connection between the north and the south
channel had been built to divert the low tide stream orientated on the channel
morphology and the moving of the tide.
The speed of the shifting of the remaining north channel decreased
immediately, but the channel kept moving towards the dike. In 1984 most of the
current of the low tide changed direction and is now running mostly in the
southern channel. The southern channel has deepened now, the northern channel
has sedimented.
At the moment the protective constructions are not endangered. A new Danger
could be an increased meandering of the south channel northwards. The
consequence would be that the connection between north and south channel,
which has sedimented over the last years, will be torn open and the low tide
stream will take the north channel. As a result the north channel will be
moving northwards again and endanger the dike. As the aerial photograph shows,
the south channel has developed a north meander. The question is now, whether
the meander will tear open and endanger the dike. Another possibility is that
a new channel, which is developing in the south at the bottom of the meander,
would reach the other bottom of the meander. Then the low tide would take the
new channel and an endangering of the dike would not happen.

5. The methodology

To analyze the complex situation in such a dynamic area, it is necessary to
use methods such as determination and presentation of the meandering channel
form in position (path of the valley), depth (level of the valley bottom) and
river sections. Furthermore there are quantitative methods necessary like
calculation of the MORAN-function (Morphological Analyzes North Sea coast)
with plans on depths - changing.
Looking for methods to answer this question, it turned out that to quantify
the consequences of the barring and to calculate morphological parameters, the
MORAN-function appears to be the most exact method (for detailed presentation
look at SIEFERT 1987; HOFSTEDE 1989 & 1991). To visualize the morphology and
to realize the geometrical form, it is necessary to use other methods. River
sections are a good possibility to show the morphology. So far, a controlling
of the precision of profiles and the method of balancing the volume of a
channel, the "Profilganglinienverfahren" has not been used. This will be
presented as follows.

5.1 Profile and the "angle of deviation"

Current velocity of the tide is an important force that forms the channel. The
knowledge of their distribution inside the cross-section, especially at the
bank and inside the longitudinal section of the channel is an important help
for the interpretation of the explanation of morphological trends, especially
with regard to the channel movement.
Usually the measurement of the current only happens selectively. Effects on
the whole area are possible to see in the morphology. Consequently, it seems

to be interesting to analyse profiles and the "Profilganglinienverfahren". This work concentrates on accuracy.

While making use of plans on depths - changing, inexactness is especially due to the non - accuracy of maps or the interpolation. The inexactness of planimetring is due to the calculation of the volume with isobaths which are smoothed out. KNOP 1963, HIGELKE 1978, PATENSCKY 1979 (sh. Franzius - Institut 1979) calculated the volume of the cross-sections, Taubert 1986 described it exactly. When using this methods the likelihood of increasing potential source of error is comparatively high. The inexactness is created by the inflexible river sections, which are orientated at fixed points. To reproduce the real river topography in the cross - sections, it is necessary to read the profile in the right angle to the isobath. If the channel meanders, the angle of the profile to the isobath changes. Consequently, the line of the reading profile is extended. The analysis of the use of profiles until know shows, that these problems were not recognized. Sometimes it had been judged by subjective criteria whether it might be possible to use profiles to balance the volume with cross-sections or not.

One possibility to find out the precision of profiles is as follows: At first one has to draw the normal profile for example orientated at the coordinates or at the morphology for one situation. For the presentation, a situation in the middle of all years was taken. This profile was called "Central Profile". It takes the position of a normal profile. Orientated along this "Central Profile" a so called "Original Profile" will be drawn, which has been laid perpendicular to the isobath. One has to pick a rotation point on the "Central Profile" in order to establish comparability between the "angles of deviation". For this example, the rotation point was chosen in the middle of all years (1980) and in the middle of the channel (for this year). This provides a possibility to find the objective "angle of deviation". In the chart 5 an example is shown in which the "Original Profile" is not orientated along the rotation point. Profile [II] shows the possibility for 1990 in which the profile is not orientated along a rotation point and only along a "Central Profile". On account of the shifting of the profile, when looking for the right angle and orientate only along the "Central Profile", it is possible to find three places for the profile. That is the reason for the rotation point. But it is possible that the chosen point is not the best point for every year. Sometimes, there would be smaller angles with other rotation points. But this drawback of the rotation point is much smaller than the incomparability of the profiles which are not orientated along the rotation point. But if the angle substantial is quite big on account of orientating the "Original Profile" along the rotation point, the profile with the smaller angle must be used. That is the reason why it is not possible to find a rotation point at very dynamic places (see interpretation).

Using this procedure, it is possible to measure the exactness of profiles and the "Profilganglinienverfahren". For example: Profile [V] shows a 45º "angle of deviation" in East/South in 1981. The extension of the length of the profile is to be calculated with cosine in the following gradation:

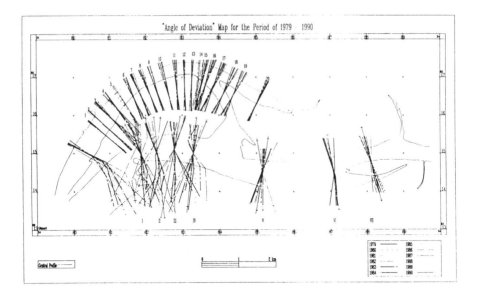

Chart 3: One example of plotting the "angle of deviation" (Profile V)

- Up to 5° it is possible to set cos. as angle.
- At 10° the length of the profile is extended by 1.5%.
- At 25° the length of the profile extended by 10%.
- At 30° the length of the profile is extended by 15%.
- At 45° the length of the profile is extended by 41%.
- At 50° the length of the profile is extended by 55%.
- At 60° the length of the profile is extended by 100%.
- At 70° the length of the profile is extended by 192%.

That means that in case of a 45° "angle of deviation", the length of the profile will be extended by 41%. Theoretically, this means that the calculation of the cross-section will result in a 41% increase. But this would only be the case if the channel is tubular. In reality this is not possible.

Comparison of Cross-Sections: Profile V
Central- and Middle Profile 1981

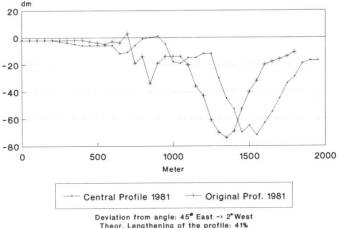

Deviation from angle: 45° East -> 2° West
Theor. Lengthening of the profile: 41%
Real enlargement: 177 sqm; 5%

Chart 4: Comparison of "Central Profile" and "Original Profile": Profile V

As shown in the chart, profile [V] differs less extreme as profiles [II] or [III] and the rotation point has been constant because of the position of the "Original Profiles". 1981 has a "angle of deviation" of 45o East/South and 2o West/North to the "Central Profile". This means that the length of the profile is extended by 41%. The cross-sections in 1981 show:

- The length of the profile had been extended.
- The undercut slope flattens.
- The cross-section increased.

In comparison:
"Original Profile" at -1.0 m: 2385.96 m^2, at 0 m: 3704.67 m^2
"Central Profile" at -1.0 m: 2459.06 m^2, at 0 m: 3881.57 m^2

The increase of the cross section is 73.1 m^2 at -1.0 m and 176.9 m^2 at 0 m. Thus the cross-section increases by 3% at -1.0 m and by 5% at 0 m despite an increase of 41% in length of the profile.
The reason behind this result is the topography and the form of the channel. Moreover, the "Original Profile" is fixed at different points because of the "angle of deviation". This can be shown with the example of profile [18]. The "angle of deviation" is 18o. This means that 1984 the "Original Profile" differs at 9o West and 1969 at 9o East of the "Central Profile". Measured at 0 meter isobath there is a difference of 220 m between the "Original Profiles" of 1981 and 1969. This shows that this procedure is not a method to calculate

cross-sections and interpret profiles, but a procedure to measure the exactness of cross-sections.

Used by coastal engineers, this procedure suggests that any drawing or interpretation of profiles should be accompanied by a calculation of the exactness using "angle of deviations". It may be suggested that up to an "angle of deviation" of 5° the cosine is equal to the angle and that this is neglectable. Up to 20° additional controls are not necessary. From 25° onwards additional control calculations are strongly suggested. Above a 40° "angle of deviation" no cross-sections can be calculated.

Furthermore for the interpretation of profiles it may be suggested to plot at least some "Original Profiles" because the lengthening of the profile may flatten the slope and broaden the valley bottom. Consequently misinterpretations are possible. It is necessary to visualize probable errors. In principle, any error can be neglected as long as it is in range of the other error probabilities.

5.2 Interpretation of area dynamics using "Original Profiles"

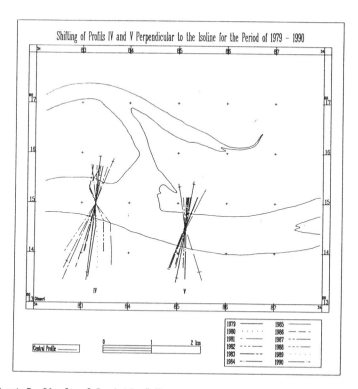

Chart 5: "Angle of Deviation" Map

There is one more possible utility in using deviation angels. By drawing a "Deviation-angle-map" it may be possible to visualize the dynamics of a certain area. Normally three profiles per meander are plotted to do a morphological analysis. But often, as the project experiences showed, three profiles are not enough to detect dynamical areas of interest. It may be suggested that there is a connection between the "angle of deviation" and a dynamical process, as it is shown in chart 5. In this "Angle of Deviation" Map the right angle had been orientated only at the north side of the northern channel, because this appeared to be the interesting side. In the southern channel, where during the period of 1969 to 1990 a meander had developed, the "Original Profile" had been laid perpendicular to the north and the south side of the channel. The "angle of deviation" in the northern cannel had a maximum of 10° in profile [17], whereas the maximum "angle of deviation" in the southern channel proved to be 48° in profile [I].

As has been described above, the "angle of deviation" is a result of the orientation of the "Original Profiles" in a right angle to the isobath. The higher the dynamics of an area, the more the isobath will change and with it the "angle of deviation". Within the research area the southern channel showed a greater "angle of deviation" and in one case (profile [II]) the rotation point could not be fixed. This is a further indication that in the southern channel there had been strong movements of the meander. Because the "Original Profiles" had been oriented along the north- and south side of the southern channel, the evolution of the slope can be observed. In most cases the "angle of deviation" in the north and the south proved to be different. Furthermore, there are points of intersection beyond the rotation angle sometimes, where the undercut slope and the slip-off slope side of the "Central Profile" are meeting. This and the different angles indicate that the northern- and the southern-slopes move northwards in batches. In between there are phases where both slopes moved parallel.

At the time the meander had developed, a parallel movement is shown at the peak while the flanks changed merely a little.

6. Further research

Further research should concentrate on correlating the results mentioned above with the turnover calculated using the MORAN-function.

The presented procedure will be reviewed as to whether it is possible to use it in analyzing the evolution of a new channel at the bottom of the meander. The hypothesis is that a new channel will cause a "angle of deviation" increase above and below the rotation point. This is because of the separation of the channel. It will be controlled through correlating the procedure and the evolution of the path of the valley.

7. Literature

Amt für Land und Wasserwirtschaft Heide, Abteilung Wasserwirtschaft, Dezernat Gewässerkunde (1986): Teilsachstandsbericht Eider. In: Büsumer Gewässerkundliche Berichte, Heft 52.

Amt für Land und Wasserwirtschaft Heide, Abteilung Wasserwirtschaft, Dezernat Gewässerkunde (1990): 1. Nachtrag zum "Rahmenentwurf für morphologische und hydrologische Untersuchungen im Eidergebiet von 1978 bis 1987 vom 28.8.1977" für den Zeitraum von 1988-2000. Büsum 1990.

Borchert, G. (1992): Analysis in Geographie. A Methodical View. In: Geo Journal, Heft 3, S.259-263.

Franzius-Institut für Wasserbau und Küsteningenieurwesen der Universität Hannover (1979): Trendanalyse der Querschnittsverformungen in der Tideeider in den Jahren 1965 bis 1977 und Untersuchungen zum erforderlichen Umfang der Peilungen in der Tideeider. Hannover 1979.

Göhren, H. (1970): Studien zur morphologischen Entwicklung des Elbmündungsgebietes. Hamburg 1970. (Hamburger Küstenforschung, Heft 14)

Higelke, B. (1978): Morphodynamik und Materialbilanz im Küstenvorfeld zwischen Hever und Elbe. Ergebnisse quantitativer Kartenanalysen für die Zeit von 1936 bis 1969. Regensburg 1978. (Regensburger Geographische Schriften, Heft 11)

Hofstede, J. (1989): Parameter zur Beschreibung der Morphodynamik eines Wattgebietes. In: Die Küste, Heft 50, 1989, S.197-212.

Hofstede, J. (1991): Hydro- und Morphodynamik im Tidebereich der Deutschen Bucht. Berlin 1991. (Berliner Geographische Studien, Band 31)

Knop, F. (1963): Küsten- und Wattveränderungen Nordfrieslands. Methoden und Ergebnisse ihrer Überwachung. In: Die Küste, Heft 11, 1963, S.1-33.

Siefert, W. (1983): Morphologische Analysen für das Knechtsandgebiet (Pilotstudie des KFKI-Projektes Moran). In: Die Küste, Heft 38, 1983, S.1-57.

Siefert, W. (1987): Umsatz- und Bilanzanalysen für das Küstenvorfeld der Deutschen Bucht. In: Die Küste, Heft 45, 1987 (Sonderdruck).

Taubert, A. (1986): Morphodynamik und Morphogenese des Nordfriesischen Wattenmeeres. Hamburg 1986. (Hamburger Geographische Schriften, Heft 42)

Wieland, P. (1990): Die Eiderabdämmung. (unveröffentlicht).

The Nösse-Peninsula of the Island Sylt / North Frisia: Hydrology and Sedimentology of Storm Surges

Belinde STIEVE[1], Jürgen EHLERS[2]
([1] Physio— und Polargeographie, Fachbereich Geowissenschaften, Universität Bremen, Postfach 33 04 40, W - 2800 Bremen 33)
([2] Geologisches Landesamt Hamburg, Oberstr. 88, W - 2000 Hamburg 13)

Introduction

Salt marshes originate from accumulation of marine sediments on tidal flats. Their generalized stratigraphy shows a near horizontal layering of seasonal deposits: more sandy in winter, more clayish in summer. Extreme flooding events leave even coarser traces, for example mollusca shells (*Littorina littorea*, *Peringia ulvae* and *Mytilus edulis*); their transport demanding higher wave energies.

Fig. 1 shows two profiles from a salt marsh region on the Isle of Sylt/North Frisia. The scale is too small to display seasonally caused differences in grain size; these layers occur with a thickness of only a few mm.

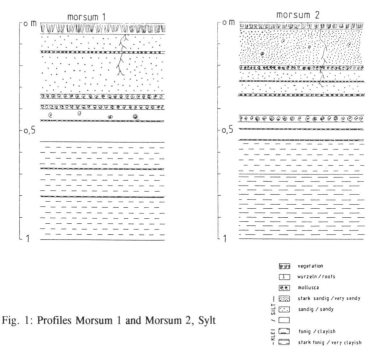

Fig. 1: Profiles Morsum 1 and Morsum 2, Sylt

This layered structure destines salt marshes to be a good coastal feature for the study of past hydrology within a region.

Storm surges play an important role in the mass movement of a salt marsh region. They accelarate both sedimentation and erosion. These aspects are studied in the project "Storm surges and their sedimentological records in salt marshes of the wadden sea; periodicy of a natural hazard" at Bremen University.[1]

The first aim is to recognize storm surge accumulation sedimentologically and geologically in salt marshes of the German Wadden Sea, and then to relate these traces to historic storminess as derived from tide gauge records for the given region. Next the defined storm surge sediment characteristics are to be applied in the investigation of older sediments, in order to extrapolate the storm surge history to times beyond tide gauge records (which exist in the region of the German Bight for some one hundred years, although for most of the stations only a much shorter period). Finally conclusions will be drawn on climatic and sea level variations over the last few hundred years.

Storm Surges on the Isle of Sylt: the Hydrology

For the whole region of the German Bight/North Sea a rise in relative sea level of 25 cm/1oo years has been calculated, this value now seems to be increasing (FÜHRBÖTER 1986, ROHDE 1988). Additionally more storm surges are recorded.

This trend is also apparent in the hydrological data from Sylt (station List). 19 year mean high water rose from o,66 m to o,78 m NN (Amsterdam level) between 194o and 1976 (values for 1936 to 1985). Mean low water in that time went down about 5 cm, thus the tidal range increased by 15 cm, with effects also on tidal wave velocities and on coastal conditions. For comparison Fig. 2 shows the 3-year, 7-year and 19-year averages, with very distinct upward orientation of all three curves:

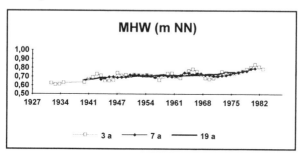

Fig. 2: Mean high water: 3-year, 7-year and 19-year averages (data from: GEWÄSSERKUNDLICHE JAHRBÜCHER, NORDSEE)

---

[1] The project is funded by the Commission of the European Community under the EPOCH-contracts "Climate Change, Sea Level Rise and Associated Impacts in Europe".

Fig. 3 depicts the High Water Events from the tide gauge List/Sylt of the past 9o years that qualify as storm surges (1,7 m above mean high water). This curve is adjusted to the rising mean sea level (the average value of 25 cm/year; with the added 1,7 m on MHW, indicated as straight line), and confirms the tendency to more intense and more frequent storm surges.[2]

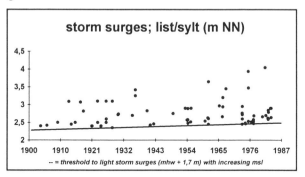

Fig. 3: HW > MHW + 1,7 m, adjusted to the rise in relative sea level, station List, m above Amsterdam level (data from: GEWÄSSERKUNDLICHE JAHRBÜCHER, NORDSEE)

The values for the years 1936, 1962, 1976, 1981 each represent the hitherto highest all time events, a definate upward development.

decade	number of storm surges
19oo - 1o	3
1911 - 2o	5
1921 - 3o	11
1931 - 4o	5
1941 - 5o	4
1951 - 6o	8
1961 - 7o	9
1971 - 8o	15
1981 - 85	(12)

---

[2] A rising reference line for MHW was chosen so the increasing reference base is taken into account. The frequently applied definition for storm surges relating to a certain level above MHW normally considers a single MHW for several decades. This would of course lead to an even greater quantitative storm surge increase.

These findings are confirmed by grouping the classified events in decades. The number of events had its first peak in the years 1921 to 193o, and has then - after a more moderate period - increased quite considerably from the 195os onward until the present. We can therefore assume that whatever influences storm surges have on the sediment budget of a region, these influences will most probably leave more distinguishable and clearer traces.

The Region

Field work is undertaken on the eastern part of the island of Sylt, the Nösse-peninsula. On the north coast the salt marsh region between the settlements of Keitum and Morsum today extends over some 15 km^2. The surface of the agriculturally used fields (cattle grazing) is slightly irregular, intersected by artificial ditches for dehydration. The coastline is formed by a 3 km long stretch of exposed salt marsh cliff (or rather microcliff) divided into bights and heads, an average height of about 65 cm with maximal and minimal heights at 1o and

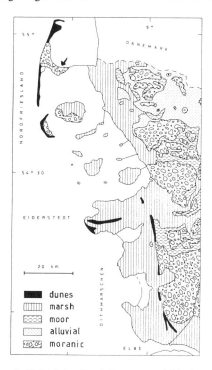

Fig. 4: The Schleswig-Holsteinian North Sea coasts. A black arrow indicates the area of field work.

95 cm. The sites are - owing to their geographical exposure - strongest affected by NW-winds. More than half the wind events (Station List) come from westerly directions, about a third from W and NW (data: METEOROLOGISCHE JAHRBÜCHER).
Results are presented from two cliff profiles (Morsum 1 and Morsum 2) and two sites further inland (Morsum A and Morsum D). The cliff sites were chosen as high cliffs with clear shell layers, the inland sites were selected as places with no history of ditches within a certain range in order to avoid a perhaps considerable mixing of sediments and therefore disturbance of layers. The salt marshes in this part of the island, which have been accumulation since the postglacial transgression are presently undergoing erosion. But the historic development also had its phases of coastal (salt marsh) advance.
The north coast of the peninsula was transgrading from the middle of the 19th century. Maps from the years 1793, 1878 and 1929 (MÜLLER, FISCHER 1938, Fig. 5a) identify this coastline advance between the settlements Keitum and Morsum: a gain between 19o m and 44o m for the total 136 years. (This means an average transgression of 1,4 to 3,2 m/year (for the period 1793 to 1878 1,2 to 3,5 m/year, in the second phase o,2 to 3,5 m/year). The values differ slightly within the region; especially in the interval 1878 and 1929 the western part, the NE-exposed flank of the young marsh, advanced with a slower rate: only 2o cm/year.
In the 20th century erosion forms the main process.[3] Topographical maps of the years 1933 and 1992 prove an average erosion of 4o cm/year (Fig. 5b). This means that on the average the coastline of 1933 was situated some 25 m further seaward than nowadays.[4]
Still today erosion dominates the situation at the sites. Field work has confirmed the erosional rates of the last decades: from October 1991 to August 1992 values between 3 and 4o cm/year were measured, the variation mainly depending on the exposition of the cliff.
Simultaneously accumulation occurs during storm surges when the salt marsh areas are being flooded. These events leave sandy material mixed with shells on the salt marsh surface. A patchy deposition leads to the irregular surface and also is the reason why salt marsh sediments are not really layered horizontally at all over distances (honever small), the characteristic feature the discontinuous layer. This demands a very cautious interpretation of sample results, as we are confronted with gaps in the profiles. This

---

[3] We assume that a high responsibility for this change in mass balance lies with the construction of the Hindenburgdamm (completion 1927), which connects the Island of Sylt/Nösse-Peninsula to the mainland and has interfered quite considerably with the currents in the Wadden Sea.

[4] It is interesting to note that for the south coast of the Nösse-Peninsula a completely different situation occured. Fig. 5a/b reveal that in teh last centureis erosion dominated, and in the last 6o years only coastal regions with artificial land gaining devices produced positive mass balances. Field work is also undertaken on sites south of the Hindenburgdamm, some 8oo m down along the coastline.

explains the differing profiles Morsum 1 and Morsum 2 (cp. Fig. 1), which are about 1oo m apart along the same coastline.

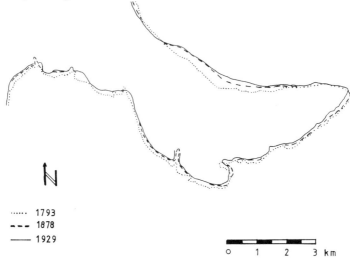

Fig. 5a: Comparison of maps: 1793, 1878, 1929 (MÜLLER, FISCHER 1938)

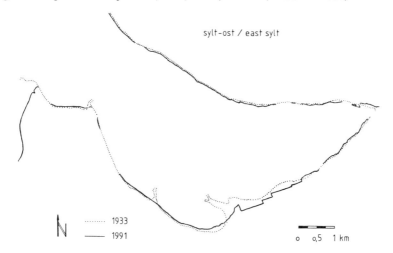

Fig. 5b: Comparison of maps: 1933 and 1992 (original scale 1:25.ooo)

In order to date the cliff stratigraphy the content of $^{137}Cs/^{134}Cs$ was measured. Peaks in radioactivity are correlated to known nuclear fall-out events in this area (1986 Chernobyl desaster, the massive nuclear bomb testing around the test ban treaty of 1963 and the Windscale (Sellafield) fallout of 1957). Defining peaks within the profile as these events, eminent shell layers can be correlated to extreme storm surges of the recent past. Thus the top 60 cm of the layered deposits are dated as represening some 80 years. (Fig. 6). This method has also been used for dating Greenland ice sheets and Mississippi delta deposits (HAMMER ET AL. 1978, REDFIELD 1972).

The Chernobyl peak is very definately definable (also because of the occurence of ^{134}Cs, which has a half life of 2,06 years). The other two defined peaks are less distinct and maybe more questionable, but as a working hypothesis for our investigations they can act as an acceptable time frame work.

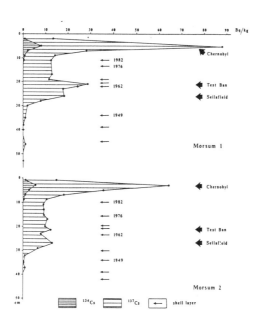

Fig. 6: Morsum 1 and Morsum 2, content of radioactive ^{134}Cs and ^{137}Cs.

To sum up the findings from dating and map evaluation we plotted both erosion and sedimentation rates in Fig. 7: a reconstruction in time over the last 60 years:

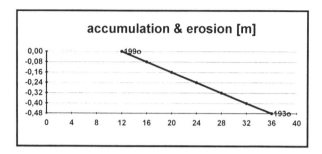

Fig. 7: The development of sites Morsum 1 resp. Morsum 2.

The rising salt marsh level which is additonally affected by the vegetational cover (promotion of sedimentation) adjusts itself to the rising sea level, the salt marsh remains above MHW.

Sedimentology of a storm surge

The general characteristics of salt marsh sediments are: median grain size in silt/clay, coarser sediments resulting from stronger wave activity, high organic content (both from terrestrous vegetation and marine microorganisms), and high carbonate-content parallel to the occurence of shells.

Marine accumulation of sediments on a terrestrial surface follows a landward gradiation, coarser material sedimentates first, i.e. during storm surges shell layers deposit close to the cliff region, these features may continue as coarse or medium sand layers further inland, but certainly in a patchy pattern.

At the feet of the profiles Morsum 1 and Morsum 2 the frontier to old tidal flat deposits is reached. The 'inland'-samples - as Fig. 5a/b indicate - should reveal salt marsh sediments in greater depths; a result of the longer time span as an island of the sites (as opposed to the tidal flat state).

At the investigated cliff profiles Morsum 1 and Morsum 2 the mean grain sizes decrease in downward direction (cp. Fig. 1). This may be the result of the described process: the coastline moved landward, thus the samples from further down in the profile accumulated, when the site was still mainland and also further away from the coastline.

Measurement of carbonate content indicates shells layers or worked shell remains. A comparison of cliff and inland samples shows that the latter are too far inland for a significant accumulation of mollusca during storm surges.[5]

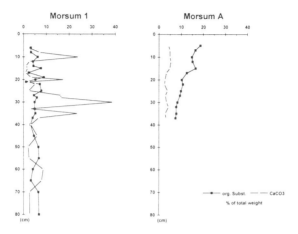

Fig. 8: Profile Morsum 1, A and D: carbonate/organic content

Organic content relates to the landhistory above msl of the area. The second parameter however are microorganisms which promote a flocculation with the finer grains, that are then deposited. More detailed carbon analysis may lead to the interpretation of phases with denser vegetation and less flooding - and therefore less storm surge activity.
Additional information is expected from grain sizes analysis. First results confirm that the inland samples are too young for visible storm surge layers, but we expect coarser grains as we drill downward. The cliff samples show characteristic peaks in the grain sizes about at parallel depths to the highest values of carbonate.

Conclusions

In the investigated area of the Island of Sylt the hydrological data reproduce the North Sea trend of an accelerated rise in relative sea level and more storm surges with higher water levels.

---

[5] Samples were taken trying to regard layers. However for the inland sites this was not possible (no visible structure), therefore 1 to 2 cm intervals were chosen.

Storm surges leave both erosional and sedimentary traces in a salt marsh region. The sediment traces are clearer visible in the immediate vicinity of a coastline. However probably also inland sites will reveal a pattern of seasonal and storm surge layering in the sediment.

In order to correlate the sediment traces to storm surges a chronological framework is fundamental. For young salt marsh deposits this can be achieved through determination of radioactive ^{134}Cs and ^{137}Cs.

The next steps for our work will include dating inland samples, taking deeper cores and also cores halfway between the two sets of sites investigated at the moment. Because of the less distinct structure of the inland sites (and also the stronger disturbance of layering through vegetaion / roots) we are investigating X-ray radiographies and thin sections along with the sediment analysis.

Literature:

DeLaune, R.D.; Patrick, Jr. W.H.; Buresh, R.J. (1978): Sedimentation rates determined by ^{137}Cs dating in a rapidly acreting salt marsh.- Nature 275: 532-533.

Ehlers, J.; Nagorny, K.; Schmidt, P.; Stieve, B.; Zietlow, K. (1992): Storm surge deposits in North Sea salt marshes dated by ^{134}Cs and ^{137}Cs determination.- J. of Coastal Res. (in press)

Führböter, A. (1986): Veränderungen des Säkuläranstieges an der deutschen Nordseeküste.- Wasser und Boden 9: 456-460.

Hammer, C.U.; Clausen, H.B.; Dansgaard, W.; Gudestrup, N.; Johnsen, S.J.; Reeh, N. (1978): Dating of Greenland ice cores by flow models, isotopes, volcanic debris and continental dust.- J. of Glaciology 2o (82): 3-26.

Müller, F.; Fischer, D. (1938): Sylt.- Das Wasserwesen an der schleswig-holsteinischen Nordseeküste. Die Inseln 2 (7), Berlin.

Redfield, A.C. (1972): Development of a New England salt marsh.- Ecol. Monogr. 42: 2o001-237.

Rohde, H. (1980): Changes in sea level in the German Bight.- Geophys. J. R. astr. Soc., 62: 291-302.

Streif, H. (1978): Zur Geologie der deutschen Nordseeküste.- Die Küste 32: 3o-49.

DYNAMIC PRESERVATION OF THE COASTLINE IN THE NETHERLANDS
how does it work out?

Roeland HILLEN

Rijkswaterstaat, Tidal Waters Division
P.O. Box 20907, 2500 EX The Hague, the Netherlands

Abstract

In 1990, the Netherlands Parliament has decided on a new national coastal defence policy. The choice for "dynamic preservation" of the coastline is essentially a choice for enduring safety against flooding and for sustainable preservation of the functions and values in the dunes. The choice for "preservation" implies that the coastline will be preserved at its 1990 position at least.

Since 1990, the concept of "dynamic preservation" has been worked out in more detail. The coastline-to-be-preserved (often referred to as the "basal coastline") has now been defined. Moreover, new ideas on innovative nourishment techniques have been and are being developed and efforts are underway to ensure the dynamic charm and quality of the dune coast, thus acknowledging the "dynamic" aspect of the policy choice.

To realize the policy choice, a set of instruments is available, including an institutional framework, a yearly budget for sand nourishments, and a coastal monitoring and research scheme.

The new coastal defence policy is not just a State affair, it is merely a collaborative effort of national, regional and local authorities in which each party assumes its own role. After slightly more than one year's experience, it appears that we can make the policy work. However, the policy can only be succesful if public awareness and support can be maintained.

1. Introduction

It is sometimes said that God created the World, but the Netherlands were made by the Dutch. Man has been living in the Dutch Delta ever since about 5000 years before present. First living on the higher elevated grounds such as beach barriers and river levees, later building mounds and dikes for protection against flooding. Today the area is among the most densely populated areas in the world, protected from the sea by natural sand-dunes and high dykes. Although more than 60% of the Dutch are living below mean sea level, the country is considered safe from flooding by the sea.

The Dutch have always been fighting the sea, often winning this struggle, sometimes losing. The last flooding disaster occured in 1953: more than 1800 death casualties and a damage of appr. 14% of the GNP. After this event the national Parliament adopted new safety standards against flooding. These standards are defined in the Water Defence Bill which provides a basic legal framework for all coastal defence measures in the Netherlands. For the coast of central Holland, for example, the sea defences (dunes and dykes) are able to withstand a storm surge level which is exceeded only once in ten thousand years on average (i.e.: the probability of exeeding this level is one tenthousandth per year). For other parts of the coast other safety standards are applied, basically depending on the economic value (real estate, infrastructure, etc) of the polderland (Fig. 1).

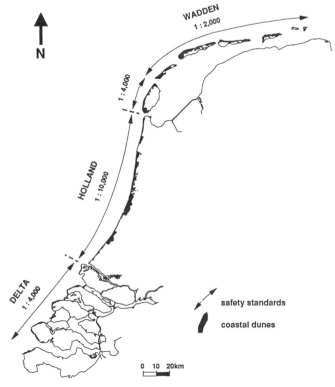

Fig. 1   *Safety standards along the Dutch coast*

Since 1990, the sea defences satisfy the safety standards. However, if no measures are taken against ongoing coastal erosion, tens of kilometres of coast will become unsafe and hundreds of hectares of valuable dune area will be lost within several decades. An accelerated rise in sea level will enhance this problem even further.

The public discussion on a new national policy for coastal defence started in the 1980's. Until that time an ad-hoc policy was followed: measures were only taken when the safety of the polderland was at stake or when special values in the dune area (e.g. drinking water supply areas, nature reserves) were threatened. In 1989, the so-called Discussion Document was presented including four policy alternatives: Retreat, Selective Preservation, Preservation and Expansion Seaward (Louisse & Kuik, 1990; Rijkswaterstaat, 1990). Benefits and costs for all policy alternatives were calculated for the period 1990-2090. After an extensive public discussion among national, provincial and local authorities, scientists, environmentalists and other people concerned with the dune and beach areas, the national Parliament decided in 1990 for the Preservation alternative. To emphasize the wish for the preservation of the natural dynamics and character of the dune coast, the chosen alternative was called "Dynamic Preservation". This policy choice is primarily aimed at enduring safety against flooding and sustainable preservation of the values and interests in the dunes and on the beaches.

Now, almost two years after the adoption of the new coastal defence policy, the parties involved are working hard to implement the new policy. Some important decisions have been made. For example: the coastline-to-be-preserved has been defined, for each coastal province a so-called Provincial Consultative Body has been formed, sand nourishment has been chosen as principal method to combat coastal erosion. But other problems arise: the increasing pressure on the coastal zone (e.g. housing, recreation, land reclamation) and the possible effects of an increased rise in sea level. Furthermore it is essential to keep the public informed. Public support and awareness are essential for a policy to be succesful.

2. The Dutch coast

The coastline of the Netherlands is appr. 350 km long; 254 km consist of dunes, 34 km of sea dykes, 38 km of beach flats and 27 km of boulevards, beach walls and the like. The width of the coastal dunes varies between less than 200 metres and more than 6 km.

The dune coast and the beach flats (occurring at the extremes of the Wadden Islands) are dynamic in character. At some locations there is sand accretion, at other locations erosion prevails. Erosion and accretion patterns also vary in time. Since the middle of the 19th century the position of the dune-foot and the high- and low-water lines is measured every year. For this purpose fixed reference poles has been set up on the beach at intervals of 200 to 250 metres. Since the middle of the 1960's the annual coastline measurements are performed through a combination of remote sensing (on-land) and sounding techniques (offshore). At every reference pole a coastal profile is measured, extending from appr. 200 metres landward of the reference pole to appr. 800 metres seaward. The result of this annual coastal monitoring is a unique data-set available for all types of coastal research and evaluation.

A typical example of the application of the monitoring data is the sand balance of the Dutch coastal system between the 8 m depth contour and the top of the first dune row (Fig. 2; Stive et al, 1990). From this figure the following general conclusions can be drawn:
1. In the North there is a structural loss of sand to the Wadden Sea;
2. For the central part of the coastline (the "segmented" Holland coast), sand is being transported from the deeper part of the foreshore to the shallower part resulting in a steepening of the foreshore;
3. In the Delta area in the Southwest, sand is deposited in front of the closure dams.

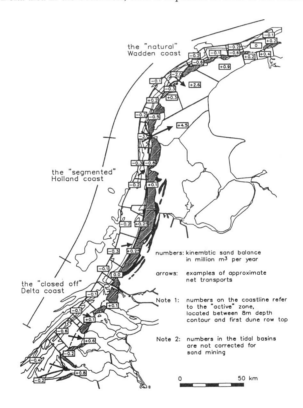

Fig. 2   *Present sand balance of the Dutch coastal system (Stive et al., 1990)*

Comparable sand balance studies have also been performed on different time and space scales. Based on that information, shoreline predictions are made indicating locations where accretion and erosion can be expected in the coming decades.

3. Dynamic Preservation

The choice for "dynamic preservation" of the 1990-coastline was based on a number of considerations: a.o. the enduring preservation of safety against flooding, the sustainable preservation of the dune area, administrative aspects and the costs for coastal defence (Rijkswaterstaat, 1990).
The most important aspect of this choice is that the coastline is to be maintained at its 1990 position. All structural erosion is to be couteracted. The 1990-coastline, called the "basal coastline", is thus a standard for the preservation policy. Every year a check is performed whether the position of the coastline still meets this standard (see section 3.1).
In line with the dynamic character of the sandy coast, sand nourishment has been chosen as the principle method to combat erosion. Sand nourishment is a common coastal defence measure in the Netherlands since the end of the 1970's. Over the years sand nourishment has proven to be an effective, flexible and financially sound method (Roelse, 1990). Prior to the policy choice of 1990, sand nourishments were mainly carried out to repair the damaged coastline at selected locations. Since 1990 the nourishments are no more repair works, but they are meant as a buffer: preventing exceedance of the "basal coastline" (see section 3.2).
Out of the four policy alternatives (retreat, selective preservation, preservation, expansion seaward), the preservation alternative was almost unanimously preferred by all parties involved in coastal defence. Especially at the request of nature conservation organisations the word "dynamic" was added to ensure the natural dynamic character of the Dutch sandy coastline. At present, efforts are underway to preserve (sometimes even re-establish) the dynamic charm and quality of the dune coast (see section 3.3).

3.1 The "basal coastline" concept

"Preservation of the 1990-coastline" and "counteracting structural erosion" looks simple enough. For hard coastal structures, such as dykes, there is no discussion on the position of the coastline. But whére is the 1990-coastline for a dune coast? And what is strúctural erosion? For these questions the concept of the "basal coastline" has been developed. The "basal coastline" is in fact the coastline-to-be-preserved. Every year a check is performed whether this basal coastline has not been exceeded.
The position of the basal coastline for a dune coast is measured for each fixed reference point along the Dutch coast. First the so-called transient coastline is determined from the results of the yearly coastal measurements. The transient coastline for a certain location and for a certain point in time is the result of a volumetric integration of the most dynamic part of the coastal profile (Fig. 3). The amount of sand on the beach and on the shallow shoreface in fact determine the position of the transient coastline.

To calculate the position of the basal coastline for a certain reference point, the position of the transient coastlines over the period 1980-1989 are plotted against time (Fig. 4). The position of the trend-line on the 1st of January 1990 is the position of the basal coastline for that particular reference point.

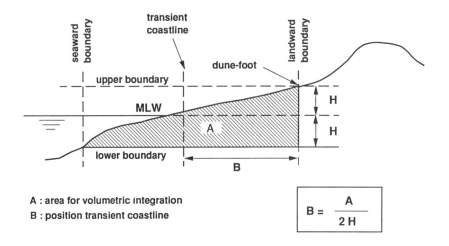

Fig. 3 *Method to calculate the transient coastline*

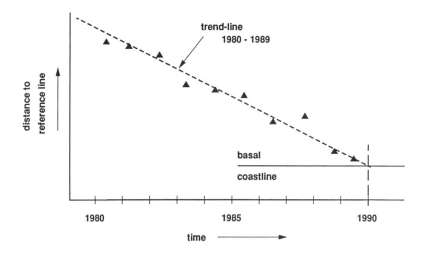

Fig. 4 *Method to calculate the basal coastline*

Thus the basal coastline for the entire coastline of the Netherlands has been calculated. The results of these calculations have been discussed among coastal morphologists and within the Provincial Consultative Bodies (see chapter 4). At the end of this year, the Minister will officially establish the position of the basal coastline for the entire country. Then the standard for the preservation of the coastline is fixed.

It is interesting to note that the results of individual storm surges do not really affect the position of the basal coastline. The concept is aimed at identifying locations with ongoing erosion in the first place. The effects of dune erosion as a result of storm surges is "filtered out" by using a volumetric approach for the calculation of the transient coastline, and by calculating the basal coastline over a period of 10 years. This implies that preservation of the basal coastline does not mean that all dune damage from storm surges will be prevented in the future.

3.2 Sand nourishment

Until 1990, many nourishment projects along the Dutch coast were carried out in or near the dune area. Such nourishments were meant to repair the coastal defence line at places where the safety against flooding was threatened. Since the implementation of the new coastal defence policy in 1990, sand nourishments are meant to maintain the basal coastline. The sand is placed on the beach, thus creating a transient coastline in a more seaward position. The nourished sand forms a buffer against the ongoing erosion and should be placed on the eroding beach before the basal coastline is exceeded.
Every year some 5 to 7 million m^3 of sand is nourished to the Dutch beaches (Fig. 5). For this purpose a yearly budget of appr. 60 million Dutch guilders (35-40 M US$; August, 1992) is available. In fact these costs can be considered the maintenance costs for the coastline. Just for comparison: the average costs for the maintenance of one km of sandy coastline is less than the average maintenance costs of a km highway.

Innovative nourishment techniques are also considered. It is likely that in the coming years a foreshore nourishment project will be carried out. A desk study has indicated that foreshore nourishment, under certain conditions, will be less expensive as compared to beach nourishment (Hillen et al, 1991). Moreover, during the execution of a foreshore nourishment, project recreational activities on the beach are not interupted. For the extensive modelling and monitoring aspects of such an innovative nourishment project, co-operation with Danish and German coastal research institutes has been established and financial support from the MAST-programme of the European Community has been requested.

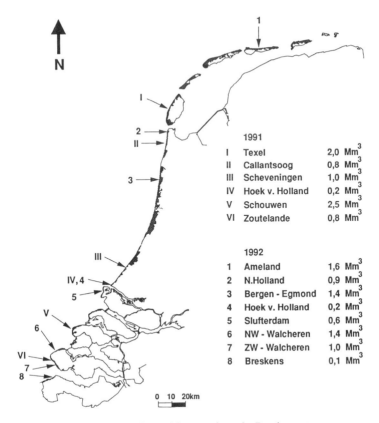

Fig. 5   *1991-1992 beach nourishments along the Dutch coast*

3.3 *Preserving the dynamics of the Dutch coast*

The dune coast of the Netherlands is of great scenic beauty and represents international biotic and abiotic values. Nature conservation organisations and ecologists fully support the policy choice of preservation of the coastline and the choice for "soft" coastal defence methods. At their request also the term "dynamic" was added to ensure the dynamic character of the Dutch coastline. Several nature conservation organisations now plead for a less strict policy with regard to the maintenance of the foredunes. Suggestions for the formation of so-called slufters (wet dune valley influenced by the tides) and dune areas with more aeolian dynamics (e.g. sand drifts, blow-outs, mobile dunes) have been presented recently (Stichting Duinbehoud, 1992).

From the viewpoint of coastal defence, there are possibilities for natural development of coastal areas, but not everywhere and unconditioned.
On the beach plains at the extremes of the Wadden islands no active coastal defences measures are carried out. Basal coastlines will not be established for these areas.
On the other hand, several dune areas (especially in the Delta area) are too narrow to allow nature development experiments.
For the remaining dune areas a less strict stabilization policy could be considered as long as the safety of the polderland is not endangered. This might imply a different management of the dune area. Presently, certain zones of the dunes are set aside to realize the coastal defence requirements, other zones are nature conservation areas or drinking water supply areas. In the future a more integrated management of the dune areas could be considered (Van der Meulen & Van der Maarel, 1989).

4. Opportunities and threats

The new coastal defence policy of "dynamic preservation" is in fact only at its infancy. A thorough evaluation cannot be given at this stage, but developments since 1990 are encouraging. A set of instruments has been developed (institutional framework, yearly budget, etc) and new developments are investigated (innovative nourishment techniques, research in the Coastal Genesis project, etc). At the request of the national Parliament a thorough evaluation of the policy will be presented in 1995.

As indicated in chapter 3, the new policy offers opportunities for new developments in the coastal zone. Opportunities for the restoration and development of nature, for the application of new coastal defence techniques, for a more integrated management approach in the coastal zone, etcetera. On the other hand there are threats, such as the increasing stress on the coastal zone and the predicted increase in sea level rise.

Some opportunities and threats are briefly considered in this chapter. It is not suggested that these are the only or the most important opportunities and threats. Besides, threats can be transformed into opportunities and the other way around. Public opinion is a key factor in this. To keep up the public interest in the new coastal defence policy is a task by itself, a task that becomes more difficult during the management phase of a policy lifecycle (Fig. 6; De Haan, 1991).

4.1 *Public opinion*

In 1990 the policy choice for "dynamic preservation" was in a way facilitated by the severe storm surges of January and February of that year. During those storm surges extreme water levels and severe dune erosion occurred and people once more realized the strenght of the sea. Directly after those events public awareness and support for coastal defence are at its maximum. But to safeguard the yearly budget for coastline preservation after several relatively calm winters is obviously more difficult.

Stages in public opinion	Phases in policy lifecycle
a. discontent	1. recognition
b. crystallization in a common need	2. formulation
c. judgement and decision	3. solution
	4. management

Fig. 6   *Public opinion and policy lifecycle (De Haan, 1991)*

Public opinion c.q. public support plays a key role in the succesfulness of a new policy. Therefore much attention is paid to inform the public through brochures, video's and the press. Some examples of discussions on "hot items" during the last two years:

- The technique of beach nourishment has been questioned by many. What people see is that much money is spent on sand which is for a part out of sight after the first storm surge. It is apparently not fully understood that sand which has been replaced from the beach to the foreshore, is not lost for coastal defence. In 1991 a brochure and video on sand nourishment has been produced which is distributed and shown at the nourishment sites. Moreover, a project to evaluate nourishment projects is presently carried out. Results are beginning to show: ecologists and policy-makers are explaining the benefits of beach nourishment to the press;
- The difference between structural erosion and incidental storm damage needs more attention. The new coastal defence policy involves counteracting of the structural erosion, but this does not mean that all damage from individual storm surges can be avoided. On this subject a press information bulletin has recently been prepared;
- In the new coastal defence policy beach flats are permitted to develop more or less without restriction. These beach flats are almost exclusively found on the Wadden Islands, more particularly at their extremes. Stopping all coastal defence measures at these locations implies more dynamics and optimal chances for nature development. For the island of Rottumeroog (the easternmost Wadden Island) the long-term consequences would be that the island would disappear into a large tidal gully. Although nobody actually lives on the island and history has proven that new islands develop in the course of time, opinion polls showed that about 80% of the Dutch did not agree to stop all defence measures. A pressure-group "Friends of

Rottummeroog" was formed and a bank donated money for the maintenance of the island. Eventually the Minister agreed to continue the maintenance on the island.

4.2 Collaboration between different authorities

The Water Defence Bill specifies that the State shall work together with the Provincial Authorities and the Water Boards to effect joint coastal defence. This collaboration is expressed in the so-called Provincial Consultative Bodies. Each party assumes a different role:
- the Water Board is responsible for the administration and upkeep of the sea defences;
- the State is responsible for the position of the coastline;
- the Provincial Authorities coordinate overall regional policy, a.o. in relation to aspects of physical planning.

Water Boards are public bodies, specially created to maintain the dykes and other sea defences and take care of water control and water quality at a local level. The interested parties (local real estate owners) select the administrators of the Water Board through a democratic process. The Water Board imposes taxes to raise the funds necessary to perform its tasks.

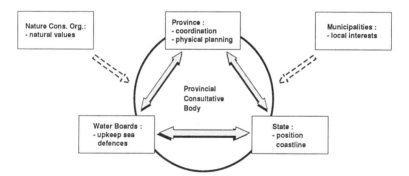

Fig. 7   Collaboration within the Provincial Consultative Bodies

In some provinces, also municipalities and/or nature conservation organisations are represented in the Provincial Consultative Bodies (Fig. 7). However, the Water Defence Bill does not specify any specific task for these authorities and organisations. Their presence merely reflects the wish of all parties to approach coastal defence matters in an integrated way. In practice, the Provincial Consultative Bodies deal with all matters relevant to the preservation of the coast, including the fight against erosion. Concepts on integrated coastal zone management could well be realized through the Provincial Consultative Bodies.

4.3 Increased sea level rise

For the evaluation of the policy alternatives in 1990, various sea level scenarios were considered. The consequences of these scenarios were quantified in terms of land loss and money (Rijkswaterstaat, 1990).
An increase in sea level rise is widely considered one of the most serious threats for low-lying countries such as the Netherlands. Studies show that if the most likely IPCC sea level rise scenario of appr. 60 cm in the next 100 years is adopted, the total costs for the Netherlands would amount about 12,000 million Dutch guilders (appr. 7,500 million US$). For the preservation of the coastline, the additional costs would be 10 million Dutch guilders per year, i.e. an increase of the present nourishment budget by 15% (Rijkswaterstaat, 1991). Both in terms of finances and know-how the Netherlands should be able to cope with an increased sea level. As compared to many of the low-lying developing countries and several small island states, the Netherlands are in fact not very vulnerable to an increase in sea level rise (IPCC/CZMS, 1992).

4.4 Coastal research and monitoring

Out of the yearly budget for coastline preservation, about 3 million Dutch guiders are spent on coastal research and monitoring. The coastal monitoring system, as briefly described in chapter 2, has proven to be extremely valuable for both research and evaluation purposes. Without these data no accretion/erosion patterns can be made and no basal coastline can be calculated.
The coastal research at Rijkswaterstaat is mainly concentrated in the Coastal Genesis project, a multidisciplinary project aimed at understanding coastal processes and predicting future coastal developments. The information and knowledge gained through this project has been of great value to evaluate the various policy alternatives.
Together with the Discussion Document, 20 Technical Reports were published in 1989. These reports can be considered as the state-of-the-art in the field of coastal defence. Based on these technical reports also the "white spots" could be identified: for example the response of the coastal system to a rise in sea level, the limited knowledge of cross shore transport processes, the "sand wave" features along the coastline that are not yet completely understood, and the processes governing transport of water and sediment in tidal inlets. Present Coastal Genesis studies are primarily directed towards these "white spots".

REFERENCES

DE HAAN (1991) Public Support, keep it awake. Paper presented at the Loughborough Conference, UK.

HILLEN, R., VAN VESSEM, P. & VAN DER GOUWE J. (1991) Suppletie op de Onderwateroever (Foreshore Nourishment). Rijkswaterstaat/Tidal Waters Division report (in Dutch).

IPCC/COASTAL ZONE MANAGEMENT SUBGROUP (1992) Global Climate Change and the Rising Challenge of the Sea. Supporting document for the IPCC-update report 1992.

LOUISSE, C.J. & KUIK, A.J. (1990) Coastal Defence Alternatives in the Netherlands. Paper # 1 in: The Dutch Coast, report of a session on the 22nd International Conference on Coastal Engineering 1990.

RIJKSWATERSTAAT (1990) A new Coastal Defence Policy for the Netherlands. Ministry for Transport & Public Works.

RIJKSWATERSTAAT (1991) Rising Waters, Impacts of the Greenhouse Effect for the Netherlands. Rijkswaterstaat, Tidal waters Division.

ROELSE P. (1990) Beach and Dune Nourishment in the Netherlands. Paper # 11 in: The Dutch Coast, report of a session on the 22nd International Conference on Coastal Engineering 1990.

STICHTING DUINBEHOUD (1992) Duinen voor de Wind, een Toekomstvisie op het Gebruik en Beheer van de Nederlandse Duinen (Dunes for the Wind). St. Duinbehoud, Leiden (in Dutch).

STIVE, M.J.F., ROELVINK J.A. & DE VRIEND H.J. (1990) Large-scale Coastal Evolution Concept. Paper # 9 in: The Dutch Coast, report of a session on the 22nd International Conference on Coastal Engineering 1990.

VAN DER MEULEN, F. & VAN DER MAAREL E. (1989) Coastal Defence Alternatives and Nature Development Perspectives. In: Perspectives in coastal dune management; pp. 183-195.

MEASUREMENTS OF TIDAL WATER TRANSPORT WITH AN ACOUSTIC DOPPLER CURRENT PROFILER AND COMPARISON WITH CALCULATIONS

M. Kolb and M. Lobmeyr
Institute of Physics
GKSS Research Centre Geesthacht
D-2054 Geesthacht, Germany

Summary:

Tidal and net transport of water were to be measured at a transect of the Elbe estuary by a vessel-mounted ADCP. Models are used for the interpolation of the measurements and for their correlation with a local water velocity and/or with the local water level. Experimental results are compared with results from a 2D depth integrated hydrodynamic model that is driven solely by the water levels of two gauges.

1. Introduction

The North Frisian wadden sea and the Elbe estuary may be seen as parts of Schleswig-Holstein's coastal area. The influence of tidal motion on the water **level** has been thoroughly observed and predicted. This is, however, not the case for the tidal water **transport**. The ever changing water motion in these regions has always caused the transport of natural sediments and affects nowadays the transport of pollutants. This motivated GKSS to conduct experimental and theoretical work on transport processes in the rivers Weser [1] and Elbe [2] and in the North Frisian Sea [3].

The present paper is based on two measurement campaigns at a transect off Bielenberg (Elbe-km 669) in 1989/90, the latter serves as an illustration. It deals exclusively with the water transport. Water motion governs the transport of dissolved pollutants and polluted SPM (suspended particulate matter) by sedimentation and resuspension [4] whose final whereabouts constituted the foremost task of the campaigns. The hydrodynamic task was to measure the flow $q(t)$ and to obtain the river's net run-off Q_R and the mean tidal flux Q_T as exactly as possible during a fairly short space of time (about 10 M2 tides). Figure 1 defines the quantities by means of a schematic electric analogue. $q(t)$ was measured by hourly transects utilizing a vessel-mounted ADCP (acoustic Doppler current profiler) [5, 6, 7].

To date, the $q(t)$ measurements have been evaluated using three approaches:

- fitting interpolation models to the data for integration according to Fig. 1,

- correlation with the velocity $v_M(t)$ measured continuously below a pontoon moored at the transect [8] in order to infer $q(t)$,

- comparison with hydrodynamic models using the water level from gauging stations as an input [9].

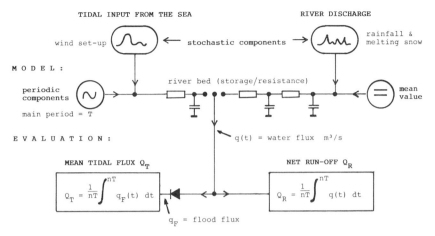

Fig. 1   Electrical analogue of the water flux in a tidal river/estuary.
(The flux $q = dV/dt$ [m³/s] is analogue to the current $i = dC/dt$ [As/s]; V = volume, C = charge)

2. Measurements

The 1.2 MHz ADCP was temporarily attached to the side of a ship which run the transect of 1.9 km within $t_E - t_S = 10$ to 15 min (t_S, t_E: times of start and end of transect, respectively) in about hourly intervals. The instantaneous water flux $q(t_i)$, $t_i = (t_S + t_E)/2$, is calculated on board according to the divergence method [10] by a simple summation [6, 7]:

$$q(t_i) = \sum_{t_s}^{t_E} \sum_{z_0}^{D} \left(g_y w_x(z) - g_x w_y(z)\right) \Delta z \, \Delta t \quad [m^3/s] \quad (1)$$

Subscript y refers to the fore/aft and subscript x to the starboard/port component of the ADCP ship/bottom - **g** - and ship/water - **w** - velocities in m/s. Δt is the time per ensemble of pings and Δz the layer size. The self-supported ADCP evaluation (Fig. 2 righthand) is used if **g** can be measured directly by the ADCP's bottom tracking facility. **g**, the vessel's speed over ground, has to be taken from navigation instruments (gyro & position system) if much resuspended sediment moves over an acoustically soft ground as was the case with the full tide running.

Monostatic ADCP's (i.e. their transducers serve to send **and** receive) cannot measure **w** in a top layer. Furthermore, they become unreliable close to the ground due to interference from side-lobes. The extrapolation up to the water surface, z_o, can be approximated by using a constant height factor with the uppermost layer. The summation down to the bottom may proceed to the full depth D if the water echo is fairly strong in relation to the bottom echo, i.e. if influence of the side-lobes may be neglected. Otherwise, a velocity profile must be fitted for extrapolation.

Fig. 3 presents typical $q(t_i)$ data from one of the so-called BILEX experiments [11]. The error in a single q-value is approximately 2000 m³/s for the narrowband ADCP technique used in 1989/90.

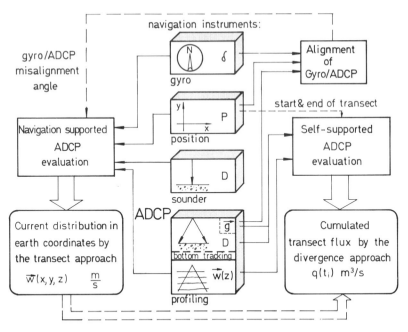

Fig. 2 ADCP measurement of water transport by a moving boat, instruments and principles.

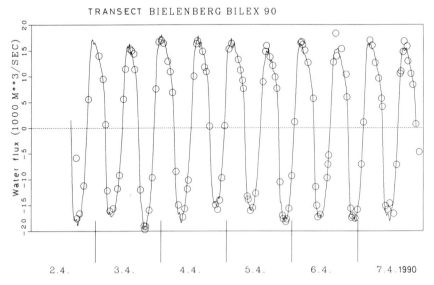

Fig. 3 Measurements $q(t_i)$ and hydrodynamic model (line).

3. Interpolation Model

A continuous integration of q(t) and an integration of its flood times only is required according to Fig. 1. For a first evaluation this was done by graphical interpolation and integration by planimeter but the results lack a reliable estimate of the eye fit's error. Therefore, a parametric "experimental model" was fitted to the measurements $q(t_i)$. A spline interpolation could not be used because data gaps corresponding to many hours occurred in the 1989 campaign.

The dominant structure of the experimental model becomes apparent when viewing the data in Fig. 3: a trigonometric term. A sum of 4 sinusoids constitutes a reasonable compromise for the number of data and parameters, and also makes physical sense. The ADMIRALITY SIMPLIFIED HARMONIC METHOD OF TIDAL PREDICTION [12] also reduces to a minimum of 4 sinusoids for the Elbe (M2+S2, K1+01, 1/4-diurnal, and 1/6-diurnal shallow water corrections).

Fig. 1 suggests further that the experimental model should have the following general structure:

$$q(t) = \text{Constant} + \text{Trend} + \text{Periodic term} = C + T(t) + P(t) \tag{2}$$

with

$$P(t) = \sum_{n=1}^{4} q_n \sin\left(2\pi\left(\frac{t}{\tau_n} + P_n\right)\right) \tag{3}$$

and

$$T(t) = q_5 \sin\left(2\pi\left(\frac{t}{t_5} + P_5\right)\right) \tag{4}$$

or

$$T(t) = q_6 t + q_7 t^2 . \tag{5}$$

Eq. (4) represents a periodic trend whose period is longer than the record length (= measurement campaign). For example, this could stem from the spring/neap cycle.

Eq. (5) represents a linear and a parabolic trend during the observation resulting from the two stochastic inputs indicated in Fig. 1. The choice between Eq. (4) and Eq. (5) is decided by the least squares criterium. Modelling routines were used to determine the 16 parameters needed with Eq. (4) [13] and the 15 parameters needed with Eq. (5) [14], respectively.

Figure 4 and Table 1 show the results of the integrations of q(t) and $q_F(t)$ which will be discussed in paragraph 6 below.

4. Data Aquisition Station

A vertical profiler positions an ultrasonic current meter 5 m below the data acquisition pontoon META [15] which is moored at the above transect about one sixth of the transect's length off the fairway. The velocity $v_M(t)$ from this sensor is continuously recorded throughout most of the year. One task of the campaigns is to establish a relation between their cross-sectional information and the data collected by META. This problem concerns all fixed data recording stations in rivers and estuaries, compare e.g. [16].

Table 1 Goodness-of-fit of the models relative to $q(t_i)$ and results of their integration for the Bielenberg transect (Elbe-km 669) during BILEX 90.

All Figures • 10^3 m³/s	SDEV σ	Mean tidal flow Q_T	Net run-off Q_R
Interpolation model:			
Eq. (2) with Eq. (4)	2.2	4.9	0.75
Eq. (2) with Eq. (5)	2.7	4.92 ± 0.13	0.59 ± 0.20
Local velocity converted by $q(t) = a + f\, v_M(t)$ from Table 2	1.6	4.8 ± 0.5	0.48
Hydrodynamic model (see below 5.)	1.3	5.82	0.57
Tidal prism model neap ... spring *		4.3 ... 6.0	- - -
River discharge measured at Neu-Darchau (Elbe-km 536) (add roughly 1/5 for the tributaries)		- - -	0.60

* The BILEX 90 campaign covered about the 1st quarter of neap-time

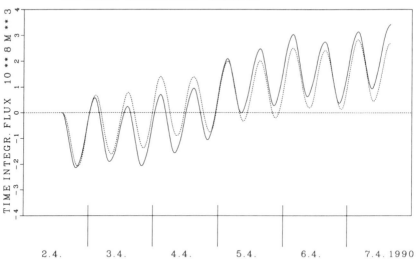

Fig. 4 Time integrated flux of the two interpolation models.
Continuous line: with Eq. (4) resulting in $Q_R = 0.75 \cdot 10^3$ m³/s.
Dotted line: with Eq. (5) resulting in $Q_R = 0.59 \cdot 10^3$ m³/s.

Spectral analyses had shown that the 4 periods of Eq. (3) were prominent constituents of $v_M(t)$ as well [17]. Therefore, some simple experimental models for the cross-sectional water flux, i.e. $q(t)$ = function $(v_M(t), h(t))$, were tested with the campaign's data, see Table 2. $h(t)$ is the water level taken as the mean value of 2 gauging stations next to the transect.

Table 2 Parameters fitted to reproduce the $q(t_i)$ by five elementary models based on the measured water level $h(t)$ and/or the local velocity $v_M(t)$:

$$q(t) = a + b\, h(t-c) + d\, h'(t-e) + f\, v_M(t) \qquad (6)$$

constant a m³/s	h, water level* factor b phase c m²/s h	h', differentiated h: factor d phase e m² h	v_M, velocity: factor f m²	goodness-of-fit x^2
19.2 ± 0.08	- 0.107 - 1.07 ± 0.0004 ± 0.002	- -	-	23
0.765 ± 0.04	- -	- 0.218 2.10 ± 0.0004 ± 0.0004	-	8
6.58 ± 0.12	- 0.033 - 0.71 ± 0.0006 ± 0.0007	- 0.167 1.91 ± 0.001 ± 0.0008	-	5.5
0.244 ± 0.04	- -	- -	15.2 ± 0.03	1.6
0.308 ± 0.03	- -	- 0.278 2.13 ± 0.001 ± 0.007	13.4 ± 0.007	1.4

* $h(t)$ = Mean of the gauges Kollmar (Elbe-km 665) and Krautsand (Elbe-km 672) each interpolated by cubic spline.
The constant a, the factors, and x^2 10^3 !

Figure 5 presents the penultimate line of Table 2 applied to the measured velocity $v_M(t)$, i.e. $q(t) = a + fv_M(t)$, (thin line) and the $q(t_i)$ measured by the ADCP (dots). It shows that the **local** velocity $v_M(t)$ measured 5 m below the water level (tidal elevation neap/spring about 2.5/3.1 m) at the META position is nearly linearity connected with the **cross-sectional** flux. The smallest x^2 results when using $v_M(t)$ complemented by the differentiated water level. (The current into a capacitor follows the differentiated voltage, compare the analogue of Fig. 1.) Furthermore, it can be seen from Table 2 that $q(t)$ is badly reproduced from **one** water level alone. The hydrodynamic model (see 5.) uses **two** water levels measured at its up- and downstream boundaries.

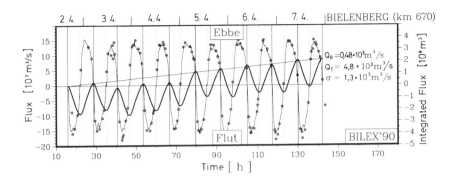

Fig. 5 Measurements q(t$_i$) and flux from local velocity (thin line) plus time-integrated flux (thick line).

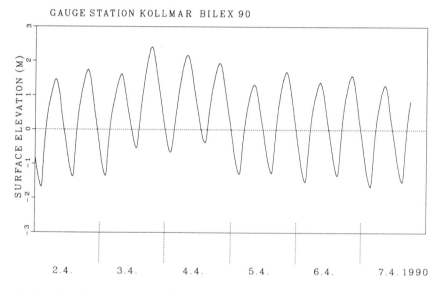

Fig. 6 Water level as measured by the gauge Kollmar (Elbe-km 667).

5. Hydrodynamic Model

The motion of a continuous medium is governed by the principles of the classical mechanics for the conservation of mass and momentum. These lead to a system of coupled nonlinear systems of differential equations for the hydrodynamic variables. In an estuary we are dealing with turbulent flow. In geophysical fluid dynamics some assumptions are applied:
1. - Boussinesq approximation leading to the hydrostatic assumption.
2. - Similarity hypothesis for the turbulent diffusion term.

Parameterizations are made for the shear stresses at the bottom and the surface.

After integrating over the vertical direction we obtain a system of differential equations for the three variables, surface elevation and the longitudinal and lateral components of the current field. These systems have to be solved numerically. The solution procedure is performed on a regularly space grid [9]. The chosen grid has a distance of 200 m between each grid point in the horizontal and lateral direction. Additionally appropriate boundary conditions must be defined: negligible flux across the solid boundaries. At the open boundaries the time series of water levels must be prescribed.

With the help of this model we have simulated a time period of 144 hours (time step 6 sec, processor time 30 min IBM). This corresponds to the time during the so-called BILEX 90 experiment from 2.4.1990 until 7.4.1990. For the boundaries time series of the values of the tidal gauges Brokdorf (Elbe-km 685) and Kollmar (Elbe-km 667) were used. Unfortunately the gauges' scales of the absolute height are known only to a systematical error of some cm. Therefore, tuning of their relative height by the model is necessary. With the data of the BILEX 89 experiment a small correction of the zero of the tidal gauge Kollmar was made. The same correction was applied for the simulation of the data of the BILEX 90 experiment.

Figure 3 shows the modelled flux (water transport m^3/sec) through the transect at Bielenberg for the time period from the beginning of an ebb period at 14:45, 2.4.1990 until the end of a flood period at 21:10, 7.4.1990. The modelled flux is compared with the ADCP measurements. A good correspondence between measured and modelled values may be found.

The net run-off is not equal for every M_2 period. The river bed itself builds up a great water storage. This storage effect can be seen by comparing the water level at the same river location over some tidal periods. Figure 6 shows the water level at the gauging station Kollmar. It can be seen that for the time period 2.4.1990 until 3.4.1990 there is an increase in the water level, subsequently there is a decrease in the water level. During the time period with increasing water level it may happen that the flood flux is greater than the ebb flux, i.e. the net run-off is directed upstream towards Hamburg. Figure 7 shows this by the time integral of the flux q(t). The time integral shows negative values for the time period 2. and 3.4.1990, that means the net run-off is directed towards Hamburg. For the following time period (4.4. until 7.4.1990) it shows positive values, that means the net run-off is directed towards the North Sea.

The integral of q(t) is very sensitive to any change of parameters. If the same simulation is performed with only a small shift in the absolute height of h(t) at the boundary, i.e. the measured values of the gauge are changed by 1 cm, then the dashed curve results for the time integral in Figure 7.

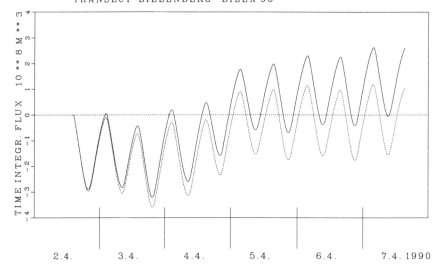

Fig. 7   Time-integrated flux of the hydrodynamic model for BILEX'90 with gauge zero tuned by BILEX'89 (solid line) and the same with gauge zero changed by 1 cm (broken line).

6. Résumé and Discussion

Table 1 lists the present results on the mean tidal flow Q_T and the net run-off Q_R obtained by integrating the ten tides of BILEX 90. The first two lines stem from model-based interpolations of the ADCP measurements $q(t_i)$. The third line represents a measured local velocity $v_M(t)$ that is converted to $q(t)$ by means of the $q(t_i)$ as done for 1989/90. Finally, results of the 2D hydrodynamic model are shown. It had been used previously for an adjustment of the 2 gauges' relative height using the $q(t_i)$ of 1989.

Only the Eqs. (2) + (5) interpolation model was fitted with a procedure providing error estimates on all parameters and results [14]. Its error for Q_R covers the three other Q_R values. However, the spread of the Q_T values is larger than the error for the fitted models on the one hand and the hydrodynamic model on the other hand. Can we obtain Q_T and Q_R independently from other sources for a comparison ? Two such values are given in the lower part of Table 1.

For Q_T we fitted the parameters of a simple "tidal prism model" from 6 calculated values near the mouth of the Elbe [18] and from former BILEX campaigns (3 values up to 1989) further upstream, where the tidal prism terminates at a point K_0 and expands linearly towards the mouth [19]. Its structure is mainly a parabola:

$$Q_T(k) = m\,k^2 + R \quad [m^3/s] \qquad (7)$$

with k = distance from K_O = Elbe-km - K_O [m]
K_O = 589/604 [km] spring/neap, end of flood current
m = 0.925/0.948 · 10^{-6} [m/s] spring/neap
R = 123/264 [m^3/s] spring/neap, offset.

The hydrodynamic model's Q_T seems to be a bit high considering the tidal prism's range. Of course, the latter corresponds to P(t) in Eq. (2) only and neglects any trend occurred in the hydrodynamic model's epoch. Still, the tidal prism model corroborates the four Q_T-values.

Table 1 shows the value measured by the official run-off gauge station at Elbe-km 536 for Q_R. It has to be augmented by the tributaries' run-off which is not exactly known for any particular epoch [18]. The agreement with the four Q_R-values is fortunate in consideration of their high relative errors which result from Q_R being the small difference of large numbers.

Acknowledgement

Professor Dr. H. Schiller kindly run the fits with Eq. (2) + (5) in Table 1 and with Eq. (6), Table 2; Dr. J. Kappenberg provided Figure 5 (both at GKSS).

References

[1] Nehlsen, A. (1982). Berechnung und Messung horizontaler Strömungsfelder in Tideflüssen. GKSS 82/E/26.
[2] Kappenberg, J. et al. (1990). Suspended Matter and Heavy Metal Transport in the Lower Elbe River. Coastal and Estuarine Studies Vol. 36, 147-152, Springer-Verlag.
[3] Boley, K. et al., Ed. (1992). Ökosystemforschung Wattenmeer - Schleswig-Holstein. Tönning, p. 110.
[4] Lobmeyr, M. and W. Puls (1991). Modellrechnungen zum Schwebstofftransport: Vergleich der Ergebnisse eines zweidimensionalen Modells mit Feldmessungen im Elbästuar. GKSS 91/E/25.
[5] Gordon, R. Lee (1989). Acoustic Measurement of River Discharge. J. Hydraulic Engng. 115, p. 925-936.
[6] Kolb, M. (1988). Durchflußbestimmungen aus Messungen mit einem akustischen Doppler-Strömungsprofilmeßgerät auf einer Stromtraverse. GKSS, Arbeitsnotiz 9. August, 9 p.
[7] Simpson, M.R. and R.N. Oltmann (1990). An Acousting Doppler Discharge-Measurement System. Proc. of the Hydraulic Eng., 1990 Nat. Conf. 2, 903-908.
[8] Bittner, K. (1991). Eine Langzeituntersuchung über Strömung und Schwebstoff im Elbeästuar. GKSS 91/E/81.
[9] Krohn, J. and M. Lobmeyr (1988). Application of two- and threedimensional high resolving models to a small section of the Elbe estuary. GKSS 88/E/52.
[10] Kolb, M. (1992). Measurement of Tidal Currents/Transports with a 1.2 MHz ADCP. Europ. Conf. Underwater Acoustics Luxembourg 14 - 17 Sept. Paper ECUA-92-53.
[11] Michaelis, W. (1990). The BILEX Concept - Research supporting public water quality surveillance in estuaries -, see [2], p. 79-88.
[12] See: Admiralty Tide Tables (1991) and User's Handbook. NP 159 A - K8 - 3/91.
[13] Ziegler, G. (1990). PC-Program FIT - SINUS in BASIC. GKSS/PE.
[14] Schiller, H. (1991). Das Fit-Programm minimize. 5 p. GKSS/DV.
[15] Fanger, H.-U. et al. (1990). The Hydrographic Measuring System HYDRA, see (2), p. 211-216.
[16] Head, P.C. and P.D. Jones (1991). Fluxes of water, solids and metals through the Mersey estuary (UK) as measured by a continuous data recording station, see [2], p. 141-146.
[17] Merseburger, Th. (1991). Meßtechnische Erfassung und Auswertung von Strömungsgeschwindigkeiten und Wassergüteparametern aus der Tideelbe. Diplom Thesis 123 p. Hamburg March 28.
[18] Duwe, K. (1989). Modellierung der Brackwasserdynamik eines Tideästuars am Beispiel der Unterelbe. GKSS 89/E/48, 184 p.
[19] Kolb, M. (1991). Flutstrom und Tideprisma, angewendet auf die Unterelbe. GKSS 91/I/8, 15 p.

IN SITU EROSION FLUME (EROSF): DETERMINATION OF CRITICAL BED-SHEAR STRESS AND EROSION OF A KAOLINITE BED AND NATURAL COHESIVE SEDIMENTS.

E.J. Houwing[1] and L.C. van Rijn[2].

1) Institute for Marine and Atmospheric Research Utrecht (IMAU), P.O.Box 80115, 3508 TC Utrecht, The Netherlands.
2) Delft Hydraulics, P.O.Box 152, 8300 AD Emmeloord, The Netherlands.

INTRODUCTION

Sea level rise causes marsh erosion through an increase in tidal flooding and an increase in wave energy. It has been observed that in the last ten years parts of the salt marshes along the coast of the Friesland and Groninger Wadden Sea Coast have been eroded due to a recent rise of the high water levels (Dijkema et al.,1990). To prevent a further erosion of the salt marshes it is important to know the processes of erosion and deposition of cohesive material. The erosion of the bed depends on the forces exerted on the bed (bed-shear stress) by wave and current motions on one side and the strength of the bed to resist these forces (bed-shear strength) on the other side. The bed-shear stress can be calculated theoretically by measuring the parameters which determine the bed-shear stress. The bed-shear strength, however, is hard to determine. As the strength of the bed is dependent on many parameters like physical-, chemical- and biological-ones, the bed-shear strength can best be measured in situ and expressed as one overall parameter.

Many different instruments have been developed in order to determine the bed-shear strength in situ. One of the problems is how to convert erosion of the bed due to a force, exerted by man or machine, into a shear force which is equal to the critical shear stress required for erosion of the bed.

In this report results are presented from the test of the in-situ Erosion Flume (EROSF) in the laboratory and in the field. By using well known functions, a circulating water motion in the Flume can directly be translated into a shear force acting on the bed. The maximum shear strength of the bed as function of the depth can be measured in the field.

Description of the In-Situ Erosion Flume

The in-situ Erosion Flume is a circulating flow system in the vertical plane. It consists of a lower horizontal test-section, two curved sections and an upper horizontal section where the flow is generated by a propeller (fig. 1). The horizontal test-section and the two curved sections have a rectangular cross-profile with a height of 0.1 meter and a width of 0.2 meter. The bottom part of the horizontal test-section is open over a length of 0.9 meter. When the Erosion Flume is resting on the sediment bed properly, the surface of the bed will be in line with the steel bottom plates of the flume on both ends of the horizontal section. The total weight of the erosion flume is about 50 Kg. The total volume of water in the flume is 0.1 m^3.

The propeller rotates at various speeds by means of an adjustable oil pressure system. The flow velocity in the horizontal section is measured by means of a small disc-type electro-magnetic flow meter (EMF) placed at 0.55 meter from the entrance of the horizontal test-section. The measuring point is at 0.075 meter below the covering plate of the test-section, which is 0.025 meter above the bed surface.

The suspended mud concentration is measured by means of an optical system, placed at 0.65 meter from the entrance of the test-section. The measuring point is 0.09 meter above the bed surface.

Bed-Shear Stress Equation

The erosion process of sediment particles of the bed is related to the prevailing bed-shear stress.
The bed-shear stress can be determined from the measured velocity profile assuming a logarithmic distribution in vertical direction, which reads as:

$$u_z = \frac{u_*}{\kappa} \ln\left(\frac{z}{z_0}\right) \qquad (1)$$

in which:

u_z = flow velocity at height z above the bed, u_* = $\left(\frac{\tau_b}{\rho}\right)^{0.5}$ = bed-shear velocity, τ_b = bed-shear stress, z = height above the sediment bed,

$z_0 = 0.033 k_s + 0.11\nu/u_*$ = zero-velocity level, k_s = effective roughness of Nikuradse, κ = constant of von Karman (0.4), ν = kinematic viscosity coefficient and ρ = fluid density.

Based on Eq.(1) the bed-shear stress can be expressed as:

$$\tau_b = \rho \kappa^2 u_z^2 [\ln(\frac{z}{z_0})]^{-2} \qquad (2)$$

During the tests it is found that the velocity distributions in the horizontal test-section are asymmetrical. This is due to the presence of the upstream bend resulting in relatively large velocities in the outer part of the bend. Redistribution of the velocity profile may take place in the horizontal section. Equilibrium flow will not be established because of the short length of the horizontal section. Therefore, the velocity distribution in the near-bed region of the horizontal section will deviate from the logarithmic distribution. Introducing a calibration coefficient α to account for this effect, and using the maximum velocity u_m at height $z = \delta$ above the bed, the bed-shear stress can be expressed as:

$$\tau_b = \rho \kappa^2 (\alpha u_m)^2 [\ln(\frac{\delta}{0.033 k_s + 0.11\nu (\frac{\tau_b}{\rho})^{-0.5}})]^{-2} \qquad (3)$$

The α-coefficient is determined, for non-cohesive material, by measuring velocity profiles above a flat bed of moving sand and gravel particles at conditions just beyond initiation of motion. The (critical) bed-shear stress at each stage of particle motion can be determined from the Shields graph shown in figure 2 (van Rijn, 1989). Four size classes (d_{90}= 320, 815, 1300 and 3770 μm) were tested for each three stages of particle motion: stage 1, 3 and 5, as defined by figure 2.

ANALYSIS RESULTS

Using the critical bed-shear stress ($\tau_{b,cr}$) and the measured velocity (u_m) at $z = \delta = 0.025$ meter above the bed and assuming $k_s = d_{90}$ for a flat bed, $\kappa = 0.4$, $\rho = 1000$ kg/m^3, $\nu = 1.31 \cdot 10^{-6}$ m^2/s (temperature 10°C); the α-coefficient can be computed from Eq.(3), yielding:

$$\alpha = (\kappa u_m)^{-1} \left(\frac{\tau_{b,cr}}{\rho}\right)^{0.5} \ln\left(\frac{\delta}{0.033 d_{90} + 0.11 \nu \left(\frac{\tau_{b,cr}}{\rho}\right)^{-0.5}}\right) \quad (4)$$

The computed α-coefficients show a clear relationship with the particle Reynolds number ($u_* k_s/\nu$, with $k_s = d_{90}$), see figure 3. The inaccuracy of the α-coefficient is ±0.1, presented as a vertical bar in figure 3. The Reynolds number expresses the hydraulic flow regime as:

hydraulic smooth for $\frac{u_* k_s}{\nu} \leq 5$, transitional for $5 < \frac{u_* k_s}{\nu} < 70$ and hydraulic

rough for $\frac{u_* k_s}{\nu} \geq 70$

As most of the experimental conditions in the Erosion Flume will be in the hydraulic smooth regime, an α-coefficient of 0.9 is recommended to determine the bed-shear stress. The velocity should be measured at a height of 0.025 meter above the bed.

EROSION OF THE KAOLINITE BED

The EROSF has first been tested in the laboratory. In the test-section an artificial bed is constructed. The bed consists of kaolinite and is formed by sedimentation in still water in the test-section. The initial concentration is 500 Kg/m³. The initial thickness of the bed is 0.06 meter. The bed is allowed to consolidate for 6 days. After consolidation the kaolinite bed is 0.05 meter thick and the top of the bed is exactly in line with the bottom plates of the test-section.
During the experiment the current velocity in the EROSF is increased in stages. During each stage the sediment concentrations are determined. The length of a stage is defined by the increase in sediment concentration (increase in erosion of the bed) towards a constant level in sediment in concentration (the erosion of the bed is arrested). The sediment concentration in the EROSF is measured with an optical sensor. At the end of the experiment the erosion depth is measured.

MEASUREMENTS

The density of the kaolinite bed in the test-section of the EROSF is determined by using the Conductivity Concentration Meter (CCM). The CCM measures the change in resistance for changes in fluid-sediment mixtures. The resistance increases (conductivity decreases) for increasing sediment concentrations. The CCM values are verified also by weighting frozen cores taken from the kaolinite bed.

Figure 4 shows the bed density as function of the depth, determined by the CCM. Results of Kuijper et al.(1989) and Parchure and Mehta (1985) are also shown.

ANALYSIS OF THE RESULTS

Initiation of erosion of the bed takes place when the bed-shear stress (τ_b), generated by the current velocity in the EROSF, becomes larger than the shear strength (τ_s) of the kaolinite bed (see also Kuijper et al., 1989; Parchure and Mehta, 1985). The bed is eroded in various stages, defined by increasing current velocities (increasing bed-shear stresses). Seven stages of erosion are determined (see fig. 5). Each stage refers to a constant shear stress acting on the bed where at the end of the stage the bed-shear stress equals the bed-shear strength.

For each stage, the bed-shear stress can be calculated according to Eq. 3, given the current velocity at 0.025 meter above the bed and $\alpha = 0.9$ for a hydraulic smooth regime.

Knowing the bed density as function of depth and the increment of the sediment concentration in suspension due to the erosion, the corresponding scour depth can be calculated. For each millimeter erosion of the kaolinite bed from the top down to the bottom the resulting sediment concentration in the EROSF is computed. It is assumed that the density of the sediment bed is constant over each millimeter.

The increment in the total sediment concentration during each of the seven stages is compared with this figure. This results in a provisional determination of the erosion depth. The depth of the bed level found at the end of each stage represents the level where the bed-shear stress equals the bed-shear strength. The shear strength of the kaolinite bed can be computed as function of depth (fig. 6).

In figure 6 the values found by Parchure and Mehta (1985) and Kuijper et al. (1989)

are shown as well. The figure according to Parchure and to Kuijper is somewhat different from the one constructed from the values measured by the EROSF. Apart from differences in bed density (fig. 4), also differences in salinity and consolidation period of the bed can cause the deviations in the measured bed-shear strength (Parchure and Mehta, 1985).

FIELD EXPERIMENTS.

In July 1992 the EROSF was preliminary tested on a mud flat in the salt-marsh area near the Groninger Wadden Sea Coast to get experience with the equipment. The salt marshes are part of the intertidal area. The bed consists of sand and mud and can be classified as a cohesive bed. During low tide the EROSF is placed carefully on the bed. The first test, on July 21, shows a continuous increasing amount of suspended sediment and by that a continuous erosion of the bed. Throughout the test, the point where the exerted bed-shear stress equals the bed-shear strength is never reached (fig. 7A). After the test (114 minutes) the bed in the test-section is investigated. It is observed that close to the edges of the test-section erosion of the bed has been taken place. It is assumed that during the positioning of the EROSF the bed was disturbed locally. In figure 7A it can be seen that only at a current $u = \pm 0.2$ (m/s) an increase is found in suspended sediment concentration. The top layer of the bed appeared to be covered by a brownish substance produced by Benthic diatoms. It is suggested that the Benthic diatoms may enlarge the bed-shear strength of the top layer of the bed (Vos et al., 1988). When the resistant top layer is eroded, the exerted bed-shear stress is too large for the underlying layers: the bed will subsequently erode more than was foreseen and without showing the different stages as defined by the tests in the laboratory.

The second test on July 30 was prepared more carefully. However, due to leakage the test could only be performed during high tide and by that the test period was shorten (fig. 7B). During the test the current velocity was constantly increased. After the test the bed in the test-section was examined. The bed was now uniformly eroded only in the latter half of the test-section. Again, only at a current speed of $u = \pm 0.2$ (m/s) the erosion of the bed starts. On this day the bed was also covered by the brownish substance produced by Benthic diatoms.

CONCLUSIONS

The exerted bed-shear stress can be determined from the measured velocity profile in the EROSF with use of eq. (3). For each stage of erosion of the bed, the critical bed-shear stress is found with use of the EROSF. The critical bed-shear stress needed for erosion is equal to the maximal bed-shear strength for that bed level when the erosion is arrested at the end of a stage.

The bed-shear strength of a kaolinite bed is dependent on the density as function of depth. The tests performed with the EROSF show good agreement with the tests performed by Parchure and Mehta (1985) and Kuijper et al. (1989). Differences found, however, can be caused by differences in salinity, in bed density and in consolidation time.

The EROSF is useful for in-situ measurements. However, the measurements are very sensitive for disturbances caused by improper use. The results of the measurements are dependent on many factors like the presence or absence of benthic fauna (Diatoms) and of the spatial and vertical differences in bed density and bed composition.

REFERENCES

Dijkema, K.S., J.H. Bossinade, P. Bouwsema and R.J. de Glopper, 1990. Salt marshes in the Netherlands Wadden Sea: Rising high-tide levels and accretion enhancement. In: Expected effects of climatic change on Marine Coastal Ecosystems (J.J. Beukema et al., eds.), pp. 173-188.

Kuijper, C., J.M. Cornelisse and J.C. Winterwerp, 1989. Research on erosive properties of cohesive sediments. Journal of Geophysical Research, Vol. 94, No. C10, pp. 14,341-14,350.

Parchure, T.M. and J.A. Mehta, 1985. Erosion of soft cohesive sediment deposits. Journal of Hydraulic Engineering, ASCE, Vol. 111, No. 10, pp. 1308-1326.

Rijn, L.C., 1989. Handbook of sediment transport by currents and waves, Delft Hydraulics, Delft, The Netherlands

Vos, P.C., P.L. de Boer and R. Misdorp, 1988. Sediment stabilization by Benthic Diatoms in intertidal sandy shoals; Qualitative and quantitative observations. In: Tide-Influenced Sedimentary Environments and Facies, P.L. de Boer et al. (eds.), pp. 511-526.

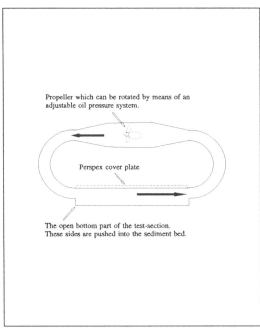

Figure 1 The in-situ EROSion Flume.

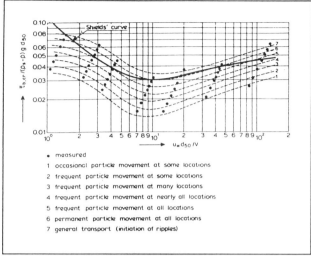

Figure 2 Initiation of motion for unidirectional flow over a plane bed (van Rijn, 1989).

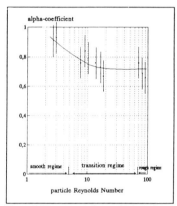

Figure 3 The computed α-coefficient plotted against the Particle Reynolds Number.

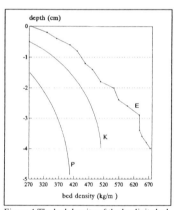

Figure 4 The bed density of the kaolinite bed. K=Kuijper et al., p=Parchure and Mehta, E=CCM.

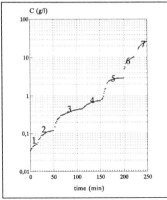

Figure 5 Seven stages of erosion of the kaolinite bed during seven stages of increasing velocity.

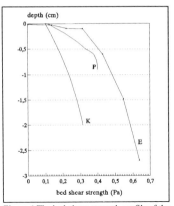

Figure 6 The bed-shear strength profile of the kaolinite bed. K=Kijuper et al., P=Parchure and Mehta, E=EROSF.

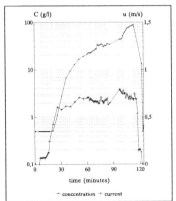

Figure 7A The current velocity in the EROSF and the resulting increase in sediment concentration on July 21.

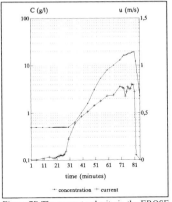

Figure 7B The current velocity in the EROSF and the resulting increase in sediment concentration on July 30.

DETERMINATION OF BENTHIC WADDEN SEA HABITATS BY HYDRODYNAMICS

Sigrid Damm-Böcker, Ralf Kaiser & Hanz Dieter Niemeyer
NLWA -FORSCHUNGSSTELLE KÜSTE- / Coastal Research Station
Norderney/East Frisia, Germany

1. INTRODUCTION

The wadden sea is a dynamic system varying in space and time. Driving forces behind the governing processes are local hydrodynamics due to tides, wind and inland discharge. Primarily they change the topography as agents of erosion and sedimentation, but secondarily they also affect the distribution of surface sediments and the living conditions of benthic habitats. There are as well direct hydrodynamical impacts on benthic fauna as indirect ones, because benthic communities are due to their specific demands adapted to different sediment fractions which zonation in Wadden Sea areas is governed by local hydrodynamics.

The goal of this contribution is to give examples demonstrating as well the zonation of Wadden Sea habitats by hydrodynamical boundary conditions as the direct impacts of hydrodynamics on the local dynamics of specific benthic animals.

2. TIDAL WATER LEVEL VARIATION

Water levels are substantial driving forces for the dynamics of the ecosystem Wadden Sea and especially of its amphibic areas. Its changing landscape due to

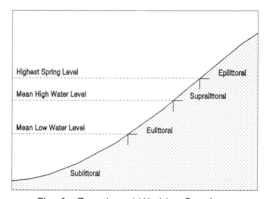

Fig. 1: Zonation of Wadden Sea Area

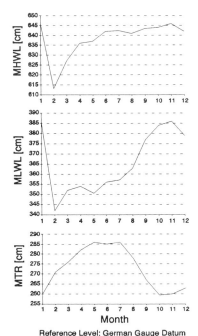

Fig. 2: Monthly mean values (time series 1982/86, gauge Bensersiel) [KAISER 1991]

water level variation in time is its most significant feature. Tidal water levels govern the vertical zonation of Wadden Sea area in sublittoral, eulittoral, supralittoral and epilittoral with reference to mean low water level, mean high water level and highest spring level (Fig. 1). The time series used for averaging individual values generally have a time scale of one to nineteen years. They are basis levels for the distribution of communities of fauna and flora referring to that zonation. But water levels are not stationary in space and time as indicated by those mean values. For certain periods of the basic time series they do not represent the whole width of variation. The borderlines of nature derived from those mean values are therefore to a certain extent artificial and do not always correspond with the existing hydrodynamical boundary conditions limiting the Wadden Sea subsections.

Mean peak water levels depend strongly on the length of the basic time series and have therefore a significant seasonal variation. Explanatory data of the tidal gauge Bensersiel in the East Frisian Wadden Sea are used here in order to demonstrate the order of magnitude of these variations for the mean monthly values of a 5 year time series of high and low water levels and tidal range (Fig. 2): The mean high and mean low water both have their minimum in February and their maximum in November. The seasonal variation in the cycle due to monthly mean values is up to 35 cm for mean high water and 45 cm for mean low water. The variation of the mean tidal range corresponds with the ones of the peak water levels and has its maximum in the middle of the year.

Transferring these water level variations into a cross-section between the barrier island of Langeoog and the mainland (Fig. 3) gives an explanatory insight into the corresponding seasonal variations of the duration of submergence, especially for the higher parts of the tidal flats (Fig. 4): Contradictory to the offshore area seasonal variations of mean high water levels have strong effects on the duration of submergence in the shallow intertidal areas of the Wadden Sea (Fig. 4). The differences become greater with increasing geodetic heights of the tidal flats.

The seasonal changes of Wadden Sea subsections due to mean water level variations is explanatory shown for tidal flat areas in and near the Ley Bay, East Frisian Wadden Sea (Fig. 5). The extension of the sublittoral and the eulittoral differ significantly between November and May whereas the supralittoral area remains constant. The areal changes due to seasonal variations of the mean low water level on a monthly basis are significant: the extension of the sublittoral in the reference area has nearly doubled comparing for instance the months of May and November. Considering these facts it is obvious that for the purpose of ecosystem research it will be insufficient to aggregate the variation of the peak water levels by mean values of one or more years neglecting the seasonal variations.

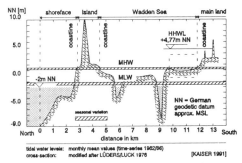

Fig. 3: Seasonal variation of mean peak water level (island of Langeoog - main land)

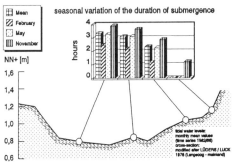

Fig. 4: Duration of submergence

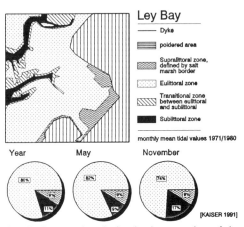

Fig. 5: Seasonal variation in the zonation of the Ley Bay

3. WAVE-INDUCED HABITAT ZONATION

Due to their origin and lifetime waves in Wadden Sea areas, sheltered by barrier islands, can be distinguished into three classes (NIEMEYER 1984, 1991):

1. Local wind waves,
2. offshore swell entering via the tidal inlets,
3. offshore generated wave systems due to onshore directed strong winds and storms and propagating via the tidal inlets.

Due to previous investigations it was established knowledge that the latter type is the most important one as well for design of coastal structures as for impacts on tidal flat and salt marsh morphology [NIEMEYER 1983]: The dynamical equilibrium between morphologically stable high-levelled tidal flats and waves of type 3 is characterized by a strong linear relationship of local wave heights and water depths which is also valid for adjacent salt marshes (Fig. 6). Theoretical studies like the construction of refraction diagrams in coordination with data from field observations and instrumental measurements confirmed the applicability of these results for instance in the Ley Bay area [NIEMEYER 1984, 1991]: For those tidal flat areas considered morphologically stable the same linear relationship could be detected as already earlier done for other parts of the East and West Frisian Wadden Sea [NIEMEYER 1983].

Corresponding to the determination of geodetic flat heights wave climate governs mainly the distribution of surface sediments in amphibic Wadden Sea areas which will be demonstrated explanatorily for wave propagation and zonation of surface sediments in the Ley Bay, East Frisian Waddensea [NIEMEYER 1984, 1991]: Waves propagate from the North Sea to the Ley Bay area via the tidal inlet Osterems splitting up into initially two dominant systems following the course of the main tributary channels Ley and Bantsbalje (Fig. 7). The first system spreads from the Ley on the tidal flat areas Pilsumer Nacken and Greetsieler

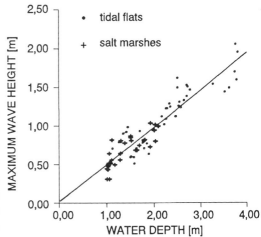

Fig. 6: Relationship of local wave heights and water depths for tidal flats and salt marshes (NIEMEYER 1984)

Fig. 7: Propagation of waves and zonation of surface sediments (NIEMEYER 1991)

Nacken being there superimposed by waves entering from offshore via the tidal inlet and its tributary gully Bantsbalje across the tidal flat area Hamburger Sand. For all other parts of the Ley Bay area waves propagating from sea via the Bantsbalje are dominating. Waves tend to follow first the course of deep gullies. The mouth of the Bantsbalje performs a nearly straightlined prolongation of the inlet's deep channel axis. Whereas waves propagating into the gully Ley will be guided mainly farther westward and only deliver subsystems to the wave regime in the Ley Bay area which are already remarkably weakened due to refraction.

The main wave system propagating into the Bantsbalje produces at least three subsystems which propagate into the Ley Bay area where they partially superimpose each other (Fig. 7): For example that one propagating further via the landward part of the Norderley gully spreads on the one hand into the bay itself and on the other hand into the tidal flat area Utlandshörner Watt superimposing there with other subsystems also entering from the Bantsbalje, but having left the deep channel further landward than the first one. That superimposition of different wave systems can effect a higher energy density with remarkable consequences. This becomes evident e.g. for the considered tidal flat area Utlandshörner Watt: The decrease of wave energy due to dissipation on tidal flats is also reflected by the zonation of surface sediments with a tendency to finer material. Wave exposed flats are sandy followed by areas with mixed surface sediments and at last by muddy flats with high silt content. The combination of a map of surface sediments [RAGUTZKI 1982] with a wave refraction diagram highlights the interaction of wave energy propagation and surface sediment zonation in that area. For example the usual landward succession of sandy, mixed and partially even muddy flats is here again interrupted by the occurrence of sandy flats. These phenomena coincide locally with the superimposition of different wave systems being an impressive example of wave effects on morphology and sedimentology of amphibic Wadden Sea areas.

The relevance of these hydrodynamical impacts on biotic zonation is evident considering the dependence of benthic communities on sediment fractions as documented for instance by MICHAELIS [1984, 1987] for the Ley Bay and the Jade Bay. For instance in the Ley Bay the amphibic areas Schweinsrücken and Buscher Heller a comparison of specific pioneer vegetation by older and newer maps shows contradictory to general trends diminishing areas of intertidal deposits and their typical pioneer vegetation [MICHAELIS 1984]. The reason is the propagation of waves with higher energy due to the deepening of the tidal gully Norderley in the meantime [NIEMEYER 1984].

4. HYDRODYNAMICS AND DYNAMICS OF MACROZOOBENTHOS

4.1 Tidal water levels and duration of submergence

Remarks considering correlations between the dynamics of abundance of macrozoobenthos and hydrodynamical impacts is very fragmentary and usually only descriptive with respect to the occurrence of storm surges [MEIXNER 1979; RACHOR 1980; MADSEN 1984; DÖRJES et al. 1986; ROHDE 1986]. In Wadden Sea areas the distribution of macrozoobenthos depends on the interrelation of hydrodynamical forces and endogenous attributes of the different species, as e. g. diurnal or lunar rhythms [ARMONIES 1991] or depth of settlement, feeding habit, strategy of reproduction and phototaxis [OHDE 1981]. The sensitivity of all these attributes to hydrodynamical impacts determines a shaping of biotic zonation, which was for instance evaluated by OHDE [1981] using the tool of areal biological thematic mapping. The importance of local duration of submergence for several species of macrozoobenthos has been mentioned by JEPSEN [1965]: The polychaete Magelona papillicornis and Nephtys hombergii and the amphipod Bathyporeia sarsi prefer biotopes with long flooding durations.

First more qualitative results gained in the Wadden Sea area of the island of Norderney which highlighted dramatic changes in the abundance of macrozoo-

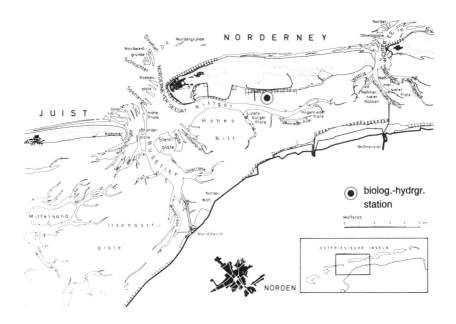

Fig. 8: Map of the Island of Norderney with biolog.-hydrogr. station

benthos due to the occurrence of storm surges [DÖRJES et al. 1986; ROHDE 1986] and deeper insight into the determination of biotic zonation of Wadden Sea areas due to hydrodynamical boundary conditions [NIEMEYER 1987; KAISER 1991] stimulated recently started field investigations on direct correlations between hydrodynamical impacts and dynamics of macrozoobenthic species at a location already used by DÖRJES et al. [1986] and ROHDE [1986] (Fig. 8). Its first tentative results are presented here by means of statistical correlation between evaluated hydrodynamical parameters and abundance of local macrozoobenthic species.

Contradictory to the results of JEPSEN [1965] the coefficients of determination r^2 indicate relations between flooding duration and the abundance of the polychaetes Anaitides spec. and Lanice conchilega and the shore crab Carcinus maenas (Fig. 9 a-c) whereas there is no evidence of a response of Nephtys hombergii and Bathyporeia sarsi to that hydrodynamical parameter. The correlations between the duration of submergence and the abundance of the polychaetes are negative whereas the one of the shore crab is positive.

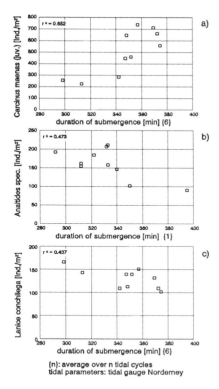

Due to correlation analysis there are significant relations between tidal peak water levels and the abundance of selected species of macrozoobenthos. For instance tidal high water levels (HWL) correlate positively with the abundance of the shore crab Carcinus maenas (juveniles) (Fig. 10a) and negatively with the polychaete Anaitides spec. (Fig. 11a).

Taking statistical analysis results into consideration it is evident that the species respond significantly different in respect of the phase lag to the hydrodynamical impact and of its repeated occurrence. This result has been evaluated by taking the parameters for one tide or by averaging them for two, four and six tidal cycles. This is a first approach to detect the sensitivity of species to individual or repeated hydrodynamical impacts without claiming systematology. In some cases the correlation is only significant for a certain parametrization due to time scale and in others its evidence differs remarkably in respect of that.

Fig. 9: Examples of species of macrozoobenthos with significant correlation between abundance and HWL

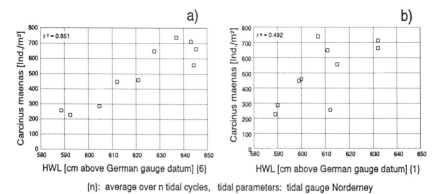

Fig. 10: Control of the correlation by the number of samples of the parameter HWL

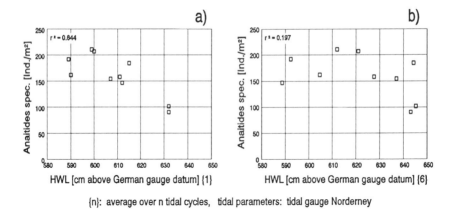

{n}: average over n tidal cycles, tidal parameters: tidal gauge Norderney

Fig. 11: Control of the correlation by the number of samples of the parameter HWL

For instance the abundance of the shore crab Carcinus maenas correlate only significantly due to changes in the high water level of consecutive tides, but not to single events (Fig. 10a + b) whereas the response of Anaitides spec. in respect of its abundance to different high tide levels is only remarkable if the singular preceding event has been considered (Fig. 11a + b).

4.2 Wave action

THAMDRUP [1935] mentions the transportation of the cockle Cerastoderma edule due to tides with remarkable set-up accompanied by eroding wave action. The results of the statistical analysis of time series of the Norderney area do not verify that observation: The abundance of that species correlate neither with high tide levels nor with any here considered wave parameter. But the data used here do not include events with remarkable set-up and strong wave action. For that reason there is no conflict with the statement of THAMDRUP [1935]. Further tests however delivered a positive correlation between the distribution of the seize of the cockle and wave energy indicating a sorting effect like that for sediment fractions as already described.

The abundance of the polychaetes Heteromastus filiformis and Arenicola marina and the mussel Macoma baltica declined dramatically after the occurrence of storm surges being accompanied by strong wave action whereas the abundance of the polychaetes Scoloplos armiger and Nereis diversicolor show no remarkable response due to those events [DÖRJES et al. 1986].

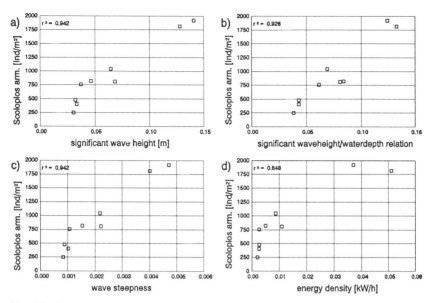

Fig. 12: Correlation between the abundance of the polychaete Scoloplos armiger versus significant wave parameters

Though the time series used here do not contain data of tides with a remarkable set-up or even of storm surges tests considering the impact of wave action on the abundance of macrozoobenthos have been carried out. For the abundance of the polychaetes Scoloplos armiger and Nereis diversicolor a strong correlation with different wave parameters exists. In order to exemplify that fact the correlation between the abundance of species and wave parameters are documented here for the polychaete Scoloplos armiger versus significant wave height H_s, relation of significant wave height and water depth H_s/h, significant wave steepness $H_s/(g \cdot T_{Hs})$ and wave energy E (Fig. 12 a-d). The wave parameters are averaged values of 12 time series of 15 minutes length for periods starting 1,5 hours before and ending 1,5 hours after high tide.

5. FINAL REMARKS

The impact of hydrodynamics on Wadden Sea ecosystems is demonstrated here by several examples. Biotic zonation of the amphibic Wadden Sea areas changes remarkably due to seasonal variations of tidal water levels and resulting duration of submergence. These areas are also determined by wave climate and resulting sorting effects of sediment fractions which mainly govern the distribution of surface sediments and respectively the habitat zonation.

First results of recent investigations on direct correlation between abundance of macrozoobenthic species and hydrodynamical impacts are presented: A number of examples make evident that duration of submergence, high tide levels and wave action determine the local abundance of certain macrozoobenthic species whereas some do not respond. Obviously the hydrodynamically sensitive species respond to the considered hydrodynamical parameters significantly different. As well as physical parametrization modifications in time averaging produce remarkable changes in the statistical quality of the correlations.

The correlation between hydrodynamical impacts and the abundance of several species of macrozoobenthos as described above base still on a rather small number of evaluated time series. Further investigations will be necessary and will be carried out in order to check and confirm the results already gained. This work will also deliver a sound basis for a deeper insight into the determination of Wadden Sea habitats by hydrodynamics.

Acknowledgements

This report bases on different research programmes sponsored by LOWER SAXONIAN MINISTRY FOR SCIENCE AND CULTURE, LOWER SAXONIAN MINISTRY FOR THE ENVIRONMENT, FEDERAL MINISTRY FOR SCIENCE AND TECHNOLOGY and FEDERAL AGENCY FOR THE ENVIRONMENT. The authors are also grateful for the support they experienced by their colleagues of the sections for coastal hydrodynamics and coastal ecology, especially Messr. G. Brandt, D. Glaser, H. Karow, H.-H. Kramer, Dr. H. Michaelis and G. Münkewarf.

LITERATURE

ARMONIES, W. [1991]: Transport of benthos with tidal waters: another source of indetermination. Commission of the European Communities: COST 647; coastal benthic ecology activity report 1988-1991, p.13-31.

DÖRJES, J., H. MICHAELIS, B. RHODE [1986]: Long-term studies of macrozoobenthos in intertidal and shallow subtidal habitats near the island of Norderney (East Frisian Coast, Germany). Hydrobiologia 142, p.217-232.

JEPSEN, v. U. [1985]: Die Struktur der Wattenbiozönosen im Vormündungsgebiet der Elbe. Arch. Hydrobiol., Suppl Elbe-Ästuar 24 (2), 3/4, p.252-370.

KAISER, R. [1991]: Hydrodynamics as ecosystem factor in Wadden Sea areas (in German). a) II. Symposium Ökosystemforschung Wattenmeer in Büsum; b) NLWA - Forschungsstelle Küste - Berichte zur Ökosystemforschung, Hydrographie - Nr. 1

MADSEN, P. B. [1984]: The dynamics of the dominating macrozoobenthos in the Danish Wadden Sea 1980-1983. Rep. Pollut. Lab. 7,p.1-35.

MEIXNER, R. [1979]: Die Fischerei auf Herzmuscheln (Cardium edule) im niedersächsischen Wattenmeer. Arch. Fischereiw. 29,p.141-153.

MICHAELIS, H. [1984]: Studies on the ecological development of the Ley Bay and to the impacts of prospected project Leyhörn (in German). Jber. 1983 Forsch.-Stelle f. Insel- u. Küstenschutz, Bd. 35

MICHAELIS, H. [1987]: Invantarisation of eulittoral macrobenthos in the Jade Bay in combination with an analysis of airial photographs (in German). Jber. 1986 Forsch.-Stelle Küste, Bd. 38

NIEMEYER, Hanz Dieter [1983]: On the wave climate at island sheltered Wadden Sea coasts (in German). BMFT-Forschungsbericht MF 0203

NIEMEYER, Hanz Dieter [1984]: Hydrographical investigations in the Ley Bay area in respect of the Leyhörn project (in German). Jber. 1983 Forschungsst. Küste, 35

NIEMEYER, Hanz Dieter [1984]: Waves and biotic zonation in Wadden Sea areas (in German). ed.: Niedersächsischer Umweltminister: Environmental precaution for the North Sea (in German). Niedersächsisches Landesamt f. Wasserwirtschaft, Hildesheim

NIEMEYER, H. D. [1987]: Changing of wave climate due to breaking on a tidal inlet bar. Proc. 20th Intern. Conf. o. Coast Eng. Taipai, ASCE New York.

NIEMEYER, Hanz Dieter [1990]: Morphodynamics of tidal inlets. CEEC (Civ.Eng. Europ.Cours.Progr.o.Cont.Educ.)-Coast.Morphol.Delft Univ.o.Techn.-Int.Civ.Eng.

NIEMEYER, H.D. [1991]: Case study Ley Bay: an alternative to traditional enclosure. Proc. 3rd Conf. o. Coast. & Port Eng. i. Develop. Countr., Mombasa/Kenya

OHDE, J. [1981]: Entstehung von Besiedlungsmustern der Makro-Endofauna im Wattenmeer der Elbe-Mündung. Dissertation an der Universität Hamburg, FB Biologie, 445 pp.

RACHOR, E. [1980]: The inner German Bight- an ecologically sensitive area as indicated by the bottom fauna. Helgol. Meeresunters. 33, p. 522-530.

RAGUTZKI, Günther [1982]: Distribution of surface sediments on the Lower Saxonian tidal flats (in German). Jber. 1980 Forschungsst. Küste, Bd. 32

RHODE, B. [1986]: Langfristige Bestandsänderungen der Makrobenthischen Fauna im Watt bei Norderney: Non-dominante Arten. Jber. 1985 Forschungsst. Küste, Bd.37, p.147-175.

THAMDRUP, H. M. [1935]: Beiträge zur Ökologie der Wattenfauna (auf experimenteller Grundlage). Meddelelser fra Kommisionen for Danmarks fiskeri - og Harundersogelser. Serie: Fiskeri, Bind 10, Nr.2 Kopenhagen, C.A. Reitzels Forlag.

TOWARDS SUSTAINABLE DEVELOPMENT OF COASTAL ZONES
Report on the activities of the Coastal Zone Management Subgroup of IPCC

Roeland HILLEN, Luitzen BIJLSMA & Robbert MISDORP

Rijkswaterstaat, Tidal Waters Division
P.O. Box 20907, 2500 EX Den Haag, the Netherlands

Abstract
In the Coastal Zone Management Subgroup of IPCC, the coastal nations of the world are developing management strategies for the coming 10 to 20 years as well as long-term strategies for adapting to climate change and sea level rise. The 1992 report of the Subgroup, which was released during the UNCED recently, includes a Common Methodology for the assessment of vulnerability of coastal areas. Up till now, the vulnerability for about 20 coastal nations has been assessed and the outlines of possible response measures for these countries have been identified. Based on the vulnerability assessment studies, comprehensive coastal zone management programmes can be developed to improve the natural and socio-economic conditions in coastal areas. The ultimate goal is to arrive at a sustainable development of the coastal zones in the world.

1. Coastal Zones of the World

The coastal zones of the world are very diverse in character. Large, densely populated and industrialized deltas alternate with long continental shorelines with lagoons, coral reefs and sandy beaches. Endless mangrove forests on muddy tropical coasts or archipelagos of thousands of islands, some of them densely populated.
The total length of coastline in the world is estimated at 1 million kilometres, of which about 400,000 km is populated. Presently about 60% of the world population lives within 60 km of the coastline. The socio-economic developments in the coastal zone are in many respects faster than elsewhere. People tend to migrate to the coastal zone, because of lack of arable land (e.g. Bangladesh), because many cities and industrial centres are located there or for recreational reasons (e.g. W. Europe, USA). Coastal zones, at the intersection of rivers and sea, often also act as sink.
The stress on the coastal zone is increasing rapidly. As a result of this also the number of problems in the coastal zone is growing: severe erosion, salt water encroachment, pollution, overexploitation of natural resources (e.g. water, fish, raw materials). If a sustainable development of coastal zones could be achieved, this could be taken as an indication that both the river catchment areas and the seas and oceans are healthy.

The problems in the coastal zones of the world are recognized by many national and federal governments. Also international organisations such as the Intergovernmental Panel on Climate Change (IPCC) devoted attention to the coastal zones. IPCC Working Group III on Response Strategies has a Subgroup on Coastal Zone Management (CZMS), jointly chaired by the Netherlands and Venezuela. This Subgroup is charged with providing information and recommendations to national and international policy centres on (1) coastal zone management strategies for the next 10 to 20 years, and (2) long-term policies for adapting to climate change and sea level rise.

Results from the IPCC Science Working Group indicate that over the next 100 years an increased rise in sea level can be expected. According to the "business as usual" scenario the best estimate for the year 2100 is a sea level of 66 cm higher than at present (Fig. 1).

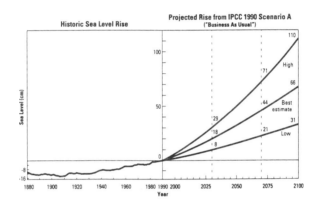

Fig. 1   *IPCC Sea Level Rise Senarios* (IPCC, 1990)

Up till now (September 1992), four reports have been published by the Coastal Zone Management Subgroup: reports on the Miami and Perth workshops (IPCC/CZMS, 1990a & b), "Strategies for Adaptation to Sea Level" (IPCC/CZMS, 1990c) and "Global Climate Change and the Rising Challenge of the Sea" (IPCC/CZMS, 1992).

In the 1990 report, three categories of response strategies are distinguised: retreat, accomodation and protect (Fig. 2). One of the conclusions of this report was:
> *It is urgent for coastal nations to begin the process of adaptation to sea level rise, not because there is an impending catastrophe, but because there are opportunities to avoid adverse impacts by acting nów - opportunities that may be lost if the process is delayed. This is also consistent with good coastal zone management practice, irrespective of whether climate change occurs.*

The 1990 report (IPCC/CZMS, 1990c) encourages coastal nations to identify coastal areas at risk as well as possible response measures. These steps should eventually lead to the development and implementation of coastal zone management plans, preferably by the year 2000.

In the 1992 report of the IPCC/CZMS a Common Methodology for the assessment of the vulnerability of coastal areas to sea level rise is presented. This Common Methodology (CoMet) has been applied in about 20 coastal nations to assess their vulnerability. The results of these country studies have been discussed during a workshop in Venezuela in March 1992 and the highlights of these studies are presented in the report of the subgroup. In Venezuela also the first results of the Global Vulnerability Assessment (GVA) were presented. In the GVA-study the effects of an increase in sea level rise for all coastal areas of the world are assessed for three items: people at risk, wetlands at loss and rice production in coastal zones.
The 1992 report has been released during the UNCED-conference in Rio de Janeiro in June 1992. It will serve as a basis for further work to be discussed in the IPCC Plenary Meeting in November 1992.

2. Vulnerability Assessment

The assessment of the vulnerability of coastal areas is one of the first steps towards better coastal zone management planning, including the response to long term changes. Coastal zone management plans provide information and possibilities for the lay-out, infrastructure and sustainable use and development of the coastal zone. Such plans can be considered as a framework for the solution of both short- and long-term problems, e.g. an accelerated rise in sea level.

To arrive at a sound cooperation between coastal nations, common problems and possible solutions are deliniated in so-called Common Methodologies (CoMet). The advantage of accepted CoMet's is the common format of information, facilitating the exchange of information and knowledge. Moreover, a "common language" is developed and the collected information can easily be assembled to be presented to international organisations and donors. The CoMet on vulnerability assessment will be discussed in section 2.1.

In section 2.2 some results of the country studies into the vulnerability of their coastal areas are presented. Most country studies are performed (or are being performed) along the lines of the CoMet. At their request, several developing nations have been/are supported by donor countries, e.g. Nigeria by the U.S.A., Tonga by Japan, Egypt by the Netherlands. The experience gained through the country studies has also been used to improve the CoMet on vulnerability assessment.

If all coastal nations of the world would have performed a vulnerability assessment of their coastal zones, a world-wide overview of the effects of an incresed sea level rise would be available. Such a "bottom-up" approach will take much time. In the meantime, there is a need for global information on the impact of an accelerated rise in sea level, and there is also a need to verify to what extend the results of the country studies are comparable. For these reasons a limited Global Vulnerability Assessment (GVA) has been carried out (section 2.3). This GVA can be considered as an attempt to realize a "top-down" approach.

Current Sea Level

	RETREAT	ACCOMMODATE	PROTECT
Buildings	Establish building setback codes	Regulate building development	Protect coastal development
Wetlands	 Allow wetland migration	Strike balance between preservation and development	 Create wetland/mangrove habitat by landfilling and planting
Crops	 Relocate agricultural production	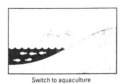 Switch to aquaculture	 Protect agricultural land

Fig. 2   *Sample Responses to Sea Level Rise* (IPCC/CZMS, 1992)

2.1 The Common Methodology

The CoMet for the assessment of vulnerability to accelerated sea level rise is directed to physical, ecological as well as socio-economic impacts in coastal areas. The CoMet defines "vulnerability" as a nation's ability to *cope with* the consequences of an accelerated sea level rise and other impacts of global climate change. The methodology outlines 7 analytical steps for a country to address the issue of vulnerability (Fig. 3). It allows countries sufficient flexibility to pay attention to their individual national circumstances and priorities.

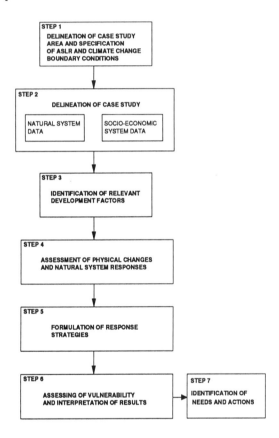

Fig. 3   *Stepwise Approach for Vulnerability Assessment*
(IPCC/CZMS, 1992)

In the CoMet three sea level scenarios are used: the present sea level, and rises of 0.3 and 1.0 metres by the year 2100. The latter two scenarios are more or less in line with the low and high estimates of IPCC (Fig. 1). It should be kept in mind, however, that sea level can also rise or fall as a result of local geotectonic subsidence or uplift, or as a result of human activities, such as the (over)exploitation of natural resources (water, hydrocarbons, etc.).

Next to factors of global climate change, the CoMet includes development factors and response options. National and local development are considered by extrapolating 30 years from the present situation. The CoMet strongly encourages coastal nations to consider a full range of response options, at least the extreme options of retreat and full protection.

The CoMet is primarily a system-approach: the natural and the socio-economic systems of a coastal area form the key of the CoMet. Besides, institutional and legal aspects are taken into account as well as the economic ability to implement mitigating measures. According to the CoMet, the most important result of each vulnerability assessment is the "Vulnerability Profile" for the country in study. In the vulnerability profiles, the results of a country study are represented according to "vulnerability classes" (Figure 4).

Based on the vulnerability profile of a country (or region), actions can be formulated to improve the natural and socio-economic conditions. To mention examples of possible actions: no more roads along eroding coastlines, a stepwise improvement of the safety against inundation, conservation measures for valuable ecological areas. Eventually such actions could be incorporated in a coastal management plan for the country. Such plans could also contain recommendations on legal and institutional aspects related to the management of the coastal zone.

IMPACT CATEGORIES	VULNERABILITY CLASSES			
	LOW	MEDIUM	HIGH	CRITICAL
People affected (#people / total) * 100%	<1%	1-10%	11-50%	>50%
People at risk (#people * probability) / 1000	<10	10-100	100-500	>500
Capital value at loss (total loss / GNP) * 100%	<1%	1-3%	3-10%	>10%
Dry land at loss (area / total area) * 100%	<3%	3-10%	10-30%	>30%
Protection/adaption cost (total cost / GNP) * 100%	<0.05%	0.05-0.25%	0.25-1%	>1%
Wetland at loss (area / total area) * 100%	<3%	3-10%	10-30%	>30%

Fig. 4 *Vulnerability Classes* (IPCC/CZMS, 1992)

2.2 Country studies

The preliminary vulnerability profiles of the country studies presented in Venezuela are given in Figure 5. For some elements of the vulnerability profiles no quantitative data are available. Based on the case study reports and expert judgement, a first assessment of these elements has been made.

For the 16 countries presented in Figure 5, the total number of people affected by an accelerated sea level rise is approximately 50 million. For the individual countries, the differences are tremendous: less than 0.5% for Mauritius versus 100% for Kiribati and one of the Marshall Islands.
According to its country profile, the Netherlands has a critical vulnerability for the number of people affected (about 10 million). However, the vulnerability for the number of people at risk is only medium (appr. 24,000), i.e. if no additional protection measures are implemented. This difference can be attributed in the first place to the high safety standards that are applied in the Netherlands. Especially for many developing countries the ratio people affected/people at risk is quite different.
From Figure 5 it can be concluded that protection/adaptation measures can be very effective in many developing nations. A problem is that financial means and possibilities in these countries are very limited. The calculated total costs for protection and adaptation for the countries of Figure 5 is about 110 billion US$. Also here, large differences occur: the annual costs for Poland are 0.02% of its GNP; for the Marshall Islands the annual costs are 7% of its GNP.

It is clear that the country studies only give an impression of the extent of the problems in the coastal zones. However, they can be very relevant tools for a sound management and planning in these areas. In other words: a first step towards sustainable, integrated planning in the vulnerable coastal zones.

2.3 Global Vulnerability Assessment

The assessment of the vulnerabilities of all resources of the world's coastal zones requires detailed world-wide information on the distribution, density and state of the resources and on the impacting hazardous events. For several elements, such as ecosystems, data on a world-wide scale are not complete. A complicating factor is also that, in order to assess the consequences of hazardous events, the response of various systems are not always known to a sufficient degree of accuracy.
In consideration of these and other constraints, it has been decided to limit the GVA-study (at this stage) to four elements of the coastal zone:
- population at risk
- coastal wetlands at loss
- rice production at change (only for Asia)
- basic protection costs (i.e. an update of the Cost Estimate presented in the IPCC/CZMS 1990 report).

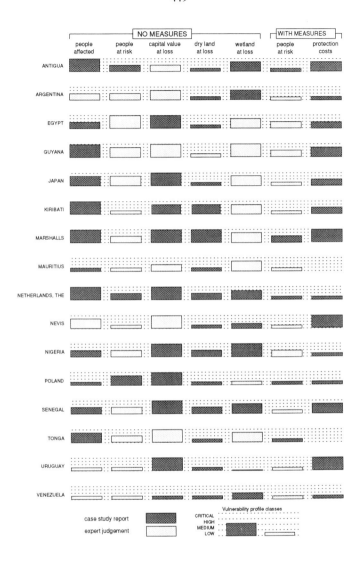

Fig. 5 *Vulnerability Profiles of Case Study Countries*
(IPCC/CZMS, 1992)

The results of the GVA-study can be summarized as follows (Delft Hydraulics, 1992):

Globally, a rise in sea level of one meter would double the number of people at risk:
- Presently, between 100 million and 200 million people are estimated to live below the annual storm surge level. Some 20% of this population experiences flooding annually.
- An additional one meter of SLR above the current mean sea level would increase the number of people subject to annual flooding, by nearly 50 percent, when coastal defense systems would be maintained at their present protecting levels. With the expected population growth for the year 2020, the number of people subject to annual flooding could double.
- If natural coastal defense systems (e.g. mangroves, dunes) decline by human impact and/or cannot migrate landward as a result of the ASLR, the number of people subject to annual flooding would rise dramatically.
- Due to differences in the regime of storm surge events, the increase of flood risk to ASLR is larger than average for small islands, the Southern Mediterranean coast, the African Atlantic coast and coast of the Indian subcontinent.
- An increase in severe storm frequency and intensity by climate changes would further increase the number of people subject to the annual flooding.

Globally, the projected impact of a one meter sea level rise on coastal wetlands is alarming:
- One third of the world's wetlands consists of coastal wetlands of great ecological and eco-nomic value (salt marshes, intertidal areas and mangrove forests).
- Coastal wetlands are presently being lost at an increasingly rapid rate worldwide. The increase in loss rates is closely connected with human activities such as reclamation, enhanced subsidence and shoreline protection, blocking sediment sources for wetlands, and the development of infrastructure, preventing or limiting wetland migration.
- An acceleration of the sea level rise would increase the rate of net coastal wetland loss. In combination with human activities a 1 m sea level rise over the next century would threaten half of the world's coastal wetlands as early as 2020. In some areas, coastal wetlands could be virtually eliminated, because their ability to migrate inland would be limited over such short timescales.
- Coastal wetland losses due to ASLR are expected to be larger than average for the South American Atlantic coast, the African Atlantic coast, the Australian and the Papua New Guinean coast.

A considerable part of the world's rice production is located in regions vulnerable to ASLR:
- Approximately 85% of the world's rice production takes place in South, Southeast and East Asia; about 10% of this production is located in areas which are estimated to be vulnerable to ASLR, such that less favourable hydraulic conditions will cause lower rice production yields, thereby affecting the staple food base of more than 200 million people.

3. Coastal Zone Management Planning and Future Activities

The assessment of vulnerability to accelerated sea level rise is by itself not a solution to problems resulting from climate change. It is merely a method to identify the problems and possible solutions (response options) in the coastal zones. Through integrated coastal zone management planning, a framework might be developed from which actions and activities can be derived to mitigate the adverse long-term effects of climate change, such as accelerated sea level rise.

In 1991, the CZMS conducted a survey among coastal nations to identify the current status of the country activities related to sea level rise and coastal zone management. From the responses of 41 countries (representing almost 50% of the world's population), the following was concluded:
* nearly 90% of the responding nations are considered vulnerable to sea level rise;
* many coastal states (around 50% of the respondents) do not currently have some form of coastal management policy;
* most countries (about 80%) do not have policies related to the impacts of sea level rise;
* appr. 85% of the respondents stated that their country would be interested in international cooperation with respect to issues related to sea level rise and coastal zone management.

The overall conclusion of the survey could be: most coastal nations are in the early stages of considering the potential impacts and responses to sea level rise and the development of coastal management programmes.

The IPCC 1990 report recommends that coastal nations should implement by the year 2000 comprehensive coastal zone management plans that deal with both sea level rise and other impacts of global climate change. The case studies show that many coastal nations do not currently have integrated coastal zone management programmes, and most do not yet have policies that specifically address the impacts of sea level rise. This is in part because the framework is not available and because of the uncertainties associated with such a long-term problem.

The benefits of integrated coastal zone planning and management are generally recognized and considered desirable as a management tool to address complex issues related to the use of coastal resources. It should be realized that a nation needs a reasonable period of time from the moment of development and implementation of a comprehensive programme till the moment that net effects can be measured. An early investment in a coastal zone management programme may, amongst others, result in the reduction of environmental damage in later years (Fig. 6).

In the IPCC/CZMS 1992 report seven recommendations for future activities are formulated:
1. coastal states should assess their vulnerability to an accelerated rise in sea level and to other potential coastal impacts of global climate change;
2. coastal states that have conducted vulnerability assessments should begin the process of planning for appropriate response strategies;

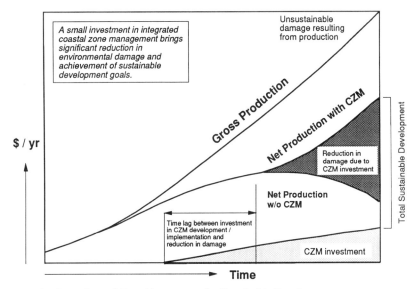

Fig. 6   *Coastal Zone Management for Sustainable Development*

3. the CZMS should continue to facilitate and build on its information network;
4. cooperation in the development of national guidelines for coastal zone management is needed;
5. the CZMS should cooperate with other IPCC subgroups to prepare an integrated assessment related to sea level rise and other coastal impacts of global climate change;
6. support for and cooperation with international organizations that are coordinating relevant data sets and implementing systematic observation and monitoring networks should be increased;
7. vulnerability to sea level rise and the need for coastal zone management are important factors that should be considered in decisions on planning and funding of coastal development, including those by international financial and development agencies.

Present IPCC/CZMS activities are for a large part directed towards the implementation of the above recommendations. Various "new" country studies into the vulnerability to accelerated sea level rise are considered. An international conference on coastal zone management planning, with preceeding regional conferences, is scheduled for November 1993 in the Netherlands. Common concepts, techniques and tools for management planning in the coastal zone (preferably in the format of a Common Methodology) are being developed and will be discussed at the 1993 conference.

The IPCC/CZMS activities must proceed, sea level rise and the stress on the coastal zones does not stop either.

REFERENCES

DELFT HYDRAULICS (1992) A Global Vulnerability Assessment; vulnerability assessment for population, coastal wetlands and rice production on a global scale; 178 pp.

IPCC (1990) Climate Change, the IPCC Scientific Assessment. WMO/UNEP. Cambridge Univ. Press; 364 pp.

IPCC/CZMS (1990a) Changing Climate and the Coast. Proceedings of the 1989 CZMS Workshop; Miami, USA; 2 volumes, 396 & 508 pp.

IPCC/CZMS (1990b) Adaptive responses to Climate Change. Proceedings of the 1990 CZMS Workshop; Perth, Australia; 246 pp.

IPCC/CZMS (1990c) Strategies for Adaptation to Sea Level Rise. Rijkswaterstaat, the Netherlands; 122 pp.

IPCC/CZMS (1992) Global Climate Change and the Rising Challenge of the Sea. Rijkswaterstaat, the Netherlands; 35 pp. + 5 appendices.

For additional information please refer to:

    IPCC/CZMS Secretariat
c/o Ministry of Transport, Public Works & Water Management
    Rijkswaterstaat, Tidal Waters Division
    P.O. Box 20907
    2500 EX Den Haag
    the Netherlands

    fax: *31-70-3282059

Mechanisms for recurrent nuisance algal blooms in coastal zones: resting cyst formation as life-strategy of dinoflagellates

Stefan Nehring
Institut für Meereskunde an der Universität Kiel,
Düsternbrooker Weg 20, D-2300 Kiel 1,
Federal Republic of Germany

ABSTRACT: The increasing occurrence of nuisance blooms and their negative ecological and economic impact has led to intensified monitoring activities. However, mechanisms of bloom formation are poorly understood. Among the most frequent producers of nuisance blooms, dinoflagellate species with a benthic cyst stage in their life cycle are prominent. The role of those resting cysts in bloom initiation, species dispersal, genetic recombination, survival of environmental stresses and as vectors of toxicity is discussed. Distribution studies of cysts in recent sediments of the coastal North Sea are presented together with a comprehensive list of cyst-forming, planktonic dinoflagellate species. Furthermore the collection, preparation and identification of resting cysts is described.

INTRODUCTION

Phytoplankton blooms are temporary phenomena and occur primarily in coastal zones, in landlocked areas, fjords and lakes. The increasing pollution of coastal waters by urban and agricultural sources, especially with nutrients, has helped to cause a worldwide increase in nuisance phytoplankton blooms over the past 20 years. Among the negative effects of such blooms there are esthetic problems of water quality (discoloration, scum-forming with all consequences on the water-based tourist industry) as well as the massive kills of fish and other marine animals, induced by oxygen deficiency or algae toxin production. Phytoplankton toxins may directly cause allergic problems to man and may directly involve the death of e.g. fish (Burkholder et al. 1992) as well as domestic animals (Nehring 1991). They may be concentrated in the food chain e.g. by mussels with fatal results for the consumers (man, marine mammals, birds). Toxins of dinoflagellate are known to cause Paralytic Shellfish Poisoning (PSP), Diarrhetic Shellfish Poisoning (DSP), Neurological Shellfish Poisoning (NSP) and Ciguatera fish poisoning which may be a hammer for a variety of aquaculture branches. At present more and more sea areas were closed down temporarily for sale of shellfish. All of these events were often associated with occurrence of toxicity in phytoplankton species that had been thought to be harmless. For ecological and economic reasons monitoring activities were intensified but the mechanism of exceptional phytoplankton bloom formation is still poorly understood.

Among causative organisms of nuisance blooms, dinoflagellate species, which often include a dormant cyst stage in their life cycle, play an important role. At present aspects of the timing, location, and duration of dinoflagellate blooms have been linked to the encystment/excystment of resting cyst-forming species (Anderson & Morel 1979, Anderson et al. 1983). Moreover there is some evidence that factors such as cyst resuspension and current transport but also cyst transport in ship ballast water may repeatedly infect areas with toxic species (Anderson 1989, Hallegraeff & Bolch 1991). Highly toxic resting cysts, containing up to 10 times the toxin of vegetative cells, may present a source of poison to organisms well after the motile species have disappeared from the water column (Dale et al. 1978, Oshima et al. 1982).

The occurrence and distribution of dinoflagellate resting cysts is subject of increasing interest because cysts represent stable populations assuring geographical maintenance in contrast to the transient bloom from which they may be derived. The fact that resting cysts can be collected and enumerated during non-bloom periods, offers a potential tool for the prediction of future toxic blooms. The purpose of this paper is to provide an introduction to modern dinoflagellate resting cyst studies.

RESTING CYSTS IN THE LIFE CYCLE OF DINOFLAGELLATES

Many plankton groups include a non motile dormant egg or cyst stage in their life history (e.g. copepods, cladocerans, tintinnids, diatoms, and dinoflagellates). Fossilized dinoflagellate resting cysts as an important group of microfossils ('hystrichospheres') are known from sedimentary deposits of 230 million years ago (Triassic) and are extensively studied by geologists for biostratigraphy (e.g. in oil exploration). More than 100 years ago, living dinoflagellate resting cysts were first observed in plankton samples of the North Sea (Hensen 1887, Stein 1883) and were occasional mentioned in the phycological literature mainly about freshwater during the first half of the twentieth century (e.g. Braarud 1945, Diwald 1938, Klebs 1912, Zederbauer 1904). Detailed investigations were, however, not carried out until the 1960s as the cyst-teca relationships of modern dinoflagellates were recognized (Evitt & Davidson 1964, Wall & Dale 1966).

Red-tide phenomena are caused by the motile biflagellated stage in the life history of dinoflagellates whereby the dominant reproductive mode is asexual fission (Fig. 1). Often towards the end of a bloom some species are also capable of sexual reproduction, forming gametes that fuse into a swimming planozygote which, in most cases, transform into a resting stage. These cysts

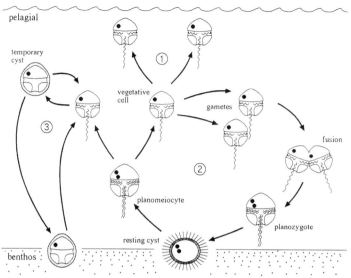

Fig. 1. Schematic diagram of basic dinoflagellate life cycle: 1) Asexual phase with motile, planktonic vegetative cell dividing by binary fission. 2) Sexual forming of non-motile resting cysts. 3) Asexual forming of non-motile temporary cysts. (●: cell nucleus).

are morphologically distinctive and differ from the motile planktonic stage. The cyst wall in most species is composed of organic matter (sporopollenin) but species of the genera *Scrippsiella* produce calcareous cyst walls (Fig. 2). Organic cyst walls are extremely resistant to natural decay and good fossilizable. The calcareous cysts and the cellulose wall of motile stages are less resistant, and they would probably not persist into the fossil record. To date, among the approximately 2000 extant dinoflagellate species more than seventy species of marine amd more than twenty species of freshwater planktonic recent dinoflagellates produce a resting cyst as part of their sexual life cycle (Tab. 1). Of these cyst-forming species, more than 20 have been known to cause red-tides and several species of *Alexandrium, Gonyaulax, Gymnodinium* and *Pyrodinium* are toxic.

Induction of sexuality in laboratory cultures is most often accomplished by sudden or gradual nitrogen depletion (Anderson et al. 1984), iron stress (Doucette et al. 1989), phosporus limitation (Anderson & Lindquist 1985) or an unfavourable temperature was also required (Anderson et al. 1985a). Sexual phenomena such as cellular fusion was, however, often mistakenly interpreted as cell division and together with the fact that only one in up to several hundred motile cells may produce cysts probably accounts for the paucity of cyst records in plankton studies.

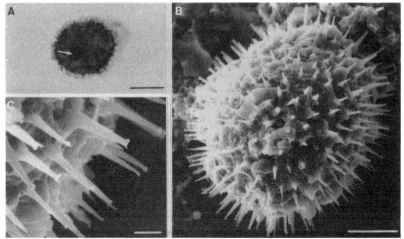

Fig. 2. *Scrippsiella trochoidea*, the most common resting cyst in North Sea sediments. (A) Living resting cyst showing red body (arrow). (B) SEM, cyst with numerous calcareous spines. (C) SEM, detail of ornament. Scale bars: 20 µm in (A), 10 µm in (B), 2 µm in (C).

In contrast to the vegetative cells, cysts have a negative buoyancy and accumulate on the sea bottom (Anderson et al. 1985b). A newly formed cyst generally has a manditory resting period during which it cannot germinate, even when growth conditions are optimal (Anderson 1980). The duration of dormancy varies significantly between species. It can last few days but also several months. When the dormancy stage is completed, germination of the cyst depends on external factors. Temperature is often cited as the major environmental factor regulating germination (Anderson 1980), but also light and oxygen as well as more subtle factors may be important (Anderson et al. 1987, Burkholder et al. 1992). In deep and relatively invariant bottom waters an endogenous circannual rhythm in cysts can control germination (Anderson & Keafer 1987).

RESTING CYSTS AS A STRATEGY IN SPECIES SURVIVAL

The formation of resting cysts has a variety of potential functions in the overall ecology of dinoflagellates:
A) The sexual reproduction of dinoflagellates is resulting in a genetic exchange which may lead to increased variation important for species survival. In this way, Anderson (1984) noted that cyst-forming dinoflagellates may maintain a viable, dormant seed population in the sediments year after year

Tab. 1.: Known recent planktonic and benthic (+) dinoflagellates producing a resting cyst (*= toxic; ?= observation of cyst forming has not since been verified).

MARINE SPECIES	REFERENCE
Alexandrium affine	Fukuyo & Inoue 1990, in Red Tide Organisms in Japan, eds. Fukuyo et al., Uchida Rokakuho Tokyo, 84-85
catenella*	Yoshimatsu 1981, Bull. Plankton Soc. Japan 28, 131-139
cohorticula*	Fukuyo & Pholpunthin 1990, in Red Tide Organisms in Japan, eds. Fukuyo et al., Uchida Rokakuho Tokyo, 88-89
excavatum*	Anderson & Wall 1978, J. Phycol. 14, 224-234
hiranoi	Kita & Fukuyo 1988, Bull. Plankton Soc. Japan 35, 1-7
leei	Fukuyo & Pholpunthin 1990, in Red Tide Organisms in Japan, eds. Fukuyo et al., Uchida Rokakuho Tokyo, 92-93
lusitanicum	Blanco 1989, Scient. Mar. 53, 785-796
minutum*	Bolch et al. 1991, Phycologia 30, 215-219
monilatum	Walker & Steidinger 1979, J. Phycol. 15, 312-315
ostenfeldii*	Dale 1977, Sarsia 63, 29-34
protogonyaulax	Montresor et al. 1991, in Abstracts 5th Intern. Conf. on Toxic Marine Phytoplankton, Newport, 86
tamarense*	Anderson & Wall 1978, J. Phycol. 14, 224-234
Amphidinium carterae*	Cao Vien 1967, Comptes Rendus Acad. Paris Ser. D 264, 1006-1008
Cachonina ? niei	v. Stosch 1969, Helgol. wiss. Meeresunters. 19, 558-568
Coolia monotis⁺	Faust 1992, J. Phycol. 28, 94-104
Diplopelta parva	Matsuoka 1988, Rev. Palaeobot. Palynol. 56, 95-112
Diplopsalis lebourae	Matsuoka 1988, Rev. Palaeobot. Palynol. 56, 95-112
lenticula	Matsuoka 1988, Rev. Palaeobot. Palynol. 56, 95-112
Diplopsalopsis orbicularis	Wall & Dale 1968, Micropaleontology 14, 265-304
Ensiculifera carinata	Matsuoka et al. 1990, Bull. Plankton Soc. Japan 37, 127-143
Fragilidium ? subglobosum	v. Stosch 1969, Helgol. wiss. Meeresunters. 19, 569-577
Gonyaulax digitalis*	Wall & Dale 1967, Rev. Palaeobot. Palynol. 2, 349-354
polyedra*	Nordli 1951, Nytt Mag. Naturvid. 88, 207-212
scrippsae	Bolch & Hallegraeff 1990, Bot. Mar. 33, 173-192
spinifera	Bolch & Hallegraeff 1990, Bot. Mar. 33, 173-192
verior	Matsuoka et al. 1988, Jap. J. Phycol. 36, 311-320
Gotoius abei	Matsuoka 1988, Rev. Palaeobot. Palynol. 56, 95-122
Gymnodinium ? breve*	Walker 1982, Trans. Am. Microsc. Soc. 101, 287-293
catenatum*	Anderson et al. 1988, J. Phycol 24, 255-262
Gyrodinium instriatum	Matsuoka 1985, Rev. Palaeobot. Palynol. 44, 217-231
resplendens	Dale 1983, in Survival strategies of the Algae, ed. G.A. Fryxell, Cambridge Univ. Press, 69-136
uncatenum	Tyler et al. 1982, Mar. Ecol. Prog. Ser. 7, 163-178
Heterocapsa ? triquetra	Braarud & Pappas 1951, Lebour. Vid. Akad. Avh. I. M. N. Kgl. 2, 1-23
Katodinium fungiforme	Spero & Moree 1981, J. Phycol. 17, 43-51
Perdinium dalei	Lewis 1991, Bot. Mar. 34, 91-106
Pheopolykrikos hartmanni	Matsuoka & Fukuyo 1986, J. Plankton Res. 8, 811-818
Polykrikos kofoidii	Morey-Gains & Ruse 1980, Phycologia 19, 230-232
schwartzii	Reid 1978, New Phytol. 80, 219-229
Prorocentrum lima⁺*	Faust 1991, in Abstracts 5th Internat. Conf. on Toxic Marine Phytoplankton, Newport, 41
marinum⁺	Faust 1990, in Toxic Marine Phytoplankton, eds. Graneli et al., Elsevier, 138-143
? pyrenoideum	Bursa 1959, Canad. J. Bot. 37, 1-31
Protoceratium reticulatum	Wall & Dale 1968, Micropaleontology 14, 265-304
Protoperidinium ? achromaticum	Popovsky & Pfiester 1990, Dinophyceae, Fischer, 272pp
americanum	Bolch & Hallegraeff 1990, Bot. Mar. 33, 173-192
avellana	Lewis et al. 1984, J. Micropalaeontol. 3, 25-43
brochii	Blanco 1989, Scient. Mar. 53, 797-812
claudicans	Wall & Dale 1968, Micropaleontology 14, 265-304
compressum	Bolch & Hallegraeff 1990, Bot. Mar. 33, 173-192
conicoides	Wall & Dale 1968, Micropaleontology 14, 265-304
conicum	Kobayashi & Matsuoka 1984, Jap. J. Phycol. 32, 251-256
denticulatum	Harland 1982, Palaeontology 25, 369-397
divaricatum	Bolch & Hallegraeff 1990, Bot. Mar. 33, 173-192
divergens	Dale 1983, in Survival strategies of the Algae, ed. G.A. Fryxell, Cambridge Univ. Press, 69-136
excentricum	Lewis et al. 1984, J. Micropalaeontol. 3, 25-34
? expansum	Hallegraeff & Bolch 1992, J. Plankton Res 14, 1067-1084
? granii	Meunier 1910, Campagne Arctique de 1907, Buleus, 343pp
? hangoei	Lewis et al. 1984, J. Micropalaeontol. 3, 25-34
latissimum	Wall & Dale 1968, Micropaleontology 14, 265-304
leonis	Wall & Dale 1968, Micropaleontology 14, 265-304
minutum	Fukuyo et al. 1977, Bull. Plankton Soc. Japan 24, 11-18
nudum	Harland 1983, Palaeontology 26, 321-387
oblongum	Wall & Dale 1968, Micropaleontology 14, 265-304
pentagonum	Bolch & Hallegraeff 1990, Bot. Mar. 33, 173-192
punctulatum	Wall & Dale 1968, Micropaleontology 14, 265-304

Table 1 cont.

Protoperidinium subinerme	Bolch & Hallegraeff 1990, Bot. Mar. 33, 173-192
thorianum	Lewis et al. 1984, J. Micropalaeontol. 3, 25-43
thulesense	Dodge 1985, Atlas of Dinoflagellates, Farrand Press London, 119pp.
Pyrodinium bahamense var. *bahamense*	Wall & Dale 1969, J. Phycol 5, 140-149
var. *compressum**	Matsuoka 1990, in Red Tide Organisms in Japan, eds. Fukuyo et al., Uchida Rokakuho Tokyo, 112-113
Pyrophacus horologium	Wall & Dale 1971, J. Phycol 7, 221-235
steinii var. *steinii*	Matsuoka 1990, in Red Tide Organisms in Japan, eds. Fukuyo et al., Uchida Rokakuho Tokyo, 116-117
var. *vancampoae*	Wall & Dale 1971, J. Phycol. 7, 221-235
Scrippsiella crystallina	Lewis 1991, Bot. Mar. 34, 91-106
lachrymosa	Lewis 1991, Bot. Mar. 34, 91-106
precaria	Montresor & Zingone 1988, Phycologia 27, 387-394
rotunda	Lewis 1991, Bot. Mar. 34, 91-106
sweeniae	Wall & Dale 1968, J. Paleontol. 42, 1395-1408
trifida	Lewis 1991, Bot. Mar. 34, 91-106
trochoidea	Anderson et al. 1985, Limnol. Oceanogr. 30, 1000-1009
Zygabikodinium lenticulatum	Matsuoka 1988, Rev. Palaeobot. Palynol. 56,95-122

FRESHWATER SPECIES	REFERENCE
Amphidinium ? cryophilum	Wedemayer et al. 1982, J. Phycol. 18, 13-17
Ceratium carolinianum	Wall & Evitt 1975, Micropaleontology 21, 18-31
cornutum	v. Stosch 1972, Bull. Soc. Bot. Fr., Mem. 53, 201-212
furcoides	Hickel 1988, Hydrobiologia 161, 41-48
hirundinella	Chapman et al. 1982, J. Phycol. 18, 121-129
rhomvoides	Hickel 1988, Hydrobiologia 161, 49-54
Crypthecodinium cohnii	Tuttle & Loeblich III 1975, J. Phycol 11 (Suppl.), 15
Cystodinium ? bataviense	Pfiester & Lynch 1980, Phycologia 19, 178-183
? cornifax	Schilling 1891, Z. wiss. Mikroskop 8, 314
Glenodinium ? emarginatum	Klebs 1912, Verh. Naturh.-Med. Ver. Heid.N.F.11, 369-451
Gloeodinium montanum	Kelley & Pfiester 1989, J. Phycol 25, 241-247
Gymnodinium ? chiastosporum	Cridland 1958, New Phytol 57, 285-287
dodgei	Sarma & Shyam 1974, Br. phycol. J. 9, 21-29
fungiforme	Biecheler 1952, Bull. Biol. France Belg. Suppl.36, 1-149
? fuscum	Bourelly 1970, Les algues d'eav douce III, N. Boubee Paris, 512pp
helveticum forma *achroum*	Skuja 1948, Symb. Bot. Upsal. 9(3), 1-399
? impatiens	Skuja 1964, Nova Acta Reg. Soc. Sci. Upsal. 18(3), 1-645
paradoxum	v. Stosch 1972, Bull. Soc. Bot. Fr., Mem. 53, 201-212
pseudopalustre	v. Stosch 1973, Br. phycol. J. 8, 105-13
Peridinium aciculiferum	Popovsky & Pfiester 1990, Dinophyceae, Fischer, 272pp
cinctum forma *ovoplanum*	Pfiester 1975, J. Phycol. 11, 259-265
forma *westii*	Eren 1969, Verh. Internat. Verein. Limnol. 17, 1013-1016
cunningtonii	Sako et al. 1984, Bull. Japan. Soc. Sci. Fish.50,743-750
gatunense	Pfiester 1977, J. Phycol. 13, 92-95
inconspicuum	Pfiester at al. 1984, Am. J. Bot. 71, 1121-1127
limbatum	Wall & Dale 1968, Micropaleontoly 14, 265-304
lubiniensiforme	Dilwald 1938, Flora (Jena) 131, 174-192
penardii	Sako et al. 1987, Bull. Japan. Soc. Sci. Fish.53,473-478
volzii	Pfiester & Skvarla 1979, Phycologia 18, 13-18
willei	Pfiester 1976, J. Phycol. 12, 234-238
wisconsinense	Evitt & Wall 1968, Stanford Univ. Publ. Geol. Sci. 12 (2), 1-15
Woloszynskia apiculata	v. Stosch 1973, Br. phycol. J. 8, 105-134
? cestocoetes	Thompson 1950, Lloydia 13, 277-299
? coronata	Wolozynska 1917, Bull. Acad. Sci. Cracovic (B) 1917, 114-122
? reticulata	Thompson 1950, Lloydia 13, 277-299
? tenuissima	Wolozynska 1917, Bull. Acad. Sci. Cracovic (B) 1917, 114-122
tylota	Bibby & Dodge 1972, Br. phycol. J. 7, 85-100

optimizing the growth and proliferation of motile cells as well.

B) This thick-walled resting stage may enable forms to survive unfavorable conditions in their environment and can be indicated as an overwintering stage. The cysts may lie dormant in bottom sediments for many years even at anoxic conditions (Anderson et al. 1987, Dale 1983).

C) Some resting cysts are themselves toxic and resuspended cyst may be then a direct source of shellfish toxicity (Dale et al. 1978). In Japan, Oshima et al. (1982) also found that the toxicity of natural cysts of *Alexandrium* spp.

was as much as 5 to 10 times that of vegetative cells germinated from the same cysts.

D) Cysts can act as seed populations, inoculating the water column with vegetative cells to initiate blooms (Steidinger 1975).

Anderson & Morel 1979 showed an *in situ* demonstration of the initiation of a dinoflagellate bloom via the excystment process. In a shallow restricted embayment cysts of *Alexandrium tamarense* were found in sediments only within the bloom area. In this stable system with reduced circulation, cyst germination initiated recurrent blooms in the overlying water.

In certain coastal and estuarine environments sediments contained only few toxic cysts but transport of suspended cysts from offshore seed beds by either periodic upwelling, storms or onshore subsurface currents to a localized site may also contribute to direct shellfish intoxication or bloom formation (Anderson & Wall 1978, Cembella et al. 1988, Dale et al. 1978, Seliger et al. 1979, Steidinger 1975, Yentsch & Mague 1979).

There is some evidence that besides to current transport human intervention, in the form of ship de-ballasting or the transfer of shellfish stocks may also be a factor to infect uncontaminated areas with toxic cyst-forming species (Anderson 1984, Hallegraeff & Bolch 1991).

The resting cyst forming dinoflagellate *Gymnodinium catenatum* was previously known only from southern Californian water. Hallegraeff et al. (1988) described the spreading of this toxic species in shellfish rearing areas in Spain, Japan and Tasmania. In European coastal waters a linering infection of *G. catenatum* from the south (Spain) to the north (France) is taken place (Wyatt 1992). A remarkable Indo-Pacific spreading of the toxic resting cyst forming *Pyrodinium bahamense* var. *compressum* between 1972 and 1984 has been documented (MacLean 1989). For Japanese waters between 1978 and 1982 *Alexandrium tamarense* and *Gymnodinium catenatum* spread into eight new areas, accompanied by shellfish poisoning (after Smayda 1990).

Besides the formation of resting cysts another strategic mechanism, i.e. the asexual formation of temporary cysts may be of importance in dinoflagellate life cycle (Fig. 1). Hypertrophic semi-enclosed, brackish basins situated behind the coast line appear to be very favorable sites for the mass development of dinoflagellates, as could be demonstrated for recurrent red tides of *Glenodinium foliaceum* in some of these ponds at the German North Sea coast (Hesse et al. in press). During and in the aftermath of these blooms, the sediment is covered by a carpet of *G. foliaceum* temporary cysts (Fig. 5A), the formation of which may be correlated due to changing environmental conditions. Temporary cysts, however, have no mandatory resting period and the protoplast of the *G. foliaceum* cyst can be devided in up to 8 daughter cells per cyst which may re-seed a second red tide. There is some evidence that the temporary cyst of *G. foliaceum* is also used as an overwintering stage. Culture experiments showed that temporary cysts of this species, which were stored in the dark at 4 °C for 6 months, germinated immediately after transfer in broad daylight at room temperature (Nehring unpubl.). The mechanism of temporary cyst encystment/excystment may plays an important role in the long term persistence of red tides in confined areas, but the occurrence of such cysts in natural waters has rarely been recorded.

Common phytoplankton monitoring activities only show the actual stock of vegetative cells. However, the life cycle of plankton organisms are undoubtedly important in many aspects of nuisance bloom formation and should be considered also in monitoring systems.

DISTRIBUTIONS OF RESTING CYSTS

Accumulations of dinoflagellate resting cysts have been observed in a variety of marine ecosystems, including offshore trenches and depressions, fjords, estuaries and shallow coastal embayments (e.g. Cembella et al. 1988, Dale 1976, Dale et al. 1978, White & Lewis 1982). The comparative distribution of

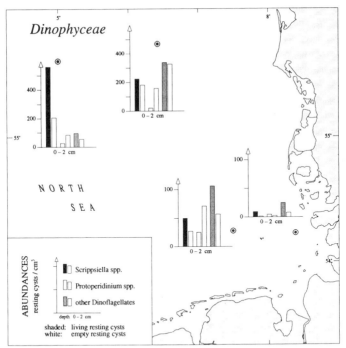

Fig. 3. Abundance of living and empty dinoflagellate resting cysts (*Scrippsiella*: 3 species; *Protoperidinium*: 7 species; other dinoflagellates: 10 species) in the topmost 2-cm of sandy mud North Sea sediments, collected beginning of October 1991.

cysts showed a general increase in cyst diversity and abundance from the inshore areas to offshore seas and is positively correlated with muddy sediments. Hydrodynamic conditions can produce high local cyst accumulations ("hot spot" for bloom initiation) e.g. convergence zones have been associated with high deposition of cysts which were found at the downstream periphery of the fronts (Cembella et al. 1988, Garcon et al. 1986, Tyler et al. 1982).

In sandy mud sediments of the North Sea the top 2-cm of offshore stations have concentrations of 583-682 living cysts/cm^3. Empty cysts constituted 34-53 % of the total cyst abundance. The inshore stations have only concentrations of 39-180 living cysts/cm^3. The fraction of empty cysts was 25-56 % of the total cyst count (Fig. 3). The dominance of *Scrippsiella* spp. calcareous cysts is remarkable, because Dale (1986) pointed out that calcareous cysts are relatively more important in tropical and oceanic rather than temperate regions.

Most of cyst distribution studies have been conducted only on qualitative status, presence or absence of a cyst species. These studies are valuable in describing the general geographic cyst distribution and can even indicate areas where nuisance blooms may be an unrecognized problem. Quantitative cyst mapping is much more time-consuming. At present only a few quantitative studies are available, mostly concerning abundances of potential toxic *Alexandrium* species (e.g. Anderson & Keafer 1985, Cembella et al. 1988, White & Lewis 1982) whereas the cyst distribution of non-toxic red-tide species is largely unknown (Nehring submitt.). In studies about the vertical distributions of cysts, the bulk was found to occur in the topmost two to three sediment centimeter (Anderson et al. 1982, Tyler et al. 1982, White & Lewis 1982). When buried in deeper sediment layers by sedimentation or bioturbation (Anderson et al. 1982), cysts may contribute little to bloom initiation, due to anoxia and other inhibitory micro-environmental factors (Anderson et al. 1987). However, cysts are able to survive for years in anoxic sediments (Dale 1983) and erosion or bioturbation bring buried cysts back to the surface. Consequently an unexpected number of cysts are present for bloom initiation.

The need for occasional dinoflagellate cyst mapping surveys is obvious, but quantification should be interpreted with caution. High cyst concentration may indicate potential seedbeds for bloom initiation, however, favorable conditions for germination are a prerequisite. Investigations of the abiotic factors of the sea area as well as studies of physiological ecology of cysts are necessary.

HOW TO COLLECT, TO PREPARE AND IDENTIFY DINOFLAGELLATE RESTING CYSTS

To document the presence of dinoflagellate cysts in sediments, a core sampler is used. Bottom samplers such as dredges or grab buckets (e.g. van Veen grab), which often lose the light fluffy material at the sediment surface, are less suitable. Best results with respect to an undisturbed sediment surface will receive by a box corer or a gravity corer (e.g. Meischner & Rumohr 1974) (Fig. 4). Sediment cores (10 cm long, 2.6 cm diameter) as subsamples can be obtained from the corer and then be stored in the dark at 4 °C until further examination.

Fig. 4. View of a modified Meischner & Rumohr (1974) gravity corer after collecting an undisturbed sediment core.

Two different methods can be used for cleaning and concentrating cysts from sediments: a standard palynological technique (Matsuoka et al. 1989) that uses harsh chemicals (hydrochloric acid, hydrofluoric acid) and a sieving technique that uses no chemicals. The palynological processing techniques can produce more concentrated samples but as a consequence all cysts are dead and additionally no information about the occurrence of calcareous cysts are available. For most biological studies in which species and living cysts are required for germination, it is appropriate to partition the top 2 or rather 3 cm of a sediment core and mix with filtered seawater. These subsamples are sonicated for one minute in an ultrasonic cleaning bath, in order to separate the cysts from organic and inorganic aggregates. Pass the suspension through a 150 μm gauze and accumulate on 20 μm gauze. The residue on the 20 μm gauze has to be rewashed and filled up with filtered seawater. Parts of this preparation may be counted on common slides using a light microscope or on Utermöhl slides using an inverted microscope.

Unknown cysts and individual cysts may be used for germination experiments to identify the species by identification of the motile thecate cell stage. To do this, pick cysts out from Utermöhl slides using a micropipette and wash twice in filtered seawater. Then cysts should be placed in small sterile incubation chambers and filled up with filtered seawater of the sample location or incubated in F/2-nutrient solution (medium). The cysts can be kept at room temperature (≈18 °C) and examined regularly for germination.

Most resting cysts are spherical, ellipsoid or polygonal with or without

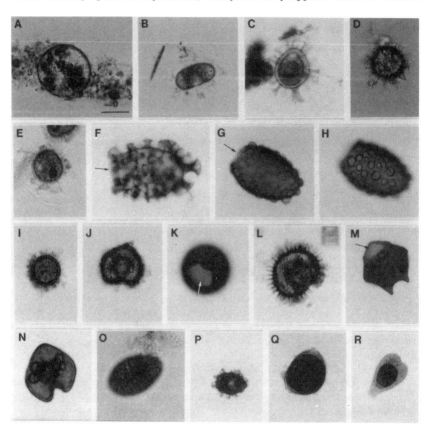

Fig. 5. Cysts of North Sea plankton organisms. **Dinoflagellate temporary cyst**: (A) *Glenodinium foliaceum*. **Dinoflagellate resting cyst**: (B) *Alexandrium* cf. *excavatum*. (C) *Gonyaulax digitalis*. (D) *Gonyaulax polyedra*. (E) *Peridinium dalei*. (F) Empty cyst of *Polykrikos kofoidii*, archeopyle (arrow). (G,H) Empty cyst of *Polykrikos* sp., archeopyle (arrow). (I) *Protoceratium reticulatum*. (J) *Protoperidinium claudicans*. (K) Empty cyst of *Protoperidinium conicoides*, archeopyle (arrow). (L) *Protoperidinium conicum*. (M) Empty cyst of *Protoperidinium leonis*, archeopyle (arrow). (N) *Protoperidinium oblongum*. (O) *Scrippsiella lachrymosa*. (P) *Scrippsiella trifida*. (Q) *Zygabikodinium lenticulatum*. **Tintinnid cyst**: (R) Heart-shaped cyst of ?*Favella* sp. Scale bar: 20 μm in (A,D,E,K,O,P), 40 μm in (B,C,F-J,L-N,Q,R).

spine-like ornamentations and range from 20 to 80 μm in diameter. The shape but also wall structure and color, paratabulation and the type of archeopyle (exit opening for germination) are important features used in cyst identification. The archeopyle is very useful in classifying the genus but the opening is not visible until excystment. Resting cysts are generally less conservative in morphology than their corresponding motile stages and therefore may be easier to identify (Fig. 5) (The asexual formed temporary cyst has no characteristics useful for identification; under non-suitable environmental condition motile cells cutting off the flagella and sometimes they shed their theca and cell membran to transform into round ball-like cells (Fig. 5A)).

A first guide to resting cyst identification has been prepared by Matsuoka et al. (1989) but not all known recent cysts are considered. At present, palaeontologists and biologists often use different names for the life history stages of the same dinoflagellates. Palynologists use a cystbased classification while biologists have developed a dinoflagellate classification system based on motile stages which is commanly used in biological cyst investigations. Unfortunately no comprehensive guide to recent resting cyst identification is available at the moment. The literature list given in this paper (Tab. 1) should provide a first introduction to cyst descriptions and taxonomy.

CONCLUSION

Many factors have been put forward to account for the development of nuisance blooms in coastal zones. Usually hydrological factors as temperature or salinity stratification and adequate nutrient and trace metal availability are held responsible for the phenomenon. The most frequent causative organisms for nuisance blooms are dinoflagellates. The role of the complex life-strategy of these forms in initiating bloom formation has not been considered sufficiently in the past. A partial explanation for the timing, persistence and recurrence of nuisance blooms may be achieved when studies of hydrodynamic, chemical and biological factors of the water column and the quantity, spatial and temporal distribution of resting cysts in the sediment are taken into account. It is suggested that the synthesis of sexuality, resting cyst formation and toxin production may be a very successful combination to balance short- and longtime variations in the ecosystem. That is why cystforming species may be of increasing importance in a changing world.

ACKNOWLEDGEMENTS: I thank Dr K.-J. Hesse for valuable discussions and comments on the manuscript and W. Huckriede for the drafting of diagrams. Part of this study is supported by the Federal Environmental Agency, Environmental Research Plan of the Minister for the Environment, Nature Conservation and Nuclear Safety of the Federal Republic of Germany (Grant 108 02 085/1), and by the State of Schleswig-Holstein. This is publictaion no.41 of the project Ecosystem Research Wadden Sea.

LITERATURE CITED

Anderson, D.M. 1980. Effects of temperature conditioning on development and germination of *Gonyaulax tamarensis* (Dinophyceae) hypnozygotes. J. Phycol. 16: 166-172.
Anderson, D.M. 1984. Shellfish toxicity and dormant cysts in toxic dinoflagellate blooms. In, Seafood Toxins, Ragelis, E.P. (ed.). Amer. Chem. Soc. Symposium Series No. 262, Wash. D.C.: 125-138.
Anderson, D.M. 1989. Cysts as factors in *Pyrodinium bahamense* ecology. In, Biology, epidemiology and management of *Pyrodinium* red tides, Hallegraeff, G.M. & Maclean, J.L. (eds.). ICLARM Conference Proc. 21: 81-88.
Anderson, D.M. & Wall, D. 1978. Potential importance of benthic cysts of *Gonyaulax tamarensis* and *G. excavata* in initiating toxic dinoflagellate blooms. J. Phycol. 14: 224-234.
Anderson, D.M. & Morel, F.M.M. 1979. The seeding of two red tide blooms by the germination of benthic *Gonyaulax tamarensis* hypnocysts. Estuar. Coast. Shelf Sci. 8: 279-293.
Anderson, D.M. & Keafer, B.A. 1985. Dinoflagellate cyst dynamics in coastal and estuarine waters. In, Toxic Dinoflagellates, Anderson, D.M., White, A.W. & Boaden, D.G. (eds.). Elsevier, Inc.: 219-224.
Anderson, D.M. & Lindquist, N.L. 1985. Time-course measurements of phosphorus depletion and cyst formation in the dinoflagellate *Gonyaulax tamarensis* Lebour. J. exp. mar. Biol. Ecol. 86: 1-13.
Anderson, D.M. & Keafer, B.A. 1987. An endogenous annual clock in the toxic marine dinoflagellate *Gonyaulax tamarensis*. Nature (Lond.) 325: 616-617.
Anderson, D.M., Aubrey, D.G., Tyler, M.A. & Coats, D.W. 1982. Vertical and horizontal distributions of dinoflagellate cysts in sediments. Limnol. Oceanogr. 27: 757-765.
Anderson, D.M., Chisholm, S.W. & Watras, C.J. 1983. Importance of life cycle events in the population dynamics of *Gonyaulax tamarensis*. Mar. Biol. 76: 179-189.
Anderson, D.M., Kulis, D.M. & Binder, B.J. 1984. Sexuality and cyst formation in the dinoflagellate *Gonyaulax tamarensis*: cyst yield batch cultures. J. Phycol. 20: 418-425.
Anderson, D.M., Coats, D.W. & Tyler, M.A. 1985a. Encystment of the dinoflagellate *Gyrodinium uncatenum*: temperature and nutrient effects. J. Phycol. 21: 200-206.
Anderson, D.M., Lively, J.J., Reardon, E.M. & Price, C.A. 1985b. Sinking characteristics of dinoflagellate cysts. Limnol. Oceanogr. 30: 1000-1009.
Anderson, D.M., Taylor, C.D. & Armbrust, E.V. 1987. The effects of darkness and anaerobiosis on dinoflagellate cyst germination. Limnol. Oceanogr. 32: 340-351.
Braarud, T. 1945. Morphological observations on marine dinoflagellate cultures (*Porella perforata, Goniaulax tamarensis, Protoceratium reticulatum*). Avh. norske Viensk. Akad. Oslo I. Mat.-Naturv. Klasse 1944 (11): 18pp.
Burkholder, J.M., Noga, E.J., Hobbs, C.H. & Glasgow Jr., H.B. 1992. New phantom dinoflagellate is the causative agent of major estuarine fish kills. Nature (Lond.) 358: 407-410.
Cembella, A.D., Turgeon, J., Therriault, J.-C. & Beland, P. 1988. Spatial distribution of *Protogonyaulax tamarensis* resting cysts innearshore sediments along the north coast of the Lower St. Lawrence estuary. J. Shellfish Res. 7: 597-609.
Dale, B. 1976. Cyst formation, sedimentation, and preservation: factors affecting dinoflagellate assemblages in recent sediments from Trondheimsfjord, Norway. Rev. Palaeobot. Palynol. 22: 39-60.
Dale, B. 1983. Dinoflagellate resting cysts: "benthic plankton". In, Survival Strategies of the Algae, Fryxell, G.A. (ed.). Cambridge Univ. Press: 69-136.
Dale, B. 1986. Life cycle strategies of oceanic dinoflagellates. Unesco Technical Papers in Marine Science 49: 65-72.
Dale, B., Yentsch, C.M. & Hurst, J.W. 1978. Toxicity in resting cysts of the red-tide dinoflagellate *Gonyaulax excavata* from deeper water coastal sediments. Science 201: 1223-1224.
Diwald, K. 1938. Die ungeschlechtliche und geschlechtliche Fortpflanzung von *Gleodinium lubiniensiforme* spec. nov. Flora, Jena 131: 174-192.
Doucette, G.J., Cembella, A.D. & Boyer, G.L. 1989. Cyst formation in the red tide dinoflagellate *Alexandrium tamarense* (Dinophyceae): effects of iron stress. J. Phycol. 25: 721-731.
Evitt, W.R. & Davidson, S.E. 1964. Dinoflagellate studies. I. Dinoflagellate cysts and thecae. Stanford Univ. Publ. Geol. Sci. 10 (1): 12pp.
Garcon, V.C., Stolzenbach, K.D. & Anderson, D.M. 1986. Tidal flushing of an estuarine embayment subject to recurrent dinoflagellate blooms. Estuaries 9: 179-187.
Hallegraeff, G.M. & Bolch, C.J. 1991. Transport of toxic dinoflagellate cysts via ships' ballast water. Mar. Pollut. Bull. 22: 27-30.
Hallegraeff, G.M., Steffensen, D.A. & Wetherbee, R. 1988. Three estuarine Australian dinoflagellates that can produce paralytic shellfish toxins. J. Plankton Res. 10: 533-541.
Hensen, V. 1887. Ueber die Bestimmung des Plankton's oder des im Meere treibenden

Materials an Pflanzen und Thieren; nebst Anhang. Bericht der Kommission zur wissenschaftlichen Untersuchung der deutschen Meere, in Kiel 1882-1886, 5: 1-108.
Hesse, K.-J., Hentschke, U. & Brockmann, U. in press. A synoptic study of nutrient and phytoplankton characteristics in the German Wadden Sea with respect to coastal eutrophication. Proc. 25th Europ. Mar. Biol. Symp.
Klebs, G. 1912. Über Flagellaten- und Algen-ähnliche Peridineen. Verh. Naturh.-Med. Ver. Heidelberg N.F. 11: 369-451.
MacLean, J.L. 1989. Indo-Pacific red tide occurrences, 1985-1988. Mar. Pollut. Bull. 20: 304-310.
Matsuoka, K., Fukuyo, Y. & Anderson, D.M. 1989. Methods for modern dinoflagellate cyst studies. In, Red Tides: Biology, Environmental Science, and Toxicology, T. Okaichi, D.M. Anderson & T. Nemoto (eds.). Elsevier, Inc.: 461-479.
Meischner, D. & Rumohr, J. 1974. A light-weight, high-momentum gravity corer for subaqueous sediments. Senckenbergiana marit. 6: 105-117.
Nehring, S. 1991. Lethal effects of a *Nodularia spumigena* bloom at the North Sea coast. Red Tide Newsletter 4 (2&3): 8-9.
Nehring, S. submitt. Dinoflagellate resting cysts as factors in phytoplankton ecology of the North Sea. Proc. 100 Years Biologische Anstalt Helgoland.
Oshima, Y., Singh, H.T., Fukuyo, Y. & Yasumoto, T. 1982. Identification and toxicity of the resting cysts of *Protogonyaulax* found in Ofunato Bay. Bull. Jap. Soc. Sci. Fish. 48: 1303-1305.
Seliger, H.H., Tyler, M.A. & McKinley, K.R. 1979. Phytoplankton distributions and red tides resulting from frontal circulation patterns. In, Toxic Dinoflagellate Blooms, Taylor, D.L. & Seliger, H.H. (eds.). Proc. Sec. Int. Conf. on Toxic Dinoflagellate Blooms. Elsevier, North Holland: 239-248.
Smayda, T.J. 1990. Novel and nuisance phytoplankton blooms in the sea: evidence for a global epidemic. In, Toxic Marine Phytoplankton, Graneli, E., Sundström, B., Edler, L. & Anderson, D.M. (eds.). Elsevier, Inc.: 29-40.
Steidinger, K.A. 1975. Basic factors influencing red tides. In, Proc. First Intern. Conf. Toxic Dinoflagellate Blooms, V.R. LoCicero (ed.). Mass. Sci. Technol. Found.: 153-162.
Stein, F. 1883. Der Organismus der Infusionstiere. III. Abteilung. Die Naturgeschichte der Flagellaten oder Geisselinfusorien. 2. Hälfte. Der Organismus der arthrodelen Flagellaten. Verlag W. Engelmann Leipzig: 30pp.
Tyler, M.A., Coats, D.W. & Anderson, D.M. 1982. Encystment in a dynamic environment: deposition of dinoflagellate cysts by a frontal convergence. Mar. Ecol. Prog. Ser. 7: 163-178.
Wall, D. & Dale, B. 1966. "Living fossils" in western Atlantic plankton. Nature 211: 1025-1026.
White, A.W. & Lewis, M. 1982. Resting cysts of the toxic red tide dinoflagellate *Gonyaulax excavata* in Bay of Fundy sediments. Can. J. Fish. Aquat. Sci. 39: 1185-1194.
Wyatt, T. 1992. *Gymnodinium catenatum* in Europe. Harmful Algae News 2, Suppl. to IMS Newsletter 63: 4-5.
Yentsch, C.M. & Mague, F.C. 1979. Motile cells and cysts: two probable mechanisms of intoxication of shellfish in New England waters. In, Toxic Dinoflagellate Blooms, Taylor, D.L. & Seliger, H.H. (eds.). Elsevier, Inc.: 127-130.
Zederbauer, E. 1904. Geschlechtliche und ungeschlechtliche Fortpflanzung von *Ceratium hirundinella*. Ber. deutsch. Bot. Ges. 22: 1-8.

INTERDISCIPLINARY INVESTIGATION OF AQUATIC COMMUNITIES USING SIZE SPECTRA

Kamenir Y.G.
Institute of Biology of the Southern Seas,
Nakhimov Av., 2, Sevastopol, 335011 Ukraine

Environmental changes (often as a result of Man's activity) are among the most acute and important problems of our time. Maximum concentrations of active masses and maximum velocities of the shifts are seen in zones of "active boundaries concentration" (Lebedev et al., 1991); freshwater and circumcontinental water bodies being important soil/air/water interfaces. Their coastal areas with "fractal" (Mandelbrot, 1982) shore line are characterized by especially high SV (i.e. surface to volume ratio). Complex multicomponent character of the medium and processes to be understood, demand integration of information, methods and efforts of specialists from many fields of science. Much success was achieved through application of mathematical models, particularly those using modern computers.

However, there is an ecosystem component well behind the others, and an important one - biota, the totality of living organisms inhabiting the water area, making use of it in many ways and transforming the inhabitance medium. Relevant biomass in coastal zones is relatively high, and of all kinds of matter the "Living Matter" (Vernadsky, 1980) - living particles and their exometabolites - shows maximum activity. So, thorough description and good understanding of its main structural and functional parameters are essential for rational use and protection of the environment (the Homo Sapiens' habitat inter alia).

The General Systems Theory methodology can be helpful in this context. It is a straight-forward top-down analysis based on a minimum model of the object as a whole. Such search for operational approaches to biota description, modelling and data acquisition was made in 1981-1991 (Kamenir, 1987, 1991) on the base of the Ideal Minimum EcoSystem (IMES; Fig. A) model.

That is a hierarchical structure of buffers, cyclic fluxes and processes which are implemented through huge number of "flow-through" elements. Stores, nonlinear characters, high branching and inertia create bufferness of the medium and equilibrium of the parameters. Such model corresponds both to the biogeochemical cycles of the biosphere (i.e. the highest known level of biological hierarchy) and the cycles of biochemical reactions (i.a. intracellular cycles, the lowest level). So, IMES is applicable to the "Living Matter" (LiM) as a whole. All in all, it resembles a vortex, a "living whirlwind". We can find analogous models in studies of the past:

the "Tourbillon Vital" (Cuvier, 1817; Vernadsky, 1924), the "Useless Machine" (Lotka, 1956), the "Microlitrosphere" (Sieburth and Davis, 1982), energy cycling (Patten, 1985), the "Local Ecosystem" (Gorshkov, 1991).

Resemblance of this structure (vortex) to the structures studied in hydro- and aerodynamics is not a remote one, but reflects profound and important aspects. Energy dissipation (heat) is an indispensible property of every biological being and structure. It is the cessation of biological motion (anabolic metabolism and the resulting heat flux) that furnishes a clear distinction between life and death (Cuvier, 1817). So, LiM as a whole is a structure existing only where there is an energy flux, i.e. a dissipative structure. Usually it exists in a specific medium protruded by energy flux from the Sun.

This substance is intensively studied in the XX century. The main structures (membranes) and biochemical components (water, DNA, RNA, ATP, protein, photopigments) common and inherent to all living nature are known now. Its structure, primarily time-and-space structure, is of great importance to us. This is a distributed parameters system, a wave-corpuscular matter. Every subsystem, every flux or pool of LiM is composed of huge numbers of heterogenous entities (membrane envelopes filled with protoplasm) where periodic processes take place, activating, synchronizing and integrating all the aggregation into a whole. So, it is a set-object whose properties can not be derived from properties of composing elements. On the contrary, parameters of the whole should be quite different, original ones. So it was made in thermodynamics with "pressure, temperature" of the gas through statistical description of huge assemblages of molecules (Vernadsky, 1980; Chislenko, 1981).

Instead of analyzing dynamics of huge sets of living beings or taxonomic subunits it is better to proceed to statistical descriptions of the whole. First, we should try to see the architectonics of the whole, the general and stable proportions of its static view. So, begin with distribution functions of the set in a specially designed multidimensional coordinate space (MCS) and 1-2-dimensional projections suitable for imaging and analysis. Such an approach is known and used in some fields of environmental science (Figs. G-L). These are statistical (regression analysis) regularities describing change in the particles' properties (allometries; Fig. B) and distribution functions of ensembles of living and nonliving "quanta" of the habitat in multidimensional space (Fig. F) or one-dimensional Size Spectra (SS; Figs. C-E, G-L).

Important is that formal quantitative parameters serving as axes of the space are suitable for Hi-Tech automatized measurement and mathematical transformations. Those are the body mass (W) or volume (V), surface (S) or size (D). Most recent are "equivalent spherical diameter" (ESD; lg D ~ 0.33 lg W) and "metabolically active biomass" and body (Wa), i.e. expressed in units of ATP, protein, "living protoplasm". So we can classify every entity as auto/heterotrophic (photopigments), living/dead (ATP), organic/unorganic (protein, DNA) etc. (Wentworth, 1931; Sheldon, Parsons, 1967; Schwinghamer, 1981; Sieburth, Davis, 1982; Peters, 1983; Yentsch et al., 1983; Kamenir, 1986, 1991; Gorshkov, 1991 etc.).

Experience of many environmental sciences and tool-making industry has shown that most simple and graphic representation of the distributions is obtained while getting from ΔX to $\Delta X/X$, i.e from X' to $(\log X)'$, using logarithmic transformations of parameters (Figs. B-L). Such axes are dimensionless that offers opportunities for mathematical calculations and comparison (e.g. coordinates of points and displacements in the hyperspace) (Fig. F). So we proceed from lists of taxonomic units (level 1) or range distributions (level 3) to MCS, i.e. to the maximum (6-th) level of the formalization (Jones, 1972).

So we can combine and integrate into unified descriptions both (living and abiotic) components of the ecosystem in question. This allows us to observe structured LiM in structured environment (Figs. C-E), intertwinned and interacting sets or vortices, where LiM is just a whirlwind among whirlwinds. Ecology then is just "the science of the habitat", of the space inhabitable for the living being (a particle, a set or a subset) under our consideration (e.g. Man, population, species etc.).

The most interesting are parameters of the media filling the space volume, first of all parameters of biological importance (such as temperature, pH). Other kinds of LiM including their statistical characters are among the main. Allometries, particularly those giving the time and space scales intrinsic for living beings (Figs. B1, B2), offer us instruments for MCS scaling and transformations.

The SS extremities are of great importance too. The small-cell (microbial) side of SS depicts the main work of the inhabitance medium maintenance, transformation and energy dissipation (Fig. I). However, as Man himself belongs to the opposite (i.e. the right) side of SS (maximum body weighters, Wm) it is very difficult for him to study the microbial side of the SS, and it is the least studied indeed (Lebedev et al., 1991).

But now, modern Hi-Tech means (laser, optics, molecular biology, microcalorimetry, electronics, cell selectors and computers) and MCS approach make this part of LiM the most suitable for study (Gnaiger, Forstner, 1983; Yentsch et al., 1983; Kamenir, 1987, 1991). The computer-operated devices in several minutes examine several millions of particles and select the most interesting of them for detailed analyses (Fig. F).

Being well adapted for quantitative comparisons, MCS descriptions gave evidence to existence of "typical patterns" (Schwinghamer, 1981; Figs. G, D, I) and typical values of integral parameters of integral natural communities (Kamenir, 1991). Those typicals may be helpful in search of general regularities. Comparison of a row of 8 aquatic communities covering about the entire water area (Ses) range (feasible for the Earth, $10^{-6} - 10^{14}$ M^2) gave evidence to the shift of the SS, and particularly, of its right extremity (Wm; Fig. G), depending on Ses. Regression equation was obtained (Kamenir, 1991):

$$\lg Wm \ (mcm^3) = a + b \lg Ses \ (m^2),$$

where $a = 8.02 \pm 1.00$, $b = 0.88 \pm 0.13$, $N = 8$, $F = 49.17$, $r^2 = 0.89$. But from the model's (IMES) logic it follows that somewhat different values ($a \sim 6$, $b \sim 1$) would be more correct. Additional analysis with help of micro- and mesocosms data could produce higher precision and give us an instrument for communities diagnstics and scaling.

There are models (Kolmogorov, 1941; Gorshkov, 1991; Kamenir, 1991) which consider the right side of the SS form and help to understand the rules of its change. Due to IMES a quantitative formal parameter can be obtained, as the community succession would give broader SS, first of all by rising Wm value (Fig. G). Then, having a "standard" (Ws, the limit to growth allowed by the environmental "niche", Ses) we could obtain a quantitative measure of a given community degradation level:

$$pW = -\lg (Wm(t) / Ws),$$

where Wm(t) is the real Wm estimate for a given moment. Using an allometry, we can derive at the regressions: $\lg \tau \sim 0.6 \lg D \sim 0.2 \lg Wm \sim 0.2 \lg Ses$, $\lg L \sim 0.5 \lg Ses$ (Figs. B-1, B-2), i.e. obtain intrinsic time and length for scaling of models of the "community to be". An example given by E.Odum (1971; Fig. 9-2), is a comparison of the forest and the microcosm successions. Such scaling could be helpful in theoretical end experimental ecology, e.g. for calculation of a "succession to be", for comparison between "normalized" equations (proceeding from t, D, W to t/τ, D/L, W/Wm).

Another important result of the comparison between 8 communities is low time-and-space variability of some integral natural aquatic community indices per unit of water area. The range of values makes ~ 23 orders for water body volumes and ~17 for water area size. For integral biomasses B_2 (17 - 194 Kcal/m^2) and Integral Surface Indices (ISI, 14 - 207 m^2 m^{-2}, a dimensionless parameter, i.e. surface of all living beings) it makes less then 2 orders (Kamenir, 1991). Taking into account the "living part of the living matter" (Vernadsky, 1980) we estimate total protein of all living organisms (B2p) ~ 10 g m^{-2}, that corresponds to B2 ~ 100 kcal. This index is better than B2 as it excludes metabolically inactive parts of organisms which make a very big part of many communities (e.g. of a forest).

It correlates well with another index, Ac (quantity of ATP of microorganisms, mgATP m^{-2}). From analysis of data published in 1964-1984 we obtained the following estimate:

lg Ac = 1.16 $\pm$ 0.55, i.e about 500 kg of living microorganisms per ha. . This estimate was confirmed by the author's data obtained during the research cruises of IBSS to the Black and Mediterranean Seas, the Atlantic, Indian and Pacific oceans (Antarctic zone included) (Kamenir, 1991).

Values of total protein (B2p) are helpful in studying the internal structure of the "living vortex". The proteins of LiM (10 g) would make a covering layer of about 10 cm thick. However, in nature the situation is quite different. Hierarhical structure of branches and the "enveloped envelopes" of closed membranes (50% composed of protein) forms a highly porous medium, "Sierpinski sponge". Huge fractals (Mandelbrot, 1982) of metabolically active surface are distributed in mats (hundreds meter thick) like a tropical forest or oligotrophic ocean. As thickness of membranes is ~ 10 nm, 10 g of protein can make ~ 2000 m^2 of membranes per m^2 of Earth.

That is about an order greater than ISI (~ 100), which includes only external membranes of all particles. This difference is evident while looking at a multilevel hierarhical structure of humane body. If the body weight is ~ 100 kg (i.e. 0.1 cubic meter with substance density ~ 1 kg/l, mean size of composing cells ~ 20 mcm, with external surface (S1) about 1 and internal surface (S2, i.e. lungs, intestines, kidneys etc.) about 200-300 m^2, we can roughly estimate integral cell (S3) and the entire bulk of internal membranes (S4) surfaces of 10^4 and 10^6 m^2, correspondingly. The membranes fractionate the same volume (the body) into hierarchy of ever-growing surfaces. As $S \sim D^2$ and $V \sim D^3$, the SV ratio is $\sim D^{-1}$ (Fig. B-4), and we can see an hierarchy of SV and the envelope sizes (D).

At every level of hierarchy (i.e. compartment fractionating) SV increases ~ 100 times. Following the turnover time logic (lg Tt ~ lg W - - lg S ~ 0.2 lg W ~ 0.6 lg D; Kamenir, Khailov, 1987) and allometry (Fig. B-1) the time scales of the microvolumes enveloped by membranes get ~ 1-2 orders less. Similar progression of logarithms (D/D ~ 2 orders, hence τ/τ ~1 order) is seen in the typical patterns of the biomass SS of benthos and plankton (Fig. G). Supposing that all membranes of LiM have about the same relative flux density, we can see a hierarchy of fluxes (vortex) or a progression of logarithms (about 2 n) of integral values. About the same is the progression of the Sm and energy dissipation SS-s (Fig. I;.look also Figs. A, G-1, G-2, B-1).

Using the SV-ratio as an ESD index, it is easy to describe (in SS form) even such complicated communities as the coastal where biomass consists mostly of macrophytes with very delicate branching and complicated geometry. Then we can follow the biota change under anthropogenic eutrophication which is among the most acute problems of coastal regions. The size spectrum shift (according to the IBSS studies and published data) (Figs. G, H), agrees well with the results of IMES analysis (Fig. G-1, G-2) where entire living matter (epiphytes included) is considered.

CONCLUSION. Summing up all written above we adopt transition from taxonomic units to statistics (distribution functions) of huge aggregations of particles in a specially designed metric coordinate space (MCS). It gives us opportunity to integrate all types of living and nonliving particles, and also some data on the "structured environment" in a uniform scheme. Logarithmically converted quantitative geometrical and biochemical characters of prime importance used as axes of MCS can provide a powerful means of description and analysis of integral natural communities.

By describing the interpenetrating and interacting ensembles of organisms and abiotic units (i.e. structured Living Matter in structured habitat) we obtain a powerful instrument for cooperation of engineers, managers and scientists of different fields. We integrate technical means of data acquisition, apparatus for mathematical transformations and scaling necessary to build up formalized quantitative, compact and graphic descriptions. Then we will become aware of complex natural objects, i.e. space filled with and controlled by life, the space we are living in. This introduces operational approaches for solving problems of aquatic environment use and protection.

REFERENCES

Chislenko L.L. 1981. The structure of fauna and flora with regard to organism size. Moscow Univ Press. {Russ.]. 208p.

Cuvier G. 1817. Le regne animal distribue d'apres son organisation. V.1. Deterville.

Gnaiger E., Forstner H. (eds). 1983. Polarographic Oxygen Sensors. Springer. 370p.

Gorshkov V.G. 1991. Ecological stability and local adaptations: The admissible limits of anthropogenic perturbation. Reprint LNPI-1701.

Jones J.C. 1972. Design Methods. Wiley. 374p.

Kamenir Y.G. 1987. Investigation of Living-and-Bioinert Matter with help of Size Spectra. Deposit N5246-B87 at VINITI. [Russ.]. 46p.

_____. 1991. Investigation of the Aquatic Systems Living Matter with help of Size Spectra. Ph.D. Thesis. Sevastopol, IBSS. [Russ.]. 214p.

_____, Khailov K.M. 1987. Metabolic characteristics and totol Surface Area of the Living Matter of the Ocean: A comparison of Size Spectra. Oceanology. 27: 492-496.

Kolmogorov A.N. 1941. On logarithmically-normal distribution law of particles formed by crushing. Repts.Ac.Sci.USSR. 31: 99-101.

Lebedev V., Aizatulin T., Khailov K. 1991. The Living Ocean. Progress.328p.

Lotka A.J. 1956. Elements of Mathematical Biology. Dover. 465p.

Mandelbrot B.B. 1982. The Fractal Geometry of Nature. Freeman.

Odum E.P. 1971. Fundamentals of Ecology. Saunders. 574p.

Patten B.C. 1985. Energy Cycling in the Ecosystem. Ecol.Model. 28: 1-71.

Peters R.H. 1983. The Ecological Implications of Body Size. Cambridge. 329p.

Schwinghamer P. 1981. Characteristic Size Distribution of Integral Benthic Communities. J.Fish.Aquat.Sci. 38: 1255-1263.

Sheldon R.W., Parsons T.R. 1967. A continuous Size Cpectrum for Particulate Matter in the Sea. J.Fish.Res.Bd Can. 24:909-915.

Sieburth J.M., Davis P.G. 1982. The Role of Heterotrophic nanoplankton in the Grazing and Nurturing of Planktonic Bacteria in the Sargasso and Carribbean Seas. Annales Inst.Oceanogr.Paris. 58(S): 285-296.

Vernadsky V.I. 1924. La Biosphere. Paris. Sorbonne.

_____. 1980. Living Matter. Nauka. [Russ].

Wentworth C.K. 1931. The Mechanical Composition of Sediment in Graphic Phorm. Univ.Iowa. 127p.

Yentsch C.M., Horan P.K., Muirhead K. et al. 1983. Flow Cytometry and Cell Sorting: A Technique for Analysis and sorting of Aquatic Particles. Limnol.Oceanogr. 28: 1275-1280.

LEGENDS TO FIGURES

A review of data available from literature. See Kamenir (1987, 1991) for references.

A. Ideal Minimum EcoSystem (IMES) - closed volume (buffer) where steady state (shading) is maintained of parameters important for stability of closed flux (Ic) of renewal of identically structured flow-through elements. Io - losses, It - compensating influx.

B. Allometries, - regression equations of the type: $\lg Y = a + b \lg X$, D - size of organisms (mcm), W - body mass or volume, $\lg W \sim 3 \lg D$; 1 - organisms generation time (h), 2 - average size of the ecosystem (m): $\lg L \sim 0.5 \lg$ Ses, where Ses - the necessary ecosystem area; 3-5 - beach sand parameters dependence on the grain size D: 3 - bacteria number (cells g^{-1}), 4 - internal surface ($m^2 m^{-3}$), 5 - oxygen consumption ($gO m^{-3} .d^{-1}$).

C. Size Spectra (SS) - distribution of the integral value (e.g. biomass) to size classes (i.e. equal increments of logarithm of D or W); $pV = -\lg B$, where B - biomass concentration (kg/kg) in soil or water; 1 - "typical pattern" of the benthos SS (Schwinghamer, 1981), 2 - limnetic plankton, 3 - plankton bloom.

D. Size distribution of species number of fish (1), its parasites (2).

E. Structured environment, i.e. Size Spectrum (relative units) of the soil particles mass (granulometry, 1-3) (Wentworth, 1931) and of the Langmuire circulation cells (4).

F. Multidimensional distribution (in a metric coordinate space) of particles (marine phytoplankton) (Yentsch et al., 1983); axes - logarithm of the photomultiplier tube signal: Chl - chlorophyll (luminescence wave length - 590 nm), PHE - phycoerithrin (540-590 nm); n - isolines of cell numbers, X,Y - the point coordinates, d - distance between two nuclei.

G. Size Spectrum evolution during succession (1 to 2; 2 to 1 - degradation), theoretical model; 4 - deviation of the benthic macrophytes under anthropogenic eutrophication (from initial position 2), 3 - additional part of community in eutrophicated conditions (i.e. epiphytes).

H. Epiphytes biomass size distribution in oligo- (1) and eutrophic (2) conditions, relative units.

I. Energy dissipation Size Spectra. D - size of organisms or the Wave length; 1 - the Sun light; the World ocean (2) and terrestrial (3) community respiration.

K. Energy distribution of the oceanic turbulence.

L. Size Spectra of soil particles mass forming the tube of the seafloor worm; 1 - young worm (W = 0.6 g), 2 - adult (4.1 g).

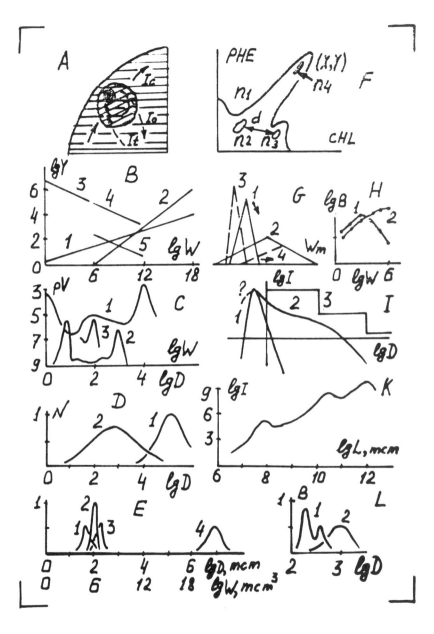

DESCRIPTION OF CHANGES IN A POCKET BEACH USING EMPIRICAL EIGENFUNCTIONS

by

N.G. Kypraios and C.I. Moutzouris

INTRODUCTION

Changes in grain-size distribution and beach profile geometry on a coastal zone occur both spatially and temporally. Wind-generated waves and the transfer of incidence wave energy to other modes of fluid motion are the main sources which drive beach changes. A knowledge of the interactions between process and response elements in a coastal zone is necessary for a better understanding of nearshore processes and coastal modelling.

A field study of sedimentary and morphological parameters was conducted on a non-tidal pocket beach of Greece under a range of wave energy conditions. In previous papers (Moutzouris and Kypraios, 1987; Kypraios, 1988) results about cross-shore and longshore sediment size distributions on the beach under study were presented. In the present paper the variability of the same pocket beach is assessed using the Empirical Orthogonal Function Analysis (E.O.F.). The resulting primary modes of variability provide an insight in the process-response under examination.

METHODOLOGY

Data Collection

The study was conducted along the Kokkino Limanaki beach (Fig. 1), which lies on the Northeast coast of Attica, Greece, at a distance of 27km from Athens. The coast is a local interruption of a continuous cliff headland, with two bounding headlands to the north and south. This kind of beach satisfies the basic requirement for a nearly constant sediment budget in the coastal zone, due to the reduced lateral exchange of sediment with adjacent beaches. Hindcast significant wave heights are less than 1.9m with significant periods less than 5.3sec. The arcuate coastline between the headlands has a length of about 400m. The width of the beach ranges from 4 to 18m.

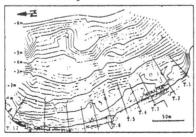

Fig.1.- Bathymetry and location of the profile lines

Measurements conducted during 16 field campaigns, from November 1985 till June 1897. The collected data include bathymetric surveys, sedimment sampling, water-level measurements, underwater photographing and some current measurements along twelve pre-defined transect lines. The data provide a detailed record of profile evolution with sediment size distribution. Spot sediment samples are limited to the upper 5cm of bed material to collect the sediment deposited during the precedent sea state in the following positions (Fig.2):
- the maximum wave run-up
- the mid-swash segment
- the plunge step (significant breaker zone)
- the transitional zone
- the landward limit of the offshore segment.

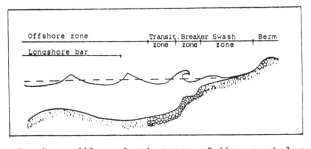

Fig.2.- Beach profile and sub-zones of the coastal zone

Grain-size parameters were computed by sieving according to the ASTM standards. Density was calculated from dry sample weight and volumetric displacement. Gross mineralogy was determined by X-ray diffraction analysis. Sea-bottom porosity was also measured in samples with standard volumetric displacement.

The grain-size distribution in the cross-shore direction is found to be related to coastal forms both reflecting the differing wave energy impact in the successive sub-zones of the nearshore zone. Coarse/fine-sized and poor/ well-sorted grains are found in zones of increased/decreased wave energy (Moutzouris and Kypraios, 1987; Moutzouris, 1988).

The longshore variation in sediment size is found to be related to the direction of wave approach. Finer grains move in the direction of the wave-induced longshore current. In addition, cross-shore sediment transport occurs which is especially influencing the pre-breaker zone. The shoreward migration of the landward limit of the sandy offshore segment and the bar crest after periods of relatively high wave-energy conditions is the most substantial result of this kind of motion. Shoreline migration and the shape of the beach also reflect the direction of wave-energy approach. Longshore sediment transport controls this equilibrium, causing erosion at the north /south and accretion at the south/north part of the beach (Kypraios, 1988).

Empirical Eigenfunctions

The statistical method used to analyse beach characteristics in this study involves the generation of sets of empirical eigenfunctions from the data (Preisendorfer, 1988). The eigenfunctions are derived from sample covariance estimates, so the structure of the functions is defined by the actual data set and does not assume some a priori functional form. The functions are "orthogonal" because they are uncorrelated over space and "empirical" because they arise from data. Each function represents a certain amount of the mean square value of the data. The eigenfunction associated with the largest eigenvalue represents the data best in a least square sense, while the second function (in rank) describes the residual mean square data best in the least square sense.

This method enables a complete description of any complex field data with a relatively small number of functions (eigenvectors) and their time dependent coefficients (principal components). The data can be represented as a linear combination of corresponding functions of time and space. Each derived component or eigenvector can frequently be identified with an actual physical property inherent in the data field. The method provides an index which measures the importance of each component within the time domain under study. Finally the analysis provides an estimate of the total percent of variance in the data set explainable by each eigenvector.

Statistical analysis of beach profile data using empirical eigenfunctions has already been used to resolve the variance structure and has produced quantitative results showing changes in profile configuration as a function of space and time (Winant et al., 1975; Aubrey, 1979; Aubrey et al., 1980; Lins, 1985; Lippmann and Holman, 1990). The results of these studies indicate great promise for the application of this statistical technique in the analysis of beach profile data, since the derived functions have a direct relation to the natural processes. The technique has also been used to study changes in nearshore sediment distribution and identify characteristic patterns of sediment movement (Ramana Murty et al.,1986; Liu and Zarillo, 1989).

In this work empirical eigenfunctions have been computed by the method mentioned above to describe the spatial and tempotal variability of the pocket beach under study as following:
- shore-parallel morphological changes
- variations in the longshore sediment size distribution
- changes in the beach profile configuration.

In the first case the migration of the shoreline, the plunge step and the landward limit of the offshore segment were examined. The data analysed consist of coastal forms positioning determined by the horizontal distances measured on the profile lines from fixed benchmarks located in the beach backshore zone. In the second case the longshore changes in the mean grain-size diameter of sediment deposited in the mid-swash zone, the breaker zone and the offshore segment were analysed. The time series of the field data were constitued on a profile-by-profile basis for each horizontal beach segment. The technique was applied to data from the 12 beach profiles (T.1-T.12) during 13 field campaigns. In the third case the variations in inshore bathymetry along the beach profiles were examined. Profiles T.3, T.6, T.12 were selected to represent the south, the central and the north beach sections respectively. The data analysed consist of water-depth readings referenced to the mean sea level recorded on the beach. The technique was applied to data collected from 5 fixed points on each profile line during 16 field campaigns. The time series were constitued on a point-by-point basis for each profile.

The spatial eigenfunctions are simply the eigenfunctions (eigenvectors) associated with the spatial covariance matrix generated from the time series. Eigenfunctions are ranked according to their eigenvalues. The ratio of an eigenvalue to the sum of the eigenvalues is expressed as the percentage of variance attributed to the mode. The eigenfunctions represent the empirical modes or characteristic patterns of variability. Associated with each mode is a modal amplitude time-series (principal component) which describes how that mode varies in time. In the present analysis the means were subtracted from time series describing variation in beach/profile data, so that the E.O.F. modes represent variation from the mean beach/ profile state. The nodes in the eigenvectors represent relatively stable, pivotal points through which sediment is transferred bidirectionally, whereas the antinodes identify places of high beach mobility and sediment redistribution. In the following paragraphs are presented the two first eigenvectors and principal components calculated from the beach survey data and sediment distribution in the breaker and after-breaker zones, as well as from the beach profile configuration. These results contribute to better understanding of the variations in the high-energy areas of the coastal zone.

RESULTS

Figure 3 shows the first two eigenvectors and the associated principal components calculated from shoreline position measurements. The dominant pattern of shoreline movement is represented by the first eigenvector mode. It contains 60% of the variance. This function indicates that the beach is oscillating round a pivotal point located in the central part of the beach, between the transect lines T.5 and T.6. A seaward migration of the shoreline (positive eigenvector weightings) in the south/north part is accompanied by a landward migration (negative eigenvector weightings) in the north/south part of the beach. This suggests that an amount of sediment eroded from the swash zone is transported alongshore and is deposited downstream. The time dependence of the first eigenvector (Principal Component I) shows positive amplitudes at the summer months and negative amplitudes at the winter months.

The second eigenvector accounts for 29% of the variance of the data. It has no zero-crossings and it thus describes an almost uniform shoreline migration along the beach. The time dependent coefficients of the second eigenvector (Principal Component II) are strongly correlated to water-level measurements (linear correlation coefficient r=0.94) conducted in the beach. This indicates that sea-level fluctuations cause shoreline movements. These movements are not related to onshore/offshore sediment transport, as far as the generated current is too weak to move sediment grains in the swash zone.

Figure 4 shows the first two eigenvectors and associated principal components calculated from the plunge step position measurements. The variance contained by the eigenvectors I and II are 72% and 17% respectively. Spatial and temporal patterns of plunge step migration represented by these eigenvectors are similar to the patterns represented by the corresponding modes of shoreline migration. Linear regresion coefficients between the two sets of eigenvector weightings or principal component amplitudes are more than 95%. It is therefore clear that the first eigenfunction mode describes alongshore sediment transfers indicating a seasonal dependence, whereas the second mode is much less significant and is related to sea level fluctuations.

To investigate the effects of shoreline and plunge step migration to sediment distribution, empirical eigenfunctions were derived also from the grain size measured in the spot samples collected in the mid-swash and the breaker zones.

In the eigenvector analysis of mean grain-size diameter along the mid-swash segment, the first mode explains 51% and the second mode 20% of the variance in the data set. The eigenvector weightings in the space domain and the associated principal component amplitudes are presented in Fig.5. The first mode of grain-size variation is linearly dependent on the first mode of shoreline migration (Fig.3) and are likely to be related to the same process. The finer grains are transported alogshore due to the energy of the wave-induced longshore currents and are deposited in the vicinity of the bounding headland. When waves attack the beach from the northeast/southeast, grain-size incrases in the northern/southern and decreases in the southern/northern part of the swash zone. Prevailing of relatively high-energy waves from the NE/SE sector during summer/winter months causes a seasonal variability in shoreline position and swash-zone sediment distribution. The second mode describes similar trends in grain-size variation occured at both ends of the beach. It possibly reflects the influence of rip-currents generated in locations of wave divergence due to the bed topography and the arcuate coastline of the beach.

In the eigenvector analysis of mean grain size diameter along the breaker zone (Fig. 6), the first mode explains 53% and the second mode 16% of the variance in the data set. The spatial and temporal dependence of the first eigenmode

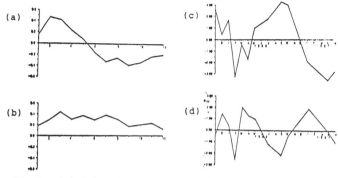

Fig.3.- (a),(b), Eigenvectors I,II and (c),(d) Principal Components I,II for shoreline position data

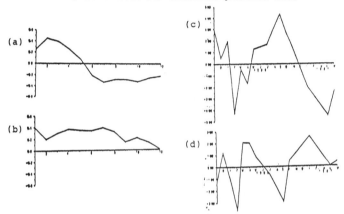

Fig.4.- (a),(b), Eigenvectors I,II and (c),(d) Principal Components I,II for plunge step position data

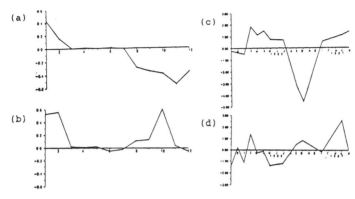

Fig.5.- (a),(b) Eigenvectors I,II and (c),(d) Principal
Components I,II for mid-swash mean grain-size diameter

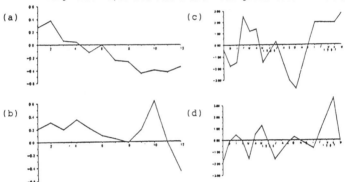

Fig.5.- (a),(b) Eigenvectors I,II and (c),(d) Principal
Components I,II for breaker zone mean-grain size diameter

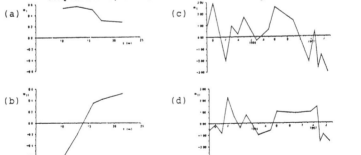

Fig.5.- (a),(b) Eigenvectors I,II and (c),(d) Principal
Components I,II for beach profile depth in transect
line T.3

are similar to that derived from plunge step position, shoreline position and swash-zone grain-size data sets. This suggests that a similar mechanism (longshore transport process) is mainly responsible for the sediment redistribution in both sub-regions (breaker and after breaker zones). The second eigenvector (Fig.6) is characterized by an almost uniform variation alongsore. It is, with positive amplitudes, associated with sediment coarsening consistent with less steep profile configuration in the breaker zone. The second principal component's plot reveals that positively weighted periods correspond to the appearance of a distinct longshore bar. The reduction in bed slope is considered to be related to the presence of nearshore sand bars increasing the level of energy dissipation and significantly decreasing the amount of energy delivered at the beachface (Carter and Balsillie, 1983). In periods of negative vector amplitudes (absence of nearshore bar) the slope of the plunge step increases and the coarsest grains move downslope. This process reduces the mean diameter in the plunge step and supports the transitional zone with the coarsest grains observed in the sampling area.

An objective procedure for use with principal components to separate "signal" from "noise" was developed by Overland and Preisendorfer (1982). The test referred to as Rule N, is designed to determine if the eigenvalues of a physical data set can be distinguished from the noise eigenvalues produced by a spatially and temporally uncorrelated random process (Lins,1985). In this study the random data set constitute a matrix with 12 spatial and 13 temporal cells. Based on these dimensions, random eigenvalues were compured and compared to physical data eigenvalues. Eigenvectors I and II pass Rule N and they are considered to be physically relevant.

A series of nearshore beach profile measurements on transect lines T.3, T.6, T.10, spanning a 21-month period have been examined for temporal variations in nearshore topography. E.O.F. analysis of the profile data indicates that most of the variation occurs in two principal modes. Fig.7 shows the eigenvectors and principal components of these modes derived from survey data in T.3. The eigenmode I explains 72%, 78% and 90% of the total variance of the profile data in T.3, T.6, T.10 respectively, whereas the eigenmode II explains 23%, 14% and 6% of the total variance of these data sets. The first eigenvector describes a uniform sediment accumulation/loss along the profile in case of positive/negative time coefficient. The associated principal component tracks closely the temporal distribution of the angle of wave incidence. When waves attack the beach from the N-NE/S-SE sector the component amplitude is calculated positive /negative. Thus, the dominant mode of neashore beach change is strongly reflects the longshore transport processes. The second eigenvector has a single nodal point indicating that at certain times sediment removed from the nearshore zone/ beachface moves onshore/offshore and contributes to beach accretion/erosion. According to the second principal component, onshore transport (negative temporal coefficients) occurs after periods of calm and relatively low-energy conditions. The increase in the variance explained by the first eigenmode from the south to the north part of the beach is accompanied by a decrease in the variance explained by the second eigenmode in the same direction. This result connotes that cross-shore sediment transport is increasingly important in the south beach section where beach protection to wave-energy impact is reduced.

CONCLUSIONS

The systematic measurement of several nearshore process and beach system variables over a 21-month period in a microtidal small-scale pocket beach under a range of wave conditions provide a data base appropriate for the study of the relationships between surf-zone process and the sedimentary and morphologic characteristics of the beach. Statistical analysis of beach data sets using empirical eigenfunctions produce quantitative results showing the changes in beach characteristics as a function of space and time. The results of this study indicate that the derived function have a direct relation to natural processes. The suggested descriptive associations between the principal component patterns and actual beach conditions have been tested by statistical significance measures and they approved to be acceptable.

The fundamental beach response along the breaker and the after breaker zones is represented by the first eigenfunction mode. It contains more than 50% of the variance, ranging from 51% in the swash-zone grain-size distribution to 72% in the plunge step position data set. It describes a longshore movement of sediment within each horizontal segment. This mode shows a distinct seasonal dependence due to the varied direction of obliquely incident high-energy waves. A landward/seaward migration of the shoreline and the plunge step is accompanied by an increase/decrease in sediment grain-size of the swash and breaker zones. Finer grains are deposited to either north or south part of the beach according to the direction of the wave-induced longshore currents. This process results in almost seasonal and spatially compatrmentalized erosion/accretion phenomena along the beach. The second mode eigenfunctions contain 16% to 29% of the variance in the data sets. They describe the influence of water level fluctuations to shoreline/plunge step position as well as the rip-currents and bed-slope induced variations in the logshore sediment distribution. Functions have no net trend indicating no net erosion or accretion on this stretch of coastline during the period of observations.

The proposed field data analysis proved to be a useful tool for studying the dynamic nature of a coastal environment especially to understand the sediment dispersal as an equilibrium response to the hydrodynamic regime. Further advance in understanding coastal processes and forward predicting beach changes is therefore dependent on the acquisition of longer, more detailed, and geographically more extensive data sets.

REFERENCES

Aubrey, D.G., 1979, "Seasonal Patterns of Onshore/Offshore Sediment Movement", Jour. Geoph. Res., Vol.84, No.C10, pp.6347-6354.

Aubrey, D.G., Inman, D.L., and Winant, C.D., 1980, "The Statistical Prediction of Beach Changes in Southern California", Jour. Geoph. Res., Vil.85, No.C6, pp.3264-3276.

Carter, R.W.G., and Balsillie, J.H., 1983," A Note on the Amount of Wave Energy Transmitted over Nearshore Sand Bars", Earth Surface Processes and Landforms, Vol.8, pp.213-222.

Kypraios, N.G., 1988, "Field Measurements on a Pocket Beach", Proc. 2nd European Workshop on Coastal Zones, N.T.U.A., Loutraki,Greece, pp. 2.39-2.53.

Lins, H.F., 1985, "Storm-Generated Variations in Nearshore Beach Topography", Mar. Geol.,Vol.62,pp.13-29.

Lippmann, T.C., and Holman, R.A., 1990, "The Spatial and Temporal Variability of Sand Bar Morphology", Jour. Geoph. Res., Vol.95, No.C7, pp.11,575-11,590.

Liu, J.T., and Zarillo, G.A., 1989, "Distribution of Grain-Sizes Across a Transgressive Shoreface", Mar. Geol., Vol.87, pp. 121-136.

Moutzouris, C.I., 1988, "Longshore Sediment Transport Rate vs. Cross-Shore Distribution of Sediment Grain Size", Proc. 21st Coast. Engin. Conf.,A.S.C.E.,Costa del Sol,Malaga,Spain, pp.1959-1973.

Moutzouris, C.I., and Kypraios, N.G., 1987, "Temporal and Spatial Grain-Size Distribution of Sediment in a Tideless Pocket Beach", Proc. Conf. "Coastal Sediments '87", A.S.C.E., New Orleans, LA, pp.1909-1924.

Overland, J.E., and Preisendorfer, R.W., 1982,"A Significance Test for Principal Components Applied to a Cyclone Climatology", Mon. Weather Rev., Vol. 110, No.1,pp.1-4.

Preisendorfer, R.W., 1988, "Principal Component Analysis in Meteorology and Oceanography", Elsevier, Amsterdam.

Ramana Murty, T.V., Veerayya, M., and Murty, C.S., 1986, "Sediment-Size Distributions of the Beach and Nearshore Environs Along the Central West Coast of India: An Analysis Using E.O.F.",Jour. Geoph. Res., Vol.91, No.C7, pp.8523-8536.

Winant, C.D., Inman, D.L.,and Nordstrom, C.E., 1975, "Description of Seasonal Beach Changes Using Empirical Eigenfunctions", Jour. Geoph. Res., Vol.80., No.15, pp.1979-1986.

BEACH DYNAMICS OF BARRED NEARSHORES: GOLD COAST, SOUTH PACIFIC AND ISLAND SYLT, NORTH SEA.

B.Boczar - Karakiewicz(1),
L.A.Jackson(2), S.Kohlhase(3) and A.Naguszewski(4)

(1) INRS - Oceanologie, Universite du Quebec, Rimouski, Canada, G5L 3A1
(2) Gold Coast City Council, Queensland, Australia
(3) Franzius - Institut, Universitat Hannover, Germany
(4) Institute of Hydroengineering, Polish Academy of Sciences, Gdansk, Poland

1 INTRODUCTION

Presented work is based on wave and bathymetry data collected on two prototype beaches: in the Gold Coast area, Australia, South Pacific and on the central part of the west-facing coast of the island Sylt, Germany, North Sea. These data are used in a mathematical model of wave-bed interactions in order to analyse the dynamics of the two nearshores, where pronounced shore-paralell bars are typical features of the bathymetry. For both analysed beach systems (Gold Coast: wave-dominated; Sylt: mixed, wave-tidal environment) the model predicts a cyclic change in bathymetry under changing wave conditions. These changes are shown to be controlled by waves and to result from a quasi-periodical exchange of sediment between the offshore bars and the visible beach. The recurrence period of beach transformation depends on the local wave climate, sediment supply and mean nearshore bathymetry. Model predictions compared with measurements show a satisfactory agreement. Conclusions presented in Sec.5 explain quantitatively the main reasons of observed beach erosion in both analyzed regions and clearly demonstrate the role of offshore bars in the nearshore dynamics.

2 BEACH AND WAVE OBSERVATIONS

2.1 GOLD COAST, AUSTRALIA, SOUTH PACIFIC

Beaches in the Gold Coast area (Fig.1) are exposed to high energy ocean waves. The amplitudes of highest waves observed in this region over to 12m and the longest periods are of 17-20s (McGrath and Patterson 1973, Gold Coast City Council 1990). The mean tidal range equals 1m.

According to observations over a period of the last hundred years, the beaches are eroded during severe storms and cyclone events but subsequently restored under mild weather conditions and swell waves. The beach system recovers under conditions that a natural buffer zone comprising the dunes is conserved and that the sediment supply from the updrift (south) across the Tweed River mouth compensates sediment carried away by the wave-induced northerly longshore transport. The Gold Coast beach system is extremely dynamic with changes occurring in several very different time scales: incident wave trains are changing during normal weather conditions in several tens of hours, the weather over a few days, the climate over years, sediment supply over decades and the shape of the seabed in days to several months. Observations and measurements show, however, that the complex beach system reaches periodically a state of dynamical equilibrium, which in turn is controlled only by two main factors. These factors are

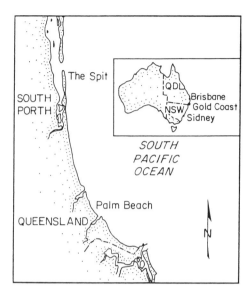

Figure 1: Location map, Gold Coast, Australia.

the incident waves and the shape of the underlying seabed from its deepwater end (extending to some 20-25m of waterdepth) up to the visible beach (Smith and Jackson 1990).

Prominent features of the seabed are large scale sand bars. Aerial photography shows evidence of sand bars in the entire nearshore of the Gold Coast area. Typically, the outer shore-parallel bar lies in a waterdepth of 6-10m and at distances of 400-500m from the mean position of the shoreline. Observations show that bars are playing an important role in the stability of the upper beach (Boczar-Karakiewicz and Jackson 1990).

2.2 ISLAND SYLT, GERMANY, NORTH SEA

In the present we will only consider the central part of the west-facing coast of the island (Fig.2). The coast is exposed to heavy wave action approaching the island during the winter month (November-January) from the NW to SW quadrant. The highest waves observed at a distance of about 1.3km from the shoreline are of 7m and the longest periods are of 11s (Dette and Fuhrboter 1976, Moutzouris 1985, Schade 1991). The mean tidal range is of 1.7-2.4m.

According to observations of coastal defence constructions dating from the end of the previous century, the beaches are eroded during severe storms, accompanied by storm surges (up to 4 m) and strong tidal currents with maximum velocities of the order of 0.8 m/s (Moutzouris 1985). Measured bathymetry (Fig.3a and 3c) and aerial photography show evidence of a large scale shore-parallel outer sand bar, which lies in a waterdepth of 3m and at distances of 350-250m from the mean position of the shoreline (Fuhrbotter 1976, Moutzouris 1985, Partenscky et al.1989,). A deep trough lying in a waterdepth of 5m separates the outer prominent bar from a small inner bar, which crest position coincides with the mean shoreline.

Observations show that the offshore bar affects the incoming wave by reducing its energy amount transfered to the upper beach. The bathymetry also reacts to waves by a seasonal change of the

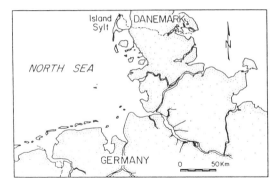

Figure 2: Location map, island Sylt, Germany.

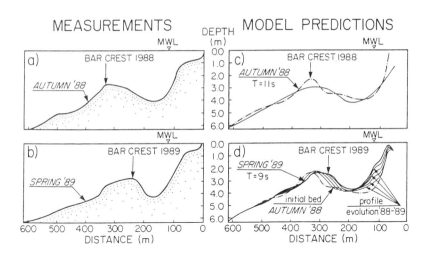

Figure 3: Island Sylt, measured and predicted seasonal bathymetry changes. Example of an observed (in 1988-89) autumn (a) and spring (b) bed profile (in 1988-89), (c) Autumn'88-bed profile (heavy line) predicted from initial topography (broken line), under extreme storm waves (peak period T=11s),(d) Spring'89-bed profile (heavy line) predicted from Autumn'88 topography (broken line), under moderate storm waves (peak period T=9s). Pointed lines show intermediate stages of the temporal evolution of the bed.

mean position of the sand bar crest. These morphological changes were measured during several seasons (Strotmann and Kohlhase 1990).

3 THE MODEL

To better understand natural beach systems of a wave-dominated environment, the effect of natural and artificial bars on their dynamics and on protection of their upper beaches, a model has been developed.

The incident wave and seabed are the only free parameters in this dynamical model which at the present stage considers the wave and bed motion to be two-dimensional (see Fig.4). The model has four constituent processes describing wave-bed interactions, namely (1) the hydrodynamics of the main inviscid water body, (2) the near-bed boundary-layer flow, (3) the sediment transport, and, (4) the evolution of the bed. The integration procedure links the four modules in two steps (Fig.5). In the first step, which corresponds to a time scale of a few wave-periods, the fluid flow and the sediment flux are calculated over a bed configuration which is assumed to be instantaneously fixed (following the spatial axis X in Fig.4). In the second step, corresponding to several hundreds of wave-periods, the temporal evolution of the seabed is calculated keeping the averaged variables of the fluid and sediment flow constant (procceding with the calculation in the τ-direction, see again Fig.4). Model's approximations are justified both on mathematical grounds and by observations (Boczar-Karakiewicz et al.1987, Boczar-Karakiewicz and Davidson-Arnott 1987, Chapalain and Boczar-Karakiewicz 1992). The time scale τ can be made explicit by using the parameters (amplitude and period) of the incident waves (Boczar-Karakiewicz and Bona 1986).

The most important aspect of the predicted evolution of the bed morphology under waves are sand bars which form from the initially featureless slope as shown in Figs.3c and 4a. The spacing of sand bars, which is typically several times the length associated to the surface wave, is a characteristic length scale induced by the nonlinear wave field and by the wave-induced sediment flux.

4 DYNAMICAL CYCLE OF BEACH CHANGES

4.1 GOLD COAST AREA, AUSTRALIA, SOUTH PACIFIC

Presented results are based on wave and bathymetry data colleted on beaches, comprising several locations: The Spit, Surfers Paradise, Burleigh Heads and Kirra (see Fig.1). These data were used in predictions, experimental verifications, and in calibrations of the described wave-bed interaction model (Boczar-Karakiewicz and Jackson 1990 and 1991). The model predicts a dynamical cycle in bathymetric changes in the nearshore under varying wave conditions (Fig.4). According to time-scale calibrations based on observations the beach cycle may be described in several very different time scales:

The time scale of the incident wave is characterised by the wave-period; the presented analysis considers a sequence of extreme and moderate storm waves with periods equal T=17s and T=10s, respectively. A system of two pronounced offshore bars under extreme storm conditions is formed in a few days (Fig.4a and 4e). A nearly uniform slope with one minor nearshore bar (model's stability state) evolves from the post-storm bars during several month of moderate storm wave activity (Fig.4b, 4c and 4d). Figure 6 showing the formation of a post-cyclone barred profile from the equilibrium pre-storm profile (see Figs 4d and 4e) in two spatial dimensions, allows to estimate sand volumes exchanged between the offshore bars and the upper beach.

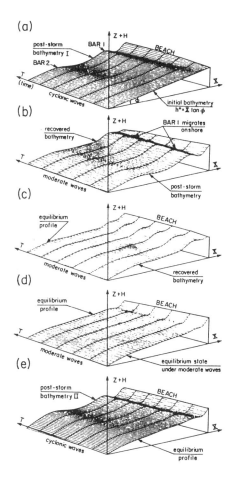

Figure 4: Gold Coast, Australia, predicted cycle of beach changes: (a) bar formation (cyclonic waves, T=17s), (b) and (c) recovery of the beach (moderate waves T=10s); (d) dynamical equilibrium (moderate waves); (e) reformation of bars (cyclonic waves).

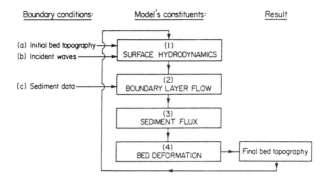

Figure 5: Scheme of the model

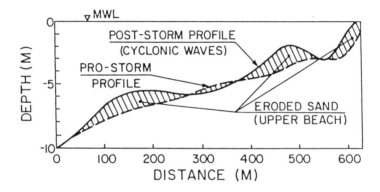

Figure 6: Formation of a post-cyclone bared profile from an equilibrium pre-storm profile (see Figs 4d and 4e) showing the estimated sand volumes exchanged between the offshore bars and the upper beach.

From results presented in Figs.4 and 6 for the Gold Coast area emerge several general conclusions applying to barred nearshores which are summarized in Section 5.

4.2 ISLAND SYLT, GERMANY, NORTH SEA

Model calculations presented in Figure 3 (Fig.3b and 3d) were carried out with a typical observed bed profile and with parameters of a simplified local wave climate (assumed to be characterized by autumn-winter peak period waves of 11s and spring-summer peak periods waves of 9s). The initial bed topography (see Fig.3b), from which wave calculations were started contained a deep trough assumed to be shaped by strong tidal currents.

The model predicts bathymetric changes in the nearshore under varying wave conditions shown in Fig.3b and 3d. These results indicate that bathymetry changes of the Sylt nearshore are correlated with seasonal changes in the incoming wave. The pronounced outer bar in his extreme seaward position (Fig.3b) is formed and maintained by autumn waves (T=11s). The corresponding autumn bed topography is transformed into an 'equilibrium' spring profile by moderate storm waves (T=9s): the outer bar crest migrates of some 50m in the landward direction (Fig.3b and 3d) and sand accumulates in the very nearshore area. A reformation of the spring bed profile by severe 11s-peak autumn waves results in a bed topography nearly identical to the primary autumn bed shown in the final stage in Fig.3b. This recurrent bed topography closes the dynamical cycle of wave-induced seasonal beach changes on the west-facing coast of Sylt.

5 CONCLUSIONS

From results presented in Figs.4 and 6 for the Gold Coast area and in Fig.3 for the island Sylt, several conclusions emerge which may be valid generally for barred nearshores in both, micro- and mezo-tidal zones:

- Waves are the main dynamical agent which predominates in the long-period cycles of beach erosion and accretion.

- The recurrence of the dynamical cycle of beach transformation is of the order of a year or longer. Its duration depends on the local wave climate and on mean morphological conditions.

- During transformation the upper beach is periodically eroded and accreted, what results from an exchange of sand between the upper beach and the offshore bars.

- For waves approaching beaches from the normal direction the model predicts the sand to move periodically in the nearshore system closed by its deepwater end (waterdepth of some 0.1 of the extreme incident wavelength) and by the upper beach. In the Gold Coast area the latter is confirmed by observations (Smith and Jackson 1990).

- Development pressures may cause severe restrictions in the natural process of long-term, periodically recurrent exchange of sediment between the underwater and visible beach. Consequently, presented results justify the offshore nourishment of sand bars which might be the most cost-effective measure of protection amd of a possible gain of volume of sediment on the visible beach.

AKNOWLEDGEMENTS

Support for this project was provided by the Gouvernement of Canada, the Ministry of Fisheries and Oceans (contract No 5203), the Natural Sciences and Engineering Research Council (Grant 276188), by the Gold Coast City Council, Queensland, Australia (Special Projects 90-92) and by the Federal Ministry of Reasearch and Technology (BFMT), Germany. This publication is a contribution of the Oceanographic Centre of Rimouski - a partnership of INRS (Institut National de la Recherche Scientifique) and UQAR (Universite du Quebec a Rimouski) operating under the suspices of the Univercity of Quebec.

6 REFERENCES

Boczar-Karakiewicz B. and J.L.Bona. 1986. Scale effects in modeling sand bar formation in coastal regions. Proc. Int. Ass. Hydr. Res., Toronto,: 46 - 61.

Boczar-Karakiewicz B. and R.D.A.Davidson-Arnott. 1987. Nearshore bar formation by nonlinear wave processes: a comparison of model results and field data. Mar.Geol.,77: 287-304.

Boczar-Karakiewicz B., J.L.Bona and D.L.Cohen. 1987. Interaction of shallow water waves and bottom topography. In: Dynamical Problems in Continuum Physics. Mathematics and its applications. (eds. J.L.Bona, C.Dafermos, J.L.Ericksen and D.Kinderlehrer) Springer Verlag. New York, Berlin Tokyo, 4: 131 - 176.

Boczar-Karakiewicz B. and L.A.Jackson. 1990. On the protection of a beach system, Gold Coast, Queensland, Australia. Proc. XXII Int. Conf. Coast. Eng., ASCE, Delft, (ed. B.Edge): 2265 - 2268.

Boczar-Karakiewicz B. and L.A.Jackson. 1991. Beach dynamics and protection measures in the Gold Coast area, Australia. Proc. X Austr. Conf. Coast. Ocean Eng., Auckland, N.Z.,: 417 - 422.

Chapalain.G. and B.Boczar-Karakiewicz. 1992. Modeling of hydrodynamics and sedimentary processes related to unbroken progressive shallow water waves. J.Coast.Res. (in press).

Dette, H.H. and A.Fuhrboter. 1976. Wave climate analysis for engineering purpose. Proc. XV Conf. Coastal Eng., Hamburg, ASCE, 2: 10 - 22.

Dette, H.H. and A.Fuhrboter. 1976. Field investigation in surf zones. Proc. XV Conf. Coastal Eng., Hamburg, ASCE, 29: 518 - 537.

Gold Coast City Council, 1990. Coastal Engineering Works, Internal Report.

McGrath, B.L. and D.C.Patterson, 1972. Wave climate at Gold Coast, Queensland, Australia. Inst. Eng. Aust.,8 Div. Tech. Papars 13 (5): 1 - 17.

Moutzouris,C.I. 1985. Coastal Processes along the eroding western coast of the island Sylt, Germany. Mitt.Franzius-Inst., Univ. Hannover. H. 60: 1 - 131.

Partenscky H.-W., S.Kohlhase, K.-F. Daemrich, H.J.Scheffer and H.Schwarze. 1988. The effect of a sand bar on wave attenuation and related sediment transport on the west coast of the island Sylt (in German). Mitt.Franzius-Inst., Univ. Hannover. H. 67: 135 - 241.

Smith, A.W. and L.A.Jackson. 1990. The siting of beach nourishment placement. Shore and Beach, 58 (1): 17 - 24.

Strotmann,T. and S.Kohlhase. 1990. Attenuation of waves passing over bars on the west coast of island Sylt (in German). In: BMFT Statusseminar: Optimisierung des Kustenschutzes auf Sylt, Kiel, 18pp.

Schade, D. 1991. Wave climate investigations based on directional wave measurements, Island Sylt, Germany (in German). Mitt.Franzius-Inst., Univ.Hannover, H 71: 211-420.

COMPARISON OF MEASURED AND CALCULATED NEARSHORE BOTTOM PROFILES

Dr. M. Szmytkiewicz

Institute of Hydro–Engineering
of the Polish Academy of Sciences
Kościerska 7, 80–953 Gdańsk, POLAND

1 Introduction

The interest in lithodynamics of coastal zone, particularly in on– and offshore sediment transport and associated sea bottom evolution, has been increasing for recent few years. The detailed review and critical analysis of cross–shore sediment transport models have been presented in Pruszak 1990.

On the basis of review of cross–shore sediment transport models, the models of Dally and Dean 1984, Kriebel and Dean 1985 and Bailard 1981 and 1982 have been chosen as the ones supposed to provide proper prediction of short–term cross–shore profile changes. Numerical algorithms secuering the cross–shore transport and resulting bottom changes computations have been worked out for each of these models.

Each of the chosen models represents a certain concept of description of sediment movement in sea. The model of Dally and Dean belongs to the group of models assuming sediment movement mainly in suspension. The model of Kriebel and Dean represents the group of global models, whereas the model of Bailard represents the group of energetic models, in which sediment transport rate is implied first of all by the asymmetry of water oscillatory motion.

The appreciation of usefulness of the chosen models for short–term bottom changes prediction in coastal zone has been made on the basis of measurements carried out at Coastal Laboratory of Shkorpilovtsi (Bulgaria) on the Black Sea.

Certain proposals of modifications of the models have been presented in Szmytkiewicz and Skaja 1990. In the case of Dally and Dean's concept the model of Engelund and Fredsoe 1976 has been proposed for calculation of so–called basic concentration C_A necessary to determine sediment transport rate. The quantity C_A is is calculated at every point of the bottom as a function of wave parameters and a depth, whereas in original Dally and

Dean's model this concentration is assumed as constant and its quantity is determined on the basis of numerical testings. For calculation of wave motion energy dissipation in all the models random wave has been assumed, as more realistic in natural conditions.

2 On– and offshore transport computation

Dally and Dean's model

The model of Dally and Dean handles sediment motion as the suspended one, especially in the surf zone, simultaneously assuming bottom layer as a particular region of sediment concentration. Within this model sediment transport rate averaged over wave period Q_{SS} is obtained by the formula:

$$Q_{SS} = \int_{-h}^{0} \overline{U}(z) \cdot \overline{C}(z) dz \qquad (1)$$

Dally and Dean divided the entire suspended sediment movement into two parts. In the lower part, the thickness of which is $-h < z < T \cdot W_s$, they assumed sediment motion to be caused by oscillatory velocities of water $\overline{U}_f$ and by return flow $\overline{U}_s(z)$, whereas in the upper part $W_s \cdot T < z < 0$ — by return flow only.

Using linear wave theory Dally & Dean determined the distance $\zeta_f(z_i)$ which is travelled by a grain until it falls to bottom

$$\zeta_f(z_i) = \frac{H}{2\omega} \sqrt{\frac{g}{h}} \sin \frac{\omega(h + z_i)}{W_S} \qquad (2)$$

from which the mean horizontal velocity $\overline{U}_f(z_i)$ per wave period yields:

$$\overline{U}_f(z_i) = \frac{\zeta_f(z_i)}{T} = \frac{H}{4\pi} \sqrt{\frac{g}{h}} \sin \frac{\omega(h + z_i)}{W_S} \qquad (3)$$

where: W_S – grain fall velocity.

The quantity of undertow $\overline{U}_s(z)$ has been determined from Dally 1980. He assumed that momentum of wave motion is transmitted shorewards between still water level and a wave crest. Hence compensatory seaward flow must exist below still water level to satisfy mass conservation law.

Dally and Dean assume the concentration profile $C(z)$ by formula of Rouse:

$$C(z) = C_A \cdot \exp\left[\frac{-15 \cdot W_S(z - z_a)}{h\sqrt{\frac{\tau}{\rho}}}\right] \qquad (4)$$

where C_A the concentration at a level z_A.

The shear stress τ is handled totally as a result of bottom friction effect τ_B and energy dissipation τ_{BT} due to wave breaking.

The determination of an optimal level z_A and corresponding concentration C_A has been an unsolved problem till now, requiring intensive investigations.

Szmytkiewicz and Skaja 1990 proposed the evaluation of concentration C_A on the basis of the model presented in Engelund and Fredsoe 1976, where strict correlation between bottom shear stress τ_0 and the concentration C_A at the bottom has been proved.

Engelund and Fredsoe, assuming a classical logarithmic velocity distribution in boundary layer and making use of Bagnold's considerations concerning bottom shear stress, determined the quantity C_A by the following formula:

$$C_A = \frac{0.65}{(1 + \frac{1}{\lambda_b})^3} \qquad (5)$$

in which: λ_b is the linear concentration at a bottom level ($z = -h$).

While calculating both energy dissipation and wave height as a function of distance from shore the random wave has been assumed. Furthermore an assumption has been made that the wave height at every point of coastal zone can be described with Rayleigh distribution. Taking advantage of the above assumption wave height and energy dissipation rate has been computed with the use of model of Battjes and Janssen 1978.

Kriebel and Dean's model

The model of Kriebel and Dean 1985 belongs to the group of models of a global character. Assuming, analogically to Swart, continuous evolution of shore profile due to variable wave–current conditions, Kriebel i Dean handle offshore transport Q_S at any point of the surf zone proportional to the difference between wave energy dissipation rates D for real profile and D_* for equilibrium profile, respectively:

$$Q_S = K(D - D_*) \qquad (6)$$

where:

K – proportionality coefficient of order of 10^{-6}

D – energy dissipation per volume unit.

The modification of Kriebel and Dean's model worked out by Szmytkiewicz and Skaja 1990 lies in employment of Battjes and Janssen's model for the description of wave field.

Bailard's model

The model of Bailard belongs to the group of energetic models originating from Bagnold's 1966 idea. Bagnold assumedd that the work P_0 done by water (or the energy E_0 derived from wave motion and transmitted to the bottom) equals, taking the effectiveness ϵ into account, the work P_e done by sediment grain moving as bedload or suspended load, i.e.:

$$P_0 \cdot \epsilon = P_e \tag{7}$$

The works of Bailard 1981, 1982 are further searching for an optimal correlation between on-/offshore sediment transport and the asymmetry of oscillatory water motion.

Bailard 1981 described the asymmetry of oscillatory velocity in the form of velocity moments series characterized by different powers and ways of averaging. As a result sediment transport rate in cross-shore direction, averaged over wave period, is given with the following equation:

$$\vec{i} = \rho C_D \frac{\epsilon_B}{\tan\phi}[\overline{|\vec{u}(t)|^2 \vec{u}(t)} - \frac{\tan\beta}{\tan\phi}\overline{|\vec{u}(t)|^3} \cdot \vec{i}] +$$
$$+ \rho C_D \frac{\epsilon_S}{W_S}[\overline{|\vec{u}(t)|^2 \vec{u}(t)} - \frac{\epsilon_S}{W_S}\tan\phi\overline{|\vec{u}(t)|^5} \cdot \vec{i}] \tag{8}$$

in which:

$\vec{i}$ – unit vector directed shorewards

ϵ_B, ϵ_S – effectiveness coefficients of bedload and suspended load, respectively

u – water particles velocity

C_D – drag coefficient

ϕ – angle of internal friction

W_s – fall velocity

β – local angle of bottom slope

Pruszak 1990, analysing Bailard's model, comes up with the conclusion that a grain moves shorebound, when $H_s < 0.9[m]$, and — seabound, if the wave is greater. The maximum of onshore transport takes place for $H_s = 0.6[m]$. In the case of severe surging, $H_s > 1.0[m]$, predicted sediment transport has the offshore direction.

3 Field measurements

The evaluation of usefulness of the models for short–term bottom change prediction in coastal zone has been made on the basis of field measurements carried out at Coastal Laboratory of Shkorpilovtsi (Bulgaria). The major advantage of this site is 230 m long pier situated perpendicularly to the shore, at which water depth attains 4 – 4.5 m. Such conditions of survey enable practically continuous access to the equipment independently on weather and are convenient to investigate all the processes and phenomena associated to on– and offshore sediment transport.

The ultrasonic gauges for depth change measurement were installed along the pier at ten 20 m spaced points. The first offshore registration point was located at the distance of about 200 m from the shoreline (at the depth $\sim 4.5m$), while the first onshore point — at the distance of about 10 m from the shore. At three points, detached 40, 100, and 180 m from the shore, wave parameters and currents 0.6 m over the bottom were registered. The detailed description of data collecting, processing and interpreting is given in Pruszak 1989.

4 Comparison of computed and measured bottom changes

Two three–day storms occured within the survey. The first one (lasting 70.5 hours) has been chosen for verification of numerical algorithms. The depth profile registered on 15 Oct. at 10.30 is depicted in Fig. 1. A single underwater bar is visible at the distance of about 50 m from the shoreline.

The depth increased nearly linearly starting from the distance of order of 80 m, with respect to the shoreline.

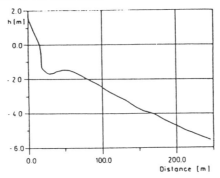

Fig. 1. Measured initial depth profile

The greatest mean wave heights observed in the measurement period were 0.95 and 1.30 m of the time of occurence 3 and 2 hours, respectively.

The computed depth changes have been compared with the measured ones after the time 1.3, 5.0, 23.0, 52.0 and 70.5 hours providing the following conclusions:

for both Bailard's models (Fig. 2)

- the calculations have provided qualitatively similar bottom changes in the direct vicinity of the shore for long periods of wave action (at least 50 hours) and the values of the coefficients $C_D = 0.01, \epsilon_S = 0.025, \epsilon_B = 0.21$.

- the coefficients $C_D = 0.5, \epsilon_S = 0.125, \epsilon_B = 0.021$ should be assumed for the computation of bottom changes in short period of wave action (up to 5 hours). However in such cases the model of Bailard 1982 provides distinct shoaling up to 100 m from the shoreline, whereas the model of Bailard 1982 forecasts bottom erosion in this region.

- neither lengthening time of wave action nor variation of wave parameters affects the character of bottom evolution. Thus if the depth increases at a certain point the lengthening of wave action results in further erosion at this point only.

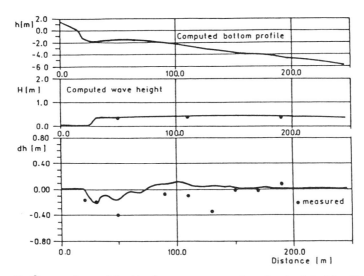

Fig. 2. Comparison of depth changes measured and calculated by Bailard's model after 70.5 hours of surge
$(C_D = 0.010, \epsilon_s = 0.025, \epsilon_b = 0.21, W_s = 0.05 m/s)$

for Dally and Dean's model (Fig. 3)

- comparatively correct description of depth changes has been obtained, particularly in the surf zone. As the wave action was lasting the beach berm was being built in the close vicinity of the shoreline, while the depth was increasing in the surf zone, like observed in nature.

- analysed process of bottom evolution is not unidirectional, i.e. erosion and accumulation can occur in turn at the same point of a profile (like in natural conditions), according to wave parameters and time of action.

- the best conformity with the measurements has been obtained assuming in computations the roughness height $K_s = 0.002m$.

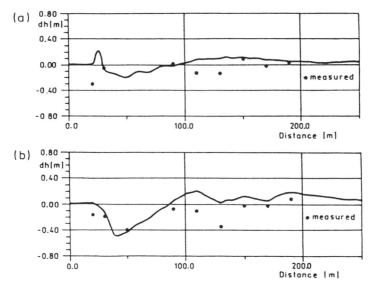

Fig. 3. Comparison of depth changes measured and calculated by Dally and Dean's model after 23 (a) and 70.5 (b) hours of surge
($K_s = 0.002, \epsilon_s = 0.040, f_w = 0.01, W_s = 0.05 m/s$)

for modified Kriebel and Dean's model (Fig. 4)

- calculated bottom changes are similar to the measured ones both in the surf zone and in the region detached more than 150 m from the shoreline for short time period of surge (up to 23 hours).

- for longer time periods of wave action (52 and 70.5 hours) calculated depth changes have been more and more different from the measured ones.

- the best coincidence has been obtained asumming the constant $K = 0.1$.

5 Conclusions

The verification implies best agreement of depth changes in successive storm phase measured and computed by the model of Dally and Dean. This model, in the Author's opinion, can be applied for computation of the short-term bottom changes in coastal zone, mainly in the surf zone. Its major advantage is that it admits of sequential erosion and accumulation at the same point of a profile, according to wave parameters. Thus, within this model, the process of bottom evolution is not unidirectional in time.

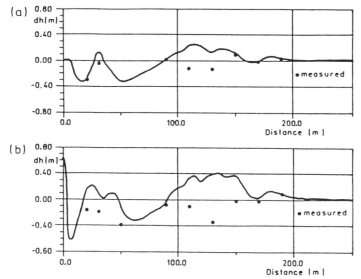

Fig. 4. Comparison of depth changes measured and calculated by modified Kriebel and Dean's model after 23 (a) and 70.5 (b) hours of surge ($K = 0.10$)

Further improvements of Dally and Dean's model ought to approach better description of velocity distributions and better description of vertical distribution of sediment concentration. The comparative analysis of the models calculating velocities in the direction perpendicular to the shore and their rough verification on the basis of measurements at Lubiatowo (Poland), Szmytkiewicz and Skaja 1990, implies that much better model for the description of vertical distribution of velocity is Svendsen's 1984 model. The description of vertical variation of sediment concentration presented in Kaczmarek 1990, particularly its innovative description in the bottom boundary layer and obtained very good agreement with measurements, proves that the employment of this description of concentration in Dally and Dean's model would improve considerably the usefulness of this model.

6 Acknowledgements

This study has been supported financially by KBN and PAN, Poland, under Programme 2 IBW PAN, which is hereby gratefully acknowledged.

7 References

Bailard J.A., 1981 *An Energetics Total Load Transport Model for a Plane Sloping Beach*, J.Geoph. Res. Vol. 86 (11)

Bailard J.A., 1982 *Modeling On-Offshore Sediment Transport in the Surf Zone*, Proc. 18-th Intern. Conf. Coastal Engng.

Battjes J. A., Janssen J. P. F. M., 1978 *Energy Loss And Set-Up Due To Breaking Of Random Waves*, Proceedings of the sixteenth Coastal Enginnering Conference Vol I

Dally W.R., 1980 *A Numerical Model for Beach Profile Evolution*, M.Sc. Thesis, Univ. of Delaware

Dally W. R., Dean R. G., 1984 *Suspended Sediment Transport and Beach Profile Evolution*, J. Waterway Port Coastal and Ocean Division Vol. 110, No 1.

Dean R.G., 1974 *Evaluation and Development of Water Wave Theories for Engineering Application*, Vol II U.S.Army, Corps of Engineers Costal Engineering Research Center. Kingman Building, Fort Belvoir, Va 22060

Engelund, Fredsoe J., 1976 *A Sediment Transport Model for Straight Alluvial Channels*, Nordic Hydrology 7

Kaczmarek L. 1990 *Non-cohesive sea bed dynamics in real wave conditions*, Ph.D. thesis, Inst. of Hydro-Engng. Polish Academy of Sciences (in Polish)

Kriebel D.L., Dean R. G., 1985 *Numerical Simulation of Time-Dependent Beach and Dune Erosion*, Coastal Engineering 9

Pruszak Z., 1989 *On- Offshore Bed-load Sediment Transport in the Coastal Zone*, Coastal Engineering 13

Pruszak Z. 1989 *Expedition Shkorpilovtsi "88*, Report, Inst. of Hydro-Engng. Polish Academy of Sciences (in Polish)

Pruszak Z. 1990 *Sediment transport and sea bottom transformation in the cross-shore direction*, Report, Inst. of Hydro-Engng. Polish Academy of Sciences (in Polish)

Svendsen I.A. 1984 *Mass flux and undertow in a surf zone*, Coastal Engng. 8, 347-365

Szmytkiewicz M., Skaja M., 1990 *The choice of numerical model for bottom changes computations in the cross-shore direction*, Report, Inst. of Hydro-Engng. Polish Academy of Sciences (in Polish)

GIS FOR SHORELINE MANAGEMENT

I H TOWNEND[+] and D LEGGETT[*]

Abstract

A geographical information system (GIS) is being used as a central tool for shoreline management on the east coast of England from the Humber to the Thames. Initially the GIS was adopted as essentially an analysis tool; to map relevant variables and analyse the inter-relationships between variables, as a means of producing interpretative maps. The adoption of a strategic approach to the planning of sea defences led to a change in focus for the application of the GIS. In addition to performing the requisite analysis, it was recognised that the GIS could play a significant role in management activities.

A key part of development procedure was the design of the data model. This specifies the way the data is structured within the GIS. In addition a range of tools were developed to extend the analytical capability, provide improved access to complex data sets and the enhance facilities for on-going management of the system. The various design considerations are reviewed in this paper, and the operational experience to-date is summarised.

La gestion du littoral sur la côte d'est d'Angleterre, par la Humber et la Thames, sont utiliser le systèm d'information geographique (SIG). En avant, le SIG est adopté pour l'analyse; pour dessiner les carte de plusieurs de variable et cherche à identifier le lier entre les variable, pour produir les carte d'inteprétation. En constatant l'échelle régionale de tel processus, les avantage d'une methode stratégique ont pu réaliser. C'est ça qui entrainer le change d'application de le SIG, pour l'analyse mais aussi pour le travail de la gestion.

Un plus important part de la dessiner, c'est la modèle de données. Ceci définir la structure de les données dans le SIG. Aussi quelque utilities ont développer de servir l'analyse et le function du gestion. Dan ce papier, le consideration de dessiner est présenter avec l'experience d'opération.

INTRODUCTION

The Anglian region of the National Rivers Authority covers one of the largest and most vulnerable coastlines in Britain stretching from the Humber in the north to the Thames in the south (Figure 1). Much of it is flat, low lying and below maximum recorded sea level.

With over one fifth of the region below flood risk level a continuing and major investment programme by the National Rivers Authority ensures the protection of three quarters of a million people and billions of pounds of investment in infrastructure and land. In the next ten years it is planned to spend about £340 million on coastal and tidal defences.

The Anglian region is protected from tidal flooding by about 1,500 km of defences. The wide range of coastal geomorphology, the underlying geology and the exposure to aggressive wind and wave action requires a variety of defence solutions for this extremely abrasive coastline. For many centuries the coast has been defended on more or less the same frontage as at present. The last major refurbishment and extension to the defences took place in the two

---

[+] Chief Engineer, Sir William Halcrow & Partners Ltd, Burderop Park, Swindon, Wilts. SN4 0QD. UK

[*] Coastal Information Officer, NRA Anglian Region, Kingfisher House, Orton Goldhay, Peterborough, PE2 0ZR, UK.

decades following the disastrous 1953 East Coast floods in which over 200 people died. Limited refurbishment and upgrading also took place in areas which were affected by a similar surge in 1978.

The coastline is environmentally sensitive. The open landscape, the numerous Sites of Special Scientific Interest and bird reserves, the Heritage Coast and the historic towns all require particular attention when considering flood protection.

The NRA, realising the problems that must be faced in the future, commissioned a study of the foreshore, which could help them to establish a shoreline management strategy based on sound scientific principles. Thus the Anglian Sea Defence Management Study emerged and is the most extensive study of coastline properties and processes to have been carried out in the UK. (NRA, 1991)

The principal outcome of the study was to establish a responsive management framework to support the ongoing development of strategy (Townend, 1990). This recognises the value of information, to develop an understanding of coastal processes, as a basis for sound management action. The GIS provides a means of managing and manipulating this information. However it is recognised that a fixed management strategy is not possible in such a dynamic environment. The classification of processes on the coast must be regularly updated based on the results of routine monitoring and/or the output of forecasts. As the coastal classification is changed, so the management response strategy may be updated and this in turn feeds back by influencing coastal characteristics (Figure 2).

This paper focuses on the application of GIS to shoreline management and in particular considers some of the issues. Upon completion of the development work, the system was put into use in April 1991 and it is now possible to provide some initial feed-back on operational experience to-date.

MANAGEMENT SYSTEMS

Following the initial task of collecting data from existing archives to establish a preliminary strategy, development concentrated on the implementation of information systems which could support the day to day management of the coast. This comprises two elements. The first is based on a Geographic Information System and provides strategic information for the whole region - this is referred to as the Shoreline Management System. The second maintains more detailed information at a local level and serves the role of integrating routine monitoring information in order to keep the regional management system up-to-date. This is a stand alone pc-based database system referred to as SANDS (Shoreline And Nearshore Data System). Together these systems service local operational needs and in particular help to establish a better awareness and understanding of coastal processes, whilst also providing the information base and analysis facilities required for regional strategic planning.

SHORELINE MANAGEMENT SYSTEM

The processes, the resource, and the uses we make of the coast all have a clear spatial dimension. By using GIS it is possible to describe real world objects in terms of their spatial description (point, line, area), their attributes (eg name, value, classification), and their relationship with other objects (ie topological relationship). The fundamentals of GIS and the concepts involved in the related issues of data structures, storage techniques, modelling (eg terrain models), data analysis, classification and most importantly data quality issues have been comprehensively set out in the book of Burrough's (1986). The specific application of

GIS coastal and shoreline management has been considered by Fleming and Townend (1989), Mitchener et al (1989) and McCue (1989), amongst others.

Functional Requirements

An initial pilot study established that the basic functionality of a GIS could meet a number of the analysis requirements. It also became evident that certain enhancements would be required to provide a more complete analysis capability. In particular the need to access time series data (eg for waves and water levels) and store spatial change (eg for coastal erosion) were of paramount importance. The ability to relate attributes to a "reference chainage" along the coast was also found to be useful for analysis, in conjunction with a range of standard statistical utilities.

The application of the GIS as a management tool envisaged a number of uses as follows:

i) Retrieval of information for a specific site.

ii) Provision of summary data for planning purposes (eg the regional extent of a feature or attribute).

iii) Preparation of graphical displays for educational and public relations exercises.

iv) Classification of the coast into management zones.

v) Sensitivity testing of the classification system.

vi) Predictive modelling to determine:
- projections of coastal change
- impact of individual schemes
- impacts from change of use
- economic consequences (eg of sea level rise)

Considering the analytical and management requirements outlined above, the capabilities that a management system needs to provide are summarised in Table 1. The selection of a suitable GIS was therefore based on its ability to meet these requirements using the proprietry tools to customise the basic system.

Data Model Design

As outlined above shoreline management application seeks to satisfy the requirements for analysis and the requirements for management. The capability of a GIS are however, dependent on the scale at which data is captured because detail cannot readily be removed or added. This may require compromise to meet the needs of different users, or alternatively multiple sets of graphics linked to a common database to meet differing presentation and analytic needs. In either case, the user requirements must be clearly specified and the data model designed to meet these needs. This has always been recognised when building large relational databases but is often overlooked when setting-up a GIS because of the apparent definition provided by real world objects. This, however, is something of an illusion because within a GIS they are represented schematically as points, line or areas (eg when mapping a river one could use the water surface area, the two banks, a centre-line, or even a point when working at very large scale). The right choice depends on how the data is to be manipulated and presented.

FUNCTION	REASON	FUNCTION	REASON
Project Management tools	Greatly facilitate the rapid set-up and subsequent control of projects developed within a GIS environment.	x-y plots of magnitude against chainage	Identified as an important requirement for studying coastal processes.
Data Security	The ability to control read/write access is vital in multi-user/multi-disciplinary GIS applications.	Access of time series data	As above
Data documenting and audit facilities	Essential if sources and definitions of the data are to be documented as an integral part of the system. Audit trails allow the data integrity to be tested.	x-y plots of time series data	As above
Feature definitions, create and edit	Standard on GIS's but these features need to be examined carefully as they can greatly influence productivity.	Regression and correlation statistics	As above
Enquiries on attributes	As above	Satellite image capture	Valuable source of data and able to play a role in long term monitoring.
Spatial analysis	Enable overlays, feature synthesis and complex spatial queries to be made.	Perspective views	Useful for visual presentations
Arithmetic transforms	Allows data attributes to be manipulated according to user defined relationships (eg $A = \log B + C^2$).	Interactive surface model	Needed for some aspects of analysis and to run process models interactively.
Areas to lines or areas and chainage	Analysis often requires mapped features to be referenced to an alongshore position on the coast. This allows changes along the coast to be studied.	Interactive use of predictive models	The ability to run numerical models interactively using information in the database, are needed for the predictive capabilities to be developed.
Summary statistics	Needed for planning purposes	Graphic attributes	Some attributes could be presented and displayed more effectively using graphics (eg geotechnical cross-sections, sea wall details, etc).
Attribute reports	The ability to construct ad hoc reports, suitable for printer output, enables specific data to be extracted from the database as required.	Measurements of areas/lengths	For analysis this facility is used from within statistical routines. For management such measurements may be required for planning or as the basis of economic appraisals.
		Plots	Screen dumps and annotated maps are essential in order to provide hard copy of the graphics.
		High quality maps	Facilities to produce colour maps are important for presentation of results.

Table 1 : Functional Requirements

Consequently, whilst the data to be included in a particular application reflects the project needs, how it is included needs to take account of some basic rules for the construction of data bases and, to a limited extent, the specific constraint imposed by the GIS being used. This process is formalised through the preparation of a **data model** which documents the format of each variable and how one data set relates to others within the system.

When bringing together large and varied data sets, it is important to ensure that consistent structures are used and that any unwarranted duplication is removed. In particular, if a relational database is to perform satisfactorily, the tables must be reduced to a form which removes unnecessary redundancy, duplication and loops. This is known as normalisation or canonical reduction and can be achieved using well established procedures. Fortunately, within a GIS this is not a particularly onerous task (although still requiring some care) because of the discipline imposed by relating data to specific real world features.

Where appropriate a feature is associated with an attribute table. This part of the model tends to vary widely depending on the nature of a particular category (or group of features). Two examples are shown in Figures 3 and 4. The first shows the coastal works category where there are many features and some interconnection between them. By way of contrast, Figure 4 shows the water levels category with only two features but each with a hierarchy of attribute tables.

In designing the data model, particular attention has been given to the time domain and the coastal location. A number of techniques have been used, to take account of attributes and graphics which can change with time. Coastal location requires features to be related to their location on the coast in an alongshore sense. This allows different features and attributes to be related to one another based on their "position" along the coast. Again this is achieved using appropriate constructs within the data model.

Clearly a detailed definition of the data model as it evolves must be maintained, if the use and integrity of the database is to be controlled. For this reason tools to provide a data dictionary were seen as essential to the development and subsequent implementation of the Data model. This includes an outline of each category, details of data sources, ownership, update requirements, type of update and defines the fields that make up each table.

Data Capture

the range of data coverage that has been included in the system is summarised as a number of categories in Table 2. Each category contains one or more features to provide a graphical representation of the different items and each feature than has data linked to it via a series of attribute tables.

Table 2 : Summary of Categories in the Shoreline Management Data Model

Forcing	Response	Character	Classification	Admin	Ecology	Water Quality
Currents Rainfall River Discharge Temperature Water Levels Waves Winds	Beach Profiles Land Subsidence Sea-bed Movement Shoreline Management	Agriculture Amenity Coastal Works Estuaries Geology Hydrogeology Industry Infrastructure Military Morphology Photographs Sedimentology Shore Ownership Topography	Coastal Strips process Units Policy Guidelines	24/5 Survey Admin Boundaries OS-10,000 OS-50,000 Raster maps	Bird Counts Bird Feeding Areas Coastal Flora Conservation Sites Fisheries Sea Banks	Sampling Points

The map data was captured using a digitising table, whereas much of the digital data was ported directly using load routines written specifically for each data set. Much care has been exercised in checking each data set for graphical and attribute completeness, as well as the integrity of the links between the two. Extensive checks were also carried out on the attributes themselves, which were passed through several independent screenings.

In certain cases some interpretation has been required either during the data gathering exercise (eg flood protection zones) or in translating the data into representative graphics in the GIS (eg coastal geology). To ensure that these considerations are not lost, relevant notes were added in the Data Dictionary.

System Customisation

To provide rapid and easy access to the data a degree of customisation has been undertaken. In addition the standard GIS analysis facilities have been extended to meet the specific needs of shoreline management. The additional utilities include:

x-y graph	-	allows various forms of graph to be generated
chainage	-	enables changes along the coast to be analysed
photos	-	enables raster scanned photographs to be displayed
graphical attributes	-	manipulates sketches and annotated cross-sections
statistics	-	provides statistical analysis on user defined queries
data dictionary	-	tools to document the data model
display	-	allows the user to explore the different data-sets, without any prior knowledge of the data structure

graphs	-	graphical displays specific to a particular type of data (eg. wind and wave roses, current vectors etc)
reports	-	enable the user to select a location and produce a report (eg all design information for a site or a summary of the data relevant to an environmental assessment)
models	-	these include; a utility to manipulate the coastal zone classification data; mapping of a hazard zone based on projection of retreat rates; and computing overtopping of all sea defences for given storm conditions.

The systems that has been established is very extensive, containing a wide range of data for over 1000 km of coast. It has been developed via a pilot system to a full implementation. During the analysis phase of the project, the ability to use and manipulate data for this purpose was thoroughly tested. The system has now been in use for almost a year, where it has also been proven as a management tool. A key aspect of this role is however, the need to keep the system up-to-date and this is the role of the monitoring programme (see Figure 2).

MONITORING PROGRAMME USING SANDS

The focus of the monitoring programme is to capture information on both the forcing and the response. Forcing comprises data on winds, waves and water levels, whereas the response component will include beach change, spit development, cliff losses, vegetation change and, over the long term, changes in the nearshore bed. These data do of course need to be supplemented by a good record of the defences and of the shoreline condition. Consequently, structure and beach inspections are also included within the monitoring programme.

Data Acquisition

The means by which data should be collected was addressed in a staged review, which included extensive discussion on requirements with Regional and District officers from the Authority. This led to proposals for a programme comprising a range of data acquisition activities, as outlined in Tables 3 and 4. An important aspect of the programme is that certain data sets, such as winds, will be obtained centrally, processed and then distributed to the Districts. At the same time, other data sets such as beach profiles, will be gathered on a local basis in each District processed and stored for local use, whilst also being sent to the centre to ensure that a formal archive is maintained.

Data Processing and Interpretation

Simply collecting the data and archiving it would have little value. One of the main objectives of the programme is to get fairly immediate feedback on shoreline behaviour and the structural integrity of the defences. This is to be aided by using a piece of software designed to capture, analyse and present the data in an appropriate format. Called SANDS (Shoreline And Nearshore Data System) it has been developed to provide the functions outlined in Figure 5. The software is PC based, sitting on a relational database, and is fully menu driven with extensive help facilities.

The principal core *function* of the system is the Diary. This allows beach inspections, structure inspection, beach profiles and miscellaneous events to be recorded in a consistent manner and against a time reference. Supporting *functions* include a map to locate measurement sites and structures and graphing facility to extract time series data such a winds, wave and tides as well as beach profiles. In addition there are various service utilities to deal with importing data, processing, reporting and archiving.

Table 3: Data Requirements and Forcing Component

Description	Format	Frequency	Analysis	Use
Wind	Velocity and direction	Hourly	• Inshore wave climate • Offshore storm climate • Extremes	To assess shoreline exposure
Water Levels	Height ref to datum	At least hourly	• Tides • Extremes • Joint wave and water level analysis	Required for inshore wave computations and allows significance of storm surge to be identified
Tidal Prism	Water levels and currents	Every 5-10 years	Estuary power curve	To define estuary dynamics and in the longer term to provide indication of changes in the hydraulic regime

Table 4: Data Requirements for Response Component

Description	Format	Frequency	Analysis	Use
Aerial Survey	Photographs	Annual	Foreshore and backshore levels	To identify key features quantify change in both a plan and cross-shore sense and to monitor both physical and ecological changes
Hydraulic Survey	Depths along profile lines	Every 5-10 years	Changes with time	Establish limit of sediment exchange on shore face and provide detailed, up-to-date bathymetry for process studies
Land Survey	Levels along profile lines	Biannual	Changes with time	Allows changes in the beach to be studied and related to forcing data (Table 3)
Inspections	Text records	Biannual	Changes with time	Observation of beach features and coastal structures can be related to processes and key events, to help establish patterns of shoreline behaviour

The implementation of well co-ordinated monitoring programmes on a regional basis, places the activity of data collection on a formal and rational footing. This approach not only provides the input for the Responsive Management Framework outlined, but greatly improves the quality of design information and may also provide an early indication of the direct impact of sea level rise.

OPERATIONAL EXPERIENCE

In April 1991 the National Rivers Authority, Anglian Region took delivery of its Shoreline Management System. The Authority had taken a conscious decision that the information collected under the Sea Defence Management Study and future monitoring data should be controlled and understood in-house. Through 1991/92 significant steps forward have been made in operational use of the system and a team has developed around the system.

The component parts of this team are highly specialised to deal with the technical complexity of the system and the data within it (Figure 6).

The team is co-ordinated by a coastal process specialist who is capable of interpreting information within the system and advising flood defence engineers about coastal processes effecting their specific schemes. There are then two main sections of the team. One half deals with the Shoreline Management Systems, SANDS and other PC databases. The Shoreline Management System is driven by a GIS specialist and assisted by an experienced technician. The other half deals with the regional monitoring programme. This includes contract management and liaison with engineers about monitoring (Table 4). This section also checks and verifies data prior to entry into the Shoreline Management System or SANDS. The team is carefully structured to ensure the regular update and supply of data and advice to other engineering staff.

The system was initially developed as a management tool but has expanded its uses into four principle areas:

(1) Flood Defence Design and Maintenance - In assessing flood defence works, the system provides an important source of data. The customised functions within the Shoreline Management System for waves and tidal currents are used frequently. Data is supplied in map form and can be output to publication standard. Data can also be supplied in hard copy form or on disk to run with site-specific models or provide quantitative backup to specific queries.

(2) Strategic Planning - The Authority is developing work undertaken in the Sea Defence Management Study to create a strategy to manage the coast into the future. This requires further development of the system particularly to define the economics of schemes through the GIS. The Shoreline Management System has already been used to identify defences requiring work in the 1992/93 corporate plan and been used in relation to coastal planning issues.

(3) Information Management - The Engineering Department holds information concerning defences including historic documents, diagrams and reports as well as CAD based engineering drawings. The system is beginning to be used to reference this data spatially and indicate who holds what information and where. The system has also been used to reference the aerial photography collected under the monitoring programme so that the relevant photography can be identified and thus located. This speeds retrieval of information not stored within the system itself.

(4) Site Specific Date - Whilst the system has regionally consistent data, it is being used increasingly to store and manipulate site specific data. This is often at higher density or frequency that regional data and is often the product of detailed design work for an engineering scheme. By archiving this data it will be possible over time to develop a highly

detailed mosaic of the Region, which will assist in interpretation of regional data as well as being and important archive facility for the future.

The Shoreline Management System has supplied an important focus to the departments activities. It has provided a central information resource which is managed by staff who specialise in coastal process, GIS, and monitoring. This has led to the standardisation of the departments activities in these fields. For example, the Shoreline Management System group now advises upon (and manages) survey contracts for site specific work,. This means all data is being collected in a consistent manner to enable comparisons throughout the Region to be made. The central resource has benefits also by creating a specialist pool which can be called upon by anyone within the department. This assists in interpretation of data and cross referencing experience from one engineer to another.

CONCLUSION

The Shoreline Management System is now an integral part of flood defence work in the Anglian Region. The value of the resource it represents and the value of work associated with monitoring is a now being realised. The first critical step to in-house control of the system has been achieved. At present the system is being reviewed to identify areas where future development is needed or where the service it provides could be improved. The Shoreline Management system to date has proven a very useful addition to the department and its role is becoming more valuable as time goes on. With the Shoreline Management System the National Rivers Authority has improved its effectiveness and efficiency in the provision of flood defence to the Anglian Region.

REFERENCES

BURROUGH P.A. 1986, Principle of Geographic Information Systems for Land Resource Assessment, Clarendon Press, Oxford.

FLEMING C A, TOWNEND I H 1989, A coastal management database for East Anglian, ASCE, Coastal Zone '89 Charleston, SC, USA.

McCUE JC 1989, The use and potential use of Geographic Information Systems in coastal zone management, MSc Thesis, Dept of Biology, Uni of Newcastle-upon-Tyne, UK.

MITCHENER W K COWEN DJ, SHIRLEY W L 1989, Geographic Information Systems of Coastal Research, Coastal Zone'89 ASCE Proc of 6th Symp on Coastal and Ocean Management.

NRA 1991, The Future of Shoreline Management, Conference Proceedings, National Rivers Authority, Anglian Region.

TOWNEND I H 1990, Frameworks for Shoreline Management, PIANC Bulletin No 71, 72-80

ACKNOWLEDGEMENT

The National Rivers Authority, Anglian Region funded the work and their permission to publish this paper is greatfully acknowledged.

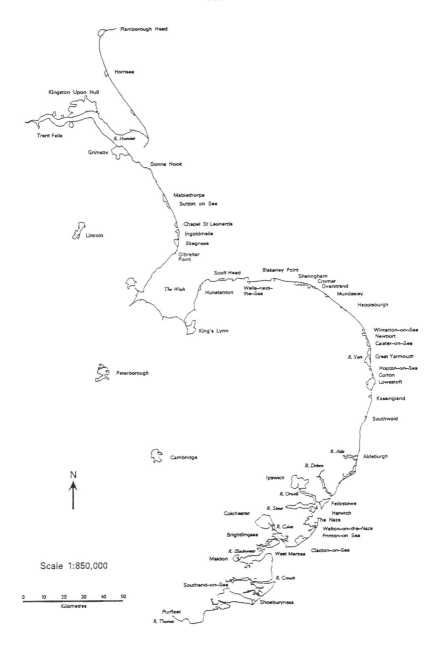

Figure 1 - The Study Area

MODEL STRUCTURE

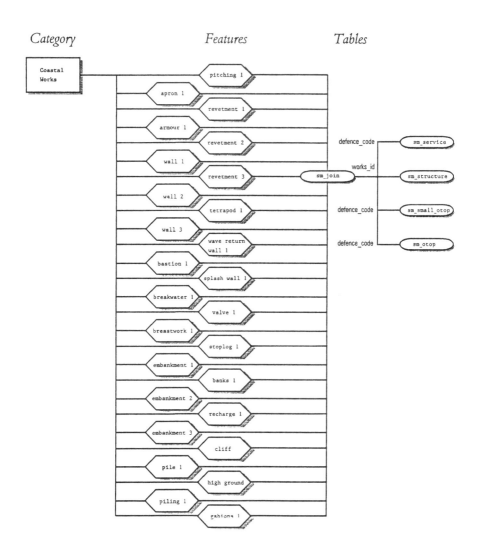

Figure 3   Coastal Works

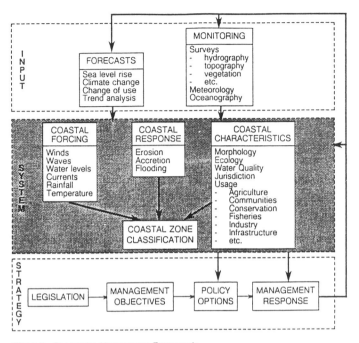

Figure 2 - Responsive Management Framework

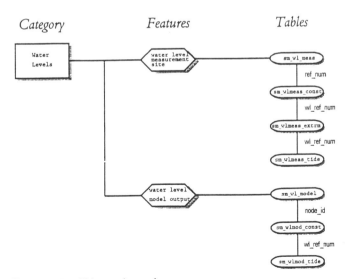

Figure 4   Water Levels

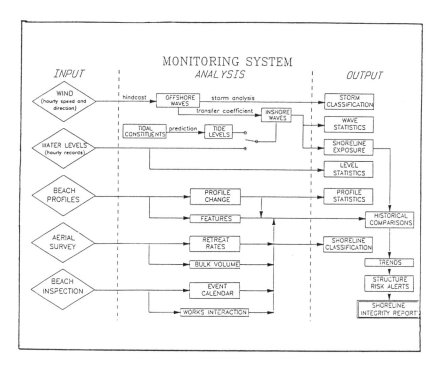

Figure 5 - Functions of SANDS

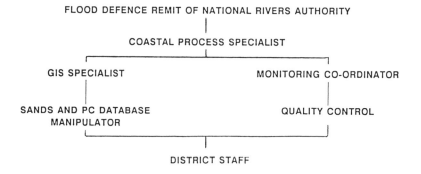

FIGURE 6 - SHORELINE MANAGEMENT TEAM

GEOGRAPHICAL INFORMATION SYSTEM FOR THE INVENTORY OF COASTAL CHARACTERISTICS IN FINLAND

by

Olavi Granö[1] and Markku Roto[2]

INTRODUCTION

The inventory of quantitative and qualitative variations of the coast has been made in the Department of Geography, University of Turku.

The object of the inventory was the 38 000 km^2 coastal belt lying between the mainland and the open sea, interspersed with islands, capes, bays, straits and open stretches of water (Fig. 1). The length of this belt from the eastern border of Finland, from the bottom of the Gulf of Finland to the top of the Gulf of Bothnia is 1 100 km. The coastal belt is widest in southwest Finland – Åland (more than 150 km).

The data of this inventory cover the lengths of different types of shore, the frequency of islands, relief amplitudes and information about the openness of shores, types of soil and land use. The data were collected by means of cartographic measurements, thus providing material covering the entire coast of Finland with the aid of which comparisons can be made. The data are stored by 5 x 5 km^2 grid cells on the geographical information system (GIS).

[1] Prof. Dr. Olavi Granö, Department of Geography, University of Turku, SF–20500 Turku, Finland

[2] M. Sc. Markku Roto, South West Finland Regional Planning Association, Rauhankatu 14 b, SF–20100 Turku, Finland

BASIC ENVIRONMENTAL CONSEPTS IN ARCHIPELAGO COAST

In the Finnish coastal environment there is fundamental difference between the concepts, coast and shore. Coast is a general regional consept, a loose expression and in which can be seen still the former features of a submerged land.

The shore on the other hand is only a narrow strip of mainland and islands which is washed by the sea, a zone between the high and low waterlines. The shore reflects the emergence of the land.

This paradox, ie. that both features of submergence and emergence can be seen at the same time, is the result of the difference between earlier and present development.

Coast

A coast is a three-dimensional space of land, water and air between the open sea and the mainland. The gradual slope of the earth's surface towards the sea and irregularities in relief cause the mainland shoreline to be sinuous and off the mainland there may be numbers of islands. This kind of coast is known as an archipelago coast.

Variations in gradient and relief result in local differences in the sinuosity of the shoreline, in the numbers of islands and the breadth of the coastal belt. In southwest Finland the surface of the earth slopes beneath the sea very gently (2'-3'). Relief amplitudes between the seabed and hill summits are large; in southern parts of the archipelago 75-150 m and farther north 50-75 m. This has led to the formation of the world's largest and densest archipelago. On the other hand, the land surface along the Gulf of Finland slopes more steeply beneath the sea (6'-7') so that the coastal belt is much narrower.

Seashore

The surface of the sea cuts across the relief of the coast and the intersections form the shores. Because of the short-term changes in water level the shores are not lines but rather strips which are at times covered by water running along the fringes of the mainland and islands.

Land uplift along the Finnish coast is more rapid than the worldwide rise in sea level (approx. 0.8 mm/a) and the shore is continually being displaced downwards. At present the displacement of the shore is largest along the coast of the Gulf of Bothnia, approx. 8 mm/a. It is slower farther south along the Gulf of Bothnia and farther east along the Gulf of Finland. In southwest Finland the displacement of the shore amounts to 4-5 mm/a and eastward along the Gulf of Finland on the eastern border of Finland it is only 2.2 mm/a (Kakkuri et al. 1985).

The width of the shore depends on the slope of shores and on the amplitude of short-term variations in sea level caused by changes in winds and air pressure. The amplitudes of sea level changes achieve their maximum values at the ends of the Gulf of Bothnia (3 m) and the Gulf of Finland (2.5 m). In southwest Finland the differences in sea level are smaller than elsewhere along the Finnish coast, less than 1.7 m (Lisitzin 1959, Mälkki 1986).

Shore length

In a coastal environment shores are factors of crucial importance. Variations in shore characteristics reflect regional variations in the coast. The method that has been used in this study to illustrate these differences is the length of different types of shore in relation to surface area (see also Granö 1960).

The length of the shore is not an unambiguous quantity. It depends on the scale and exactitude of the map and on the method used to measure the length of the shoreline. Many researchers have drawn attention to the special cartographic characteristics of the shoreline. Steinhaus (1954) noted that the measured length of the shoreline, just as the length of other empirical curves, increases in a certain mathematical proportion with increasing scale. The phenomenon came to be known as Steinhaus' paradox of length.

The comparability of results obtained by measurement can be estimated with the aid of Perkal's empirical curve measurement technique (1958). The line is measured by moving a circle of given radius and drawn on a transparent overlay on both the land and the water side of the shoreline. The centre of the circle draws a thin strip around the shoreline, which is known as an epsilon area. The surface area of this strip is then measured and divided by its width. In this way the length measurement obtained corresponds to the length of the original line on the map and the generalisation used in the measurement can be expressed as the length of the radius of the circle (epsilon value).

Mandelbrot (1967) studied the problem of measuring and determining the shoreline with the aid of a fractional model. In his view the shoreline is not a one-dimensional line nor a two-dimensional plane. The number of dimensions is not a whole number but lies somewhere between one and two. He suggested as the value for the dimensions of the shoreline along the coast of Britain the figure 1.25. A special feature of fractions is that they repeat themselves inwards infinitely (self-similarity) and they cannot be measured or drawn in complete detail.

INVENTORY METHODS

The length of the shoreline was measured from the 1:20 000 Basic Map of Finland by squares of 5 x 5km^2. This method has been used for the first time in the study of the coast of south Finland (Granö 1960).

The Basic Map was chosen for the map because it is the largest-scale available map showing the whole of the Finnish coastal belt. At the same time this scale provides the best basis for collecting data concerning man's functional environment. A 5 x 5 km^2 square was chosen as the basic unit since it corresponds to the visual landscape in archipelago conditions and is well adapted to illustrating regional variations.

Measurements were made using a digitiser connected to a computer. The coordinates of the shoreline were digitised for use with the measuring program, which calculated the length of the line. Coordinates were drawn for the program when the cursor had progressed an average of 2 mm on the map.

The results obtained by means of this digitiser method were compared with those obtained with the aid of a curvimeter. The results from the digitiser method proved to be 0–10% greater depending on the sinuosity of the line measured. The increase in length shows that with the digitiser method the exactitude of the details of the shoreline improved. The method was also compared with measurements made with a pair of dividers. Dividers with a two-millimetre interval gave the same result on average as the digitiser method. When the digitiser measurements were compared with Perkal's measuring techniques (1958), it was noticed that the digitiser results corresponded to the one-millimetre epsilon results. This epsilon value represents a 2 mm-broad strip, which on a map with a scale of 1:20 000 corresponds to a belt of shore 40 m in width.

The measurements were rounded to an exactitude of 0.1 km and stored on DOS floppy disks. The lengths of the shoreline in the provinces, regional planning areas and communes were calculated by generalising the boundaries according to the grid cells. The boundary between land and sea drawn on the map was considered to be the shoreline. Shorelines of lakes and rivers were excluded from the measurements. In estuary areas local conditions were taken into consideration in estimating the boundary between the river shoreline and the seashore. The shoreline of bays that had been formed into lakes during land uplift was measured in those cases where there still existed a direct link with the sea on the map. Dams and embankments were measured as a part of the shoreline. The measurements were taken from as recent maps as were available.

The shore types were classified according to the deposits of the shore. Five different categories were used: rock; till; gravel and sand; silt, clay and marsh; artificial.

With the aid of the 1:20 000 Basic Map of Finland it was possible to reliably measure the length of the shore in each grid cell. Port constructions, embankments and dams were classified as artificial shores. Geological maps and and other sources showing soil composition and geomorphology were used in addition to the Basic Map.

In practice, the work was conducted so that the total length of the shoreline was first measured, then the length of rocky shoreline, the length of artificial shoreline, the length of gravel and sand shoreline, the length of grass-covered silt, clay and marsh shoreline and the remaining length of shore was interpreted as till shore according to the primary deposit. This means that it constitutes a very heterogeneous group, ranging from stony shores to grass-covered shores made up of finer sediments.

When the length of the shoreline was measured, the number of islands was also counted. The islands were divided into two categories: those with an area of more than 1 hectare and those of less than 1 hectare. The smallest size of island shown on the 1:20 000 Basic Map of Finland is about 5 ares.

The difference between the highest summit and the greatest depth (relief amplitude) and the information concerning the material on highest summit, the openness of shores (fetch) and land use were also included to the grid cell data.

RESULTS

The total length of the Finnish seashore is 39 139 km. The Finnish seashore can be classified on the basis of shore types as follows:

Shore type	length (km)	percentage
Rocky	16 394	41,9
Till	16 284	41,6
Gravel and sand	1 870	4,8
Silt, clay and marsh	4 090	10.4
Artificial	501	1,3
Total	39 139	100,0

There are large variations in the predominant type of shore encountered in different parts of the coast. The following table shows the lengths (km) of different types of shore by regional planning area. Kymenlaakso, East Uusimaa, Helsinki and West Uusimaa regional planning areas are located on the coast of Gulf of Finland (from east to west) and Satakunta, Vaasa province, North Ostrobothnia and Lapland on the coast of Gulf of Bothnia (from south to north).

Regional planning area	rocky	till	gravel, sand	silt, clay	artificial
Kymenlaakso	370	799	270	155	76
East Uusimaa	583	1 007	185	256	23
Helsinki	183	186	48	55	86
West Uusimaa	1 856	1 174	228	439	19
Soutwest Finland	6 809	3 643	343	1 325	53
Åland	5 952	1 535	74	372	8
Satakunta	152	1 604	65	161	38
Vaasa province	418	5 585	207	684	67
N. Ostrobothnia	64	419	368	481	71
Lapland	8	332	83	198	60
Total (km)	16 394	16 284	1 870	4 090	501

In the coastal belt of Finland there are 73 021 islands of which 20 423 with an area more than 1 hectare.

The resulting grid cell maps showing the regional variation of coastal characteristics have so far been published in the 'Atlas of Finland' (1986) 'Relief' (Folio 121) and 'Water' (Folio 123). These maps show the length of shore, predominant shore type, relief amplitude, summit material, percentage of rocky shore and number of islands by 5 x 5 km^2 grid cells along the Finnish coastal belt.

The map (Fig. 1) 'Fragmentary nature of the Finnish coastal belt as shown by length of seashore in relation to surface area' and the map (Fig. 2) 'Characteristics of the shore along the coast of Finland' are based on the grid cell data and give an overall view of coastal characteristics in Finland.

The map 'Shores in Southwestern Finland' is also based on the grid cell data and shows the length and character of seashore by municipality in the SW Finland regional planning area (Granö and Roto 1991).

CONCLUSION

Characteristic morphological features of the coast of Finland are: the large size of the whole coastal belt (approximately 38 000 km^2), the great length of the shoreline (39 139 km), the great frequency of islands (73 021), the fragmented nature of sea and land areas, the latitudinal and parallel zonation of the coastal environment.

By using the grid cell data structure and computer-aided geographical information system (GIS) it is possible to analyse the interrelationships between the morphological characteristics, the different rates of land uplift, the different amplitudes of water level changes and the different rate of wave action and so produce maps and models illustrating the regional variation and zonation of the coastal belt (Granö and Roto 1989a and 1989b).

References

Granö, Olavi (1960). Die Ufer der Südküste Finnlands. Fennia 83:3.

Granö, Olavi and Roto, Markku (1986a). The Coast. Atlas of Finland, 5th edition, Folio 121, Relief. National Board of Survey and Geographical Society of Finland.

Granö, Olavi and Roto, Markku (1986b). The Shoreline. Atlas of Finland, 5th edition, Folio 132, Water. National Board of Survey and Geographical Society of Finland.

Granö, Olavi and Roto, Markku (1989a). The duration of shore exposure along the emerging Finnish Coast. Journal of Coastal Research 5.

Granö, Olavi and Roto, Markku (1989b). Zonality in the Finnish Coastal Environment. Essener Geographische Arbeiten 18.

Granö, Olavi and Roto, Markku (1991). Map of Shores in Southwestern Finland. University of Turku. Development project for the archipelago. Department of Geography.

Kakkuri, J. & Vermeer, M. (1985). The study of land uplift using the third precise levelling of Finland. Reports of the Finnish Geodetic Institute 85:1

Lisitzin, E. (1959). The Frequence Distribution of the Sea-Level Heights along the Finnish Coast. Merentutkimuslaitoksen julkaisu 190.

Mandelbrot, Benoit (1967). How long is the coast of Britain? Statistical Self-similarity and Fractional Dimension. Science 156.

Mälkki, Pentti (1986). Sea areas. Physics. Atlas of Finland, 5th edition, Folio 132, Water. National Board of Survey and Geographical Society of Finland.

Perkal, Julian (1958). On the length of empirical curves. Michigan Inter-University, Community of Mathematical Geographers. Discussion 10. Translated by R. Jackowski 1965.

Perkal, Julian (1958). An attempt at objektive generalization. Michigan Inter-University, Community of Mathematical Geographers, Discussion 10. Translated by R. Jackowski 1965.

Steinhaus, H. (1954). Lenght, shape and area. Colloquium Mathematicum 3.

Fig.1 Fragmentary nature of the Finnish coastal belt as shown by length of seashore in relation to surface area (Granö and Roto 1989a).

Fig.2 Characteristics of the shore along the coast of Finland (Granö and Roto 1989b).

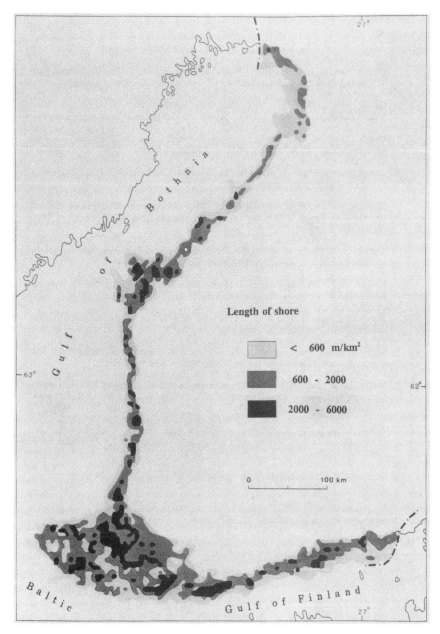

Fig.1 : Fragementary nature of the Finish coastal belt as shown by length of seashore in relation to surface area (Granö and Roto 1989a)

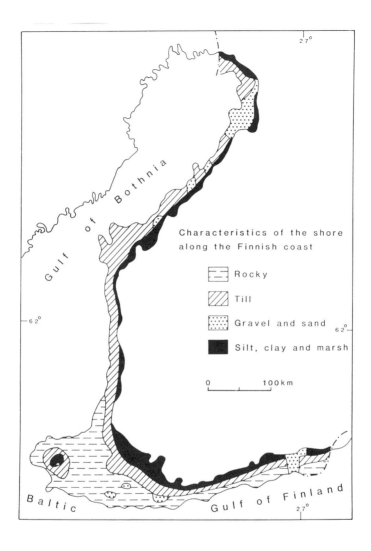

Fig. 2: Characteristics of the shore along the coast of Finland (Granö and Roto 1989b)

A GIS-supported sensitivity analysis. Implementation of results from ecosystem research.

M. Stock[1], D. Boedeker[2], U.-H. Schauser[3] & R. Schulz[4]

Abstract

As part of an applied ecosystem research project in the Wadden Sea Nationalpark of Schleswig-Holstein an interdisciplinary approach has been taken to study the impact of touristic activities and natural vegetation succession on breeding birds on beach-barrier systems. The study area is situated at the most significant beach-barrier system off the westcoast of Schleswig-Holstein. This area holds the most important breeding population of Kentish plover *(Charadrius alexandrinus)*, at the same time attracting millions of tourists every year, leading to many conflicts between nature conservation and recreation.

In order to quantify the extent of human impact on the breeding success of Kentish plovers, detailed investigations on population dynamics have been conducted since 1989. The ornithological data were incorporated into the existing Geographic Information System GIS-WEST. The combined data were used to analyse and document recreational impacts on coastal breeding birds.

The sensitivity analysis enables the authors to formulate certain objectives for the protection of the area. They can be used by the National Parc authorities to work out detailed management plans.

Introduction

As part of an applied ecosystem research project in the Wadden Sea Nationalpark of Schleswig-Holstein (Leuschner & Scherer 1989) an interdisciplinary approach has been taken to study the impact of touristic activities and natural vegetation succession on breeding birds on beach-barrier systems. In order to quantify the effects of these parameters a detailed investigation on population dynamics of the Kentish Plovers (Charadrius alexandrinus) - as a case study - has been conducted in 1989 in St. Peter-Ording on the Eiderstedt peninsula,

---

[1] Nationalparkamt, Am Hafen 40a, 2253 Tönning, Germany

[2] Internat. Naturschutzakademie Insel Vilm, O-2331 Lauterbach/Rügen, Germany

[3] Geogr. Institut der Universität Kiel, Ludwig-Meyn-Str. 14, 2300 Kiel 1, Germany

[4] Inst. für Haustierkunde, Universität Kiel, Am Botanischen Garten 9, 2300 Kiel 1, Germany

Schleswig-Holstein.

The area is situated at the only nearshore beach-barrier system off the mainland coast of Schleswig-Holstein (Fig. 1). A description of the study area can be found in Schulz & Stock (1992, 1993). The natural beauty of the extensive system of primary dunes, natural sandy saltmarshes and beaches attracts millions of tourists every year. At the same time the area holds the most significant breeding population of Kentish Plovers in Central Europe (Schulz & Stock 1991). Existing touristic facilities on the beach (extensive parking lots, restaurants, etc.) lead to conflicts between recreation and nature conservation (Boedeker & Schauser 1990). Results from ornithological investigations (Schulz & Stock 1991, Stock 1992) clearly illustrate the outstanding significance of the area for breeding birds of beach-barrier systems and have lead to protection measures in the most important part of the breeding area. The primary dunes and parts of the open sandy beaches close to the vegetated zone were closed for the public during the breeding season. An extensive wardening of the area was necessary. Data on the impact of touristic acitivities and the natural dynamics of the beach-barrier system on the distribution and the breeding success of Kentish Plovers were incorporated into the existing Geographic Information System GIS-WEST (Schauser et al. 1992). The data were used to run a GIS-supported sensitivity analysis. The main aim was to check the applicability of the existing GIS-WEST to formulate objectives for the protection of the area and to demonstrate the use of the Geographic Information System as a practicable tool for the National park authorities to work out detailed management plans.

Study area, methods and data handling

The coast off St. Peter-Ording is build up by an extensive system of barrier beaches which are gradually migrating eastward (Ehlers 1988). On the mainland, dunes vegetated by Pine forest, have formed. The beach-barrier system is composed of a mosaic of beach ridges, sandy flats, primary dunes, which occur mainly within restricted sites, elongated lagoons developing behind the ridges, and saltmarshes. The beach barrier is dissected by a system of small creeks and gullies. The study area of about 120 ha (Fig. 1) includes saltmarshes on sand, bare sandy areas and primary dunes on the youngest beach ridge.

Data on phenology, distribution, habitat choice and breeding success of Kentish Plovers and on habitat parameters were collected from 1989 to 1991 mainly by standard procedures, but a detailed description of the methods used can be found in Schulz & Stock (1992, 1993). The bulk of the bird data are based on observations of individually marked birds. The

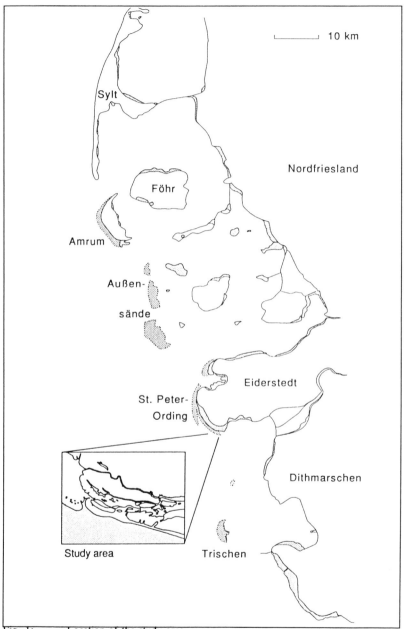

Fig. 1: Location of the study area

position of the clutches was measured by means of a DECCA navigation system. Vegetation succession was monitored with the help of aerial photography interpretation and ground-checking. The touristic impact was measured by counting and mapping the number of tourists (walking and sun-bathing persons were considered separately) on about 40 days from April to July each year. Counts were made during peak-times of touristic activity as well as on certain days in 30 min. intervals.

Data analysis was carried out within the Geographic Information System GIS-WEST, which is installed on IBM-compatible PC (386). All procedures were run with the software package PC-ARC/INFO. In order to analyse tourist and clutch data, we generated a raster map (grid cell width: 25 by 25 m). For the sake of manageable data handling we used point data files, which were referenced to the center of each grid cell, to store and to compute the information. Vector maps were generated to store and analyse the vegetation data.

Results and discussion

In a first step we analysed the occurence of different habitat types in the entire study area over the three years of our study (table 1). Sandy saltmarshes rank first and cover more than 50% of the area. There was a slight increase from 50.6 to 55.5% over the three years. Bare sands, often covered with a shell layer, and tidal flats rank second. The proportion of bare sand in the study area was decreasing from nearly 19% to nearly 15% over the study

Tab. 1: Number of clutches and number of tourists (sun-bathing people) related to habitat types.

Habitat type	area (%)			clutches (%)			touristic impact (%)		
	1989	1990	1991	1989	1990	1991	1989	1990	1991
primary dunes	11.1	11.7	9.3	63.3	75.4	59.0	18.1	43.7	37.6
sandy saltmarsh	50.6	51.1	55.5	6.1	7.9	16.7	18.5	11.2	7.3
sparse vegetation	0.5	3.1	2.9	-	12.3	7.7	-	-	-
bare sand	18.6	16.4	14.7	22.4	4.4	16.7	45.7	40.9	53.3
tidal flats	19.2	17.6	17.6	5.1	-	-	17.7	4.3	1.8
Total (ha resp. no)	121.5	121.5	121.5	98	114	78	276[x]	843[y]	836[y]

[x] April - Mai counts [y] May - June counts

period. Primary dunes cover less than 12% and were also decreasing. Sparsely vegetated sands, colonized by Salicornia europea or algae mats, cover only 0.5% in 1989, but were increasing to nearly 3% in the following years. Fig. 2 summarizes the vegetation succession in % of the total cover over the three years. There seems to be a trend towards an ongoing loss of open or sparsely vegetated habitats, while at the same time the sandy saltmarsh habitat with its more or less closed plant cover is gaining ground.

The percentual distribution of clutches and the percentual occurence of tourists (only

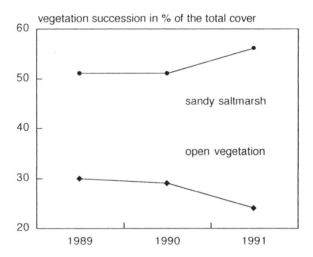

Fig. 2: Vegetation succession in the study area from 1989 to 1991, represented in % of the total area.

sun-bathing persons were taken into account) related to the different habitat types are also shown in table 1. Both Kentish plovers and sun-bathing tourists prefer the primary dunes and the bare sand areas. Throughout the study period 59 to 75% of the clutches were found within the primary dunes. These dunes and comparable vegetation zones with a plant cover of about 30% can be considered to be the most prefered breeding habitat for Kentish plovers (Schulz und Stock 1993). Bare sands rank second and with the exception of 1990 more than 16% of all clutches could be found here. The remaining habitats play a minor role as breeding places. The open sandy areas are also the most prefered sites for sun-bathing persons, followed by primary dunes. Thus both habitats can be considered as the most threatened beach zones in the study area leading to competition between both user-groups. The touristic impact on the three main habitat types is shown in figure 3. There was a steep increase in the use of primary dunes by sun-bathing persons from 1989 to 1990. Bare sands were highly prefered during the first year of study but showed a slight decrease in the next year. Sandy saltmarshes are less favoured and the touristic impact decreased more and more. Due to the closure of the main part of the area in the 1991 breeding season, the touristic impact in primary dunes decreased from 1990 to 1991, whereas the pressure in the remaining bare sand areas increased.

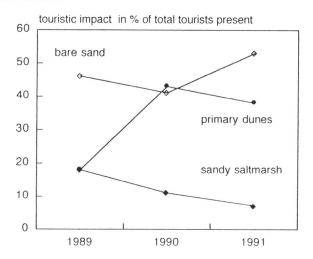

Fig. 3: The impact of tourists in the study area from 1989 to 1991, represented in % of total tourists present.

Competition between plovers and tourists

In a second step we investigated the interrelationship between areas used by breeding plovers (clutches) and areas used by sun-bathing persons. For this purpose we looked at single raster cells (grid width: 25 by 25 m). We only selected grid cells which at least contained one clutch. The result of this analysis is shown in table 2. Only between 2.6 and 5.1 % of all grid cells considered were used by both sun-bathing persons and breeding plovers. The bulk was used either by plovers or by sun-bathing persons. This clearly indicates that a mixed occurence of both user groups is in most cases impossible. A competition for space occurs.

Tab. 2: Interrelationship between grids with breeding plovers and grids with touristic impact (sun-bathing people)

year	grid cells used by birds or tourists (1)	grid cells used by birds and tourists (2)	% of 1
1989	332	17	5.1
1990	267	7	2.6
1991	227	8	3.5

Tab. 3: Nearest mean distance (m) between grid cells with clutches and grid cells with sun-bathing people. (n) = number of grid cells

year	total grids		excluding grids used by birds and tourists	
1989	51	(73)	67	(56)
1990	97	(81)	106	(74)
1991	86	(91)	100	(54)

In a further step we calculated the nearest mean distance between grid cells with clutches and grid cells with sun-bathing persons. This analysis allowed us to answer the question how a buffer zone between breeding plovers and tourists in the area, and thus how an effective closure of the breeding area should look like. We ran the calculation with all grids which were included in the former analysis. Nearest mean distances for all cases and for those excluding grids that were used by birds and tourists simultaneously are presented in table 3. This exclusion was done on the assumption that a situation where birds breed within the same grid cell that tourists use for sun-bathing is an extremely "forced" situation for the birds, resulting from lack of breeding habitat, and therefore should not be included in the analysis. Looking at all grid cells, the mean nearest distance over time varies between 51 and 97 m. Considering only those cases excluding simultaneously used grid cells, the distance varies between 67 and 106 m. These results indicate that an undisturbed buffer zone of at least 100 m around a clutch is needed by breeding plovers.

The spatial dimension

The spatial distribution of the different habitats of the study area in 1989 and 1991 is shown in Figs. 4 and 5. Sandy saltmarshes comprise the largest area; primary dues, being the most prefered breeding habitat of the Kentish plover, bare sands, tidal flats and sparsely vegetated areas are also represented. The spatial use of the area by plovers and tourists in relation to habitat distribution in 1989 is shown in Fig. 6. The interrupted black line marks the protected core zone of the National Parc. Widely cross-hatched squares represent grid cells with observed touristic acitvities. More densely cross-hatched squares represent grid cells containing clutches and densely cross-hatched squares represent grid cells with simultaneous use by birds and tourists.

The westernmost part of the primary dunes up to the edges of the sandy saltmarsh are heavily used by tourists. The people are widely spread over the area and the separated primary dune in the west is extensively used by sun-bathing people and covered with beach-chairs. Clutches of Kentish plovers are mainly found in the extensive primary dune complex in the eastern part of the study area and on open spots within the sandy saltmarsh. In several cases clutches occur in combination with sun-bathing tourists (black solid squares). This presentation indicates clearly that large parts of the study area cannot be used by plovers because of the widely spread presence of tourists.

The situation in 1991 (Fig. 7) is influenced by two phenomena. Firstly a protected area was designated during the breeding season. The solid black line marks the politically negotiated compromise between nature management and touristic requirements. Secondly, a distinct change in habitat type has taken place (Figs. 4 and 5). Large parts of the primary dunes and other slightly vegetated areas are now covered by closed vegetation and will change to a sandy saltmarsh (eastern part of the primary dune complex). Due to the closure of the potential breeding area, the impact of touristic activities in the western part of the study area was reduced (Fig. 7). Parts of the formerly blocked dunes are now colonized by plovers. As a result of the change in vegetation cover in the central part of the primary dune area, the birds moved further east and a concentration of clutches can be found in the totally undisturbed easternmost areas.

Management implications

The Kentish plover is one of the most threatened wader species in the Wadden Sea area and breeding numbers have declined dramatically during the last centuries (Becker & Erdelen

1987, Stock 1992). The breeding population in St. Peter-Ording in 1990 accounted for about 235 breeding pairs representing nearly 50% of the German breeding population (Hälterlein & Behm-Berkelmann 1991).

Primary dunes and similar sparsely vegetated sandy habitats are the most prefered breeding places not only in our study area. In general these habitats form a very small proportion of the coastal habitats along the Wadden Sea coast of Schleswig-Holstein. Apart from a small stretch at Westerhever (about 12 km north of St. Peter-Ording), the study area is the only location with a natural beach-barrier system on the mainland coast. Because of progressive loss of suitable and undisturbed habitats for Kentish plovers, which is due to natural vegetational succession, coastal protection measures and an intensive use of these areas for recreational purposes, a more effective protection of the remaining area is required.

As we could demonstrate, tourists and plovers compete for space in primary dunes and on bare sandy spots. Only in a very small number of cases Kentish plovers bred in close vicinity to sun-bathing tourists. With the help of the software package ARC/INFO we were able to calculate a minimum buffer zone which is at least necessary to separate breeding plovers from sun-bathing tourists. The Geographic Information System furthermore enabled us to depict the result of a closure of the breeding terrain. In comparing the situation in 1989 (before the closure) and 1991 (after the closure) (Figs. 5 and 6) it emerges, that a demarcation line has a positive effect on the spatial use of the area by tourists, but this is not sufficient. A buffer zone of at least 100 m is urgently required. The graphic representation furthermore showed the supplemental effect of the vegetation succession on the reduction of the potential breeding area. From our results it can be concluded that a long-lasting protection of the remaining breeding area of Kentish plovers can be obtained only by declaring the entire beach-barrier system of St. Peter-Böhl (which comprises the southeastern part of the St.Peter-Ording beach barrier system), "Zone 1" of the National Parc. Closure of only isolated parts of the breeding terrain does not guarantee long-term protection of this endangered breeding bird, which requires a highly dynamic habitat.

It can also be concluded that the existing Geographic Information System GIS-WEST represents a practicable tool for the National Parc authorities to illustrate the results of scientific research and to establish management plans on the basis of a GIS-supported sensitivity analysis.

Acknowledgements

This work was financially supported by the Federal Environmental Agency and the State of Schleswig-Holstein as part of the Project: "Ecosystem Research Wadden Sea". This is contribution No. 44.

References

Becker, P.H. & M. Erdelen (1987): Die Bestandsentwicklung von Brutvögeln an der deutschen Nordseeküste 1950-1979. - J. Orn. 128: 1-32

Boedeker, D. & U.-H. Schauser (1990): A G.I.S. for the coast. Experiences from Schleswig-Holstein (West Germany). - In: Quélennec, R.E., E. Ercolani & G. Michon (eds.): Littoral 1990. Comptes rendus du 1er symposium international de l'Association européenne:588-595

Ehlers, J. (1988): The Morphodynamics of the Wadden Sea. Balkema, Rotterdam, 397 S.

Hälterlein, B. & K. Behm-Berkelmann (1991): Brutvogelbestände an der deutschen Nordseeküste im Jahr 1990 - Vierte Erfassung durch die Arbeitsgemeinschaft "Seevogelschutz". - Seevögel 12: 47-51

Leuschner, C. & B. Scherer (1989): Fundaments of an applied ecosystem research project in the Wadden Sea of Schleswig-Holstein. - Helgoländer Meeresunters. 43: 565-574

Schauser, U.-H., D. Boedeker, S. Matusek & H. Klug (1992): The use of a Geographic Information System for the Wadden Sea conservation in Schleswig-Holstein. Neth. Inst. Sea Res., Publ. Ser. No. 20: 281-283

Schulz, R. & M. Stock (1991): Kentish plovers and tourists - conflicts in a highly sensitive but unprotected area in the Wadden Sea National Park of Schleswig-Holstein. - Wadden Sea Newsletter 1/91: 20-24

Schulz, R. & M. Stock (1992): Seeregenpfeifer und Touristen - Der Einfluß der touristischen Nutzung von Strandgebieten auf die Ansiedlung und den Bruterfolg des Seeregenpfeifers. - ÖSF Bericht 1989-1991, WWF-Wattenmeerstelle, Husum und Nationalparkamt, Tönning. 69 S.

Schulz, R. & M. Stock (1993): Kentish Plovers and Tourists: Competitors on Sandy Coasts? - WSG Bull- Suppl. (in press)

Stock, M. (1992): Ungestörte Natur oder Freizeitnutzung ? - Das Schicksal unserer Strände. - In: Prokosch, P. (Hrsg.) Ungestörte Natur - Was haben wir davon ? WWF-Tagungsbericht 6: 223-249

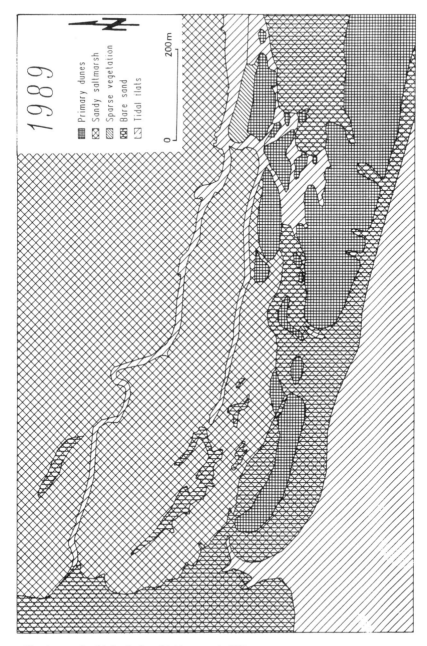

Fig. 4: Spatial distribution of habitat types in 1989

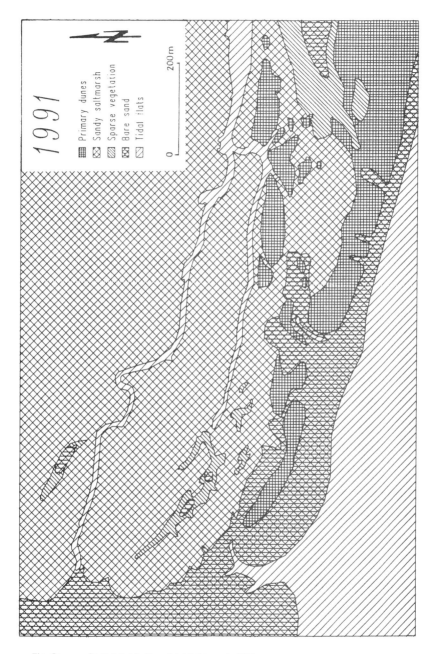

Fig. 5: Spatial distribution of habitat types in 1991

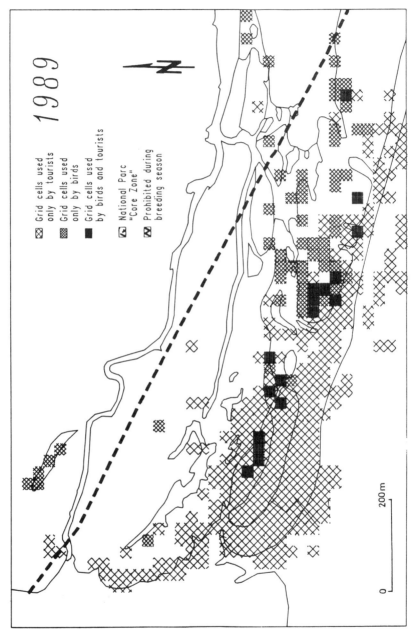

Fig. 6: The spatial use of the study area by plovers and tourists in relation to the different habitat types in 1989

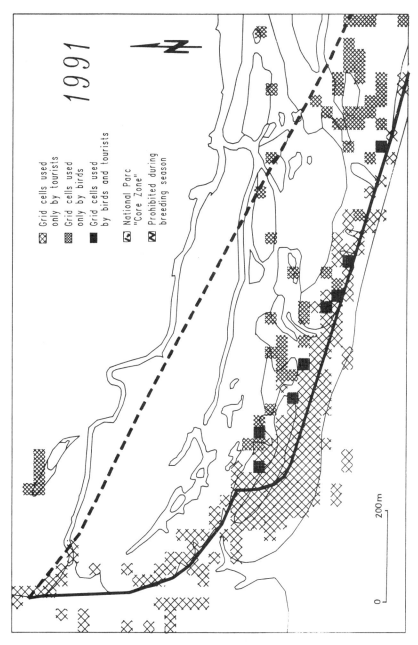

Fig. 7: The spatial use of the study area by plovers and tourists in relation to the different habitat types in 1991

THEMATIC MAPPING AND SENSITIVITY STUDY OF MUD FLAT AREAS IN THE GERMAN WADDEN SEA

K.H. van Bernem [1], M. Grotjahn [2], J. Knüpling [3], H.L. Krasemann [1], A. Müller [1], L. Neugebohrn [4], S. Patzig [1], G. Ramm [3], R. Riethmüller [1], G. Sach [5], S. Suchrow [4]

The results of many years of field research and laboratory experiments have shown that there are significant habitat-dependent, long- and short-term consequences of oil pollution in the Wadden Sea. The dependence is determined by both abiotic and biotic parameters and results from the interrelationships among toxicity, mobility, and persistence. These findings provided a basis for the development of a spatially and temporally differentiated vulnerability study of the tidal flats to help formulate methods to combat oil pollution. Necessary prerequisites for this study included a new, uniform, methodical cartographical survey and the development of a concept for evaluation.

Even though narrow limits were set for the cartographical survey, possible aspects of general ecological monitoring were to be included and integrated using a comprehensive data processing system. This "Wattenmeerinformationssystem" -WATiS- with the database "Wattenmeerdatenbank" -WADABA- should make it possible to recover detailed data to determine vulnerability to oil in individual cases with the necessary degree of simplification. It should also make information available for the needs of the national park authorities and to serve scientific purposes.

The main tools of the underlying technical concept of WATiS are a relational database for data storage and geographical information systems (GIS) for the handling of thematic maps. Much concern has been put on the close connection between the locally distributed GIS and the coordinating central database, to guarantee an automatic data exchange and a consistent data set within the WATiS (Fig. 4).
To combine data from different research groups all data are well documented. The data structure has been designed in close cooperation among the research groups and the staff of the information system.

For the cartographic survey the following independent categories were established:
Saltmarsh, sediment, benthos (macro-, meio-, and microphyto-) and juvenile fish populations. Data for the categories avifauna and mammalia were prepared from

literature studies at the "Inselstation der Vogelwarte Helgoland" as part of a special program.

After completion of a feasibility study on the tidal flats between the rivers Weser and Elbe and in the Wadden Sea between the island Spiekeroog and the mainland conducted from 1983 through 1986, the assessment of the entire part of the German Wadden Sea was completed in 1992 under the direction of the GKSS Research Centre Geesthacht. The work was financed in concert by the following institutions: Umweltbundesamt Berlin, Nationalparkverwaltung Niedersächsisches Wattenmeer, Landesamt für den Nationalpark Schleswig-Holsteinisches Wattenmeer, Sonderstelle der Küstenländer "Ölunfälle See/Küste and the GKSS Research Centre Geesthacht.

Data on the content of heavy metals and chlorinated hydrocarbons in tidal flat sediments at selected stations have been compiled since 1989 in a special study, also directed by the GKSS.
The methods used for the analyses are TXRF, AAS and INAA for heavy metals and GC for chlorinated hydrocarbons.

Scheme of data taken during the annual surveys:

Topography: surveys of official institutions

Avifauna and mammalia: official censuses of birds (nesting, non-nesting, molting) and common seals

Saltmarshes: floral surveys of species and community types on transsects with 1000m between

Ichthyofauna: juvenile fish and brown shrimp surveys in spring and autumn on selected eulitoral areas (2m beam trawl; 80 tows/season)

Benthos
Macrofauna: species and abundance (1 station/km^2)

Meiofauna and microphytobenthos: species and abundance (selected stations)

Sediment: determination of grain size, shear strength, redox conditions and water content (1 station/km^2)

Habitat description and depiction of borders

Evaluation of the data obtained

Content of chlorinated hydrocarbons and heavy metals: sediment of selected stations

Short description of methods:

Thematic mapping
Eulittoral zone

A network of stations has to be established over the entire area of the annual cartographical surveys. There is one station/km^2. Samples are taken at all stations for quantitative analysis of the macrofauna and abiotic parameters. In addition stations within connected areas that can be considered representative of the region and dispersed stations that were selected according to to the homogeneity of the habitats are sampled quantitatively for identification of the meiofauna and microphytobenthos. Some of these intensively sampled stations are investigated monthly to record seasonal influences.

The characteristics of the individual habitats at and between the stations are recorded according to a standardized format. The locations of all stations and the habitat borders are plotted using the Decca navigation system.

As an example for the depiction of individual results, the depth and quality of the oxidazing layer and the dispersal of *Arenicola marina* and *Lanice conchilega* are shown in areas of the North-Frisian islands (Fig. 1 and 2).

Supralittoral zone

The mapping of the salt marshes is undertaken with emphasis to meet the demands of programs to combat oil pollution – a complete depiction of all communities is not required. Starting from the dike or base of the dunes, transsects about 1 km apart are plotted across the saltmarsh at angles of about $90°$ to the coastline. From among the vegetation types occurring along these lines, several are selected for floral surveys on areas of $1m^2$ (species and dominance as a percentage of coverage). The levels of these locations are then determined relative to normal datum. The identified community types are grouped together as blocks so that the ecotone along lines of increasing elevation can be recognized.

Sensitivity study
Evaluation

For all categories of the study: Saltmarsh; benthos, including juvenile fishes; and avifauna together with mammalia, syn- and autecological criteria were established to permit estimates of the sensitivity to oil of the individual species.

This sensitivity of the species is interpolated either according to the biotic communities, such as the saltmarsh, or as in case of the benthos, in relation to the abundance of the species and various sediment parameters. As an example, figure 3 shows the distribution of the individual class values "benthos-sediment" on areas between the islands Sylt and Pellworm.
Indices are calculated for all categories, and their ranges are divided into four sensitivity classes.
To summarize these findings, the class values for all categories are placed on a scale from 1 to 12 with increasing vulnerability.
This permits an evaluation of temporal and spatial differences in the sensitivity of habitats based on the detailed data in the WATiS-system under the continual supervision of experts.

Summary

A methodologically uniform inventory of the entire German Wadden Sea will be completed in 1992. It is based on quantitative and descriptive data and includes a documentation of temporary aspects. Making use of the synchronously prepared WATiS data processing system, a basis will be established for the designation of representative sampling areas as part of a monitoring program for the entire region as well as for the direct application of preventive and remedial measures combatting oil pollution accidents and for environmental protection perhaps on the basis of quality objectives yet to be developed.

[1] GKSS-Forschungszentrum Geesthacht, Max-Planck-Straße, 2054 Geesthacht, FRG

[2] Forschungsstelle für Insel- und Küstenschutz, An der Mühle 5, 2981 Norderney, FRG

[3] Institut für angewandte Biologie, Alte Hafenstraße, 2163 Freiburg/NE

[4] Institut für angewandte Botanik der Universität Hamburg, Marseiller Straße, 2000 Hamburg, FRG

[5] Forschungsinstitut Senckenberg, Schleusenstraße 39a, 2940 Wilhelmshaven, FRG

EXAMPLES OF A THEMATAIC DEPICTION OF SELECTED DATA:

1. Depth and quality of the oxidazing layer and the occurrence of reducing conditions on the sediment surface

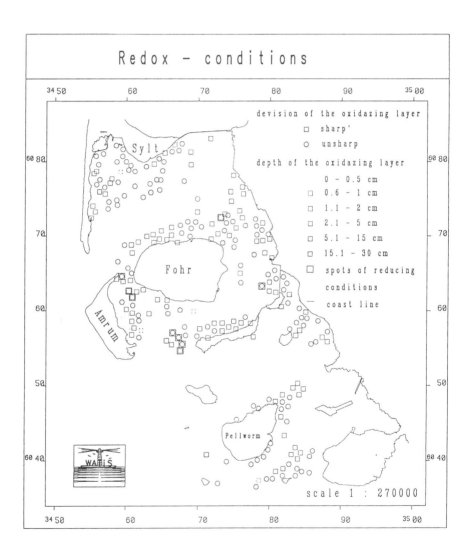

2. Dispersal of Arenicola marina in 5 classes of abundance and the occurrence of Lanice conchilega at sampling stations in areas between the islands Sylt and Pellworm

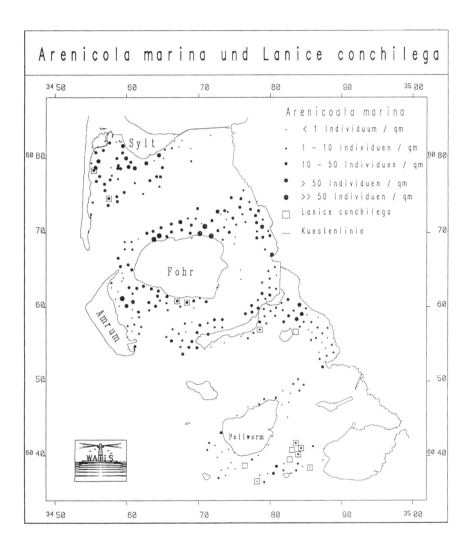

3. Distribution of the individual class values - "benthos-sediment" (1-4, with increasing sensitivity) on areas between the islands Sylt and Pellworm

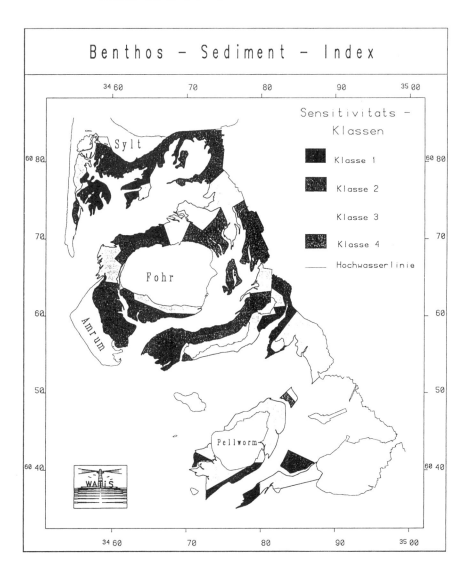

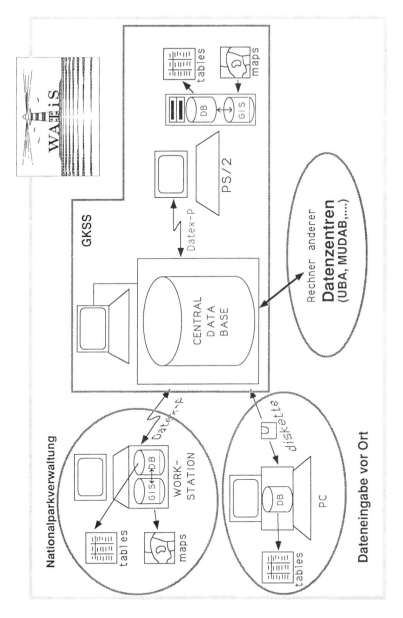

4. Technical Concept of the Wadden Sea Information System WATiS

GIS-APPLICATION ON WADDEN SEA AREAS
by
Wolfgang Liebig, Coastal Research Station, Norderney
East Frisia / Germany

In the last years more and more computers have been used to produce and handle maps. The required extensive tools, consist of the soft- and hardware, are called Geographical Information System (GIS). A GIS is a tool for storing and manipulating geographic informations in a computer. Once maps are in the computer, you can ask questions of the database and manipulate, analyze and display geographic information.

In this paper it is explained how GIS is used at Coastal Research Station, Norderney. Three examples are going to show some investigations of the Wadden Sea where the GIS database "Ostfriesland" is implemented. With aid of contour, sedimentical, biological and hydrological maps several overlays, volumes, contourlines and 3D-surfaces are computed.

At first the "Project Leyhörn" presents an example of change of morphology at two states, 1976 and 1990. The "Project HISWA" shows a connection of the mathematical waveheight model HISWA with GIS. In the "Project WADE" volumes and cross-section areas are computed.

PROJEKT "LEYHÖRN"

In the 80th the peninsula "Leyhörn" (7) was built to improve the water management situation around the "Ley Bay" located in the north-west of Lower Saxony. In order to check the interactions of the morphology this **GIS** application is carried out. To compute 3D-surfaces of the investigation area different maps of the morphology (1976, 1990) are digitized (figure 1, figure 2).

Furthermore a clip of investigation area has been made to explore the morphology around the head of Leyhörn. Changes of the morphology as difference of both states 1976 and 1990 is showed as 3D-surface (figure 3). Therefrom contourlines of changes are calculated (figure 4). This points out a remarkably change of the tidal channel, moving to the north. In addition removed and filled sediment volumes are computed.

PROJECT "HISWA"

Since 1989 the mathematical wave propagation model "HISWA" (Hindcasting Shallow Water Waves) (8) is used at Coastal Research Station, Norderney. The wavefield is described at every grid point (figure 5) within the model boundaries for given morphology and wind and the waves at the up-wave boundary. The required morphology for HISWA is digitized and prepared by **GIS** to transfer into HISWA. In reverse order the waveheights calculated by HISWA are transfered into GIS using its powerful commands to store, update and analyze all forms of geographically referenced informations.

In the plots below distributions of waveheights for the investigation area Norderney are pointed out. The waveheights are plotted in connection with sediment and contourlines (figure 6, figure 7). Furthermore the distribution of waveheights is covered over a three-dimensional surface of the investigation area. Finally a database query was done to overlay waveheights with different sediments like mudflat, mixed-flat, sand-flat and shell bank (figure 8).

PROJECT "WADE"

Within the scope of the German-Dutch research project "WAdden sea morphological DEvelopment" (WADE) investigations are carried out to obtain parameters describing the interactions between the hydrological impact and the morphology. Based on digitized surveys from the map-database "Ostfriesland" (figure 9) volumes and cross sectional areas are computed for different intake areas and waterlevels.

To compute volumes and profiles some 3D-models are calculated from the digitized morphology by utilization of **GIS**. In this case it is done for the inlet "Norderney" (figure 10). Furthermore statistical analyses are used to develop parameters to show physical relationships, e.g. between water level and intake area (figure 11) or basin volume (figure 12).

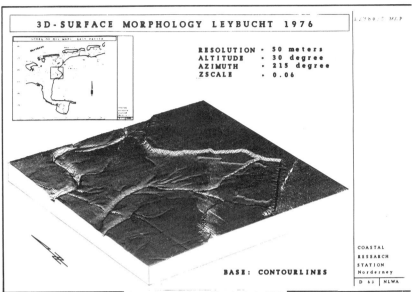

Figure 1 State of morphology of Leybucht in 1976 without new dyke 'Leyhörn'

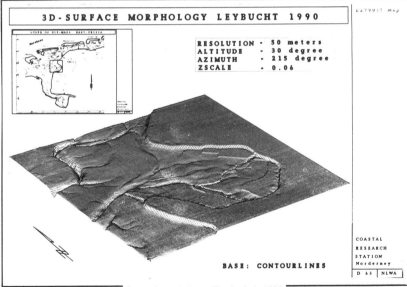

Figure 2 State of morphology of Leybucht in 1990 with new dyke 'Leyhörn'

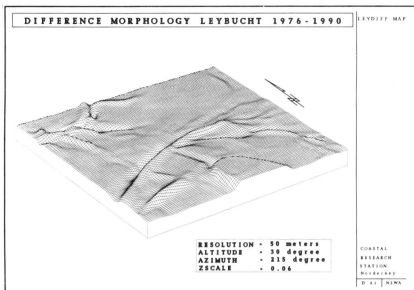

Figure 3 Difference of the states of morphology in 1976 and 1990 showed as 3D - surface and calculated from section above

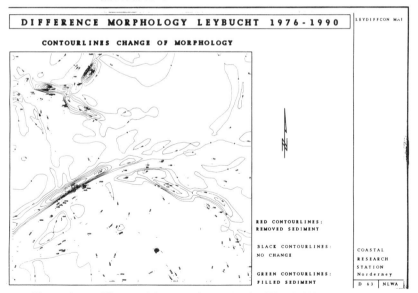

Figure 4 Difference of the states of morphology in 1976 and 1990 showed as contourlines and calculated from 3D - surface left

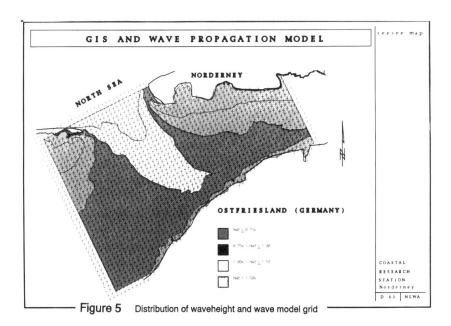

Figure 5 Distribution of waveheight and wave model grid

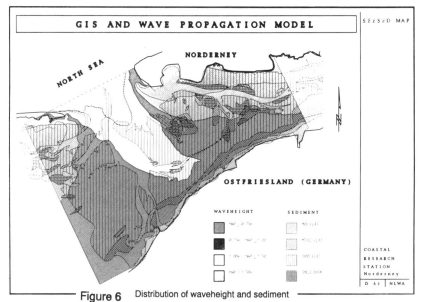

Figure 6 Distribution of waveheight and sediment

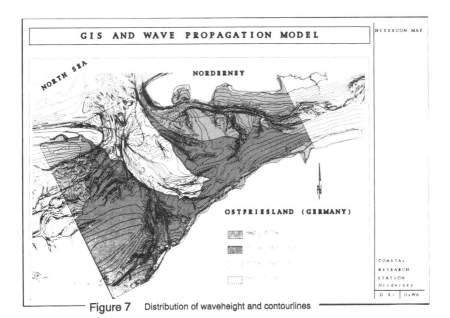

Figure 7 Distribution of waveheight and contourlines

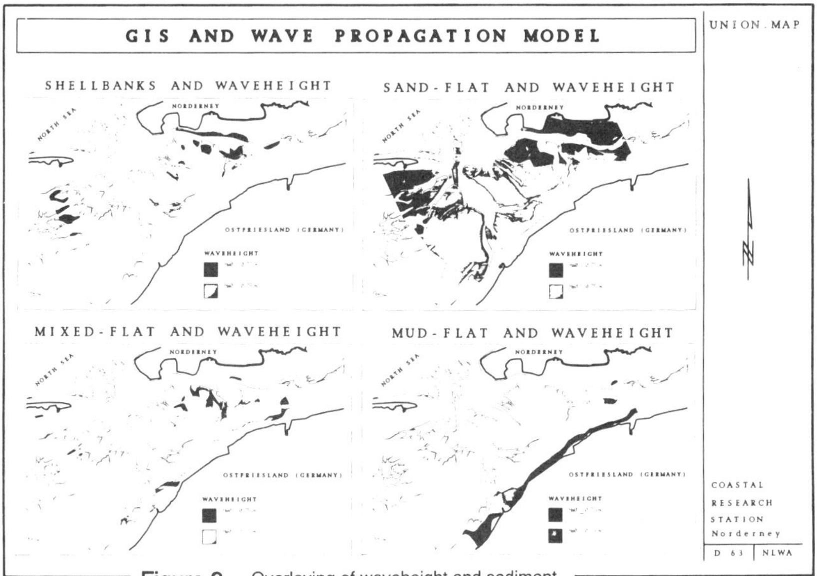

Figure 8 Overlaying of waveheight and sediment

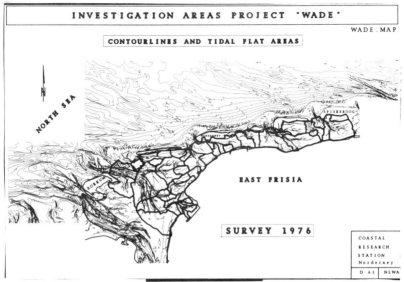

Figure 9   General survey tidal inlets " Ostfriesische Inseln "

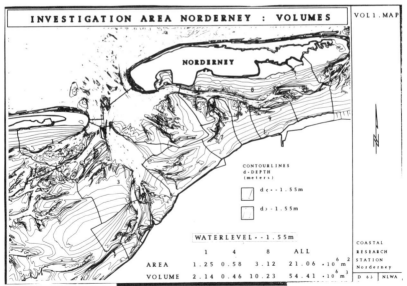

Figure 10   Inlet " Norderney " and computed volume with water level -1.55 m [NN]

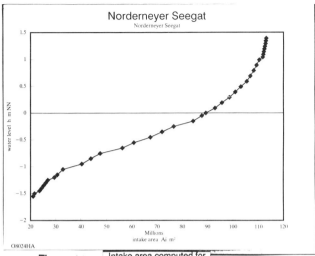

Figure 11 — Intake area computed for different water levels

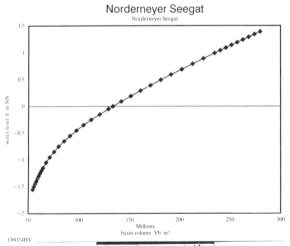

Figure 12 — Basin volume computed for different water levels

References

(1) Proceedings of the European ARC/INFO User Conference
Rotterdam 1991

(2) EGIS '92, Third European Conference and Exhibition on Geographical Information Systems
Munich 1992

(3) Proceedings of the International ARC/INFO User Conference
Palm Springs 1991

(4) Rhind, D., and H. Mounsey: "Understanding GIS"
Taylor and Francis, London 1989

(5) Geographical Information Systems, A New Frontier
Proceedings of the International Symposium on Spatial Data
Zürich 1984

(6) Coppock, J.T., and K.E. Anderson: "International Journal of Geographical Information Systems (IJGIS)"
Taylor and Francis, London 1987

(7) Jahresbericht der Forschungsstelle Küste, Norderney 1982

(8) Holthuijsen, L.H., Booij, N., and Herbers, T.H.C., 1988:
"A prediction model for stationary, short-crested waves in shallow water with ambient currents"
Coastal Engineering, Vol. 12, pp. 23 - 54

SHORELINE CHANGE SIMULATION AND EQUILIBRIUM SHORE–ARC ANALYSIS OF EMBAYED MUDDY COAST

Qiu Jianli, Department of Geography
Hangzhou University, Hangzhou 881224, P.R.China

ABSTRACT

In this paper one–line theory for simulating shoreline changes and equilibrium shore–arc theory are initially applied to the shoreline form and dynamic studies of embayed muddy coasts. The applications are based upon some characteristics of the coasts in the study area. A suspended load transport rate replaced the equation of the littoral sand transport rate on a sandy beach and is uniformly used both in the simulation of shoreline changes and the analysis of shore–arc. Assuming the suspended load that makes a net contribution to the shoreline dynamic states for a long time is carried into the bay from the left and right sides by longshore currents, which are caused by the waves along two main wave directions, and taking a profile thickness that is far larger than the experiential value in a sandy coast model, the shoreline dynamic states of Yeyashan area, Zhoushan Island, are simulated. Then the equilibrium shore–arc expressions of this shore section and Beilun area are deduced. The theoretical shoreline forms are coincident with actual shoreline configurations. The conclusion on the shore dynamics is also identical with those from other studies. It showed the effectiveness of the theoretical methods presented and provided quantitatively geomorphological data for exploitation of these shorelines.

INTRODUCTION

The shoreline form is related to wind and wave vectors acted on the shore, and the geometrical forms of equilibrium or stable shore–arcs which generally are present in nature have in close relationships with wave vectors. These have been confirmed by numerous researches on sandy coasts (King , 1972; Komar, 1976; Komar, 1983; Bakker, 1970; LeBlond, 1972; Price *et al.*, 1972; Muir Wood *et al.*, 1981; Silverster, 1974; Xue *et al*, 1980). At present, one–line theory model developed for predicting shoreline changes has come into actual uses (*e.g.* Hanson, 1989) . For muddy coasts, however, the whole and unified theoretical formulas which can be used both for simulating the shoreline changes and determining the equilibrium shore–arc forms have not been reported up to now.

In fact, the basic assumptions in the available shoreline simulating method and shore–arc theory can be applied to any coast that has been a dynamic equilibrium condition in the profile. Moreover, the shoreline configurations of many muddy coasts, their deformation and relation with the wave conditions are coincident to a considerable degree with the predicted results from shoreline evolution theory for sandy coasts (e.g.Qiu and Feng, 1987) . Therefore, it should have active methodological significance to explore the applicability of the available theory to muddy coast study. Basing on these considerations and some differences between muddy and sandy coasts, this paper properly revises the present shoreline change model so that it can be applied to embayed muddy coast, and describes the results from the modified model taking the shorelines of Yeyashan and Beilun areas as examples.

METHODS

The fundamental assumptions of the one—line model that the bottom profile does not change in time and the bottom moves only in a certain limiting depth are suitable for a muddy coast which the profile has reached equilibrium state. And so the continuity relationship in the model is. For this reason, we only need to change the sand transport equation in order to use the one—line model for simulating shoreline changes of a muddy coast.

For reflecting the average conditions of the factors that change with tides, the line where the shore and the annual mean sea level meet may be as a simulating shoreline. This really assumes that the net contribution of longshore transports under several tidal level conditions takes place under the mean sea level state over a long period of time. This assumption is rational because the profile adjusts itself in whole profile thickness and shoreline change of a shoreline cell only depends on the volume change of sediment within this cell width.

In preliminary modeling to shoreline changes only longshore sediment transport can be taken into account. Generally, the longshore sediment movement close to a shore is chiefly caused by the wave—induced longshore currents. This phenomenon has been also confirmed through the analysis of dynamic conditions in the embayed muddy coasts adjoined tidal channels in the eastern Zhejiang region (Qiu and Feng, 1987), although the local wave activetives are not strang. However, for muddy coasts the sediment is mainly drifted in suspended pattern, thus the original sand transport equation is no longer applicable. As initial simulating to a muddy coast, we use a suspended load drift equation:

$$Q = \frac{S \overline{V}_l A}{a'} \tag{1}$$

where A is the total cross—sectional area of the nearshore region depended on breaking wave conditions and the bottom slope; a' is the pore space factor; S is the average volume concentration of suspended sediment and $\overline{V}_l$ is the longshore current velocity. Placing the longshore current relationship of Komar (1976) into equation (1), we have

$$Q = Q_o \sin\alpha_b \cos\alpha_b \tag{2}$$

where Q_o is a longshore suspended load transport amplitude and α_b is the angle of breaking wave crests to the shoreline. In order to reflect the mean conditions of the local dynamics and sediments, the annual mean values of the measured sediment content of the water and the wave data are used in the transport equation.

Based on the same theoretical assumptions and using equation (2), an equilibrium shore—arc formula can be deduced:

$$X = \frac{Q_o}{K}(\sin\alpha_b - \frac{2}{3}\sin^3\alpha_b)$$

$$(0 < \alpha_b < 45°)$$

$$Y = \frac{Q_o}{K}(\cos\alpha_b - \frac{2}{3}\cos^3\alpha_b - \frac{1}{3}) \tag{3}$$

where K is the change rate of longshore suspended load transport along the shoreline; X and Y are longshore and offshore coordinates, respectively ; the coordinate origin is placed at $\alpha_b = 0$ on the shoreline; and y axis is parallel to the ray of breaking wave. It may be seen that the equation (3) approximately is representative of a circular arc in the range of assuming α_b values, the radius of which mainly depends on Q_0 / K ratio. If saying Q_0 represents the dynamic and sediment conditions , then K the topographic condition .Using the contrast of the both, it will be possible to determine the shoreline curvature and to explain the shoreline dynamic state. But one should explain the shoreline curvature not only from Q_0 because the variation of K is a function of the shoreline orientation and the wave direction.When comparing the shore-arcs of different regions or those in the same region but different times and analyzing the effect of wave in some direction on the shoreline, one must remember that the selection of coordinate origin should be same as above. Additional description of computer modeling of shorelines can be found in Komar (1983) .

RESULTS

1.Numerical simulation of the shoreline changes in Yeyashan area

Yeyashan embayment is located in the southwest of Zhoushan Island, Zhejiang Province (Figure 1) .The average trend of the shoreline is 324 degrees.The width of intertidal zone outside of the present seawall is 148 meters , the offshore slope is about 1.3 percent and the median diameter of the tidal flat sediment is about 14.3μm .From the topographic surveying data , the change amplitude of the bottom profile has been very small, and there is a stable point of it.This strictly conforms to the basic assumption of one-line model .

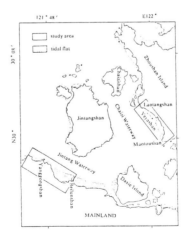

Figure 1   Locations of the study sections
(the present seawall is not marked)

The directional distribution of the wave factors (Figure 2) indicates that there is no a obvious unidirectional longshore current in this shore section. The wave action in the average direction thus is only considered.We take the annual mean wave height Hrms= 0.2 meters, wave period T = 2.2 seconds, $S = 3.4 \times 10^{-4}$ and $a' = 0.57$ from the local measurements.The depth of profile closure D = 16 meters (profile study period is 25 years) is used, which is far

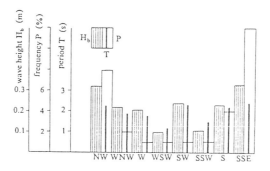

Figure 2 Typical data of the breaking waves

		tidal flat width (m)	147	105	109	172	140	147	133	119	130	144	158	70	175	207	210	200	161	158	119	148	
		total (m)	27.9	6.3	1.9	23.6	16.6	10.1	0	6.7	7.9	30.4	40.0	0	16.2	30.7	18.3	17.1	21.0	32.6	29.3	17.7	averages (m)
		mean (m)	1.39	0.32	0.10	1.18	0.83	0.51	0	0.33	0.40	1.52	2.00	0	0.81	1.53	0.92	0.86	1.05	1.63	1.46	0.89	
simulated time (yr.)	20	annual shoreline advance (m)	1.20	0.46	0.30	0.89	0.79	0.46	0	0.30	0.65	1.30	1.35	0	0.76	1.19	1.03	0.95	1.12	1.47	1.48	0.83	mean (m)
	16		1.26	0.42	0.21	0.98	0.82	0.47	0	0.30	0.59	1.39	1.56	0	0.80	1.29	1.01	0.91	1.10	1.51	1.48	0.85	
	12		1.34	0.36	0.10	1.09	0.84	0.49	0	0.30	0.50	1.49	1.81	0	0.83	1.41	0.98	0.87	1.08	1.58	1.48	0.87	
	8		1.44	0.30	0	1.24	0.85	0.52	0	0.32	0.35	1.58	2.13	0	0.84	1.59	0.91	0.83	1.04	1.66	1.46	0.90	
	4		1.55	0.20	0	1.43	0.83	0.54	0	0.38	0.15	1.67	2.53	0	0.82	1.34	0.79	0.79	0.98	1.77	1.44	0.93	

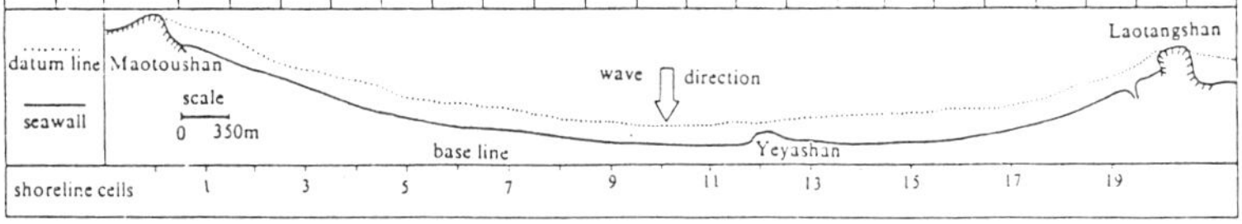

Figure 3 Simulated results of the shoreline changes along the Yeyashan area

$(\Delta t = 1 \quad Day, \ \Delta x = 350m)$

larger than that of many sandy shore models and may be related to that the material is finer and tidal effects are more serious for the muddy shore. Based on SSE and NW waves, respectively, and the local shoreline orientations, the left and right boundary's suspended load amounts into the simulating section are calculated. In addition, it is assumed that the original shoreline can not retreat due to it just being a firm seawall. The simulated results for a total period of 20 years are shown in Figure 3. It can be seen that the average advance rate is only 0.89m *per annum* and is reducing at a rate of 1 cm *per annum*. The seventh cell shoreline may change into slow advance in the 21st year, the stability of Yeyashan cell shoreline could be maintained still for a few ten years at least.

2. Equilibrium shore-arc analyses

The wave directions in the study shore section change with seasons, and the total time of shore-arc development is greatly longer than the replacement period of wave directions. Therefore, the waves in different directions can be considered to act on the shoreline simultaneously, and just for this reason the shore-arc which approximates to or fully is a symmetric form can be developed.

We can take a coordinate system, the origin of which is located at the head of the bay and y axis is positive in the offshore direction. This means that the wave crests are parallel to the shoreline at the head of the bay and α_b increases towards both sides of the bay. Using Q_o, equation (3) and the actual shoreline, K value can be calculated. If a shoreline has been a state of equilibrium, its K should approximately be a fixed value. For the shoreline of Yeyashan area, we take the depth datum line as the natural curved shoreline for the purpose of avoiding the affection of artificial seawall. The caculated results indicate that the mean $K = 5.56$, thus the equilibrium shoreline expression of Yeyashan area is deduced:

$$X = 13045.47 \ (sin\alpha_b - \frac{2}{3}sin^3\alpha_b)$$

$$Y = 13045.47 \ (cos\alpha_b - \frac{2}{3}cos^3\alpha_b - \frac{1}{3}) \qquad (\alpha_{b\ max} = 17°) \qquad (4)$$

The comparison between equation (4) and the nature is shown in Figure 4a. The theoretical arc length is 7295 meters and the actual length is 7280 meters. The convenient index of curvature (King, 1972) $c/p = 11.1$ (c is the chord length; p is the perpendicular to the shore at the mid-point of the chord), it also is close to a possible index ($c/p = 15$) that shows the equilibrium character of the beaches. These results as well as the shoreline change data obtained above indicate that the shoreline of this area has basically reached an equilibrium state and been quite stable.

Because all waves in any direction tend to balance with the shoreline, it is clear that we can use the equilibrium shore-arc theory to analyze the effect of some given waves on the shoreline. As illustrated in Figure 4 a (the thick spot line), the waves from the west would increase the shoreline curvature and in turn probably cause obvious erosion near the Yeyashan. This is more important for the shore protection.

Assuming that Q_o of this area in 1885 was roughly same as the present, K in equation (3) is calculated ($K = 9.89$) for the shoreline at that time. This means that the advance rate of the shoreline has reduced by 44 per cent since then. The decreasing rate is about 1.3cm *per annum*. The data about the shoreline dynamic states obtained from the historical shoreline research and the tidal flat sedimentation measurement are very close to the results above.

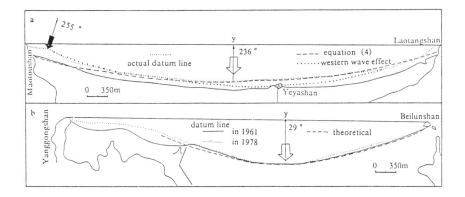

Figure 4 Theoretical and natural shore—arcs

For the purpose of analyzing the shoreline dynamic difference resulted from the topographic conditions, a shore—arc to the west of Beilunshan which both dynamic and sediment conditions are close to those of Yeyashan area is taken as a contrast. It is assumed that Q_o is constant, then K (Beilun area) = 15.51 ($\alpha_{b,\ max} < 35°$). The comparison between the theoretical shore—arc and the natural shore—arc is shown in Figure 4b. The c / p index of this section only is 8.6, hence its stability is not as high as Yeyashan section. This conclusion has been confirmed by the shoreline development study (Feng and Wang, 1984). In addition, the theoretical shore—arc in Figure 4b is deduced from the depth datum line in 1961 and the shoreline development in the west end almost coincides with the theoretical prediction. To the west of Yanggongshan other shore—arc examples may be found, including one with K = 28.67 and c / p = 5.5 and another in disequilibrium. Through the contrast study on shore—arcs which develop in different embayments or at different curved forms (concave or convex) as well as in the same embayment but at different time, it will be possible to draw up some indexes which are used for determining shore—arc types and their evolutive rates.

CONCLUSIONS

1. The shoreline configuration of embayed muddy coast can be related to the characteristics of the local breaking waves. Through the proper revision, the simple one—line model and the shore—arc theory are effective for the study of muddy coasts.

2. If there is no a unidirectional longshore current for a long time, a shoreline of embayed

muddy coast may form a circular arc configuration that is more or less symmetric relatively to the normal line at the embayment top.

3. The embayment shorelines of Yeyashan and and Beilunshan regions have been in equilibrium forms, the stability of the former is better than the latter but their progression rates are all very small. These results provide important geomorphological basis for the development and protection of the shorelines.

REFERENCES

Bakker, W.T.; Klein–Breteler, E.H.J., and Roos, A., 1971. The Dynamics of a coast with a groyne system. *Proceedings of 12th Coastal Engineering Conference*, ASCE, N.Y., PP.1001−1019.

Feng, H. and Wang, Z., 1984. Development of coastal lanforms and coastline changes in Beilun Harbor. (in Chinese) . *Literary Selections of Beilun Harbor*, Science and Technology Commission of Zhejiang Province, H.Z., PP2−21.

Hanson, H., 1989. Genesis−A Generalized shoreline change numerical model. *Journal of Coastal Research*.5 (1), pp.1−27.

King, C. A. M., 1972. *Beaches and Coasts* (2ed.) . Edward Arnold Ltd, London. pp.364−379

Komar, P.D., 1976. *Beach Processes and Sedimentation*. Prentice−Hall, Englewood Cliffs, N.J., pp.168−226, pp.249−287.

Komar, P.D., 1983. *Handbook of Coastal Processes and Erosion*. CRC Press, Inc. Florida, pp.191−216.

LeBlond, P.H., 1972. On the formation of spiral beaches. *Proceedings of 13th Coastal Engineering Conference*, ASCE, N.Y., PP.1331−1344.

Muir Wood, A.W. and Fleming, C.A., 1981. *Coastal Hydraulics* (2ed.) . The Macmillan Press LTD. London, pp.128−159.

Price, W.A.; Tomlinson, K.W. and Willis, D.H., 1972. *Proceedings of 13th Coastal Engineering Conference*, ASCE, N.Y., pp.1321−1329.

Qiu, J. and Feng, Z., 1987. The Dynamic state of muddy beach in Beilun area. *Journal of Hangzhou University* (Natural Science Edition), (in Chinese), 14 (4), H. Z., pp.502−511.

Silverster, R., 1974. *Coastal Engineering*, II., ESPC. N.Y., pp.127−152.

Xue, H.; Gu, J. and Ren, R., 1980. *Coastal Dynamics*. (in Chinese), Peoples Communication Publishing House, Beijing, pp.377−406.

COASTAL AND NEARSHORE EROSION AT VEJRØ, DENMARK : COMBINED EFFEECTS OF A CHANGING WIND-CLIMATE AND NEAR-SHORE DREDGING

CHRISTIANSEN, C. (Department of Earth Sciences, University of Aarhus, Ny Munkegade Build. 520, 8000 Aarhus C., Denmark), CHRISTOFFERSEN, H. (Geoscandic A/S, Asmusvej, 1, 8530 Hjortshøj Denmark) & BINDERRUP, M. (Geological Survey of Denmark, Thoravej 8, 2400 København NV, Denmark).

ABSTRACT

Shoreline, cliff and near-shore erosion during the period 1922-1988 on the island Vejrø, Denmark has been evaluated from aerial photoes and marine surveys. During 1922-1954 shoreline erosion was strong (30 m) on the southern part of the island while the western and eastern part had accumulation (up to 40 m). During 1954-1988 shoreline erosion was pronounced on the eastern (66 m) and northern (30 m) part of the island. Cliff erosion was up to 19 m on the northern part and insignificant on the southern part of the island. In the 20 km² near-shore area the surveys showed a total gain of 2.1×10^6 m³. There was, however, a loss of 1.2×10^6 m³ in connection with the retreat of a partly submarin spit stretching out from the NW part of the island. It is concluded that the observed erosion partly is due to a changing wind-climate with higher frequencies of strong winds. It is also due to dredging activity. The activity has changed bottom topography and thereby induced new wave-refraction patterns. Steep slopes in the dredging hollows do not allow the sediment to return to the littoral zone after periods of erosion.

INTRODUCTION

The use of raw materials from the seabottom in Denmark is regulated by law (Fredningsstyrelsen, 1979). It is not allowed to extract raw materials inside a 300 m zone from the nearest coastline and it is not allowed to extract material up to 20 cm in diameter inside a 500 m zone from nearest coastline if the resulting depth is bigger than 1/50 of the distance to the coastline. In vulnerable areas further regulations can be imposed.

It is wellknown that nearshore dredging might influence the littoral sedimentbudget (Komar, 1976, 1983; Dean and Maurmeyer, 193; Bird and Schwarts, 1985 and Carter, 1988). According to Carter (1988) dredging hollows might act as sedimenttraps and thereby reduce the inshore sedimentbudget. In spite of this there are only few case studies on the effects of nearshore dredging on coastal development.

Another factor that might influence the coastal sediment budget is a change in longshore sediment transport. Changes in the dominating winddirection and/or changes in the frequencies of storms can be responsible for a change in the coastal equilibrium (Komar, 1976 and Christiansen et al., 1985). Christiansen and Bowman, 1990 showed that long-term beach and shoreface changes in the NW part of Denmark were responses to a changing wind regime. Also, according to the Bruun rule (Bruun, 1962), a rising sea-level might change the equilibrium profile and thereby induce coastal changes.

STUDY AREA

The study area covers 20 km² in the central part of Kattegat and includes the island Vejrø (Fig. 1). The island covers an area of 5×10^5 m² and consists of a clayey morainic core with a narrow beach and active cliffs on both the northern and the southern side (Binderup, 1991 a,b). Moraines are also highlying in the nearshore. In

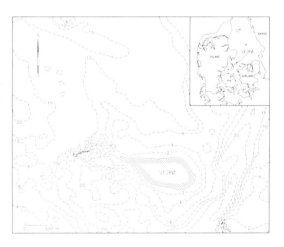

Figur 1 Locational map showing study area.

many places they are only covered by a thin carpet of residual or marine sediments. Depressions between former morainic hills have beeen filled by Late-glacial and Holocene marine sand and gravel. These later deposits have for a long time been exploited for raw materials. Raw materials in the area were mapped by the Danish Raw material Survey (Skov- og Naturstyrelsen, 1987).

The Kattegat is microtidal with a tidal range of 30-40 cm. These small sea-level variations are by far exceeded by wind induced variations, which can be up to +1.7 m above DNN (Danish Ordnance Datum) and down to -1.5 m below DNN. Westerly wids give high and easterly winds give low sea-levels. The mean sea-level is rising by 0.5-1.0 mm y^{-1} (Christiansen et al., 1985).

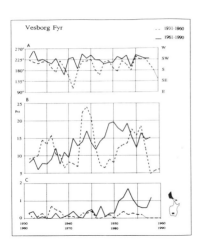

Figur 2 a) Yearly average wind direction and frequencies of wind velocities exceeding b) 6 Beaufort and c) 9 Beaufort.

There has been a change in in windclimate during the period studied. Wind exceeding 6 on the Beaufort scale have been steadily more frequent since 1960 and winds exceeding 9 on the Beaufort scale has increased drastically since 1980 (Fig. 2). As consequences of the change in windclimate both yearly max. and min. waterlevel has got more extreme during the last 15 years (Geoscan A/S, 1991).

METHODS

Coastal changes was mapped using old maps and sequential air-photoes covering a period of 66 years. Winddata from the nearlying (22 km) lighthouse Vesborg was obtained from Kristensen & Frydendahl (19-91). Wave parameters was esti-mated using the methods outlined in Christiansen et al. (1992). Waverefraction patterns and sediment mobility was found using the program REFRACT (Geoscandic,

1990,1991). Seabottom erosion and accummulation was found by comparing hydrograhic surveys from 1981 (Farvandsdirektoratet, 1981) with new detailled mapping using Trisponder system for positioning. Grain-size analysis were carried out using conventional sieving and pipette techniques.

RESULTS

SHORELINE AND CLIFF CHANGES

Coastal changes in the period from 1022 to 1954 was examined by comparisons of maps. The main conclusions are that coastal retreat was strong (20-30 m) along the southern part of the island while the eastern part of the island had coastal progradation of up to 35-40 m. Also the western part of the island had accumulation. The subaerial part of the spit was prolonged by 30 m. The wind climate in this period was characterized by yearly averaged wind directions from the SSW. The mid 1930'ies and 1940'ies had high frequencies (15-23 %) of wind velocitites > 11.4 m/s (Fig. 2).

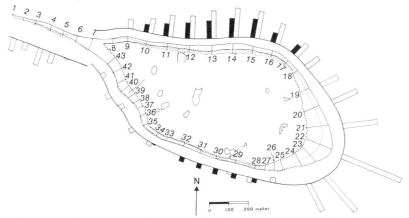

Figur 3 Histograms of coastal changes 1954-1988. Shoreline changes are in white and cliff changes in black. Numbers refer to cross-section positions.

Coastal and cliff changes in the period 1954-1988 were examined using aerial photoes. The wind climate during this period was characterized by yearly averaged wind directions from the SW. Frequencies of wind velocities > 11.4 m/s has increased steadily since 1970. The frequency of winds > 19.3 m/s has increased drastically since 1980 (Fig. 2).

Fig. 3 shows the shoreline and cliff changes 1954-1988. The spit has diminished in width by 7-17 m. The spit has been fluctuating during the years, but erosion has been accellerating since 1969-1975. This coincide with the raise in in the frequency of wind velocities > 11.4 m/s. Along the northern part of the island erosion has been between 9 and 30 m. The eastern part of the island had shoreline retreat of up to 66 m, whereas there has been only minor changes along the southern part of the island. In total there has been a loss of 50208 m^2 along the 3400 m shoreline. Accumulation has been only 645 m^2.

Cliff erosion in the period was concentrated on the northern shore. On cross-section 13-14 and 15 (see fig. 3) cliff retreat was 19 m. This averages 0.5 m/y. Totally, a loss of 193319 m^3 was found in the north facing cliffs. Maximum cliff retreat on the southern part of the island was 6 m (0.15 m/y). Here the volumetric loss was only 18864 m^3.

NEARSHORE CHANGES

By comparison of detailed old (1981) charts and the 1991 surveys a total gain of 2.1x10^6 m^3 was observed in the 20 km^2 study area. However, some areas had losses. The loss was mainly concentrated inside the 6 m depth contourline around the island Vejrø and to the west and north of the island. The loss was specially pronounced on the submarin part of the spit. Fig. 4a show the topography in 1981 and fig 4b the topography in 1991. Between these two years the spit had a loss of 1.2x10^6 m^3. Approximately half of this loss can be explained by dredging activity as seen in the dredging hollows on fig. 4b.

The new topography induces new wave refraction patterns. However, the refraction pattern is also influenced by the changing windclimate in that the observed increasing wind velocities from westerly and northerly directions result in higher waterlevels and bigger depths whereas increasing wind velocities from easterly and southernly directions result in lower waterlevels and smaller depths.

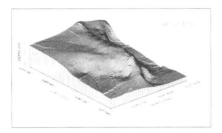 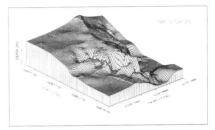

Figur 4 Nearshore topography in the spit area. a) 1981. b)1991.

By comparing wave refraction patterns based on the 1981 topography and windclimate and patterns based on the 1991 topography and windclimate it can be observed that the refraction pattern has changed and that waves now are breaking much closer to the coastline. Also, the inshore area inside which bottom sediment can be mobilized by wave activity has been considerably enlarged. Fig. 5 a,b show the the paths for sediment transport. It can be seen that before dredging activities had removed parts of the spit (Fig. 5 a) the sediment could be transported across the spit from both sides. After the dredging the sediment transport is now only from south to north across the spit. The steep slopes in the dredging hollows do not allow the sediment to return to the spit in periods with westerly winds.

DISCUSSION

Allthough the island Vejrø is situated in a low energy environment, the observed rates of cliff erosion (0.5 m y^{-1}) can be compared to rates from much more exposed environments. McGreal (1977) observed retreat rates of 0.4 m y^{-1} in the glacial sediments of Northern Ireland sea cliffs.

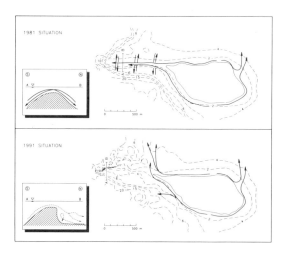

Figur 5 Paths of sediment transport. a) 1981. b) 1991.

There has been only insignificant erosion on the island Kyholm situated less than 10 km to the west of Vejrø. On the contrary, Christiansen et al. (1981) showed that this island had two periods (1930'ies and 1960'ies) of major coastal progradation in connection with dieback in eel-grass. Compared to Vejrø, the island Kyholm has much smaller fetches towards the south. However, it is supposed that the major reason for the difference in coastal developmnet between these two islands is that no near shore raw material exploitation takes place around Kyholm.

According to Bruun (1962) the erosion could, in theory, be a response to the long term rise in mean sealevel of 0.5 mm y^{-1}. However, coastal erosion due to long term sea level rise is normally hidden by effects of short term fluctuations (Carter, 1982; Christiansen et al., 1985). Also, such effects of sealevel rise can be more than balanced if the sediment budget is positive (Christiansen et al., 1985).

Effects of sand and gravel dredging may be long lasting. Geoscandic A/S (1991) estimated that with the present rate of cliff erosion it would take more than 200 years to balance the loss in the near shore area. One major reason was that the small grain sizes in the morainic clay in the cliff would be carried away in suspension and therefore not contribute to the near shore sediment budget. Valentin (1954) also shoved that only about 3% of the sediment eroded from the boulder clay cliffs of the Holderness reaches the sandy spit of Spurn Head to the south.

ACKNOWLEDGEMENTS

Thanks to the Skov -og Naturstyrelse for permission to publish the present study. This paper is also part of the GEOKAT project.

REFERENCES

Binderup, M., 1991 a: Kystændringer på Vejrø 1922-1988. Skov -og Naturstyrelsen. Unpubl. report. 28pp.

Binderup, M., 1991 b: Kystændringer på Vejrø 1954-1988. Skov -og Naturstyrelsen. Unpubl. report. 14pp.

Bird, E.C.F. & Schwartz, M.L., 1985: The world's coastlines. Van Nordstrand Rheinhold, New York.

Bruun, P., 1962: Sea-level rise as a cause of shore erosion. ASCE, Jour. Waterw. Harbour Div., 88, 37-50.

Carter, R.W.G., 1982: Recent variations in sea-level on the north and west coasts of Ireland and associated shoreline response. Proc. R. Ir. Acad., 82B, 117-187.

Carter, R.W.G., 1988: Coastal environments. An introduction to physical, ecological and cultural systems of coaslines. Academic Press.

Christiansen, C. & Bowman, D., 1990: Long-term beach and shoreface changes, NW Jutland, Denmark: Effects of a change in wind direction. In: Beukema, J.J et al. (eds.): Expected effects of climatic change on marine coastal ecosystems. Kluwer Academic Press. p. 113-122.

Christiansen, C., Christoffersen, H., Dalsgaard, J. & Nørnberg, P., 1981: Coastal and near-shore changes correlated with die-back in eel-grass (Zostera marina, L.). Sedimentary Geology, 28, 163-173.

Christiansen, C., Møller, J.T. & Nielsen, J., 1985: Fluctuation in sea-level and associated morphological response: Examples from Denmar. Eiszeitalter u. Gegenwart, 35, 89-108.

Christiansen, C., Zacharias, I. & Vang, T., 1992: Storage, redistribution and net transport of dissolved and sediment-bound nutrients, Vejle Fjord, Denmark. Hydrobiologia (in press).

Dean, R.G. & Mauermeyer, E.M., 1983: Models for beach profile response. In: Komar, P.D. (ed.): Handbook of coastal erosion and processes. CRC Press. p. 151-165.

Farvandsdirektoratet, 1981: A-blad 5550/1040. Farvandsdirektoratet.

Fredningsstyrelsen, 1979: Råstofindvinding på havbunden. Lovgivning-geologi-teknik. Fredningsstyrelsen. København.

Geoscandic A/S, 1990: The wave theory behind the Geoscandic program REFRACT. Hjortshøj. 4pp.

Geoscandic A/S, 1991: Vejrø NV-rev. Skov -og Naturstyrelsen. Unpubl. report. 20pp.

Komar, P.D., 1976:Beach processes and sedimentation. Prentice Hall.

Kristensen, L. & Frydendahl, K., 1991: Danmarks vindklima fra 1870 til nu. Havforskning fra Miljøstyrelsen Nr. 2.

McGreal, W.S., 1977: Retreat of cliff coastline in the Kilkeel area of county Down. Queens University of Belfast (unpubl. thesis).

Skov -og Naturstyrelsen, 1987: Havbundsundersøgelser: Samsø-nordøst. Miljøstyrelsen.

Valentin, H., 1954: Der Landverlust in Holderness, Ostengland, von 1852 bis 1952. Die Erde, 3, 296-315.

Short and long term variations in vertical flux and sediment accumulation rates in a semi-enclosed bay at the frontal zone between the North Sea and the Baltic Sea.

Lund-Hansen[1], L.C., Floderus[2], S., Pejrup[3], M., Valeur[3], J. and A. Jensen[4].

[1]Institute of Geology, Aarhus University, 8000 Aarhus C, DK, Denmark. [2]Department of Physical Geography, Uppsala University, Box 554, S-751 22, Sweden. [3]Institute of Geography, Copenhagen University, 1350 Copenhagen, DK, Denmark. [4]Danish Hydraulic Institute, Agern Alle 5, 2970 Hoersholm, DK, Denmark.

ABSTRACT
The stratification in Aarhus Bay is due to changes in salinities, as waters originating in both the North Sea/Skagerrak and the Baltic Sea are recognised in the bay. The vertical flux of suspended particulate matter in Aarhus Bay is inhibited by stratification, but increases in periods of high wind speeds. Measurements of vertical flux in periods of inflow of high saline waters increased the vertical flux due to high concentrations of suspended matter in the water. Vertical fluxes have been measured by means of sediment traps as averages over 1-2 weeks, and have been compared to fluxes measured pr day by a multitrap.

INTRODUCTION
Marine sediment accumulation rates are important sedimentological parameters, which have been investigated on a broad scale with different methods and aspects. Investigations on physical properties of parameters influencing sediment accumulation rates in coastal areas, have dealt with transport of suspended particulate matter (Abraham et al., 1981; McCave, 1984), erosion of bottom sediments by currents (Lavelle, et al., 1978), and rates of resuspension (Gabrielson and Lukatelich, 1985; Demers et al., 1987). A study by Lund-Hansen and Skyum (1992) showed that the distribution of suspended particulate matter in Aarhus Bay was strongly affect-

ed by the changes in hydrographic conditions. In the present study, changes in sediment accumulation rates and vertical flux are investigated in relation to hydrographic conditions as salinity, stratification, concentration of suspended particulate matter, and current speeds.

STUDY AREA

The study area is the semi-enclosed Aarhus Bay, in the southwest part of Kattegat (Fig. 1). Hydrographic measurements and measurements of sediment accumulation rates have been carried out at a single position in the bay (56° 09.10 N, 10° 19.20 E). The bottom sediments at the position consists mainly of silt and clay and contain approximately 10 % of organic material (Lund-Hansen and Skyum, 1992). The water depth is 16 m. The bay has a tidal range of 40 cm, and covers an area of 250 km^2, and contains a volume of water of $5*10^3$ km^3 (DHI, 1980). The hydrographic front between the highly saline North Sea/Skagerrak waters (32-34) and the less saline Baltic water (15-20)

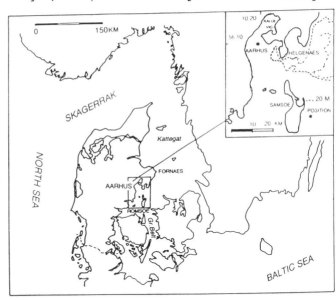

Figure 1. Location of study area and position.

is placed in the Kattegat (The Belt Project, 1981). The stratification in Aarhus Bay is fully dominated by the changes in salinity, due to prevailing outflow from the Baltic Sea or inflow originating in the North Sea/Skagerrak. Due to changes in hydraulic and meteorological conditions in the Kattegat and the Baltic Sea area, related south or northwards movement of the front are relatively frequent (The Belt Project, 1981).

METHODS

Two different systems of sediment traps were deployed at the position. One trap system consisted of a string trap with traps measuring the vertical flux at 9 different levels above the seabed (0.3, 0.5, 0.8, 1.0, 2.0, 4.0, 6.0, 8.0 and 10.0 m). The string trap consisted of polyethylene pipes, closed at the lower end and mounted on a wire, at each level above the seabed. The wire was held in position by a concrete block at the bottom and a buoyancy buoy near the surface. The pipes have an inner diameter of 8.0 cm and an inner length of 40.0 cm - an aspect ratio of 5, which is considered to be the optimal ratio for measuring vertical flux (Hargrave and Burns, 1979). A more detailed description of the string trap system is given by Valeur et al., (1992). The string trap was recovered by divers at 16 days intervals on average (range 6-39 days). The second trap - the multitrap - measured pr day the vertical flux 1.5 mab by automatic shifting of 21 cans - one for each day. The multitrap was recovered by taking it on board the ship and replacing the cans. Formalin was used in the cans to prevent decomposition. The multitrap is described in detail by Honjo and Doherty (1988). In the laboratory samples were filtered using preweighed (0.1 mg) Millipore filters (0.45 µm) and dried for 24 hours at $60°$ C. By use of the Pb-210 method for age determinations (Robbins, 1987) sediment accumulation rates have been determined. Continuous (10 min) measurements of salinity, temperature, current direction and speed were obtained by a mooring Aanderaa instrument (RCM 7), placed 1.0 m above the bottom at the position. The string trap system was deployed between February 1990 and May 1991, and the multitrap was deployed between November 1990 and June 1991.

RESULTS AND DISCUSSION

Advection between Aarhus Bay and Kattegat

Fig. 2 shows the relation between bottom current direction and current speed at the position, compiled for the period between 11 November 1990 and 23 May 1993. Two characteristic directions with relatively high current speeds are recognised - $140°$ and

330°. Compared to the geometry of Aarhus Bay (Fig. 1) 140° is towards Helgenaes from the position, whereas 330° is towards the position from Helgenaes. However, rela-

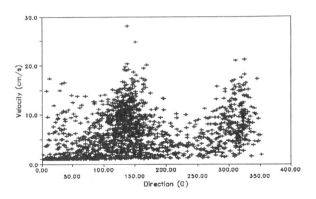

Fig. 2 Current direction and speed.

tively high current speeds between 0° and 90° are also recognised, and are probably due to a coastal current originating in the southern part of Aarhus Bay (Fig. 1). The current directions between 100° and 200° are the most frequent directions. This is due to a low response time of winds between east and south. These wind directions produce a barotrophic field between the Baltic Sea and the North Sea/Skagerrak as waters in the Kattegat are moved northwards (The Belt Project, 1981), and thereby lowering the water level in Aarhus Bay. Southward movements of the front in the Kattegat are due to persistent and relatively strong winds between southwest and west which develop a barotrophic field from Kattegat towards the Baltic Sea (Skyum and Lund-Hansen, 1992). Bottom current directions about 330° in Aarhus Bay are correlated with increasing salinities (Lund-Hansen et. al, 1992). To the south Aarhus Bay is enclosed by islands and shallow water areas (Fig. 1), and corresponding to Fig. 2, it is seen that the exchange of bottom water between Aarhus Bay and Kattegat, mainly takes place at the deep-water entrance at Helgenaes.

Vertical flux and sediment accumulation rates

For comparing the string and multitrap system a period (19 February - 23 May 1991) of concurrent measurements have been studied. The average vertical fluxes (g $m^{-2}*day^{-1}$) measured by the string trap system at the levels 1.0, 2.0, and 10.0 mab and the vertical flux (g m^{-2}*day^{-1}) measured pr day with the multitrap 1.5 mab are shown in Fig. 3. It also shows salinity, and current speed (cm s^{-1}). The sediment accu-

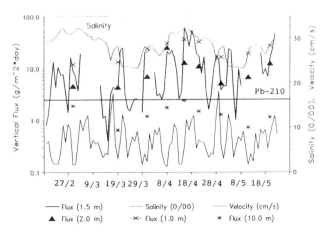

Figure 3. Flux, salinity and current speed.

mulation rate measured by the Pb-210 method, which is a mean of 50-75 years (Robbins, 1978), is depicted as a constant in Fig. 3. In general, it is seen that the vertical flux increases towards the seabed, with a mean vertical flux of 1.4 (g m^{-2}*day^{-1}) at 10.0 mab, 10.0 (g m^{-2}*day^{-1}) at 2.0 m, and 23.2 (g m^{-2}*day^{-1}) at 1.0 mab. The mean vertical flux 1.5 mab measured by the multitrap is 11.7 (g m^{-2}*day^{-1}). Valeur et al. (1992) showed that the vertical distribution of the measured fluxes at the 9 levels increases exponentially towards the seabed. The sediment accumulation rate at the position as given by the Pb-210 method is 2.5 (g m^{-2}*day^{-1}), which is relatively low compared to the average vertical flux at 1.0, 1.5 and 2.0 mab, whereas the vertical flux at the 10.0 mab is lower. Sediment traps measure a gross sediment accumulation rate, which is here denoted the vertical flux, in opposite to the sediment accumulation rate given by Pb-210, which is a net sediment accumulation rate (Lund-Hansen, 1991; Valeur et al., 1992). The difference between the vertical flux and the rate determined by the Pb-210 method is primarily due to resuspension of bottom material by surface waves and current shear stress (Valeur et al., 1992, Lund-Hansen et al., 1992). Increased vertical flux measured at 1.5 mab by the multitrap are associated with periods of maximum current speeds (16 March and 29 April), and the increases in current speed are associated with a significant decrease in the salinity between 15 and 17 March and between 26 and 30 April (Fig. 3), and thus waters flowing out of the Aarhus Bay (Fig.

2). These changes in salinity are associated with movements of the hydrographic front in the Kattegat (Lund-Hansen et al., 1992).

Increased vertical fluxes measured both by the string and the multitrap are associated with the increase in salinity in the period between 26 March and 8 May (Fig. 3). Relatively low current speeds are recognised in the period, whereas high current speeds are recognised in connection with the outflow about 28 April. A vertical flux measured by sediment traps is given by $C*w_s$ (g $m^{-2}*day^{-1}$) - a concentration C (kg m^{-3}) of suspended particulate matter and a settling velocity w_s (m s^{-1}) of the particles (Dyer, 1986). The variations in the vertical flux in this period are thus considered to be due to a higher concentration of suspended particulate matter of the inflowing waters. Provided that particle size is approximately constant. In combination with another study between 9 and 11 April 1991, at the same position, the concentration of suspended particulate matter of the bottom waters was 10-15 mg l^{-1} and 4-5 mg l^{-1} of the surface waters (Lund-Hansen and Skyum, 1992). On average, the concentration of suspended particulate matter in Aarhus Bay is approximately 1-2 mg l^{-1} (Lund-Hansen et al., 1992), which shows that the inflow enhance the vertical flux in the Aarhus Bay. However, between 19 February and 16 March, waters with high salinities are also recognised, but no similar related changes in vertical flux are recognised (Fig. 3). This inflow of high salinity waters (29-31) started about 2 February, whereas the outflow started about 1 March (Fig. 3). It is supposed that suspended particles during the stagnant period had settled out prior to this study period, which explains that increased vertical fluxes are not associated with increasing salinities. The in- and outflow between 26 March and 8 May is due to a barotrophic forced southward movement of the frontal zone in the Kattegat (Lund-Hansen and Skyum, 1992). The current speeds during the inflow are relatively low, whereas high current speeds are recognised during the outflow. The average vertical flux measured pr day at 1.5 mab follows the vertical fluxes measured in the string traps, although there are significant variations at 1.5 mab, which, to some degree, are positively correlated with the variations in the current speed

(Fig. 3). However, during the inflow relatively low current speeds are recognised, which also indicated, that the inflow waters contain comparatively higher concentrations of suspended particulate matter.

Density difference and wind energy
The average wind energy transfer to the water (E_w) have been calculated by : $E_w = kC_d\rho_a U^3$, where k is a wind factor ($1.8*10^{-2}$), C_d a drag coefficient ($1.1*10^{-3}$), ρ_a the air density (1.2 kg m^{-3}), and U the average wind velocity (m s^{-1})(Kullenberg, 1977). Data on wind speeds were obtained from Fornaes Lighthouse (Fig. 1). The mean density difference ($\Delta\sigma_t$) between surface and bottom waters in Aarhus Bay has been calculated on basis of weekly CTD-measurements at the position. E_w is shown against ($\Delta\sigma_t$) in each of the 28 periods of trap deployment, and approximate time of year is also shown.

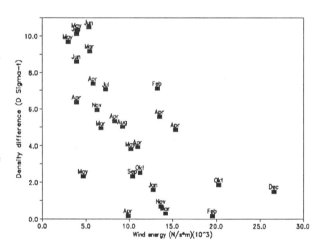

It is seen that Figure 4. Wind energy and density difference.
a strong negative correlation (r = - 0.69) exist between wind energy transfer (E_w) and density difference ($\Delta\sigma_t$), which indicates that density difference, i.e. the stratification of the water is roughly a function of wind energy transfer. However, a seasonal component is recognised with relatively high wind speeds during winter and season, and low during summer and spring. The hydrography of Aarhus Bay is primarily controlled by the changes in the transitional zone between the North Sea and the Baltic Sea, where outflow of low salinity waters (15-

20) from the Baltic Sea occurs in spring followed by compensating return flow of high saline waters (28-30) during summer (Lund-Hansen et al., 1992). Both outflow and return flow enhance the density difference between surface and bottom waters, which thus shows that part of the negative correlation (Fig. 4) is not due to erosion of the layer of stratification in the Aarhus Bay, but merely due to the hydrographic changes in the transitional zone. However, periods of E_w above $1.2*10^{-4}$ (J $s^{-1}*m^{-2}$) and low density differences are supposed to represent periods where the wind energy have broken down the layer of stratification by entrainment (Lund-Hansen et al., 1992a). In these periods of no or weak stratification the vertical flux of suspended matter in the water column is enhanced, and eventually resuspension takes place.

The present study showed that during inflow of waters with high salinities, originating in the North Sea/Skagerrak, vertical fluxes in the Aarhus Bay increased due to higher concentrations of suspended particulate matter in these waters. These waters contain about 3-6 (mg l^{-1})(Eisma and Kalf, 1984) and reaches Aarhus Bay as a bottom current resuspending material from the bottom during transport. In periods when Aarhus Bay is dominated by low saline waters originating in the Baltic Sea Lund-Hansen et al. (1992) showed that the vertical flux was low, which is supposed to be due to a general lower concentration of suspended particulate matter in the Baltic Sea. Yemelyanov (1975) give a range of 0.4-12 with a mean of 3.0 mg $^{-1}$ af concentrations. And also that the waters from the Baltic Sea reaches Aarhus Bay as a surface current, without any bottom contact.

CONCLUSION

The stratification in Aarhus Bay is due to changes in salinity, and Lund-Hansen et al.(1992a), showed that the vertical flux in general was inhibited by density difference between surface and the bottom waters, i.e. stratification. The present study showed that in periods of inflow of water origination in the North Sea/Skagerrak the vertical flux was enhanced due to higher concentrations of suspended particulate matter in the water. Variations in vertical flux were recognised by the string trap, whereas much more detailed changes in the vertical flux were recorded by the multitrap, where increased vertical fluxes could be positively correlated to periods of high current speeds.

ACKNOWLEDGEMENTS

This study was a part of the HAV-90 program initiated by the Danish Agency of Environmental Protection, who are acknowledged for providing the funds for the investigations.

REFERENCES

Abraham, G., Adrian, G. van O. and Verboom, G.K., 1981. In: Transport models for inland and coastal water. Proceeding of a symposium on predictive ability. (Ed.) H.B. Fischer.

Demers, S., Therriault, J.C., Bourget, E. and Bah, A., 1987. Resuspension in the shallow sublittoral zone of a macrotidal estuarine environment: Wind influence. Limnol. Oceanogra., 32, 327-339.

DHI, 1980. Danish Hydraulic Institute - Aarhus Bay, 145 pp. (In Danish).

Dyer, K., 1986. Coastal and Estuarine Sediment Dynamics. John Wiley & Sons.

Gabrielson, J.O. and Lukatelich, R.J., 1985. Wind-related resuspension of sediments in the Peel-Harvey estuarine system. Estuarine and Coastal Shelf Science, 20, 135-145.

Hargrave, B.T. and Burns, N.M., 1979. Assessment of sediment trap collection efficiency. Limnol. Oceanogr.,24, 1124-1136.

Honjo, S. and Doherty, K.W., 1988. Large-aperture time-series sediment traps; design objectives, construction and application. Deep-Sea Research, 35, 487-498.

Kullenberg, G., 1977. Entrainment velocity in natural stratified vertical shear flow. Estuarine and Coastal Marine Science, 5, 329-338.

Lavelle, J.W., Young, R.A., Swift, D.J.P. and Clarke, T.L., 1978. Near-bottom sediment concentration and fluid velocity measurements on the inner continental shelf, New York. Journal of Geophysical Research, 83, 6052-6062.

Lund-Hansen, L.C., 1991. Sedimentation and sediment accumulation rates in a low-energy embayment. Journal of Coastal Research, 4, 969-980.

Lund-Hansen, L.C. and Skyum, P., 1992. Changes in hydrography and suspended particulate matter during a barotrophic forced inflow. Oceanologica Acta (In print).

Lund-Hansen, L.C., Pejrup, M., Valeur, J. and Jensen, A., 1992. Variations in sedimentation rates at the North Sea - Baltic Sea transitional zone. Oceanologica Acta (Submit).

Lund-Hansen, L.C., Pejrup, M., Valeur, J. and Jensen, A., 1992a. Vertical particle flux related to wind energy transfer and stratification in a marine Bay. (Paper submitted at the ECSA 22 Symposium, Plymouth).

McCave, I.N., 1984. Erosion, transport and deposition of fine-grained marine sediments. In: Stow, D.A.V. and Piper D.J.W. (Geol. Soc. London Spec. Publ. 15): Fine-grained sediments: deepwater processes and facies.

Robbins, J.A., 1978. Geochemical and geophysical applications of radioactive lead. In: Nriagu (ed.). The Geochemistry of lead in the environment. Elsevier/North-Holland Biomedical Press.

Skyum, P. and Lund-Hansen, L.C., 1992. Barotrophic and baroclinic forcing of a semi-enclosed bay at the frontal zone between the North Sea and the Baltic Sea. Geografiske Annaler (In print).

The Belt Project, 1981. Evaluation of the physical, chemical, and biological measurements. The National Agency of Environmental Protection, Denmark.

Valeur, J., Pejrup, M. and Jensen, A., 1992. Vertical sediment fluxes, measured by sediment traps. Limnol. Oceanogra. (Submit).

West, J.R., Oduyemi, K.O.K., Bale A.J., and Morris, A.W., 1990. The field measurement of sediment transport parameters in estuaries. Estaurine, Coastal and Shelf Science, 30, 167-183.

Yemelyanov, Y.M. and O.S. Pustelnikov, 1975. Compositions of river and marine forms of suspended matter in the Baltic. Geochemistry International, 12, 195-208.

Measurement of morphological changes by means of a sand-surface-meter

by

J. Straube

Research Engineer at the Leichtweiss-Institut für Wasserbau,
TU Braunschweig, Beethovenstr. 51a, 3300 Braunschweig, Germany

ABSTRACT

Quasi-continuous measurements of bed surface variations by means of a digital sand-surface-meter and measurements of related sea state parameters have been carried out in the Baltic Sea. Results of seasonal behaviour and storm events are presented with respect to the temporal and spatial variaton of morphological changes and the influence of wave-induced velocities on the change of the sea bed.

INTRODUCTION

Measurements of waves and wave-induced currents as well as measurements of morphological and sedimentological changes have been taken at the Baltic Sea near SCHÖNBERG/HOLSTEIN since 1990 in an interdisciplinary research project by the AMT FÜR LAND-UND WASSERWIRTSCHAFT KIEL (ALW), the GEOLOGISCH-PALÄONTOLOGISCHES INSTITUT DER UNIVERSITÄT KIEL (GPI) and the LEICHTWEISS-INSTITUT FÜR WASSERBAU DER TECHNISCHEN UNIVERSITÄT BRAUNSCHWEIG (LWI).

At the measuring site KALIFORNIEN/SCHÖNBERGER STRAND, four steel piles which are located in different water depths along a shore normal profile carry the wave gauges and the current meters. Fig. 1 shows the measuring location KALIFORNIEN with the position of the measuring piles. On Fig. 2, the related underwater profile together with the location of the sand surface meters is illustrated.

The underwater profile of the measuring site is surveyed extensively every 4 to 6 weeks and additional surveying will be done e.g. after severe storm conditions. Despite of the short time intervals between the surveys, these measurements may lead to wrong interpretations of the morphological changes if the time-scale of the change is much shorter than the time-scale of the survey (MANZENRIEDER, SNIPPE 1991). For this reason, quasi-continuous measurements with digital sand surface meters that operate at a time interval of 1 or 2 hours offer additional information about instantaneous sea bed changes and may complete and support the results of the surveys.

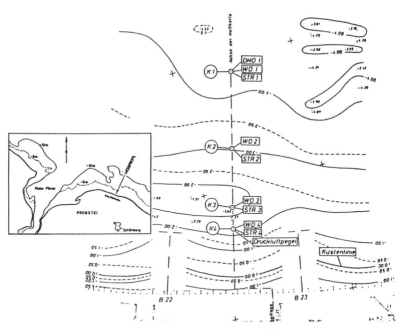

Fig. 1: Location of the measuring site KALIFORNIEN/SCHÖNBERGER STRAND

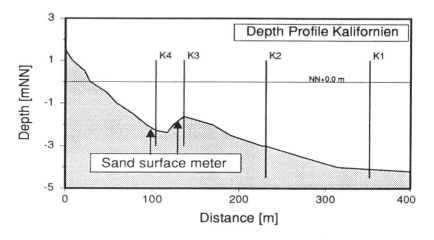

Fig. 2: Depth profile of measuring site KALIFORNIEN with location of measuring piles and sand surface meters

MEASUREMENTS

The digital sand surface meter consists of a rectangular steel rod with 61 optical sensors emitting infrared light. The distance between the sensors is selective either 1 cm or 2.5 cm, which results in a total measuring length of 60 cm or 1.5 m respectively. Depending upon the desired accuracy and upon the expected bed level changes, the appropriate sand surface meter can be chosen. Sensor control, battery supply and memory board in watertight housing make the system independent of any wiring and thus a universal use can be achieved. The system is washed into the ground by divers up to the middle of the measuring area and does not need additional fixation e.g. in a sandy soil. With a measuring interval of 1 hour, the system can operate up to 42 days. At the end of the measuring time and after removing the sand surface meter the data on the memory board can be read out by the aid of a Personal Computer at the landstation. The system can then be reinstalled immediately after a change of batteries and some easy maintenance work e.g. cleaning of the system has been done.

TEMPORAL AND SPATIAL VARIATION OF THE BED SURFACE

From July to the end of September 1991, several measurements of morphological changes were taken close to the measuring piles K3 and K4 (see Fig. 2). The data were analyzed and the development of the sea bed over time was plotted. From the plots covering the whole measuring interval, general tendencies in the morphological development can be seen as well as periods of extreme changes, that correspond with intensified sea state conditions.

In Fig. 3 (top), the development of the sea bed at the locations K3 (near the bar crest) and K4 (landward of the trough region) is shown for the measuring interval 16.09.1991 - 30.09.1991. Fig. 3 (bottom) shows for the period of maximum morphological activity (26.09.-27.09.) the measured offshore and longshore velocities and the related sea bed changes.

It can be seen from Fig. 3 (top) that the sea bed near the station K3 remained upon a constant level most of the time during the measuring interval, except for small changes of approximately 5 cm. During the storm event at the 26./27.09. with a maximum significant wave height of 1.2 m , a decrease of 15 cm was observed. After the end of the storm, sedimentation started until the former level of the sea bed was reached. Similar observations have been made for storm events during other measuring intervals.

The record near the station K4 showed a quite similar behaviour, even though at the beginning of the measuring interval a decrease of approximately 2 cm was observed. The difference between the stations K3 and K4 may be explained by the different resolution of the two used sand surface meters (K3: $\Delta s=2.5$ cm; K4: $\Delta s=1.0$ cm). The erosion during the storm reached 10 cm. After the storm, the sedimenta-

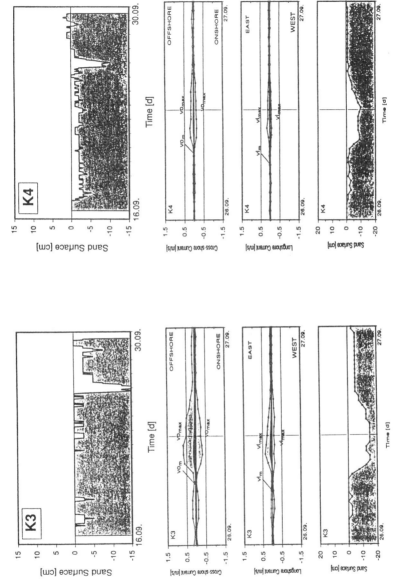

Fig. 3: Temporal and spatial variation of bed surface and measured wave-induced velocities from storm event 26./27.09.1991, KALIFORNIEN

tion up to the original sea bed level was faster at the station K4.

Both stations showed that the original sea bed level after the storm was reached again only 20 hours after the beginning of the erosion. Morphological changes in such a short time cannot be observed by surveying. This shows the great importance of completing the data from surveys by measurements with sand surface meters.

From Fig. 3 (bottom), it can be seen that with decreasing longshore and offshore velocities (measured 20-50 cm above the ground) the sedimentation started until the original level was reached again. At the beginning of the erosion process, maximum velocities between 20 and 50 cm/s were measured. Even though velocities at station K4 were almost half of the size as at station K3, the erosion rates of 15 cm (K3) and 10 cm (K4) did not differ in the same magnitude.

MORPHOLOGICAL CHANGES AND WAVE-INDUCED VELOCITIES

Fig. 4 shows wave data and observed morphological changes of a storm event that was recorded at the end of June, 1991. After the beginning of the storm, an erosion of 15 cm over 7 hours was observed. During this time, the maximum wave-heights and wave-induced velocities occurred. After the storm peak, erosion stopped and sedimentation with a rate of 40 cm over 36 hours followed. While the erosion process was quite similar to the one observed at the 26./27.09., the resedimentation did not stop when the original bed level was reached again. Comparing the time series of wave-induced velocities it can be seen that along with the sedimentation of up to 40 cm stronger maximum velocities in cross- and longshore direction were measured, whereas at the 26./27.09. velocities decreased a few hours after the storm peak.

Several explanations for the strong sedimentation have been discussed. Besides of the possibility of sedimentation of material from either onshore or offshore, the migration of the longshore bar could have been another possible cause. A first analysis of morphological data from the GPI seems to support this assumption, but further investigation has to be done before a final judgement can be made.

Besides of the still unknown origin of the sedimented material, it is of interest to analyze the relationship between the morphological changes and the related wave-induced velocities. The time series of the sand surface from Fig. 4 can after some "smoothing" be divided into several characteristical sections with specific erosion/-sedimentation rates in [cm/h]. For each section, the mean bed surface change rate and the mean velocities in cross-shore and longshore direction can be calculated. Table 1 shows the results for the erosion/sedimentation period from the 28.06., 2:00 a.m. to the 29.06., 10:00 p.m..

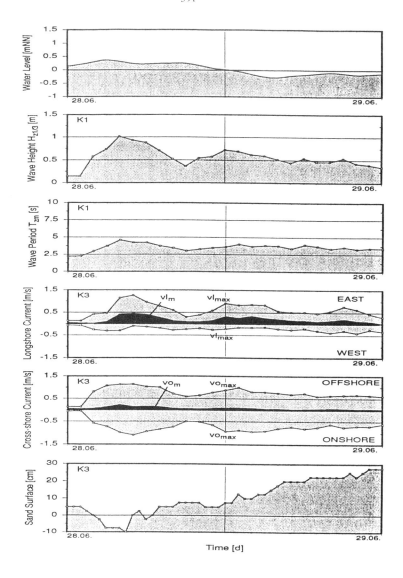

Fig. 4: Beach erosion/sedimentation and wave-data from storm event 28./29.06.1991

Time	Bed level change ds/dt [cm/h]	mean maximum velocities [m/s]		mean velocities [m/s]		
		v_{lmax} [+)]	v_{omax} [+)]	v_{lm} [+)]	v_{om} [+)]	v_{result}
02:00-09:00	-2.1	0.68	1.03	0.17	0.14	0.22
09:00-14:00	3.0	0.48	0.70	0.11	0.07	0.13
14:00-23:00	0.0	0.57	0.72	0.13	0.05	0.14
23:00-09:00	1.3	0.80	0.94	0.27	0.08	0.28
09:00-22:00	0.6	1.00	1.06	0.37	0.14	0.40

+) v_l = longshore, v_o = on-/offshore

Table 1: Mean bed level change and mean velocities for characteristic erosion/sedimentation sections, storm event 28./29.06., Station K3,

The results from Table 1 are plotted in Fig. 5. High values of maximum and mean velocities correspond well with high erosion/sedimentation rates as could be expected. The magnitude of the maximum velocities, which are mainly responsible for the initiation of sediment movement, was 0.70 to 1.00 m/s for the highest erosion/-sedimentation rates.

The magnitude of the maximum and mean cross-shore velocities is similar for erosion and for sedimentation. The "symmetrical" development of the cross-shore velocities with high maximum values and low mean velocities indicates that these were probably dominated by the oscillating wave-orbital velocities. The longshore velocities were mainly directed eastward (see Fig. 4 and 5) and mean velocities during sedimentation were much higher than during erosion. The resulting mean velocity is mainly influenced by the longshore mean velocity.

CONCLUSION

Digital sand surface meter with a data sampling interval of 1 or 2 hours give valuable information about instantaneous morphological changes that complete and support the results of surveys. Additional informations from closely located wavegauges and current meters allow investigations about the magnitude of wave-induced velocities at the beginning of sediment movement and may help to understand the process of sediment movement in the coastal zone.

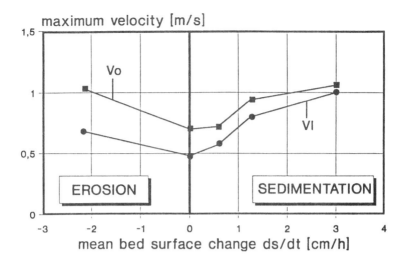
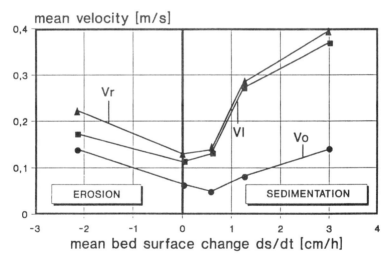

Fig. 5: Maximum and mean wave-induced velocities during erosion and sedimentation period of storm event 28./29.06.1991, Station K3, KALIFORNIEN

ACKNOWLEDGEMENTS

The research project "VORSTRANDDYNAMIK EINER TIDEFREIEN KÜSTE" is supported by the BUNDESMINISTERIUM FÜR FORSCHUNG UND TECHNOLOGIE and the KURATORIUM FÜR FORSCHUNG IM KÜSTENINGENIEURWESEN. Field assistance was provided by the TAUCHGRUPPE DER UNIVERSITÄT KIEL.

REFERENCES

ALW KIEL et al.	1991	Forschungsvorhaben Vorstranddynamik einer tidefreien Küste, 1. Zwischenbericht, Zeitraum 01.07.-31.12.1990
MANZENRIEDER, H. SNIPPE, B.	1991	Kontinuierliche Beobachtungen der Gewässersohle mit einem Sandstandpegel, Die Küste, Heft 52, 1991
LERCH, D. MANZENRIEDER, H. WITTE, H.-H.	1990	Ein digitaler Sandpegel zur permanenten Beobachtung der Gewässersohle, HANSA, Nr. 1/2 1990

EROSION/ACCRETION SYSTEM OF SOUTH BALTIC COAST DURING THE LAST 100 YEARS

Elżbieta Zawadzka-Kahlau - Instytut Morski
Długi Targ 41/42
80-830 Gdańsk
Poland

1. INTRODUCTION

Changes of the Polish coastline during three periods: 1875 - 1979, 1960 - 1983 and 1971 - 1983 were measured by means of cartometric investigations. The determined in this way changes of position of coastline (and for 1960 - 1983 of dune foot) allowed defining accumulative and erosive forms of various length along the Polish seacoast.

The alternating sequence of erosive bays and accumulative protrusions along the coast shows that there exists an erosion-accretion system, and allows to define the trends of change and scale of variability of distinguished elements of the system during the last 100 years. Obtained results allow also to forecast the changes of given elements of the coastal system.

2. CONDITIONS OF COASTAL CHANGE

Observed contemporary processes of Polish coast change are connected with geological/geomorphological preconditions and present geographical conditions of the region.

On the area formed by the North-Polish icing, dune forms started building up after the first phase of the Litorina transgression. Intense development of dunes took place during a temporary lowering of sea level which occured after transgression maximum, abt. 5 - 6 thous. years B.P. (Rosa 1967). The South Baltic spits were formed in diverse dynamic situations. In conditions of proceeding transgression, deposition took place in the southern part of the Vistula Spit and in the region west of Piasnica. When sea level became stabilized, dune complexes at the Rega outlet, along the stretch from Ustka to Lake Sarbsko spit, and east of Piasnica were formed. The root of the Vistula Spit was built in conditions of depositionary regression (Rosa 1980).

Predispositions of the substratum had significant influence on the development of dune forms. Most of the spits are located in areas of ice-marginal valleys, some of them in post-lob depressions (Rosa 1980). The substratum of the dune coasts is built of Pleistocene (moraine tills and fluvioglacial material) and Holocene (peat and lake formations) material.

Holocene dunes take up 80% of the length of the contemporary Polish coastline, the remaining 20% are cliffs formed mainly by Pleistocene formations (Fig. 1). Untill the middle of the XIX-th century lithodynamic processes along this coast were of nearly natural character. The main obstacles to littoral transport were formed by river outlets and elevations of the older substratum, which also constituted sources of material for the transport.

In general, the main characteristics of natural preconditions and hydrometeorological factors have been nearly constant in the times directly preceding the analized period and during the last 100 years.

Variable wind fields cause a differentiation of current fields. E.g. when wind blows from the south, in the coastal zone of the Koszalin Bay currents are directed south-west, while east of Jaroslawiec they are directed eastwards. With north winds, currents diverge west- and eastwards in the region of the Leba elevation (Jankowski 1979).

Mean storm wave heights on the South Baltic are within 1.7 - 1.9 m. In storm conditions, at wind speed over 15m/s, H5% is 4 - 6 m. Storm surge at the coast reaches 1.5 - 2.0 m above mean sea level (Jednorał 1987).

East of the Koszalin Bay, the net longshore transport in the shallow water zone is eastwards, along the west part of the coast the net transport is westwards (Fig. 1) (Cieslak 1987).

Hydrodynamic processes occur on the background of long-term processes connected with the neotectonics and eustatics of the region. The west and east part of the Polish coast is characterized by a slight downward movement of abt. - 1 mm/year. The central part, in the region of Łeba elevation, rises at a rate of abt. +0.2 mm/year (Wyrzykowski 1971).

Contemporary transgression, which is superposed on the regionally differentiated neotectonic displacements, has been observed along the Polish coastline-since the XIX-th century. Increase of relative sea level is, depending on area, between 1.8 and 2.3 mm/year (Dziadziuszko 1987). In the central part of the coast, the relative sea level was rather stabilized.

Danger to land appeared first along the west coast, where the longest stretches (of even 25 km length), characterized in the period 1837 - 1875 by a prevalence of erosion, were observed (Hartnack 1926). Along the east coast dominated accumulative processes, which however occured parallel with erosion (Hartnack 1926). Probably domination of earosion along the west coast was related to earlier influence of sea level rise, at simultaneous downward movement of the land and the rather thin layer of littoral cover sediments.

Since the middle of the XIX-th century, natural conditions became disturbed by antropogenic factors. Along the west coast small ports were built. Endangered stretches of coastline became protected by hydrotechnic structures. In effect of increasing erosion, in the period 1870 - 1970, gradually along about 20% of the coastline hydrotechnic protective structures were built. 80% of these is located on the west coast.

During the investigated period 1875 - 1979, erosion occurred along 60% of the coastline. For the period 1960 - 1983 the percentage of eroded stretches grew to 69%, and to 72% for period 1971 - 1983. In the last period more intense erosion is observed along the east coast, where it proceeds at a 20% higher rate than on the west coast. Results of measurements of coastal changes along the west coast indicate that erosion has become stabilized at average rate of -0.4 m/year, among others due to construction of a second line of defence at dune/cliff foot. The coastal protection system limited the amounts of freely moving sediments. This could be one of the reasons for the increased intensity of erosion in the eastern region. Additionally, during the last decades in this region a quicker rise of relative sea level is observed.

Technical ingerence into the natural course of coastal processes - often realized on an emergency basis - has, at least at some places, resulted in consolidating lithodynamic trends negative to coastal stability (Semrau 1980).

3. INVESTIGATED MATERIAL AND METHODOLOGY

For evaluating the changes of coastline, basic 1;25000 German topographic maps from the period 1875 - 1922 and Polish topographic maps from the period 1908 - 1979 were used. The dates of the German maps were verified thanks to the kindness of the Staadsbibliothek Preussischer Kulturbesitz in Berlin. For the period 1960 - 1983, coastal changes were taken from 1:2500 plans of the technical belt of the Maritime Offices, from 1:5000 photointerpretation topographic plans, and from 1:2000 and 1:5000 cadastral plans.

For measurements on 1:25000 maps, a 500 m (Neiczew, Pelczar, Szeliga, Ziolkowski 1989), and on 1:2500 plans a 100 m quantization step was assumed. On the maps, within each of the scales, a common reference base was established. The position of investigated morphological lines was detrmined relative to this base.

Changes of coastline position in thus defined points and in known times are the basis for calculating the rates of displacement (in m/year) of the investigated lines. The coastline change curve for 1875 - 1979 (Fig. 2) was verified by measurements on plans of the technical belt and through analysis of survey measurements from the period 1958 - 1978 (Zawadzka 1989).

4. CHANGES OF COASTLINE

Analysis of the 100-year coastline change curve allowed to distinguish three classes of length of erosion or accretion stretches of coastline. 200 stretches with various direction of changes were found, in that in class I (length ≥ 4 km) - 44.6%, in class II (2 - 3.5 km) - 37.5%, and in class III (0.5 - 2.0 km) - 19.7% of length of analized coastline. During the last 100 years erosion prevailed on 60.8% of the analized coastline, accretion - on 39.2% (Fig. 2).

The relative position of characteristic forms and erosion/accretion bistructures allowed to distinguish, within the class I and II stretches: a single erosion bay or accumulative protrusion, pair of erosion bays or accumulative protrusions, pairs of divided stretches with different trends, and pairs of connected stretches with different trends (Fig. 3). The main feauture arising from the analysis is the alternating position of erosive and accumulative stretches along the coastline, and the lack of stretches showing clearly visible lithodynamic equilibrium.

Erosion processes, dominating in the 100-year balance of changes, resulted in a retreat of the Polish coastline at a mean rate of - 0.08 m/year. The erosion processes were accompanied by a partial restoration of the land area equal to abt. 71% of the area lost on erosive stretches. Erosion processes, which are of dominating importance in the balance of coastal changes, took place within class I stretches. Within them, in the hundred-year period, average rate of erosion was -0.53m/year, and of accretion was 0.59 m/year. Theses stretches occured with no clear regularity, but maintained a fairly stable position on the coastline.

The position on the coastline of stretches of different length classes depends on complex land-based conditions, such as the general orientation of the coastline, geological build, reserves of transportable (sandy) material, on variable hydrodynamic processes, and on the influence of natural and man-made obstacles to the transport of sediments.

The quasi-wavy character of the coastline change curve is related to the migration of coastal sediments and to the influence of the changes of wave and current energy in the coastal zone.

In the period 1960 - 1983, erosion intensified within stretches of all classes. This is indicated by the growth of the rate of coastline retreat, which for the period 1960 - 1983 is -0.39 m/year, and for 1971 - 1983 is -0.81 m/year. Restoration of land area reduced to 22.5% for 1960 - 1983, and for 1971 - 1983 to 18.9%. Rate of erosion on class I stretches determined for the hundred-year period was during 1960 - 1983 -0.46 m/year, and during 1971 - 1983 -0.79 m/year. In years 1960 - 1983 accretion along class I stretches decreased to 0.05 m/year. These values may be considered as the basic indicators of growing intensity of erosion of the Polish coast and of increasing danger to the coastal zone and lying behind it low land areas.

The alternating position of stretches with opposite trends and the relatively stable location of main morphological forms are the basic features of the coastal system. This allowed to analize the variability of structurally higher levels of the system, that is the erosion/accretion bistructures formed by stretches of length class I and II. For 75% of the analized bistructures, the eroded part is east of the

part showing accretion. 28 erosion/accretion bistructures, distinguished for the hundred-year period, covered 50% of the coastline.

Within the erosion/accretion bistructures the rate of coastline displacement was -0.01 m/year in the hundred year period, -0.30 m/year for 1960 - 1983, and for 1971 - 1983 it was -0.99 m/year. The increasing rate of destruction of the coast within the bistructures points to a negative balance of sediments in the coastal zone.

The similar course of lithodynamic processes along parts of coastline or within the bistructures was the basis for defining higher elements of the coastal system, i.e. the lithodynamic units. Hydro- and lithodynamic phenomena with higher uniformity indicate the approximate boundaries of lithodynamic units. Groups of units with different rates of change, are characterized in the hundred year period and the 1960 - 1983 period by alternating directions of coastal change. During 1960 - 1983 erosive processes intensified in the group of units characterized by a prevalence of accretion in the hundred-year period. This is connected with the general intensification of coastal erosion, which gradually covered nearly all morphodynamic regions distinguished along the Polish coastline (Fig.4).

In the Bay of Gdansk, where accretion prevailed in the hundred-year period, in 1960 - 1983 erosion intensified in two morphodynamic regions. Morphodynamic regions of the open sea coast with prevailing accretion in the hundred-year period (Pomeranian Bay), variable erosion (Jaroslawiec - Miedzyzdroje) or relative equilibrium (Wladyslawowo - Jaroslawiec) passed into a phase of dominating erosion. Accretion occuring in the east part of the Pomeranian Bay became weaker.

The erosion/accretion system of the Polish coastline, which in the hundred-year period is characterized by a small rate of change, in years 1960 - 1983 became dominated by erosive processes proceeding at medium rate, and in years 1971 - 1983 - at high rate. A negative balance of sediments of the coastal system is observed also at the foot of dunes and cliffs (Musielak 1988, Subotowicz 1982, Zawadzka 1989).

5. SUMMARY

Assuming that geomorphological conditions and hydromoteorological influences before the investigated period and during the last hundred years remained stable, it is considered that the most probable reason for the intensification of erosive processes is the gradual sea level rise. Data on coastline change correlated with average change of sea level during the last hundred years (a mean rise of 10.6 cm) show that increase of sea level influences the rate of erosion occuring within the distinguished elements of the system, and along the whole Polish coastline.

The growing level of the erosive base acts on all elements of the erosion/accretion system, which is characterized by alternating position of stretches with erosion and accretion, connecting into erosion/accretion bistructures and groups of lithodynamic units. Lithodynamic processes occuring at lower levels of the system influence the course of changes within morphodynamic regions and the whole coastal system. Sea level change is the cause of irretrievable loss of a part of the sandy sediments from the nearshore zone due to increased transverse transport.

During the last hundred years, basic elements retain stable trends; erosive stretches are characterized by a growing rate at which this process proceeds, on accumulative stretches accretion becomes weaker. Elements of the system which showed equilibrium or weak accretion in the hundred-year period, passed to medium or strong erosion. (Fig. 5).

The erosion/accretion system was formed in conditions of natural course of processes. Defining the elements of the system and determining the direction of changes in the last hundred years, the measure of which is the rate of movement of the coastline, allows to forecast the variability of these elements and of the whole system in the antropogenic phase of erosive transgression development.

LITERATURE

Cieślak, A. et al. (1987): Dynamic trends of coast and nearshore bottom along the Polish coastline (in Polish). Instytut Morski, WW-IM 2783 Gdansk.

Dziadziuszko, Z., T. Jednoral (1987): Changes of sea level along the Polish coastline (in Polish). Studia i Materialy Oceanologiczne 52, Dynamika morza 6, Gdansk.

Jednorał, T. (1987): Waves on the South Baltic (in Polish). Studia i Materialy Oceanologiczne 52, Dynamika morza 6, Gdańsk.

Jankowski, A. (1979): Diagnostic model of wind and density currents on the Baltic Sea (in Polish). Doctor dissertation, Instytut Geofizyki Warszawa.

Hartnack, W. (1926): Die Kuste Hinterpommers... Greiswald.

Kielmann J.: Grundlagen und Anwendung eines numerischer Modells der Geschisten Ostsee. Ber. Inst. Meereskunde Universitat Kiel 87a, b.

Musielak, S. P. Madejski (1988): Dynamics of seacoast on stretch Swinoujscie - Dzwirzyno (in Polish). OPGK Szczecin.

Nejczew, P., M. Pelczar, J. Szeliga, J. Ziolkowski (1989): Investigations of changes of Polish coastline on XIX-th century maps (in Polish). Polish Cartographic Conference, Gdansk.

Rosa, B. (1967): Morphologic analysis of South Baltic bottom (in Polish). Torun.

Rosa, B. (1980): On spits of the South Baltic coast (in Polish). Peribalticum 1, Ossolineum, Gdansk.

Semrau, I. (1980): Lithodynamic processes along selected stretches of the Polish coastline. Processes of erosion in the regions of ports and coastal protection structures (in Polish). Report No. 141, Instytut Morski, Gdańsk.

Subotowicz, W. (1982): Lithodynamics of Polish cliff coasts (in Polish). Ossolineum, Gdańsk.

Wyrzykowski, T. (1971): Map of recent absolute velocities of vertical movements of erth's crust surface on the territory of Poland. Polish Academy of Sciences, Committee of Geodesy, Warsaw.

Zawadzka-Kahlau, E. (1989): Morphodynamics of selected dune coasts (in Polish). Studia i Materialy Oceanologiczne 55, Brzeg Morski, Ossolineum, Gdansk.

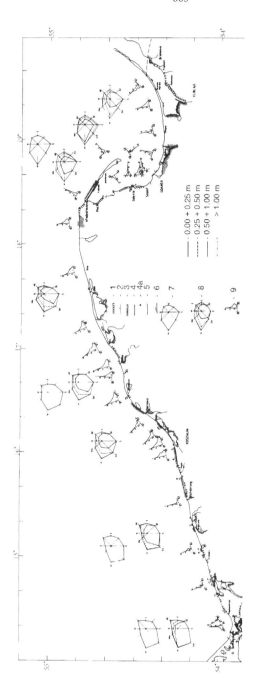

Fig.1 Meteorologic, hydrometeorologic, lithodynamic and hydrotechnic conditions of South Baltic coast.

1 - cliff; 2 - port breakwater; 3 - groynes; 4 - revetment, light type; 4a - revetment, heavy type; 5 - dyke; 6 - areas in danger of flooding by storm surges; 7 - wind frequency (in %); 8 - mean wave height frequency (in %) in 5 m water depth; 9 - littoral transport direction and intensity

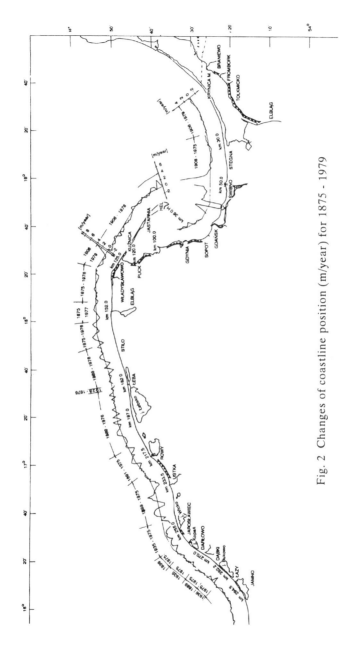

Fig. 2 Changes of coastline position (m/year) for 1875 - 1979

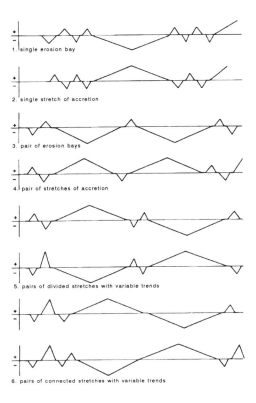

Fig. 3 Diagrams of typical erosion-accretion bistructures on Polish coast.

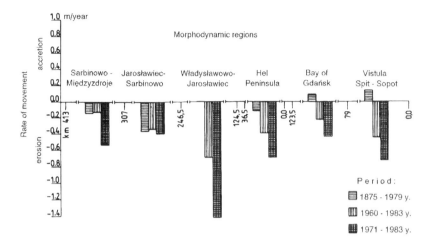

Fig. 4 Average rates of coastline movement for the last hundred years.

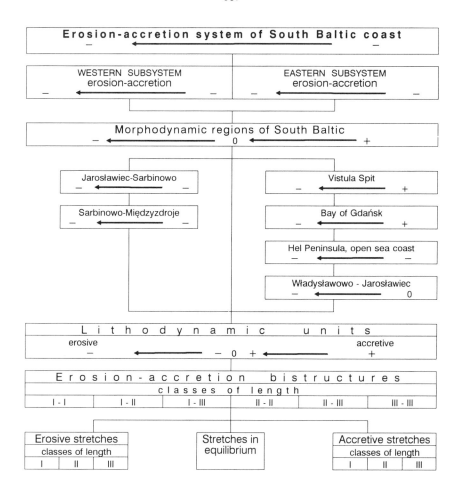

Fig. 5 Diagram of structure of erosion-accretion system of South Baltic coast in period 1875 - 1983

Near-shore sedimentation and pollution influence on the ecosystem of the Baltic Sea in Lithuanian boundaries

Dr. Olegas Pustelnikovas, Institute of Geography, Vilnius
(Lithuania)

Analysis of sedimentation processes of the Baltic Sea during the recent epoch of the industrial activities and agriculture developement becomes more actual. It is especially important in the near-shore zone with high recreational potential and maximum human influence.

Problems of sedimentation to a higher or lower degree used to be solved only on the basis of analysis of bottom sediments as its final product (1-4). In the last decades the direction of suspended matter investigations (5-8) and determination rates of sedimentation by methods 14 C and excessive 210 Pb (3,10-14) has developed. The use of the last method is epecially important in determining human influence on sedimentation in deep sea zone, where different rates of sedimentation (0,63-2,5 mm/year) were determined (11).

The near-shore sedimentation includes the zone of active waves up to 20-25 m depth. It represents the final differentiation phase substance of the abrasional-erosional, eolian and alluvial origin, enriched by material of human activities. It is necessary to know the total budget of sedimentary material in the Baltic Sea in order to calculate its components. Moreover, the last one is considered as a result of dynamic geochemical bilance between the annual inflow and outflow of this material (15). Among the total inflow (outflow) (217,3 mln t of sedimentary material) the main part the bilance is determined by biogenous material, participation of which in the process of sedimentation is connected with its rotation and like security (activity) of the organisms, for only 1,6% of the material from the total income is accumulated on the bottom. Obvious predominance of the terrigenous process must be pointed out and the analysis of the rates of near-shore sedimentation studied. It can be rendered among rather difficult tasks.

The analysis of the lithodynamical data of the suspended matter in the range of 2-meter nearbottom waters in the areas Lubiatowo (Poland), Sambian Peninsula (Kaliningrad district) and Nida (Lithuania) (16-19) allowed us to estimate the role of sedimentary material mobilisation on the rate of bottom erosion processes. The decisive role in these processes belongs to the geological structure of the shore, bottom and hydrodynamical stage of the sea (8).

The zone of sandy riffs (abreviation - ASR) and the other part of the off-shore zone (abreviation - OPZ) are clearly distinguished in the suspended matter distribution during small and stormy waves.

The high concentrations of suspended matter in ASR during small (waves hight up to 1,2 m) are determined by transportation of particles from the depth of 7-8 m towards the shore. The silty-pelitic material prevails in its composition. During the stormy waving (hight of waves up to 2,8 (and more) m) the material is washed out from the riffs by compensation seaward currents. In this case the obvious predominance of sandy particles is observed. The distribution of suspended matter in ASR is determined by the weighing and wash-out of bottom sediments, mixing of the water column, inflowing material of beaches and eolic origin and transportation of the sediments by longshore currents. The speed of such currents often surpasses during storms 30-50 (average 10-15) cm/s (20).

The wave influence on the bottom in the OPZ decreases, the entered suspended material is tranported further under the residual influence of compensational and drift currents. It is testified by increasing quantity of glauconite and silicaceous particles - as indices of the wash-out of the old quarternary and Mezo-Cenozoic deposits of the shore area. The coarse particles of suspended matter, as a rule, accumulate at the depth of 15-20 (and more) m. Deeper a sharp increase of the quantity of silty and pelitic material is observed. The quantity of sandy paticles sharply decreases from 75% in ASR (depth 6 m) to 5% - at the depth 26 m of OPZ profile.

The changes of current directions (20) in the seaward slope of ASR (depth 6 m, 13 and 17,5 m on 180° in many cases determine a rather stable character of distribution of the total quantity of suspended matter and the amount of sandy fraction in them. This is noticeably reflected on the mineral contents and granulometrical types of suspended drifts. Those fluctuations, apparently, promote the wash-out of bottom at the depth of 10 m, where the thickness of a sandy fractions layer increased, as well as the quantity of goodgrained particles of quarz, fieldspars, glauconite etc. The regular decrease of zones sites in which the sandy fractions and certain minerals are still clearly transported upon the waves influence can be observed up the depth 8-9 m. Deeper more dispersed-silty-pelitic material and in its composition - a number of light minerals - quarz, mica, biogenic carbonats etc., are transported only by longshore currents even with no high velocity. Those are incorporated into a superficial layer of the currents and enter the deep-water zone of sedimentation.

Our date shows that hydro-, lithodynamical processes in the ASR are by 4-7 times more intensive than in the OPZ. During stormy time in comparison with calm waves period the quantity of suspended matter in near-shore zone increases by 9-14 times. Thus, ASR is the main source of suspended matter, while OPZ predominantly the area of its transportation and partial sedimentation, especially during stormy period of the sea.

Having looked through the regularities of suspended matter dynamics in the near-shore zone we shall try to make bilance calculations with certain assumptions. First of all, 60 % of the time is assumed as the time of small waves. The concentration of suspended matter in the 2-meter nearbottom layer these conditions makes up to 5 mg/l (in the surface layer - 2 mg/l), and with calculation of its increase during stormy time, we get annual average concentration of suspended matter of 20 mg/l. In case the relation between sandy-gravel and silty-pelitic fractions of suspended matter is changed we get 50% of the last one of the total quantity in ASR, but it is assumed to take 60% of the whole suspended matter in the 2-meter thick nearbottom layer. Finally this material is carried outside the limits of the shore sedimentation zone.

The area of the bottom, which occupies the shore sedimentation zone, is about 150.000 km². It annually contains 10,8 mio. t of sedimentary material in 2 meter layer. In the remaining layer of the water of the shore zone 21,3 mio. t and 6,3 mio. t material of river solid discharge can be found, where as the annual quantity of sedimentary material in the shore sedimentation zone is 38,6 mio. t. Exactly in this zone 12,5 mio. t, accumulate forming the drift of the wave-field. We must take into consideration that almost 70% of this material (sandy fractions) is sedimented in ASR. The calculated indices of the income (256 t/km²) and accumulation (83 t/km²) give general evaluations of the processes of sedimentation on the bottom. Having sandy deposits in the ASR making up only 20% of the shore zone, the indice of accumulation in it is equal to 513 t/km². Then we get that the velocity of recent sedimentation in the ASR makes up 29,5 cm/1000 years and in the whole shore zone -15 cm/1000 years. It's practically 6-10 times less than the analogical velocity in the deep-water sedimentation zone of the Baltic Sea.

Analysis of the mineral composition of suspended matter shows us, that in shore zone the material of erosional and eolic origin exists and in the OPZ a significant place is taken by glauconite, carbonats etc. This date clearly prove the income of material as a result of the wash-out of the shores and the bottom of the south and the south-eastern part of the Baltic Sea, formed by Quarternary and Mezo-Cenozoic rocks. The same complex minerals typical to Paleozoic rocks we discovered on the shores of north-eastern shores of the sea (in Riga and Finnish bays). It enabled us to single out the terrigenous-mineralogical provinces of suspensions, which reflect the source of sedimentary material.

The above presented general features of sedimentary material differentiation in all shore zone, change in the Lithuanian coastal zone. These changes are connected with drifts from Nemunas-river (lagoon Kursiu Marios); in the zones of their impact the fields of coarse silt as type of deposits are developed. This influence along with geological structure and bottom relief is clear enough on the distribution of silty and pelitic fractions. Generally, near the coast sandy drifts are distributed. The presence of the stone-pebble zone and morainic deposits on the depth 3-25 m from Klaipeda to Liepaja created the deficit of drifts for its longshore flow transportation towards north. Lithuanian coasts include the areas of calm and moderate sedimentation (Melnrage-Giruliai and Nemirseta-Sventoji), abrasional-cliffy, strongly abraded and abrasional step-like, moderately abraded areas (Giruliai-Karklininkai-Nemirseta) (21). The width of beaches of the accumulated areas changes from 30-40 m to 150-170 m, layer of sandy deposits - from 0,2-0,5 to 1,5-2 m. In abrasional areas the width 10-20 m, with sandy deposits layer on the seperate "islands" of sandy-gravel deposits - only 10-20 cm.

The hurrican storm with a wind velocity up 30 m/s in 19. November 1981 endured an accident of Gibralta tanker "Globe Asimi". To the stormy sea about 17.000 t of black oil was spilled, which at first flowed inside the lagoon Kursiu Marios, then - along the coast towards the north. Detailed analysis of consequences of this wreck is given in (22, 23). So, here we can speak only about its influence on litho- and morphodynamics of the coastal zone.

The black oil-spill and its liquidation were the main reason of expectionally heavy consequences of storms on the eastern coasts of the Baltic in January 1983. The removal from the shore zone about 600.000 t of sand-black-oil mixture broke the regime of the coast, created the drifts deficit, in the longshore flow, sharply changed the shoreline. That in its turn conditioned a subsequent washing-out of beaches even same parts of accumulated areas. Generally, the coastal zone of Lithuania (partly, of Latvia, too) last about 1 mio. t of beaches and dunes sand, also material of the morainic cliffs. They were abraded and transported with stormy waves towards underwater part of the coastal slope.

The data of investigations of 1980-1989 show us, that the influence of wastes from Mazeikiai Oil Processing Factory upon the investigated ecosystem, in case careful requirements are observed, is not high on the whole. Here, however, attention must be paid to the negative influence of discharge of municipal water of Palanga town and of different objects worsening the sanitary-epidemiological stage of the Baltic near-shore. Episodic or constant pollutants of the water in this coast area are fresh water flows from lagoon Kursiu Marios. Improvement of recreational zone of Lithuania will be promoted by the construction of the 2nd line of deep sea throw out of cleaned waste water of the Mazeikiai complex and Palanga-town. Disposal of such waters into deep water zone would actually produce no influence on sedimentation processes.

Sedimentation in the lagoon Kursiu Marios as transitory basin-accumulator is by 30 times higher as in the near-shore zone of the Sea (10, 24). These huge (2,9-3,6 cm/1000 years) rates, apparently, determined decreasing of lagoon area during the last 50 years on 35 km^2, e.g. on 0,7 km^2/year.

Though the lagoon is shallow (average depth 3,8 m), but geomorphology of its bottom is rather complicated. The deepest depression of the lagoon can be found in south and central parts of the lagoon. Depths more than 3 m make up only about 50% of its area. If we consider the average rates of sedimentation as 3,2 m/1000 years, then we assume, that after 1000-1500 years the lagoon of Kursiu Marios in its recent state will cease to exist. The decrease of the depth will speed up the natural processes of eutrofication. It, of course, will lead to sharp changes of its area sizes. Apparently, the seperate near-delta lakes would occur. An example of which is now delta-lake Kroku Lanka. River Nemuas would change its course, the mouth would continue towards west-south-west. In our opinion, the first indices of such processes are observed in Morskoye (Sarkuva) settlement, where sharply with the speed up to 1,7 m/year the coastline is abrasioned. Processes of deep wash-out of Kursiu Nerija spit will start in this district. Finally, those processes would be completed after 2-3.000 years. The new lagoon-lakes such as Leba, Gardno etc. on the Polish coastal zone would be formed. This process, of course, will be very difficult. The data of the newest and recent movements of the Earth crust (25) confirmed it. Thus, date of O. Yakubovskiy by mareographic observations during the last 85 years (1886-1970) show us, that Earth crust decreased from 0,6 $\pm$ 0,3 (Liepaja) and 0,8 $\pm$ 0,3 (Klaipeda) to 1,2 $\pm$ 0,3 mm/year in

Baltijsk (Pilau) area. S.Pobedonostsev shows that Kursiu Marios lagoon decrease 1-2 mm/year. As we see, the decrease is evident on the Baltic Sea and lagoon Kursiu Marios coastline distance of only 100-200 km. Moreover, the northern part of this area decreases by 2 times slower than the southern one. As a result - the River Nemunas will change its course towards south and west (now settlement Morskoje (Sarkuva), the drainage towards north will stop. This prognosis is possible only in the case, if, as V.Gudelis thinks (26), "recent epoch of slow decrease will not with time be changed by a period of insignificant elevation". At present the role of the changes of level in the world ocean, is not clear and as we know the Baltic Sea is it constituent part.

References

1. Pratje, O., 1956. Die Bodenbedeckung der südlichen und mittleren Ostsee. - Dtsch. Hydr. Z. B.1

2. Gudelis, V., Emelyanov, E. (eds.), 1976. Geology of the Baltic Sea. - Vilnius, "Mokslas", p. 386.

3. Lisitzyn, A., Emelyanov, E. (eds.), 1984. Geological history and geochemistry of the Baltic Sea. - Moscow, "Nauka", p. 248.

4. Emelyanov, E., Lukashin, V. (eds.), 1986. Geochemistry of the sedimentary process in the Baltic Sea. - Moscow, "Nauka", p.248.

5. Pustelnikov, O., 1974. Geological characteristics of the basin and mechanism of the bottom sediments formation in the Baltic Sea (by data of the suspended matter studies). - Diss., Vilnius University, p. 22.

6. Pustelnikov, O., 1975. Particulate organic matter and its sedimentation on the bottom of the Baltic Sea. - "Okeanologija", vol. 15, 6, p. 1040-1045.

7. Pustelnikov, O., 1976. Absolutwerte der Masse an Sedimentmaterial und das Tempo rezenter Sedimentation in der Ostsee. - Beiträge zur Meereskunde, H.38, Berlin, S.81-93.

8. Pustelnikov, O., 1980. Some features of shore and deep sea sedimentation and bilance of sedimentary material in the Baltic Sea. - Thesis of 12th Baltic Oceanogr. Conf., Leningrad.

9. Kuptzov, V., Lisitzyn, A., Zeldina, B., 1982. Sedimentation rates of the Riga-Gulf bottom sediments by data of the radiocarbon method. - "Okeanologija", vol. 22, 4, p. 616-619.

10. Jankovski, H., Pustelnikov, O., Dushauskiene-Duz, R., Lijv, N., 1991. Comparison of sedimentation rates and scales of accumulation of heavy metals in the upper (0-1 m) bottom sediments layer of Finnish Bay and Kursiu Marios lagoon. - Thesis of Conf. "Ecobaltic - 92", Kaliningrad.

11. Pustelnikov, O. (ed.), 1992. Biogeochemistry of the semiclosed bays and human factor of the Baltic Sea Sedimentation. - Vilnius, "Academia" (in press).

12. Voipio, A. (ed.), 1981. The Baltic Sea. - Amsterdam-Oxford-New York, Elsevier Sci. Publ. Comp. p. 418.

13. Voipio, A., Niemisto, L., 1974. Studies on the recent sediments in the Gotland Deep. - Merentutk. Julk., 238, p. 17-32.

14. Voipio, A., Niemisto, L., 1979. Sedimentological studies and their use in pollution research. - ICES C.M. 1979/C, 46, 10 (mimeogr.).

15. Pustelnikov, O., 1982. Near-shore sedimentation in the Baltic Sea. - "Baltica", 7, Vilnius, "Mokslas", p. 137-145.

16. Antsyferov, S. et al., 1980. Distribution of suspended sediment over coastal zone of Lubiatovo. - Polish Academy of sciences, Institute of hydroengineering, Gdansk. Hydrotechnical Transactions, vol. 41, Warszawa-Poznan, p. 211-228.

17. Pustelnikov, O., 1980. Lithological characteristics of suspended sediment under various hydrodynamical conditions. - Ibidem, p. 229-242.

18. Pustelnikov, O. et al., 1977. Suspended drifts in the transit section of the South-Eastern part of the Baltic Sea coast (1. Quantitative distribution of drifts on upper part of shelf at the calm and stormy sea.) - J. of Acad. of Sciences Lithuanian, ser. B, vol. 5(102), Vilnius, p. 155-165.

19. Pustelnikov, O., Stauskaite, R., 1979. Ibidem (2. Granulometrical and mineralogical composition of the drifts at the calm and stormy sea), vol. 1 (110), p. 133-151.

20. Zeidler, R. et al., 1980. Hydrodynamical conditions in the coastal zone under "Lubiatovo-76" programme. - In "Hydrochemical Transactions", vol. 41, Warszawa-Poznan, p. 67-80.

21. Gudelis, V., Janukonis Z., 1977. Dynamical coast classification and zonation of the coastal zone of the south-eastern part of the Baltic Sea (2. Zonation of coastal zone). J. of Acad. of Science Lithuanian, ser. B, vol. 4 (101), Vilnius, p. 135-145.

22. Pustelnikov, O., Nesterova, M. (eds.), 1984. Environmental influence of black oil spill in the Baltic Sea. - Vilnius, p. 150.

23. Simonov, A. (ed.), 1990. Accident of tanker "Globe Asimi" in Klaipeda Harbour and its ecological consequences (Investigation results according to interdepartmental programme). - Moscow, Gidrometeoizdat, p. 232.

24. Gudelis, V., Pustelnikov, O. (eds.), 1983. Biogeochemistry of the Kursiu Marios Lagoon. - Vilnius, p. 149.

25. Recent movements of the Earth Crust, 1973. - Collection of articles, Tartu, p. 430.

26. Gudelis, V., 1982. Newest and recent movement of the Earth Crust in the south-eastern shore of the Baltic Sea. - "Baltica", 7, Vilnius, p. 179-186.

CLASSIFICATION AND CHARACTERIZATION OF CONTEMPORARY COASTS OF ESTONIA

Kaarel Orviku

Institute of Geology, Estonian Acadamy of Sciences, Tallin

The subdivision of the Baltic Sea coast has been carried out by a few investigators. The most complete classification of the coasts in the Baltic as a whole, belongs to V. Gudelis (1967) who has generalized the results obtained through coastal investigations in the states surrounding the Baltic Sea. In his scheme V. Gudelis presents the Baltic Sea coasts and shores. As a basis he uses the classification of shores and coasts of the world ocean acknowledged by Ionin et al. (1964). However, not all of his standpoints can be entirely agreed with. For example, the whole southern coast of the Gulf of Finland from Osussaar Island in the west up to Narva Bay he classifies as glint coast. To our mind (Orviku, K., Orviku Kaarel 1961) and considering the results of the recent studies this stretch of the coast should be subdivided into four types or subtypes which will be discussed in more detail below.

In his popular work P. Hupfer (1978) attempts to classify the coast of the Baltic Sea. Accurately enough he has classified only its northern coasts which he regards as belonging to the skaren type.

In his monography K. Orviku (1974) has dedicated a whole chapter to the classification of coasts. Below will be brought the main principles of the classification supplemented with the recent study results. Classifying the contemporary seacoasts K. Orviku proceeded mainly from the morphogenetic classification of the coasts elaborated by Ionin et al. (1964) and by Kaplin (1973). The above-named authors have subdivided all contemporary coasts, regardless of their belonging to oceans, inner seas (e.g. the Baltic) or large lakes, into three groups. As a criteria serves the main agent responsible for the formation of coastal relief forms and deposites:

I - The coasts where the effect of the waves has been negligible; the character of the coasts has been determined by the initial relief and rocks.

II - The coasts whose formation is not resulting from wave action.

III - The costs formed under the effect of prevailingly wave processes.

According to this subdivision the Estonian seacoasts are reffered to the third group, i.e. to the coasts formed under the effect of wave action prevailingly. A great part of the coasts of Estonia have undergone only slight changes due to wave action and they are classified as straightening coasts. These are the southern coasts of the Gulf of Finland from Spithami Cape in the west up to the settlement of Aseri, and a great part of the coasts of the West-Estonian Archipelago. In those areas the straightening of the primary contour of the intended coast has not been observed. There have preserved peninsulas and the bays cutting deep into the land and island. Prominent accumulative coastal formations such as barrier beaches a.o. are also lacking. In view of the above, the coasts of the kind should have been classified as abrasion-accumulation-embayed coasts. In some coastal areas within the limits of the above-mentioned type with inconsiderable tilt in the wave shade the primary relief of the coast has practically not been changed. The contour of the shoreline is a result of the unevenesses of the primary relief of erosional and accumalative formations formed under the effect of the continental ice sheet of the last glaciation. On those dynamically little active shores wich are subjected either to weak abrasion or insignificant accumulation of fine-grained drift material silty shores are developing.

The two stretches of the coast - coast of the Northern part of Kopu Peninsula, and the second one of the Liivi Bay - are believed to belong to the straightened abrasion-accumulation type.

For the Estonian area the straightened accumulation coasts are not as common as for the Southern and Central Baltic. It is represented by one short stretch of sandy beach in Narva Bay, east of Merikyla.

Between Aseri and Merikyla the line of the North-Estonian (Baltic) Glint runs in the immediate vicinity of the contemporary coastline. Due to intensive wave action which has been observed on the southern coast of the Gulf of Finland in the area of the Baltic Glint for several millennia (at least since the Ancylus stage of the Baltic Sea), a wide abrasional platform came into being in front of the glint and the line of the glint has straightened coast type.

However, for the most part the glint scarp, up to 56 m in height at Ontika, has attained its dynamic maturity through time. It means that at present the glint scarp remains mostly out of the reach of the wave action and is dying out or has already died out. At the foot of the scarp coastal dabris is covered by permanent forest vegetation. Whereas at the glint foot there was formed a beach. The contemporary shores have turned into accumulation ones. The glint scarp is episodically subjected to the wave action in limited areas only, e.g. at Paite. It means that this stretch of the coast at the present time should be referred to the straightened type and term as accumulation coast with the dying out cliff.

As a whole, the Estonian coasts, the straightened stretches included, suffer the deficiency of sediment drift. For this reason the coasts which have already reached the state of maturity are from time to time subject to abrasion. Most frequently these are the sandy beaches as the least resistant.

The development of coasts has been strongly affected by the lithology of the parent rocks. This is most prominent while speaking about the most common coasts - straightening abrasion - accumulation - embayed coasts. On the Estonian coast these are represented by two varieties with the first one being related to the areas of the distribution of glacial and fluvioglacial formations and deposits on the fore-glint lowland between Aseri and Viimsi and in the southern coasts of Saaremaa and Hiiumaa.

The other variety is associated with the area of the blocks and ancient erosional valleys in the Pre-Quaternary bedrock. On the southern coast of the Gulf of Finland it is also spread on the Northern border of the North-Estonian Plateau within the boundaries of the so called glint bays and headlands. This area starts in the east with Viimsi Peninsula and continues up to Osmussaar Island in the west. The second one is found on the Northern coast of Saaremaa Island. This variety of the straightening coast should have been termed as abrasional - accumulation coast of glint bays.

The structure, dynamics and development of the Estonian seacoast as a whole are closely connected with the geological and geomorphological history of the territory. This has been repeatedly outlined (Orviku, Orviku, 1961; 1974 et al.). The ancient relief, rocks and deposits have affected the development of both the contemporary coast and ancient shore formations wide - spread in the area under consideration.

The contemporary coast of Estonia is shoaly. The degree of slopeness varies within wide limits and exerts, therefore differently influencing the character and intensity of morpho-lithodynamics of the shore as well as its development. However, on the contemporary shore the tilt of the primary relief slope does not attain the altitude which would enable one to assume the existence of deep water coasts.

The shoaly coast is subdivided into low and bluff shores. According to the tilt of the primary relief, geological character of initial rocks and prevailing shore processes the author has distinguished the following shores:

1 - cliffed shore - an abrasional bluff in resistant Paleozoic rocks;
2 - scarp shore - an abrasional bluff in brittle loose Quaternary deposits;
3 - rocky shore - an abrasional sloping shore in resistant Paleozoic rocks;
4 - till shore - an abrasional sloping shore with a protective cover of bouldres;
5 - gravel shore - an accumulation shore with beach ridges of along-shore and on-shore drift or derived from local bedrock;
6 - sandy shore - an accumulation sandy beach with a ridge of foredunes or dunes;
7 - silty shore - an accumulation shore with inconsiderable accumulation of silty-peletic sediments;
8 - artificial shore - the shore with man-made coastal protection constructions.

All above-listed shores may display a variety of natural patterns according to which stage of development they have reached depending on the exposure to prevailing currents, the time of exposure and several other factors (Orviku, Sepp, 1972). The rich coastal vegetation both on the shore and submarine coastal slope, the occurence of bulrush reed is an indication of dying out of any of the above named 8 shore types.

Different coasts and their varieties are of uneven distribution in consideration of the whole length of the shoreline. On Estonian coast for example, the type of straightening abrasion-accumulation-embayed ones are of the widest distribution. At the same time each coast type has its characteristic combination of shores. Below a short characteristic of litho-morphodynamic peculiarities and regularities governing the distribution of shores within some of the coast types will be given.

The straightening abrasion-accumulation-embayed coast which in its typical form occurs within the limits of Lahemaa National Park is characterized by active development of the contemporary shore due to intensive wave action. In dependence on the character of ancient deposits, tilt of the primary relief and direction of prevailing currents on the capes there prevail abraisonal shores in the form of small abrasion scarps eroded into till, fluvioglacial deposits and varved clays and in places even into gravel- and shingle deposits. Although on abrasional coasts formed in loose Quaternary deposites the protective cover from coarse-grained components develops rather rapidly. A small amount of loose material (pebble-gravel) resulting from abrasion of rocks accumulates in the vicinity of the abraded areas along the edge of the bays where it forms sandy beaches also on the shallow sea floor. But the gravel beaches are rather rare making up only about 10% of the total length of the shoreline of that coast type. The finest particles (pelite) are transported in a suspended state farther and they accumulate in the deep-water parts of the bays. The increased storm activity, especially the increasing frequency of extremely strong storms has favoured the activization of coastal processes in recent years. For example, at head of Eru Bay (southern coast of the Gulf of Finland) there accumulates the sandy-silty material resulting from the abrasion of till, fluvioglacial and other sediments in the area of Kasmu and Parispea peninsulas. The pelitic matter accumulates in the flat south-western part of the bay. In this area the coastal vegetation (mainly bulrush and reed) is widely distributed.

As for the length, the next place occupy the straightening abrasion-accumulation-embayed coast of bays and headlands on the southern coast of the Gulf of Finland and on the northern part of islands Saaremaa and Muhu (West Estonian Archipelago). This type of coasts is first of all characterized by distribution of cliffed shore on the dependence of the carbonate bedrock and sandstones in places. The abrasion of carbonate rocks yields rather abundant shingle and gravel material which is subjected to alongshore drift and accumulates in the form of gravel and shingle beach ridges. On the contemporary coast the accumulation forms are represented by spits, asymmetric forms, terraces a.o. As a result of the abrasion of sandstones there is formed a rather great quantity of finer material which accumulates for the most part at the heads of the bays and supplies the sandy beaches with loose material. It should be pointed out that the so-called glint bays are for the most part buried valleys of big rivers whose floor lies some tens of meters below the contemporary level and they have been filled up with loose Quaternary deposits and drift material (Tammekann, 1928; a.o.). For this reason at the heads of the contemporary bays within the boundaries of the ancient buried valleys there are excellent preconditions for the formation of wide sandy beaches.

For example, in Lahepere Bay between the capes of Pakerort and Tyrisalu the following characteristic alternation of the beaches is traceable. An active cliff with relatively unresistant Lower Odovician sandstones in its lower part is spread on the cape of Pakerort. The wave-cut notches in it account for the frequent collapse of the higher lying carbonate rocks. At the foot of the cliff the carbonaceous material is crushed and subjected to longshore drift along both sides of the peninsula. It accumulates along the rim of Pakri peninsula forming there a gravel and shingle beach. The finer abrasion products (sand, coarse silt) move farther where they supply at the head of Lahepere Bay the accumulation sandy beach with loose material. Owing to the large quantity of sandy material in the area of accumulation at the head of bay there develops a wide sandy beach with 3-4 submarine ridges and foredune belt.

The above shows, that both cliffed and gravel shores (abrasion-accumulation system) of the contemporary coast are most common for this type making up about 50% from the total length of the shoreline in the area of the distribution of the above-mentioned coast type.

Artificial shores are unevenly distributed on the Estonian coast. They are connected first of all with big seaports (Tallinn, Parnu a.o.) and for example, correspondingly, in the middle part of the southern coast of the Gulf of Finland they make up only less than 5% of the whole length of the coastline.

Among the others the artificial shores may be differentiated on the southern coast of the Gulf Finland as well. For example in Tallinn Bay the length of the jetties belonging to the Tallinn Port totals 4 km. In addition at the head of the bay in connection with the reconstruction of the highway leading from the city to the Olympic Yachting Centre the shallow coastal part of the sea floor was artificially filled and a wave-breaking wall of about 2.5 km in length constructed along the coast in 1980 (Martin, Orviku, 1988). The beach at Pirita, in the immediate vicinity of the Olympik centre was reconstructed by 1980 - on account the sand transported onto the beach it turned higher and wider. Since 1988 the Narva-Joesuu sandy beach has been artificially supplied on account of the loose material taken from the bay bottom.

Besides the coasts of the Estonia abound in small fishery harbours, which to a certain degree affect the natural regime of the development in their immediate vicinity as well.

REFERENCES

Gudelis, V. (1967)   The morphogenetic types of the Baltic Sea-coasts. Baltica. Vol.3. Vilnius. 123-145 (in Russian)

Hupfer, P. (1978)   Die Ostsee - Kleines Meer mit grossen Problemen. Leibzig. 152 p.

Ionin, A.S., Kaplin, P.A., Medvedev, V.S. (1964)   World Ocean seashores and seacoasts types, their classification and aspects of their division into districts. Theoretical problems of seashore evolution. Moscow. 19-32. (in Russian).

Martin, E., Orviku, K. (1988)   Estonian SSR. Artificial structures and shorelines by Kluwer Academic Publihers. 53-57.

Orviku, K., Orviku Kaarel (1961)   On the geological development of the Estonia Seacoasts. Proceedings of the Estonian Academy of Sciences. Geologia, VII, 187-202 (in Estonian).

Orviku, K., Sepp, U. (1972)   Stages of geological development and landscape types of the islets of the West-Estonian Archipelago. Geographical Studies. Tallin, 15-25.

Orviku K. (1974)   Estonian seacoasts, Tallin, 112 p. (in Russian)

Tammekann, A. (1928)   Das Relief und die Abflussverhältnisse in Estland (2. Baltische hydrologische und hydrometische Konferenz). Tallinn 5 p.

THE PRESENT STATE OF ESTONIAN COASTS AND EVOLUTION TENDENCIES

Kaarel Orviku

Institute of Geology, Estonian Academy of Sciences, Tallinn.

The Baltic Sea coastline of Estonia is about 3780 km long, of 1240 km are on the mainland, and two thirds on about 1500 km islands.

The great lenght and proximity of the coastline allows the majority of Estonians to be considered coastal people or islanders. The sea provided Estonians with food since the time when our forefathers settled in these northern areas. Evidence of ancient settlements has been obtained from archaeological excavations in the coastal area.

One of the oldest and most studied is the Pulli settlement, occupied about 7500 years ago in the vicinity of Parnu on the coast of the Yoldia Sea.

The coast has always attracted men, been attractive for the settlement, even though it has sometimes proved to be a difficult environment (Schwartz, 1976). The sea has provided food, and a means to travel to neighbouring and distant countries. When the coastline changed its position, due to the advance or retreat of the sea, man has moved in accordance with it. In primeval ages, when buildings were light, this way was easy, but now that the buildings are more durable, and constructed to last for centuries, it is more difficult to make such movements.

As long as man used only small boat harbours, such as inlets, sheltered areas beside promontories, or primitive landing-piers there was little disturbance of natural processes. With the growth of large modern harbours there have arisen many problems. In choosing a place for a harbour one should keep in view how, at the least expense, to construct a refuge for ships without causing deterioration of the surrounding environment. Unfortunately, this principle has not always been kept in view.

Since 1919 coastal theory has reached a high level, but mistakes are still quite frequent in applying the principles of the science in practice. There seems to be an internal contradiction of some kind. To my mind it is, on the one hand, related to the seemingly simple and tangible dynamics and evolution of beaches, while on the other hand, the great diversity of natural processes in the shore zone complicates the understanding of the coastal environment.

For example it is difficult to measure processes in a stormy sea, but strong storms have enormous effects on the beach.

Fortunately for Estonian geologists, problems of hydrotechnical constructions on our coast have been rather rare, but we still have some sad experience.

Systematic studies of the litho- and morphodynamics of the present coast were not initiated in our country until the end of the nineteen-fifties. The aim was achieved in 15 years, and the results presented in a monograph (Orviku, 1974). However, at present

we must confess that despite our relatively good knowledge of the evolution and dynamics of the shore, much work has still to be done to solve specific questions.
What do we actually know about the evolution and dynamics of the Estonian coast? Kessel (1967) and Orviku (1974) distinguished abrasion shores such as cliffs cut in limestones and sandstones, scarps in moraines and other unconsolidated sediments, from accumulation shores, grouped into sub-types according to their sediments: shingle, sand and silt. The evolution of each type of coastal area depends upon many processes. Although sea waves serve as the main process, one has to consider the geological structure and developmental history of Estonian territory as a whole. In recent years the present shore has been modified considerably by human impact.

Estonian coasts are characterized by low, gently sloping topography, formed during the last retreat of continental ice. Relative relief changes over short distances, but changes are small. The unevenesses of ancient (Cambrian to Devonian) rock outcrops result in irregularities along the coast, and there are variations in the nature and thickness of the mantle of Quarternary glacial and postglacial deposits that are reflected in coastal outlines. In places where Quarternary deposits were thin, they have often been destroyed by the sea, exposing Paleozoic limestones, sandstones and clays to the action of waves. Thus the diversity of shore types is accounted for by the ancient relief, the lithology of coastal outcrops and the pattern of unconsolidated surficial sediments.

For example, cliffs have developed where the prequarternary relief was steeply sloping, beneath a thin layer of Quarternary sediments. Abrasional scarps occur on morainic shores related to glacial relief forms. Where glacial and fluvioglacial sediments have been subjected to erosion, material has been generated to produce accumulative forms such as sands and gravel beaches. Dunes have formed behind some sandy beaches, especially during successive oscillations of the Baltic Sea. Due to slow uplift of the Earth's crust the Estonian coast and beaches are of emergent character; as the ancient relief emerges from the sea, islands come into being, and bays become shallower. As morainic shores emerge, the finer sand and gravel is washed out to from nearby beaches and spits, leaving scattered boulders or bouldary ridges (morainic ness). In general, near-shore sea floor deposits are of silt and clay, so that continuing emergence leaves bouldery and sandy shores stranded behind muddy deposits, which are colonized by reeds and rushes to form fringing fens.

However, the Estonian coast has not always been an emerging coast. During the Holocene there have been several transgressive stages of the Baltic Sea. These stages are clearly distinguished in the structure of coastal formations (Kessel and Raukas, 1967). As a rule, the latter stages are bolder and larger in area. Within their boundaries coastal formations, such as beaches, spits and dunes are present. Erosional features such as bluffs also formed during these periods of marine transgression.

The Estonian coast began to emerge about 5-6 thousand years ago and precise relevelling indicates a slow uplift of the earth's crust (Vallner et al., 1988).

Despite the emergence, there has been erosion of Estonian sandy beaches in past decades, and in place the sea is advancing again.

Erosion of sandy beaches is a world-wide problem and several causal factors have been identified (Bird 1985). These include a rising global sea level, increasing storm activity and diminishing sand supplies, as well as widespread human impacts.

Since 1954 Estonia has been subject to at least six (in 1954, 1967, 1969, 1975, 1982, 1985) extremely strong storms of the kind that, according to statistics, should occur only once in a hundred years. In autumn and winter period the westerly and southwesterly stormy winds raise the water up to 2 m above its summer level: in Leningrad it rises over 3 m. The effects of storm waves then extend landwards far beyond the average coastline. While the erosional effects of a single storm are usually compensated by acceleration during subsequent calmer weather, a succession of severe storms of the kind seen in Estonia in recent years results in more permanent recession of the coastline and re-shaping of coastal and near-shore profiles. As there is little evidence of a rising sea level in Estonia over this period, beach erosion appears to be largely due to recent increased storminess in the eastern Baltic Sea. With each storm the amount of sand on the beaches has decreased, and after 3 or 4 storms the backshore dunes were subjected to erosion and cliffing. Subsequent storms have continued this process. The Estonian coast is characterized by deficiency of sandy sediments and those present are in its main, composed of very fine material $Md = 0,1$ mm, which is very readily removed by winds and waves.

In winter, an ice fringe develops on Estonian coast. A steady ice cover puts an end to wave action for the winter period, but in spring the ice is broken and pushed on to the coast by heavy winds, to pile up as hummocks 10-15 m in height. At the same time, blocks of ice scour the sea floor to depths of 5 m, where divers have seen deep furrows left by ice. Ice pushed on to the shore loosens scours sea-floor sediment, driving large boulders up from the sea bottom. It may also damage buildings on the coast. From the other point fast ice protects the shore from strong winter storms. During the last 4-5 years the winters have been extraordinary warm and fast ice was formed in sheltered bays only. Therefore its protecting effect was negligible.

On some sectors of the coast, a terrace composed of shingle ridges has been built. As the coast has emerged, the older ridges are left inland, higher than these near the present coastline. Some individual large beach ridges probably result from severe stormy periods in the past. Severe storms have thus occurred in the more distant past and are likely to occur again in the future.

Measurements of the growth of a spit with typical shingle ridges have shown that in the past 10-15 years the length of the spit has increased about ten times faster than during the previous 10-15 years, indicating the recent increase in storm activity.

The Estonian coast of the Baltic Sea is thus a natural laboratory, where in a relatively restricted area it is possible to make detailed studies of coastal evolution and dynamics, as a basis for the management of our coastal resources.

References

1. Bird, E.C.F. (1985) Coastline changes, Wiley Interscience. New York. 219 p.

2. Kessel, H., Raukas, A. (1967) The deposits of the Ancylus and Littorina Sea in Estonia. Tallinn. Valgus. 135 p. (in Russian).

3. Orviku, K. (1974) Estonian seacoasts. Tallinn. Valgus. 135 p. (in Russian).

4. Schwartz, M.L. (1976) Man has lived in harmony with the shoreline - can he again? In XXIII International Geographical Congress Symposium "Dynmaics of shore erosion". Tbilisi. p. 230-231.

5. Vallner, L., Sildvee, H. and Torim, H. (1988) Recent crystal movements in Estonia. Journal of geodynamics, 9. p.215-223.

ENVIRONMENT AND DYNAMICS OF THE KALININGRAD COAST OF THE BALTIC SEA

Boldyrev V.

The Kaliningrad coastal zone includes the native coasts of the Sambian peninsula (Samland) and the adjacent Kursh and Visla Spits (about half of the length of each). Due to its jutting into the sea position and west storms predominance there is an irreversible jut demolation to both sides of almost all abrasion material, mostly sand. The Kursh and Visla Spits are mainly composed of these materials. The Visla alluvium contributes to the latter too. According to our estimate there are 4 billion m^3 of sand in the Kursh Spit.

Because of the predominance of the sand drift beyond the Sambian peninsula shore zone there is an intensive abrasion and recession of the coast at the speed of 0.5-1.5 m a year, and in case of extreme violent storms - from 7-8 m to 12-15 m during one such storm only. The abrasion process affects 12 km of the root part of the Kursh Spit too. In the pre-war time the whole coast from the cape Jaran (Brüsterort) as far as Baltijsk (Pillau) was demolished.

The abrasion and recession process of the Sambian peninsula coast and the Southern part of the Kursh Spit has been inherited since the first phase of the Holocene transgression. Only since the beginning of the litorina phase the coast has receded from 3.2 km to 4.5 km. It is at this distance from the coast that at depth of 19-27 m the peninsula is fringed by an ancient 7-8 m high abrasion escarp, whose formation began during one of the phases of the Ancylus transgression and was over at the beginning of the Litorina transgression. (Blazhchishin A., Boldyrev V. and others. - Ancient Shorelines and Shore Formations in the South-Eastern Part of the Baltic Sea. - Baltica, 7, Vilnius, 1982, p.57-64).

In the process of coast demolation and during the glacial loams wash out large boulder fields were formed at the bottom. They blocked the bottom from a further abrasion and consequently the feeding of the coastal zone goes actually at the expense of the coastal abrasion. As there is not enough sand feeding material for the beaches, they are badly developed and on certain parts there are no beaches at all. There is an average sand defecit of 500-700 m^3 per 1 km.

With the potential annual alongshore sand transport capicity at Svetlogorsk (Raushen) of 500 000 m^3, the actual transport capicity is only 180 000-200 000 m^1. With such a sand transport deficit the pre-war beach protecting constructions (walls and groynes) proved uneffective. The sand beaches in front of the walls are completely washed out and the intergroyne "pockets" are not refilled with sand and so the coast recedes. Since the construction of the last groynes in 1927 the coast, they are to protect, has receded from 25 to 38 m. The steel grooved groynes of 1937 have also failed to preserve the coast (Lashenkov V. - Coast Protecting System of the Kaliningrad Coast of the Baltic

Sea. - Natural Fundamentals of Beach Protection., Nauka, Moskow,1987, p.154.165). So, artificial alimentation and sand replenishment as well as inwash of wide sand wave-extinguishing beaches have been recognized as the most effective fundamental beach protecting methods.(Boldyrev V. - Main Principles of Fortification of the Sea Sand Shores. - Baltica, 7, Vilnius, 1982, p.203-213, Boldyrev V.,v Gretschitchev E. and others - Principles of Shore Protection of Kaliningrad Sea Coast of the Baltic Sea. Baltica, 7, p.187-194; Schulgin J., Morosov L. -Schutz der sandigen Meeresufer mittels des künstlichen Strandbaus. Baltica, 6, 1977, p.99-109.)

An example of such effective beach protection was a mass sand feeding coming from the amber open pit soil discharges. In the postwar period alone the shore zone at Jantarnoe (Pulmniken) has received more than 50 million m^3 of the amber pit discharge feeding. As a result, wide, 30 km long sand beaches, with a rampart-like shore dune in their rear part, were formed south of Jantarnoe. They protect effectively the former active abrasion ledge.

Part of the sand reached the Visla Spit and considerably contributed to the sand feeding of the beaches and the shore dune.

The transfer of sand from the amber pit to the north coast will start in the near future. At present an experimental construction is underway, including cutting off and terracing of the landslide coastal slope and an inwash of a sand wave extinguishing beach at the expence of the cut off material. The first stage of the project is almost completed. Along the shore slope 5 berm-steps with sea-oriented gentle slopes were formed. More than 2.5 million m^3 of sand have been cut off and dropped into the sea and a 150 m wide (maximum) and 2 km long beach has been formed. The research data of the 4 years testify to the fact that the cut off sand material drifts eastword, and in the eastern part of the Svetlogorsk cove along 1.5 km. 45-50 m wide sand beaches were also formed and they proved considerably resistant to storm abrasion.

The cut off material is distributed unevenly along the profile of the shoreline: in the above-water part of the beach about 10%, about as much goes down to over the 10 m depth, and as to the 80% left - 50% are distributed between the water edge and 5 m isobathes, and 30% between 5-10 m isobathes. This tendency of sand distribution is observed on the 10 km long and 3 km wide area.

The second stage of the experimental construction envisages pumping over the cut off sand material by the 7 km long pipe and an aggratation of a 100 m beach in the Svetlogorsk cove along 1.5 km.

For the damaged parts of the shoreline, which are subject to severe abrasion the scientific production association "Baltberegozaschita" has worked out and successfully put into life some origional types of wave-extinguishing constructions which function rather effectively west of Zelenogradsk (Kranz) and at the root part and 10th km of the Kursh Spit. This year the building of a third type of a wave extinguisher will be started: a sloping wave extinguisher with a promenade.

At present out of 145 km of the shoreline 50 km are being intensively damaged. We believe that the best way to solve the beachprotection problem in our case, is artificial sand alimentation for abrasion areas - with about 1 million m^3 annually. We expect a firm shoreline stabilization in the course of 20 years at that rate of sand feeding.

The research data of the last 3 years point to a possibility of building new large ports with the annual freight turnover of more than 10 million tons in Primorsk cove (Fischhausen) of the Visla Bay, at Jantarnoe, as well as a remodelling of the port Pionerski (Neukahren). This can be done in such a way that the new ports will not destroy the natural equilibrium but will improve the local coastal zone.

COASTAL MANAGEMENT IN POLAND

Andrzej Cieslak - Instytut Morski, Dlugi Targ 41/42
80-830 Gdansk, Poland

This paper concentrates mainly on coastal protection. However, it must be always remembered that coastal management is much more than just coastal protection, though this last may be its most important part and objective.

1. SHORT CHARACTERISTIC OF POLISH COASTLINE

The total length of the Polish coastline, including banks of internal sea waters (the Szczecin and Vistula Lagoons) is abt. 843 km. In that the length of the seacoast is 500 km, of the banks of the Szczecin Lagoon is 241 km, and Vistula Lagoon - 102 km.

1.1. Seacoast

Regardless of type (dune/cliff), the seacoast nearly on its whole length is systematically retreating at a rate of 0.2 - 1.5 m/year. An exception are the regions of Swinoujscie, Niechorze - Mrzezyno, Lake Lebsko, tip of Hel Peninsula and Vistula outlet where the coast is accreting or remains fairly stable. Short stretches of accretion, forced by the presence of breakwaters, exist on the west sides of the ports at Darlowo, Ustka, Leba and Wladyslawowo.

Erosion processes are accompanied by gradual decrease of beach width and hight, and degradation, in places even complete disapearance, of dunes. Observations and measurements point to generally growing erosion trends. Cartometric investigations (Zawadzka 1990, 1992) indicate that the average rate of seacoast retreat during the last 100 years gradually increases. Reasons for this situation may be connected with:
- increased frequency of appearance of very high storm surges and long duration of states of high filling of the Baltic Sea,

- cumulated during the years influence of significant reduction and qualitative change of longshore transport downstream of ports,
- pollution of sea environment which resulted in the disappearance of various sea plants; in the past the plants contributed to stabilization of the foot of the active profile.

The state of the coast, specially along cliff coasts, is additionally worsened by bad water and sewage management. Incorrectly aligned pipelines, discharge of sewage- and rain-water directly into the soil or onto cliff slopes significantly increase cliff crown retreat. Often they are the direct cause of large landslides. Appearance of larger amounts of water on cliff slopes is also caused by devastation of vegetation on cliff crowns and hinterland. Plants retain a significant part of rain- and ground-water.

The total length of seacoast along which hinterland is endangered and intervention is required is abt. 80 km.

It is expected that on top of this rather unfavourable situation will come the earlier mentioned results of the greenhouse effect. This will result in quicker erosion and coastline retreat, and appearance of danger in many new places where everything seemed under control.

1.2. Lagoon banks

Similarly as along the seacoast, increased erosion is observed along the banks of both lagoons. However, no numbers can be given because of lack of data. It is supposed that the larger erosion rates are caused by deepening of the straits leading to the sea. In effect high water levels, caused by storm surges on the sea, are more frequent and reach larger values.

At the same time, because of increased exchange with the sea, salinity of lagoon waters has grown. Together with increased pollution by communal and industrial waste-waters that results in gradual disappearance of reed fields which up to this time were the best protection of the banks.

A large part of the banks of both lagoons is formed artificially by dykes. Dykes along the Szczecin Lagoon are too low, especially in the light of the expected greenhouse effect influences. In some places on the Vistula Lagoon lacking foredyke slopes should be built.

1.3. Banks of estuaries and coastal lakes

The banks of coastal lakes and estuarial parts of rivers and adjacent areas are in danger of flooding in result of long-lasting storm surges on the sea. Existing incomplete data indicate that in effect of the surges, water level in the lakes can rise nearly 1.5 m above mean level. In estuaries storm surges, specially when they occur simultaneously with freshets on the river, may result in high water levels in the estuary. Observed levels in estuarial stretches reached 2.16 m above mean level.

Inflow of sea water into coastal lakes and estuaries may cause periodical salination of ground-water.

1.4. Technical coastal protection

At present hydrotechnical structures protecting the seacoast take up 22% of its length, i.e. abt. 126 km, and are situated in 20 regions. In some regions systems of protection comprising several types of structures have been built. The total length of all kinds of structures along the seacoast (excluding artificial nourishment) is abt. 150 km. It is estimated that their devastation, mainly by waves and ice, reaches 20 -30%.

Nearly 88 km of dykes have been built along the Szczecin Lagoon, and 49 km along the Vistula Lagoon.

Most of technical protection works carried out during the last 40 years was forced by the postwar state of the coast, its protection and state of hinterland development, and by later decisions concerning the location of various investments. Unhappily, the Maritime Offices, which in Poland are responsible for coastal protection, had no real influence on these decisions.

For many years, various types of pile groynes and seawalls were the basic types of coastal protection structures used along the seacoast. At present, in principle no new groynes are built - except groynes filling in gaps in larger groyne systems. Heavy seawalls are constructed only in places where this is absolutely necessary. Soft protection is preferred, i.e. beach nourishment and light revetments situated at the dune foot or lately - even inside the dune.

1.5. Biotechnical protection

Biotechnical protection is carried out systematically within the technical belt by the coastal protection service of the Maritime Offices. The work involves various dune-forming techniques, planting dune grass, bushes and trees, and cultivation and maintenance of the protective vegetation, including protective woods.

Developed during many years of experience and constantly improved methods of biotechnical protection seem to fulfill satisfactorily their tasks, specially along the dune coasts. However, methods of biotechnical protection of cliff coasts and underwater part of foredyke slopes on the lagoons still must be worked out. Some attempts have been made recently on the Szczecin Lagoon.

Pressure of functions other than protection on the coastal zone is a grave danger to biotechnical protection systems. In areas with intense recreation and touristic exploitation, the vegetation is subject to significant degradation, specially in the direct neighbourhood of the technical belt. Proper development of biotechnical protection and protective woods behind the dunes can be seriously impeded by air pollution from nearby industry and by improper management of industrially exploited woods on old dune areas neighbouring with the technical belt.

1.6. Pollution of the coastal zone

The sea in the coastal zone is polluted through rivers carrying waste-waters, through direct sources of waste, by sea transport, from the atmosphere, etc. About 88% of pollutant load comes into the coastal zone through rivers. The remaining 12% comes from direct outlets, in that 10% from coastal towns and villages and 2% from industry. During the last 20 years the state of coastal waters steadily deteriorated. This, as stated earlier, probably has indirect but significant influence on coastal erosion processes.

1.7. Legal status before 1991

Protection of the seacoast and of the banks of internal sea waters was the responsibility of the Maritime Offices in Szczecin, Slupsk and Gdynia. All protection works were carried out on a narrow strip of land encompassing the beach, foredune/cliff and hinterland from a few to abt. 1000 m inland, and in the neighbuoring waters. However, permits for other than coastal protection kinds of use of this area were not issued by the Maritime Offices but by other local or governmental authorities. Maritime Offices had very little influence on these decisions - in most cases they only could express their opinion. This state of law resulted in many wrong and costly - even disastrous - use and location decisions realized against the judgement of the maritime administration.

2. FUNCTIONS LOCATED IN THE COASTAL AREA

Main functions located along the coast and in its direct hinterland are the following:

PROTECTION - protection of land against damage or destruction by the sea, soil and water protection, protection of hinterland against flooding and direct influence of the sea climate. These functions are concentrated in the area of the coastal zone (fig. 2) and technical belt (fig. 3).

RECREATION AND HEALTH - area of most popular summer holidays; because of climatic value and mineral waters is important for therapeutics. These functions concentrate in the coastal belt.

ECONOMY - sites of ports, industry, agriculture and forestry, spawning grounds. These functions are situated in the whole width of the coastal zone, coastal belt and hinterland.

HOUSING - attractive area of urban development.

All these functions are in conflict with each other. Complete elimination of the conflicts is impossible. However, it should be stressed that recreation, health, economical and housing functions can be properly fulfilled only on the condition that the protection functions are correctly performed. Therefore the coastal zone must be in such a state that it shall ensure safe and healthy use of the coastal belt and its hinterland.

3. MAIN FACTORS TO BE CONSIDERED IN COASTAL MANAGEMENT

Research and practice seem to indicate that:

1) The coastal zone, which contains the most active part of the sea bottom, the beach and dunes or cliffs, is the most dynamic, constantly changing and vulnerable area of the country. In fact it is a transient zone between sea and land, and its state depends heavily on processes occuring in the sea. However, land originated processes are also of significance.

2) Processes occuring within the coastal zone are characterized by large spatial dimensions. Any change or ingerence introduced into the coastal zone is felt (mostly with a negative effect) along stretches of coast of the length of several tenths - sometimes over one hundred - kilometers. At the same time the sea directly influences the land environment (in Poland abt. 3 km inland). In some cases this influence may be felt at a distance of over 40 km from the sea (estuaries, lagoons, coastal lakes). The time scale of processes occuring along the seacoast exceeds the lifetime of one generation. Minimum time that should be considered is abt. 50 years.

3) Because of the greenhouse effect, during the next 100 years the sea level shall rise, and changes in air mass circulation are highly probable. In result during annual storms the sea level along the Polish coastline will be about 1.4 - 1.5 m higher than it is now, i.e. it will be at least as high as during the storms of January 1983 and 1988/89. These were nearly catastrophic storms of 1 in at least 100 year probability. In the future, even during average storms significant areas of hinterland will be for some time below sea level.

4) The coastal zone is very susceptible to degradation, even to devastation, caused by direct and indirect human influence. At the same time it is under constant pressure of industry, urban development and other forms of exploitation resulting from local interests.

Among others, it results from the above that Coastal protection activities must have depth, both sea- and land-wards.

Maintaining a proper state of the coast, such that it would guarantee correct fulfilling its protective functions, requires comprehensive, large-scale, long-term and systematic planistic, technical and ecological activities, which would take into account the specifics of the coastal zone and the possibilities of coastal protection. It also requires withstanding some local pressure.

The activities mentioned above by far exceed areas of jurisdiction of any one administrative unit, their time scale exceeds the scale of local planning, and the necessary funding - especially in the light of the expected greenhouse effect - exceeds the possibilities not only of local coastal communities but also of voivodships.

An important factor in formulating the legal frames and strategy of coastal protection is the coast awareness of people living in coastal areas. In Poland nearly all the population living in coastal communities is land-oriented. Only lately people start to think about the beaches and dunes, but mainly as areas of economical activity, not as areas providing safety. This state of awareness, among others, is the result of prior practice, in which the people had no contact with solving coastal problems. They have become used to the idea that the problems are solved for them. But participation in solving means participation in responsibility, and requires sufficient knowledge and understanding from the decision makers.

Therefore, at present stage, people must participate in decision making, so that they have an inducement and chance to learn. But on the other hand the participation should be such that the risk of forcing wrong (i.e. often - especially in the long-term - disastrous) decisions would be minimized.

4. THE NEW ACT OF PARLIAMENT

On July 1-rst, 1991 a new Act of Parliament "On the Sea Areas of Poland and on Maritime Administration" comes into life. According to this Act a coastal belt is established and
"The coastal belt comprises:
  1) the technical belt - which is the zone of interaction of sea and land; it is an area set apart for maintaining the coast in appropriate state according to requirements of safety and environmental protection,
  2) the protective belt - which covers the area in which human activity has direct

influence on the state of the technical belt."

The Act establishes also ranges of competence (fig. 3) of the maritime administration and of other governmental or local authorities. In short:

- On the sea: Maritime administration is fully competent and responsible.

- In the technical belt: Maritime administration is as before responsible for programming and execution of coastal and environmental protection, but also issues permissions for all kinds of use or construction in the belt except permits for water use; these last are issued by other authorities but have to obtain agreement of the appropriate Maritime Office. On all permits issued by the maritime administration, opinion of local authorities is required.

- In the protective zone: All construction or land- or water-use permits issued by local or governmental authorities which may have influence on the state of the technical belt, also regional development plans at all stages of elaboration must be agreed upon with maritime administration.

The Act does not say anything about ownership of land, it only establishes the frames of acceptable use and administrative competence over the coastal belt.

5. REQUIRED DIRECTIONS OF FURTHER RESEARCH AND LEGAL WORKS

- Development of a coastal protection oriented monitoring system and Data Bank for the whole Polish coastal zone; this work is already started.

- Development of models for long-term (tenths of years) and large-dimensional forecasts of coastal processes. This task may require a changed parametrization of the coastal zone.

- Establishment of safety standards for the technical belt, and introducing them into practice by Act of Parliament.

- Determination of necessary widths of the technical belt and protective zone, and legal establishment of the boundaries of these areas.

- Obligatory (by law) introduction of coastal variability and coastal protection problems into planning of coastal region development.

- Improvement and development of "soft" and low-cost coastal protection methods.

LITERATURE

Cieślak A., W. Subotowicz [editors](1986): Long-term program of coastal protection and maintenance of beaches (in Polish). Internal Report of the Maritime Institute, No. WW-IM 3631.

Cieślak A., W. Subotowicz (1987): Report on the state of knowledge about the Polish seacoast and its protection (in Polish). Inżynieria Morska, No. 2.

Cieślak A., et al. (1989): Actualization of the long-term program of coastal protection and maintenance of beaches (in Polish). Internal Report of the Maritime Institute, No. WW-IM 4403.

Zawadzka E., et al (1990): Determination of coastline changes - in Polish. Internal Report of the Maritime Institute, No. WW-IM 4571.

Zawadzka E. (1992): Erosion/accretion system of South Baltic coast during the last 100 years. Eurocoast '92, Kiel.

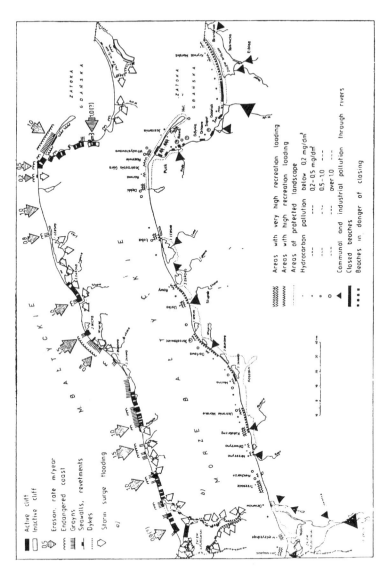

Fig. 1 State of seacoast (Cieślak, Subotowicz, 1987)

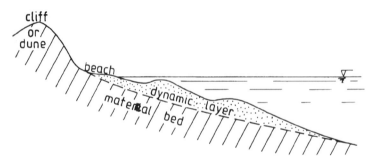

Fig. 2 Cross - section through coastal zone

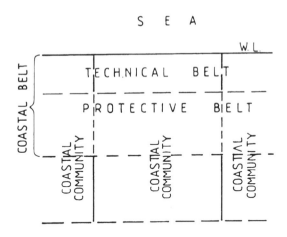

Fig.3 Administrative division of coastal zone and hinterland

Experience from shore protection of the Hel Peninsula

By
Tadeusz BASIŃSKI

Polish Academy of Sciences,
Institute of Hydroengineering,
Gdańsk, Poland.

1 Introduction

Hel Peninsula is a famous recreation area at the Polish coast and belongs to the landscape park. There are two towns Hel and Jastarnia, and three villages: Chałupy, Kuźnica, Jurata (Fig. 1). The railway and the road passes along the Peninsula. The harbours of Hel and Jastarnia are situated at the Puck Bay side. The seaward coast of the Peninsula is strongly eroded by sea although some protection works have been started in 1946. This coast is very low; 45% of its area is situated below the ordinate +2.5 m, i.e. within the reach of storm waves running up simultaneously with high sea level. There is also no possibility to draw back from the abrasion. Protection of the Peninsula against breaking through or overflowing creates not only economical but also social problems.

2 Morphological and hydrodynamical data

- Area od the Peninsula – 32.3 km^2;
- length – 34.0 km;
- maximum width – 2.9 km;
- minimum width – 0.1 km;
- length of shore line – 74.4 km;
- 0.44 km^2 land for 1 km of shore line;
- seabed slope (depths 0–20 m) from 0.003 at the base, till 0.06 at the headland of the Peninsula;
- storm surges $\Delta h_{1\%}$ – 1.48 m above the MSL;
- wave conditions along the Peninsula differ considerably;
- longshore sediment transport to SE is 70–200 thousand m^3/yr.

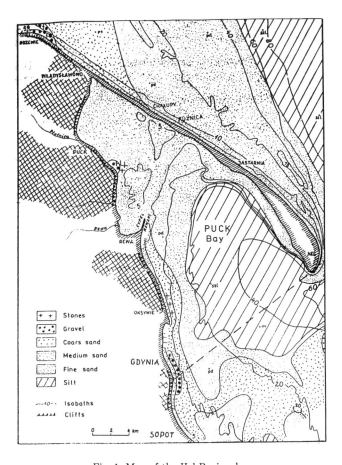

Fig. 1. Map of the Hel Peninsula

3 Shore protection by groynes and seawalls

- Protection works in larger scale were started in 1945.
- The actual state of protective construction from the open sea side consists of 420 m of concrete seawall in the vicinity of Władysławowo harbour, 162 groynes on the 12 km distance Władysławowo – Kuźnica, 1500 m of sand dike with seaward slope protected by rubble mound and concrete sleepers in the Kuźnica region, and about 3 km of artificial beach on the 4th and 13th kilometer of the Peninsula.

- First protection constructions (1948) caused further disorder in natural coastal processes disturbed by the Władysławowo harbour. As the consequence erosion processes concentrated in some places of the shore. Later protection caused transfer of erosion from one place to another. Finally all protection works between Władysławowo and Kuźnica were hazardous; it was done with no plan, with no consistent preventing action.
- Groynes were constructed to facilitate sand accumulation on the beach. It happens in case while waves approach the coast at an angle. From the observations and measurements made at the Peninsula it can be concluded that:
    - directly after construction groynes cause accumulation of beach and in the nearshore area of several tens of meters width;
    - after initial accumulation shore line goes back to its mean position and starts to stabilize. Beaches formed by groynes in the spring season diminish the dune erosion during autumn – winter season;
    - the erosion created at the ends of the group of groynes is harmful. This happens also in case of discontinuity in the system of existing groynes;
    - constructed groynes did not manage to stop dune abrasion, but successfully delayed it.
- Seawalls were used to stabilize the position of dune foot. From the experiences gained for the Hel Peninsula it is possible to conclude the different types of sea walls in different ways;
    - concrete sea wall protects effectively the area of Władysławowo harbour;
    - non-existent now bulkheads filled with the concrete blocks have turned out to be undurable, ineffective, and even harmful, so useless for the open sea;
    - sand dike protected with rubble mound and sleepers (in Kuźnica region) is a temporary construction, which may protect areas situated behind the dike during one storm, but with high damages of rubble mound and dike. To maintain the dike it is necesssary to keep a wide beach. In this case its protection role is limited to extreme storm conditions only.

The existing coastal structures were not able to prevent the dune erosion during severe storms and at high water levels. The increasing sand deficit have forced us to look for other shore protection methods.

Beginning from 1980 different technologies of artificial nearshore and beach nourishment have been developed and tested.

4 Artificial nearshore and beach nourishment

The effectiveness of artificial beach nourishment is proved in many published papers. Nevertheless, a direct supply of the sand to the beach or dunes area is not always possible. Tehnological problems arise, due to the difficulty of crossing the breaker zone, usually

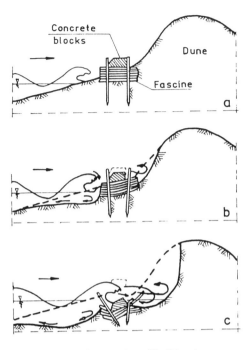

Fig. 2. Destruction of bulkheads

existing between the borrow and reclamation area. The breaker zone is too shallow to use large trailing suction dredgers, while small size dredgers, of smaller draft, cannot operate in the rough wave conditions. There is also a risk of using long, floating pipelines from the dredgers to the shore, as a sudden storm is able to destroy the installation. A rather expensive solution are pile moorings, platforms or mooring boys, which enable a connection of the dredger to a flexible pipeline. The short flexible part is joined to a stable pipeline, layed on the sea-bed, up to the beach.

The direct supply of sand to the beach is very costly and requires a complex technology. A simpler solution is a nourishment of the nearshore zone. It may be applied in following conditions:

- the necessary volume of sand is inconsiderable, comparing to the cost of a pipeline,
- the supply should be executed immediately, due to dangerous erosion, thus, there is no time for constructional works,
- the material cames from routine dredging of a neighbouring navigation channel and there is an alternative of wastage by deposition in deep water,

– in developing countries, when the realisation of a proper technology is not possible.

Following question should be answered in analysing that solution:

- Does the nourishment of the nearshore bottom really protect the dune?
- If yes, what is the effectiveness of that method in relation to a direct nourishment of the beach?

The efficacity of the artificial nourishment of the nearshore zone is determined by:

- its protective effect thus, the rate of wave dissipation caused by the artificial bank in relation to the dissipation over the natural bottom configuration,
- the stability of the artificial shoal,
- the cost of supply of 1 m^3 of sand.

4.1 Technology „A"

The nourishment by direct dumping with hopper barges in the nearshore zone (fig.3) was realised in 1980 and 1981. Altogether 440 x 10^3 m^3 of sand was reclaimed. 290 x 10^3 m^3 in 1980 and 150 x 10^3 m^3 in 1981 (Mierzyński, 1985). The fill site 850 x 150 m was placed 150 – 300 m from the shore line, at water depth 4 – 6 m. The borrow material was taken from a routine dredging of the navigation channel of Władysławowo harbour, from a depth 5 – 9 m. The dredged material was finer than that of the fill site. A volume 90 x 10^3 m^3 of sand, thus only 20% of the supplied volume, was found on the sea bottom one year after the nourishment, (Mierzyński, 1985).

Fig. 3. Technology „A" – Direct dumping in the nearshore zone

4.2 Technology „B"

The technology B (fig.4) was applied in years:

1984 – a supply of 136 x 10^3 m^3 sand
1985 – a supply of 208 x 10^3 m^3 sand
1986 – a supply of 72 x 10^3 m^3 sand
1987 – a supply of 99 x 10^3 m^3 sand

The material was taken from dredging the navigation channel to Władysławowo harbour. The works were performed during several summer months. It happend that due to heavy wave conditions (sea state > 2°B) the supply had to be suspended for several days. The trailing suction hopper dredger transported about 1000 m^3 of sand every 8 times per day from the borrow to the reclamation area. The dredger pumps ejected the stream of water sand mixture to an area distance about 20 – 35 m from the bow – depending on: the wind velocity and direction, the output capacity of the pumps, the relation of water to sand in the mixture (Basinski, 1985).

A part of the supplied sand reached the beach after 1.5 month. During 2.5 months the beach widen by 20 m over a distance of 400 m. On the lee-side of the reclamation area ($\sim$ 400 m) an erosion of the beach appeared.

During one year the artificial sand banks moved some hundreds meters along the shore. The sea bottom profiles became shallower. At a distance of 1 km from the nourished spot (lee-side), the beach was eroded by 10 m.

One year after the operation 17% of the supplied sand was found on a 300 m long part of the nourished area, but over the whole tested section (2.1 km long $\sim$500 m wide) about 200.000 m^3 of accumulated sand was measured. This material consisted probably partly of the artificial supply, partly of a material due to natural coastal processes.

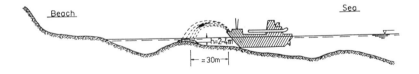

Fig. 4. Technology „B"– Rainbow system

4.3 Technology „C"

The direct beach supply started in 1989. It will be continued every year, till 1993. The borrow area was in a part of Puck Bay (fig. 5), sheltered from wayes. A small trailing suction dredger was used, of a draft of about 1.2 m. Fine sand was pumped to a distance of 2 km, first through a floating and further through a stable pipeline, crossing the peninsula. The works went on, non–stop, during several summer months in a few different places.

Below are described the results of the nourishment in 1989 at the region of Kuźnica village. Over a distance of 1845 m of the beach 652×10^3 m^3 sand was supplied (345 m^3/per m). Taking into account the flow–down coefficient, ($\sim$0.62), 407×10^3 m^3 of sand remained on the beach and its close vicinity (221 m^3/per m) (Shore ..., 1991). On the site where no natural beach existed, an artificial beach was built, $\sim$ 2 m high and 80 m wide.

Fig. 6 shows the changes of a central profile of the supplied area. During the year a considerable amount of the beach fill was transported toward SE and deposited on neighbouring bottom regions. In 1990 the nourishment was renewed. The results are under consideration. Actually (spring 1991), a wide beach maintained in the region.

Fig. 5. Technology „C" – Pumping directly to the beach zone

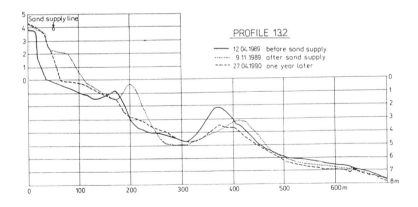

Fig. 6. Profile variations after sand supply (Basiński, Szmytkiewicz, 1991)

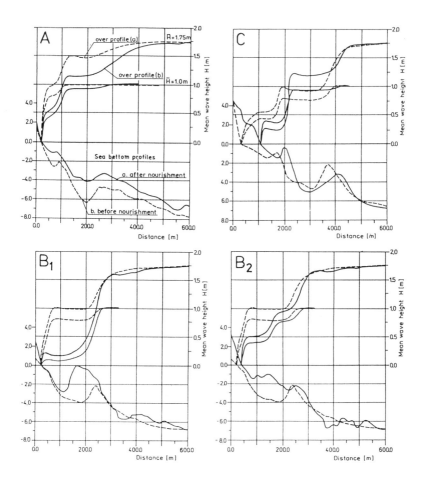

Fig. 7. Transformation of wave height over different nearshore profiles. A) technology „A"; B) technology „B" during (1) and after (2) sand supply; C) technology „C" (Basiński, Szmytkiewicz, 1991)

4.4 Dissipation of wave energy

The energy dissipation, as well as the wave height changes with distance from the shore, were determined under the assumption that the wave heights in every point of the coastal zone (including the nearshore zone) follow the Rayleigh distribution. The results for all three technologies are shown in Fig. 7.

Wave run-up was calculated for storm wave ($H = 1.75$ m, $T_p = 7.0$ s) and sea water level $+ 1.0$ m over M.S.L. Then a percent of wave energy approaching at the dune foot, over different nearshore and beach profiles, was estimated.

4.5 Conclusions

- The most effective appeared technology „C". It caused a 100% wave energy dissipation and fully protected the dune from erosion. After one year 43–46% of the fill was left on the beach and in the nearshore zone. Due to longshore transportation of the material the neighbouring sea bottom area become shallower. The results were even better after the nourishment in 1990.

- Technology „B" was the second as the effects are concerned. During the nourishment about 99% storm wave energy could be dissipated over the new artificial sea bottom and beach profile (83 – 88% over the natural profile), nearly 100% of dissipation could occure after the nourishment during the first storm season. The supplied material moved with velocities of 2 m per day on shore and of 0.5 m per day longshore. After one year 17% of the supplied sand remained on a limited area of the nourishment but the vicinity area became considerably more shallow.

- The application of technology „A" resulted in a 97–99% of wave energy dissipation (82 – 89% in natural conditions). The majority of the supplied sand was dispersed. After one year 20% of the fill remained on place.

References

Basinski T. (1985), *Experimental technology of artificial nearshore zone nourisbment* (in Polish), Inżynieria Morska No 6.

Basiński T., Szmytkiewicz M. (1991), *Effectiveness of different artificial nearshore and beach nourishment technologies*, Proc. of the 3–rd Internal. Conf on Coastal and Port Eng. i Develop. Countries. Mombasa.

Manual on artificial beach nourishment (1986), Center for Civil Eng. Research, Codes and Specification, Report nr 130, Delft.

Mierzynski S. (1985), *Experimental artificial nearshore zone nourishment on the Hel Peninsula* (in Polish), Materiały na sesję naukową, Instytut Morski, Gdansk.

Shore Protection System on the Hel Peninsula (1991), Stage 1990 (in Polish), Prace Instytutu Morskiego, Gdańsk.

Skaja M., Szmytkiewicz M., Tarnowska M. (1987), *Geomorphological effects after artificial nearshore zone nourishment at Kuźnica* (in Polish), Inżynieria Morska No 6.

ENVIRONMENTAL ASPECTS OF MINING CLASTICS MATERIAL FROM THE SEA BOTTOM

Szymon Uscinowicz

State Geological Institute, Branch of Marine Geology,
81-740 Sopot, St. Polna 62, Poland

Gravel and sand mining from the sea bottom may have a negative influence on the marine environment; the sea bottom and the watt layer, and in fact on the marine biocenosis.
Investigation carried out during 1988/89 on the Slupsk Bank / Southern Baltic / gave first data on the scale marine environment changes caused by gravel mining. Within the 1 km x 1 km test field detailed observations and oceanographic, sedimentologic and biologic measurements were made.
Oceanographic investigations aimed at determining the temporal and spatial scale of change hydrologic parameters during some days directly before the mining operation, and after defining the hydrologic background of the area - at determining the deformation of physical fields caused by gravel mining in the inspected area.
Among others the variability of the temperature and of light attenuation coefficient / light extinction / fields during 3 days before exploitation, and also distributions of these parameters 2 hours before exploitation, during the mining and 2 hours after stopping operations were determined. Perturbations of temperature and light attenuation coefficient / light extinction / fields occuring during gravel mining were recorded by towing a set of sensors at 1 m depth below sea surface. The profiling was done along profiles perpendicular to the dredgers course, directly after its passing / behind the dredger's stern /.
Observed changes of the light attenuation coefficient and of temperature exceed significantly earlier observed background fluctuations. The width of the disturbance zone did not exceed 50 m. Profiling repeated several minutes after stopping dredger operation and measurements carried out 2 hours later showed that both the temperature field and the extinction field quickly return to equilibrium state / L.Gajewski, Sz.Uscinowicz, 1991 /.
Sedimentological investigations included:

- determination of the content and variability of suspensions in the sea water,
- determination of concentration and grains' sizes of mineral suspensions in water flowing from the dredger and falling onto the sea bottom,
- repeated observations of sedimentary structures around points / markers / installed in the sea bottom,
- repeated measurements of bottom elevation changes at the datum points / markers /
- repeated observations and measurements of troughs left by the mining operations.

Average suspension content in water was: in May 1988 - 1.3 mg/l, in August, 1 day before exploitation - 2.0 mg/l. Suspension concentration in water flowing from the dredger during operation was between 1510 mg/l and 11250 mg/l, depending on exploitation intensity. In the granulometric composition of suspension fractions 0.25-0.125 mm and 0.125-0.062 mm dominate. Material smaller than 0.063 mm occured in the same quantities as in the mined field, i.e. the percentage was below 1%.

Measurements of suspension concentrations in sea water, made during mining and 1 hour after stopping operations, showed that the material falls very quickly on the sea bottom and does not propagate further than 50 m to both sides of the dredger's course. Similar results were obtained from light attenuation coefficients measurements.

The amount of suspensions falling onto the sea bottom was measured by means of 16 traps placed on the sea bottom along two profiles, 8 traps along each profile. The traps were placed at distance of 50, 150, 300 and 500 m to both sides of the dredger's course. The thickness of the sand settling on the bottom of the mined troughs was 0.5 to 1.0 cm, which means that the suspended material settles at a rate of 7500 to 15000 g/m^2. In traps placed at a distance of 50 m from the dredger's course settlement of 1.29 to 1221.52 g/m^2 was recorded.

At distances exceeding 50 m the amount of suspended material falling to the bottom decreased very quickly. The grain size of suspended matter falling onto the bottom is similar to the material flowing off the dredger.

Direct, repeated several times observations of the bottom and measurements at the datum point showed that where on the bottom surface are course sands with an addition of gravel, a layer of at least 0.2 m thickness is mobile, equal to the maximum hight of ripplemark crests.

In areas where on the bottom surface coarse gravel is present and there are no ripplemarks, no accumulation or erosion processes have been observed during one year of measurements. In areas where on top of the gravel are fine sands, the process of ripplemark generation is accompanied by local accumulation and erosion processes. Maximum observed during one year accumulation was +0.15 m and erosion -0.15 m. These processes are connected with the propagation / migration / of fine sand fields / sand pathes and probably sand ribbons / over the gravel deposits.

Observations of troughs left over the mining were carried out directly after the mining operations and then after exploitation.

Inital trough depth was 0.2 and 0.7 m. After 2.5 months the depth of partly silted up troughs was 0.1 m to 0.15 m. Traces of the troughs were visible also after 9 months and their depth was 0.04 m to 0.12 m. Observation of bench marks driven into the troughs and analysis of grains' size distribution of sediments indicate that the troughs have been filled up partially by material sliding down from trough edges and partially by fine sand migration along the bottom surface.

Biological observations aimed at determining the qualitative changes in the biocenosis of the Slupsk Bank caused by gravel mining. Investigations carried out during the last twelve years showed that the hitherto exploitation of aggregates has not disrupted biocenosis equilibrium. Benthos and periphyton of the Slupsk Bank has a high ability for developing on postmining excavations. A year after ending mining operation, fairly numerous macrozoobenthos and proliferous development of microperiphyton were found in the excavations / G.Okolotowicz 1991 /.

In view of the planned mining of gravel and sand in the Polish part of the Southern Baltic it will be necessary to monitor the exploited areas in order to obtain knowledge about longterm effects in hydrological, sedimentological and biological aspects and to prevent possible ecological damage.

References

1. Gajewski, L., Uscinowicz, Sz., 1991 - Hydrologiczne i sedymentologiczne aspekty exploatacji kruszywa na Lawicy Slupskiej, Inzynieria Morska i Geotechnika. No.4. Gdansk.

2. Okolotowicz, G., 1991 - Biologiczne aspekty exploatacji kruszyw z Lawicy Slupskiej. Inzynieria Morska i Geotechnika. No.3. Gdansk.

SEDIMENT VARIABILITY - AN IMPORTANT ELEMENT IN THE EVOLUTION OF TIDAL FLATS

EITNER, V. & RAGUTZKI, G.
Forschungsstelle Küste
- Niedersächsisches Landesamt für Wasser und Abfall
An der Mühle 5
W 2982 Norderney (Germany)

Abstract

Sediment distribution and properties should not be considered as time and space invariant because of the complex interaction of hydrodynamics, morphology, biology and sedimentology. There are many dynamic conditions which produce spatial and temporal variation of sediment distribution. In reality, these changes often reflect natural deviations of a dynamic equilibrium.

Sediment samples taken monthly from 100m^2 areas in the tidal flats of Niedersachsen (Lower Saxony), reveal the problems relating to temporal and spatial variability. Sediment samples taken with 6 cm cylinders, were used to examine short-term sediment redistribution. Spatial variability was determined from the differences between five samples taken from each area. A mean for each area was used to analyse temporal variability. Samples cored to a depth of 12 cm, display the characteristics of longterm changes, because short-term sediment redistribution is predominantly restricted to the upper millimeters/centimeters.

In the past, the aspect of spatial and temporal variability was often neglected in sedimentary mapping. This led to an incorrect picture of a static sediment distribution.

This analysis presents a more complex picture of sediment dynamics. The variability is an essential component of the development of the tidal flats. The type of sediment analysis and hence the scale of the survey, will depend on the required approach to coastal and ecosystem protection.

Introduction

Sediment distribution on tidal flats is subjected to many external influences, of which the hydrodynamic, morphological and biological conditions are considered to be the most important. The dynamic evolution of sediments should be placed into the foreground of research because of these complex ecosystem interactions.

In the past, the large changes of sediment distribution were not often recorded. Although it has always been known that sediments are subjected to steady redepositions, these changes were only analysed by morphological studies - especially when coastal protection and shipping interests were affected, for example, severe dune or beach erosion and silting-ups of shipping routes. These processes were considered in so far as technical and financial capabilities would allow. Securing of the west-spits of many east Frisian islands by solid seawalls, revetments and groins should be regarded as such an attempt. In the long-term, dynamic processes cannot be controlled by these static measures. Therefore more natural forms of coastal protection were found, e.g. artificial beach nourishment.

Information about the transport processes and morphological changes can be infered from a knowledge of the sediment composition and its properties. Variability was not often given with mapping of sediments. The maps which were made during the geological survey, give the impression that the sediment distribution was as unchangeable as a geological formation. Only recent sedimentary maps have represented temporal changes. Herewith the next argument arises because the spatial variability was mostly neglected in this context. The spatial variability could be confined by cluster or other statistical analysises but the result was patchy sedimentary maps which are only hard to correlate with hydrodynamic conditions.

The neglect of variability may be acceptable when a rough sediment classification e.g. sand, silt and clay has been used. The problems of variability have to be considered, if finer classification will be made. The finer the grain analytical resolution is, the more the variability will be placed into the foreground. The sampling style is due to the required approach of sediment analysis, because a sample is only representativ for a certain area within certain variability boundaries.

Therefore sediment dynamic changes - a succession of erosion, sedimentation and redeposition - are the unalterable natural processes of tidal flats. The change of sediment distribution is known as natural variability, which comprises of both a temporal and a spatial component. Moreover, errors can be made during sampling, preparation or analysis, which will influence variability further.

The Temporal and Spatial Variability of Tidal Sediments

Sedimentological and soil-physical parameters were determined monthly in an attempt to analyse temporal and spatial variability of tidal flat sediments in seven areas within the tidal flats of Niedersachsen (Lower Saxony) during 1991 (fig. 1). These areas are situated along two profiles in the backbarrier tidal flats of Norderney Island and Spiekeroog Island. It was also possible to determine the influence of along-shore groins on sedimentation conditions as well.

Five sediment samples were taken monthly using a 6 cm-cylinder in each 100 m^2-area. The spatial variability was determined from a calculation of standard deviation and range. A mean of these data was used to analyse temporal variability which may be also defined as the difference between maximum and minimum of the monthly means. Moreover, samples cored to a depth of 12 cm show characteristic long-term changes, since short-term effects are mostly restricted to the upper millimeters/centimeters. The penetration and shear resistance was determined in the field, and the grain size distribution, water content, bulk and dry bulk density, ignition loss, carbonate content and organic content of the sediment samples were measured in the laboratory.

In addition, the sediment variability can be revealed by only two parameters: grain-size

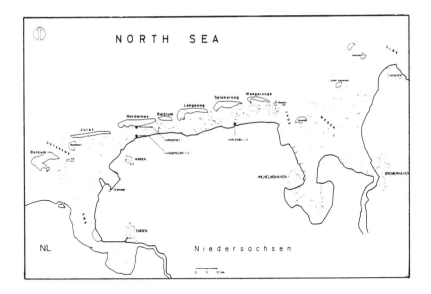

Fig. 1: Location map

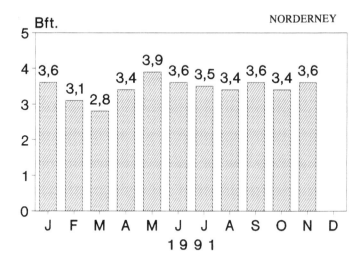

Fig. 2: Wind forces in 1991 (in Beaufort)

distribution and organic content.

Grain-Size Distribution

The grain size distribution was determined by wet sieving analysis. The grain-size distribution and temporal/spatial variability of the silt and clay content show how different sedimentation conditions may be developed in along-shore groin fields (fig. 3, 4). The content of silt and clay is, as a rule less than 50 % in the area Mandepolder 1 (tab. 1). But the area of Harlesiel 1 shows values about 20-30 % higher. These differences are due to the fact that the Mandepolder areas are exposed to stronger hydrodynamic forces than the Harlesiel-areas because they are situated closer to a tidal inlet.

Area	Temporal Variability (mean) -6cm -12cm		Difference of anual maximum and minimum	Spatial Variability (standard deviation)	Spatial Variability (range)
Mandepolder1	10 %	12 %	40 %	3 - 12 %	7 - 33 %
Mandepolder2	50 %	26 %	75 %	1 - 20 %	1 - 56 %
Mandepolder3	17 %	8 %	55 %	1 - 21 %	4 - 56 %
Harlesiel 1	15 %	20 %	40 %	2 - 16 %	7 - 39 %
Harlesiel 2	20 %	40 %	35 %	2 - 13 %	6 - 35 %
Harlesiel 3	7 %	9 %	17 %	2 - 7 %	4 - 17 %
Norderney	3 %	3 %	7 %	1 - 2 %	1 - 5 %

Tab. 1: Temporal and spatial variability of mean of silt and clay content in 1991

This fact is revealed in fig.5 and 6, in which a different evolution of sediment compositions can be recognized. The mean silt and clay contents are quite high (about 50 %) at the beginning of the year. But soon they reduce to ca. 30 %. This short-term covering by finer grained sediments reflects the calm sedimentation conditions (fig. 2). This decrease of almost 20 % (temporal variabilty), contrasts with a spatial variability of standard deviation of 2-13 % (tab. 1). A similar temporal decreasing grain size can be recognized in the area Harlesiel 2 (the backbarrier tidal flats of Spiekeroog Island) within the course of the year, but to a lesser extent. The spatial variability of both areas is similiar, but the temporal ranges are quite different.

The spatial variability of Mandepolder 2 is rather high in the beginning of the year. At this time, the silt and clay content is also high. The reverse can be recognized in Harlesiel 2. Here, the spatial variability increases towards the end of the year.

In the beginning of the year, the area Mandepolder 3 (an area some distance from the coast) shows an high spatial deviation range of the silt and clay content which strongly

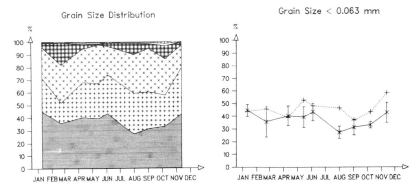

Fig. 3: Mandepolder 1

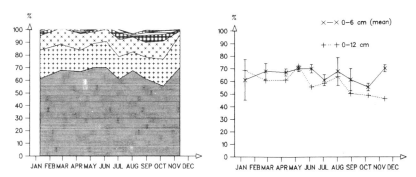

Fig. 4: Harlesiel 1

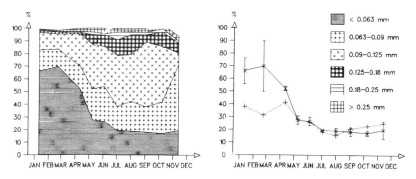

Fig. 5: Mandepolder 2

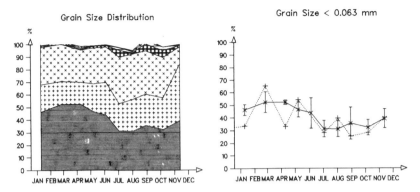

Fig. 6: Harlesiel 2

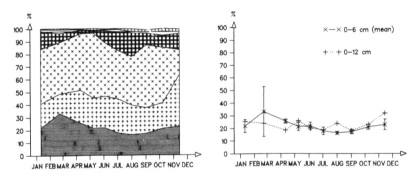

Fig. 7: Mandepolder 3

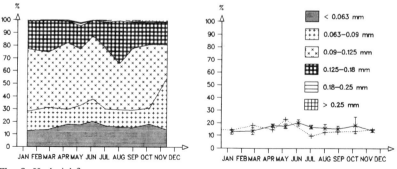

Fig. 8: Harlesiel 3

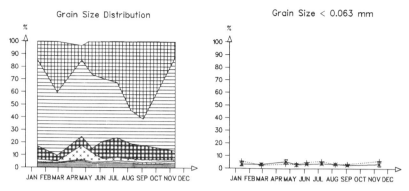

Fig. 9: Norderney

decreases (fig. 7). The comparable area Harlesiel 3 is characterized by a coarser grained sediment composition (fig. 8). The spatial variability can be typified as small, but it is even less in the area Norderney (fig. 10) because of the extremly small proportion of grain sizes < 0.063 mm.

Generally a decrease of spatial variability of silt and clay content can be recognized in the summer months. The calmer hydrodynamic conditions of this season cause a reduced intensity of redeposition which produces a smaller deviation range.

The highest temporal variability can be recognized in the area adjacent to the along-shore groins (Mandepolder 2 and Harlesiel 2). The areas of the groin fields are characterized by little variance which is in turn less on areas with higher sand contents (tab. 1).

The temporal variation of samples cored to a depth of 12 cm (showing long-term developments) is less than that of the 6 cm-samples, in Mandepolder areas. The opposite situation is developed in the Harlesiel areas (tab. 1). This shows once more the complex problems which appear in this analysis.

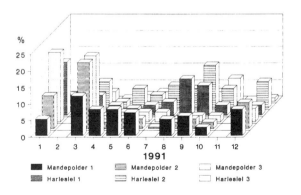

Fig. 10: Spatial variability - grain-size < 0.063 mm

ORGANIC CONTENT

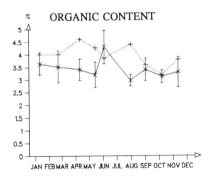

Fig. 11: Mandepolder 1

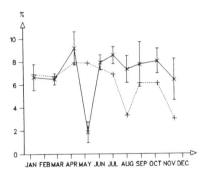

Fig. 12: Harlesiel 1

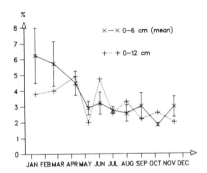

Fig. 13: Mandepolder 2

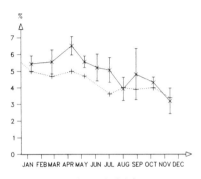

Fig. 14: Harlesiel 2

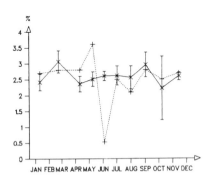

Fig. 15: Mandepolder 3

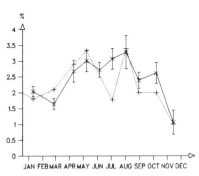

Fig. 16: Harlesiel 3

Organic Content

The organic content, which is mainly transported in suspension is deposited during calm sedimentation conditions. Therefore the highest organic content can be expected in the groin fields. But there are differences between the areas Mandepolder 1 and Harlesiel 1, which could be already recognized in the grain-size distribution previous (fig. 11, 12). In the latter case there are organic content values of about 10 %. But there are not only differences in absolute value but also in variability (tab. 2, fig. 18).

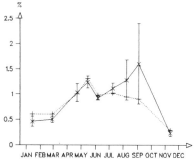

Fig. 17: Norderney

The annual variation of the organic content in the areas in front of the groins, Mandepolder 2 and Harlesiel 2, is quite similiar as are their grain-size distribution. This demonstrates the close connections between organic content and grain size, especially silt and clay content. There is similiar situation in terms of the spatial variability which decreases as a trend in area Mandepolder 2 and increases in area Harlesiel 2 (fig. 18).

Area	Temporal Variabilty (mean) -6cm	-12cm	Difference of annual maximum and minimum	Spatial Variability (standard deviation)	Spatial Variability (range)
Mandepolder1	1.3 %	1.4 %	3.2 %	0.2 - 0.7 %	0.7 - 1.9 %
Mandepolder2	4.4 %	5.1 %	6.2 %	0.1 - 1.8 %	0.3 - 4.6 %
Mandepolder3	0.8 %	3.1 %	2.7 %	0.2 - 1.0 %	0.4 - 2.7 %
Harlesiel 1	7.3 %	4.6 %	10.9 %	0.6 - 1.9 %	1.3 - 5.2 %
Harlesiel 2	2.6 %	1.4 %	5.1 %	0.3 - 1.5 %	0.9 - 4.1 %
Harlesiel 3	1.6 %	1.6 %	3.0 %	0.2 - 0.5 %	0.5 - 1.4 %
Norderney	1.1 %	0.7 %	2.8 %	0.1 - 0.8 %	0.1 - 2.2 %

Tab. 2: Temporal and spatial variability of organic content

It is not possible to make relevant statements on the variability of the organic content in the areas Mandepolder 3, Harlesiel 3 and Norderney (fig. 15-17). In the Mandepolder areas the temporal variability of the 6 cm-samples is generally smaller then of the 12 cm-samples. The contrary situation is developed in Harlesiel areas. Here the similiarity between the results of grain size distribution and organic content can be seen again, but this time in an inverse effect.

Conclusions

This paper considers the high spatial and temporal dependence of sediment distribution. They tend to complicated sediment distribution hypotheses at any location, as well as showing where further developments can be made. Taking single samples only once does not allow a seperation to be made between temporal and spatial variability.

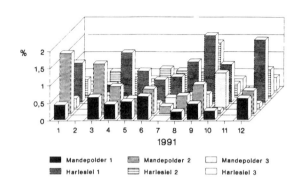

Fig. 18: Spatial variability - organic content

Moreover, the paper reviews the scale of the variation of sediment distribution on tidal flats. A change in the sediment composition is usual and not necessarily a sign of changing external - maybe even anthropological - influences. The results of Mandepolder 1 and Harlesiel 1 illustrate that coastal protection measures influence sedimentation conditions. Groin fields reduce the turbulence of currents and especially waves. They support a building-up of sediment up to a certain degree which is influenced by a lot of external factors. The sediment supply, the height of the groins in relation to the tidal flats and the hydrodynmic conditions all play an important role. A comparison of the groin field areas confirm this presumption, because both areas are characterized by different sediment distributions. Natural sedimentation conditions without a filtering of the wave influence (i.e. outside of the groin fields) are typical of areas Mandepolder 2 and Harlesiel 2. The temporal and spatial variability of these areas reveals the dynamic sediment transport and exchange without the hindrance of groins.

With respect to environmental protection another conclusion can be drawn. Natural processes change sediment distribution because of their dynamics and therefore also biotopes without endangering the ecosystem. The ecosystem of tidal flats is a highly dynamic system that reacts on changes - as far as they do not exceed the natural variation - in a dynamic way.

Acknowledgments

The studies were financially supported by the Umweltbundesamt and the Federal State of Niedersachsen (Lower Saxony).

BEACH CUSP GRANULOMETRY: A STUDY OF BEACH CUSP SEDIMENT GRAIN-SIZE STATISTICS.

PETER JANCA KRISTENSEN (Dept. of Earth Sciences, University of Aarhus, Denmark),
GEORGE GHIONIS (Dept. of Marine Geology, University of Patras, Greece) and
CHRISTIAN CHRISTIANSEN (Dept. of Earth Sciences, University of Aarhus, Denmark).

Abstract

Three series of sediment samples from beach cusps of different cusp spacing from the Gulf of Kyparissia have been examined.

Grain-size statistics for two of the series C and U Folk & Ward and moment parameters were computed based on sieving technique (½ Φ-interval). Series V was analyzed using settling-tube technique. In addition all series was fitted by the hyperbolic distribution.

Although simple and of low resolution grain-size spectra maps clearly showed the trends obtained by the more sophisticated computations of hyperbolic parameters.

The parametric comparisons showed extensive shore parallel differences between Folk & Ward and moment parameters, specially the shape parameters (skewness and kurtosis). The extensive variation of shape parameters mainly originated from the bimodal nature of grain-size distributions. All parameters of central tendency showed a systematic variations between cusp horns and bays, in contrary to the values of sorting. Only the fitted hyperbolic distribution showed systematic variation i sorting of the grain-size distribution of horn and bay sediments systematically.

The mineral composition of the sediment indicates that the distributional bimodality originates only from shape and size of the single grains, i.e. not by density differences.

The spatial distribution of beach cusp sediment settling-velocity indicated the presence of underwater deltas seawards of the bays. It is proposed, that beach cusp sediment datasets should be reinterpreted using the settling-tube technique combined with the hyperbolic distribution.

Introduction

Beach cusps are morphologic features occurring in the swash zone. They appear as seaward pointing promontories separated by embayments. Underwater deltas may be formed seawards of the embayments (Kuenen, 1948). The spacing between to successive horns, λ_c, varies between one 10th of a meter to more than 60 m (Komar, 1976). The difference in elevation between horns and bays varies from a few centimeters to more than one meter (Gorycki, 1973) The slope of the swash zone, β_s, is usually greater at horns than at bays, but no correlation between λ_c and β_s has been shown (Russels & McIntire, 1965).

The most studied theme is without doubt the formation and occurrence/frequency of the cusps (e.g. Guza & Bowen, 1981). Although the majority of articles gives thorough investigations both on natural beaches and in laboratory basins (e.g. Bagnold, 1940; Antia, 1989) some facets of the *beach cusp story* still needs elucidation. Numerous authors have performed grain-size analysis on beach cusp sediment (e.g. Pyökäri, 1982), but no systematism in the grain-size distribution seems to be present (Kristensen, 1991).

The purpose of the present study is to through more light upon the sediments in which beach cusp develop. Knowing that beach cusps can develop in a great variety of sediments (e.g. Russels & McIntire, 1965), and that the mineralogic and size composition of the sediment is a product of the local geology and processes the aim of the present study is to find out whether the scale and location invariant parameters of the hyperbolic distribution provide for more information than traditional parameters.

Study Area

Facing the Ionic Sea the Gulf of Kyparrisia is situated on the western coast of Peloponnesos, Greece (Figure 1). The approximately 80 km semicircular shaped Gulf is oriented in a NW-SE direction and bound by rocky headlands. The prevailing wind directions approach from NW, SW and W resulting in theoretical maximum wave periods of respectively 6.7s, 11.6s and 9.25s. The tidal wave amplitude of 20 cm classifies the area as microtidal.

Three major morphologic regions can be distinguished on aerial photographs (Ghionis & Ferentinos, 1988). Using the Australian beach stage model (Wright & Short, 1984) the northern part is dissipative, the central part predominantly

Figure 1 Location of the Gulf of Kyparissia on the western coast of Peloponnese, Greece.

intermediate while the southernmost parts mostly occur in reflective beach stage. The different energy regimes throughout is reflected in the mean grain-size of the sediment giving fine sand ($M_z = 3$ Φ) at Katakolo in the north and very coarse sand ($M_z = 0$ Φ) in the south at Kyparissia (Pannagos et al., 1976). Analogous the mean beach slope steepens from 1-2° to 5-7° from north to south. The submarine topography (distance to 20 m and 50 m depth contour) shows similar trends. Sandy material is present in the coastal zone to a distance of about 200 m from the shoreline.

Methods

Three series of beach sediment were collected from the swash zone representing different aspects of beach cusp sediment. **Series C** were collected successive on the apex of the horns and in the middle of the bays. **Series U** were collected in a shorenormal transect over the apex of a relict mega cusp, and the **series V** were collected in a grid over an active beach cusp. The series thus represent both the shorenormal and shoreparallel as well the spatial variation of beach cusp sediment.

All samples were collected using a 75 mm ∅ corer driven into the sediment to a depth of 50 mm. The bulk sample was dried at 110 °C and split into 15 - 20 g samples. Two laboratory methods were used to determine the grain-size distribution. Sieving of split samples for 20 min in a vibrator using ½ Φ-interval (series C and U) and settling-tube analysis of split samples of 2-5 g (series V).

Grain-size statistics were computed by the method of moments (Griffiths, 1967) and the Folk & Ward (1957) parameters were calculated as well. In addition all samples were fitted by the hyperbolic distribution using the method in Christiansen & Hartmann (1988a) (the parameters are explained in the appendix).

Selected split samples were pulverized in an agathe mortar for determining the mineral composition using X-ray diffractometry.

Unrefined balance readings were used to construct *grain-size spectra maps* (GSS-maps) according to Kristensen (1991). This presentation do not presume any distribution of the grain-size population.

Sieving was performed at the University of Patras, settling-tube analyses at the Christian-Albrechts-Universität in Kiel and X-ray diffraction at the University of Århus.

Results

According to Folk & Ward (1957) the **series C** samples are medium grained sand, well sorted, very negative skew, very platykurtic (using the parameters from the method of moments computed on Φ-scale).

To testify the proficiency of the various parameters to reflect the morphology all parameters were plotted as a function of sample number representing successive cusp horn and bays. The measures of size mean and M_z and of sorting sdv and σ_I gave similar results (Table 1) whereas the shape parameters (SK/Sk_I and KU/K_G) parameters showed comprehensive variability.

The hyperbolic distribution has two parameters of central tendency (ν and μ). ν, which is regarded as the typical (peak) grain-size showed the same relative picture as mean and M_z, but tended to assign the samples a smaller diameter (in mm).

SERIES C	BAY mean ± sdv	HORN mean ± sdv
mean	1.71 ± 0.14	1.50 ± 0.15
sdv	0.49 ± 0.04	0.52 ± 0.04
skewness	-0.41 ± 0.12	-0.30 ± 0.23
kurtosis	0.63 ± 0.30	0.37 ± 0.54
M_z	1.72 ± 0.14	1.50 ± 0.16
σ_I	0.50 ± 0.05	0.53 ± 0.04
Sk_I	-0.08 ± 0.05	-0.07 ± 0.06
K_G	1.05 ± 0.08	1.02 ± 0.07

Table 1 Mean values of moment and Folk & Ward parameters of cusp horn and bay sediment samples.

The degree of sorting can be expressed in several ways using the hyperbolic parameters. Unfortunately no verbal classification exists to describe the sorting. δ, κ, τ and ζ are all sorting parameters, but of these only δ was able to show the systematic variation, that horn sediment is poorer sorted than bay samples (Figure 2). Neither sdv nor σ_1 showed this trend systematically. The extreme value of δ for sample No 8 is a consequence of the very near normal distribution of the grain-size population of this sample. Two of the samples (No 13 and 14) are labeled respectively intermediate bay and horn because they were less distinct comparing to the rest of the horns and bays. It also

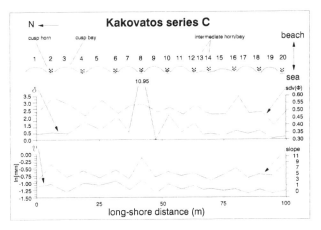

Figure 2 *The upper part shows location of samples series C. In the lower part the curves represents from above sdv(Φ), δ, slope (°) and ν (ln[mm]).*

appears from Figure 2 that both ν and δ was able to distinguish this delicate change in morphology. The slope of the intermediate horn (sample No 14) was even somewhat smaller than the two adjacent bays.

The domain of the hyperbolic distribution known as the shape triangle (a scatterplot of ξ and χ) has been used to separate different sedimentary environments (Hartmann, 1988) as well as erosional/depositional areas (Barndorff-Nielsen & Christiansen, 1988). When plotting the samples of series C it appears

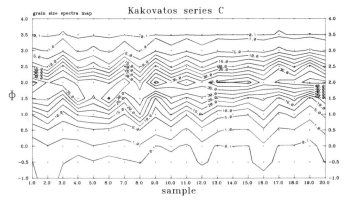

Figure 3 *Grain-size spectra map for samples series C. Location of tha balance readings are marked with dots. Note the systematic sinusoidal trend of the contour curves.*

that all samples are hyperbolic positive skew, and all placed in the predominantly depositional part of the triangle. The bay samples tend to be concentrated in the lower part of the triangle reflecting more normality.

The **GSS-map** for series C (Figure 3) shows, that the contour curves from 1.0 % to 35.0 % of the fine flank (upper) all have a sinusoidal pattern. This is also the case for the coarse flank, but less distinct. The sinusoidal curvature reflects the relative coarser nature of the horn samples (even numbers). The peaked percentage values of the bay samples (odd numbers) combined with the steep gradient of the flanks illustrates a better sorting in bays than on horns. In addition the 0.1 % and 1.0 % contour curve of the coarse flank shows a rhythmic pattern (at samples 20, 16, 12, 8, 5 and 2) with a greater spacing ($2\lambda_c$) than the horn-bay variation ($\frac{1}{2}\lambda_c$). The periodical variation of $2\lambda_c$ may reflect an other event than the $\frac{1}{2}\lambda_c$ periodicity, although this is not reflected in any of the sorting parameters. This may indicate a secondary edge-wave of different wavelength.

The **series U** sediments are characterized by coarse grained sand, moderately sorted, very positive skew and very platykurtic (using the parameters of the method of moments computed on Φ-scale).

Both the mean grain-size (mm-scale) and the sdv increased landward, i.e. the sediment is coarser and has a poorer degree of sorting landward. This is reflecting decreasing energy flux as the sea level is lowered during the time the storm is reduced.

The shape parameters showed extensive variation between respectively sk_I and SK as for k_G and KU, albeit both SK and KU showed distinct decrease respectively increase trends in landward direction. A great variability was observed in the very sensible sorting parameters (δ, κ, τ and ζ).

φ and γ expresses the slope of respectively the fine and coarse asymptote of the fitted hyperbola. These parameters tended to increase landward, i.e. resulting in more narrow distributions of the grain-size distribution going landward reflecting a better sorting in contrary to sdv.

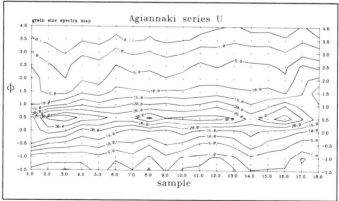

Figure 4 Grain-size spectra map for samples series U. Note the trend of the contour curves towards the lower left corner and the different increment of the gradient of the two flanks.

The **GSS-map** for series U (Figure 4) visualize the trends in both mean grain-size as well as in φ and γ. The 0.1 % to 15 % contour curves are displaced approximately 1 Φ in the coarse direction going landward (to the left), i.e. the mean grain-size increases landward. The 10 % to 30 % curve on the fine flank and the 1 % to 25 % on the coarse flank are more spread seawards than landward, which illustrates a better sorting in landward direction. In addition it can be seen, that the gradient of the two flanks increases landward with different rates, as also observed in the φ and γ trends.

Comparisons between sieving and settling-tube analyses were performed on selected samples and resulted in the following trends, when settling is used instead of sieving: **1)** the mean size increases, and ν and mean gave different trends, **2)** sorting parameters sdv, δ, ζ and κ indicated poorer sorting. The distribution in the central part (τ) tended to be more narrow indicating a better sorting in the central part, and **3)** a considerable better fit of the hyperbolic distribution to the grain-size population.

The use of settling-tube analyses has been discussed widely (e.g. Sengupta & Veenstra, 1968; Brezina, 1979) and one of the advantages using this technique is the great amount of datapoints provided for each sample resulting in a more accurate computation of both moment and hyperbolic statistical parameters (Lund-Hansen & Oehmig, 1992).

The characteristics of the grain populations of **series V** were determined exclusively using the settling-tube technique. The samples have a velocity specter of -0.5 to 3.5 Ψ (~ 1.4 to 11.3 cm/s). Most of the samples showed bimodality with a dominating coarse modal. The mineral composition is limited to only three different species: calcite, quartz and plagioclase with calcite as the overall prevailing mineral. Quartz to

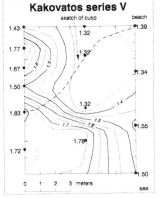

Figure 5 Above: location of samples series V. Right: Spatial distribution of the typical settling velocity (ν) on the sampled active beach cusp. Unit is ln[cm/s]; 1.72 ln[cm/s] ~ 5.58 cm/s and 1.32 ln[cm/s] ~ 3.74 cm/s.

calcite ratios were in the range of 0.37 to 0.63. Hence, the observed bimodality is caused solely by the shape and size of the individual grains and not by differences in density.

The typical (peak) settling-velocities expressed as ν varies between 3.72 and 6.25 cm/s. In general the settling-velocities are greatest seawards and decreases landward. Settling-velocities are greatest on the horn and lowest on the flank of the cusp. The relatively high settling-velocities seawards of the bay might indicate the presence of a underwater delta (Figure 5).

The parameters of sorting (δ, ζ, τ, κ and sdv) showed no systematic variation in the sorting of settling-velocities, but there is a tendency, that sorting is poorer landward both for the horn, flank and bay row. There is also a tendency, that the samples increases in hyperbolic positive skewness landward.

The parameters of kurtosis (KU and ξ) do not shown any systematic trends, but horn samples tends to be plotted in the upper left part of the group in hyperbolic shape triangle. All samples were hyperbolic positive skew.

Discussion

The results above demonstrate that grain-size parameters shows great variability in two ways. Firstly, the applied computation methods yield different interpretations of the sediment. Secondly, within the various methods some of the parameters do not directly reflect the morphology observed and therefore the morphological conclusions to draw are limited.

Several reasons for the apparently chaotic parametric response to the morphology can be claimed. The main uncertainties are sampling method, laboratory procedure and the selection of statistical approach to the grain-size distribution.

The collection of individual samples must be done within the active layer representing the actual dymanic event. Chafetz & Kocurek (1981) observed a vertical variation of M_z in the sediment of cusp horn from 1 Φ to -2.8 Φ from surface to a depth of 50 mm. The sampling method has a great influence on the estimation of hyperbolic parameters (Christiansen & Hartmann, 1988b). The depth of the active layer varies in time and space under alternating wave energy regime (breaker type/high and angle of approach). Bryant (1984) found that 2 g surface samples were representative for both mean and sdv in a radius of 5 m.

The laboratory routine used might as well have influence on the computed parameters. The use of ½ Φ-interval instead of ¼ Φ-interval reduces the amount of datapoints by 50 % and results in an inaccuracy in the computation of both moments (Kennedy et al., 1981) as well on hyperbolic parameters (Christiansen & Hartmann, 1988b). In case of bi- or polymodal distribution the error increase (Hansen, 1974; Fieller et al., 1984) because the change from the fine to coarse modal often is seen within one half Φ. In addition the use of non-calibrated sieves have influence on hyperbolic parameters (Dalsgaard & Sørensen, 1985).

In contrary the use of settling-tube reduces the possible maximum errors remarkable (Sengupta & Veenstra, 1968). Using settling-velocities instead of "equivalent" grain-size results in a better environmental discrimination (Reed et al., 1975) and a better fit of the hyperbolic distribution (Lund-Hansen & Oehmig, 1992). The fact that the results of series V do not show any distinct systematism may be a result of the small amount of samples representing the same event.

The work of Flemming (1964) dealed with the distribution of the shape of the grains of beach cusp sediment. The bay sediment consisted predominantly of plate shapes in contrary to the sphere shaped grains of the horns. Flemming (1964) also noticed, that sediment of high settling velocity was deposited in the seaward part of the cusps. Observations of the mobile active surface layer consisted of coarse fractions of same density, which states that the shape of the individual grains are very important in cusp sediment. From the present work and the results of Flemming (1964) it follows, that the use of settling-tube combined with the hyperbolic distribution is a suitable method for the analyses of beach cusp sediment

The use of different statistical approach to the grain-size distribution has been discussed widely. The great differences of the parameters of 3rd and 4th order in the calculation of moments and Folk & Ward (1957) parameters is caused by the rough approximation of the Folk & Ward (1957) parameters to the moment measures. The Folk & Ward (1957) parameters do not take to account the fraction below the 5 % and above the 95 % percentiles and thus results an incorrect picture of the sediment characteristics.

It was also seen that trends in sorting (δ and sdv of series U) varies with the used computation method.

To proove any systematism in the grain-size statistics of beach cusp sediment cannot be done by comparing the results of the foregoing work of other scientists. It is recommended that future work should be done using settling tube analyses combined with fitting of the hyperbolic distribution. Comparisons of beach cusp, mega cusp and giant cusp sediments should be done using the location and scale invariant parameters of the hyperbolic distribution.

Appendix

Φ is defined as -$\log_2$ (grain-size [mm]), Ψ as -$\log_2$ (settling-velocity [cm/s]).
The parameters of Folk & Ward (1957).

M_z	graphic mean grain-size
σ_I	inclusive graphic standard deviation
Sk_I	inclusive graphic skewness
K_G	inclusive graphic kurtosis

Parameters of the method of moments.

mean	arithmetic mean
sdv	standard deviation
SK	skewness
KU	kurtosis

Parameters of the hyperbolic distribution (Figure 6).

ν	nu	typical (peak) grain-size
μ	mu	parameter of central tendency
δ	delta	sorting parameter
κ	kappa	sorting parameter ($\kappa = \sqrt{\phi\gamma}$)
τ	tau	sorting parameter
ζ	zeta	sorting parameter
φ	phi	slope of fine/slow asymptote
γ	gamma	slope of coarse/fast asymptote
χ	chi	approximated value of the hyperbolic skewness
ξ	ksi	approximated value of the hyperbolic kurtosis

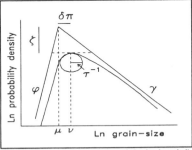

Figure 6 Geometric interpretation of the log-hyperbolic parameters.

Acknowledgments

The authors wish to thank Dr. Reinhard Oehmig, Christian-Albrecht Universität, who kindly made the settling-tube analyses and Hanne Birch Madsen for running the x-ray diffractions analyses.

References

Antia, E. E., 1989: **Beach Cusps and Burrowing Activity of Crabs on a Fine-Grained Sandy Beach, Southeastern Nigeria.** *J. Coast. Res. Vol 5. No 2. p 263-270.*

Bagnold, R. A., 1940: **Beach Formations by Waves: Some Model-Experiments in a Wave Tank.** *J. Inst. Civ. Eng. Vol 15. paper No 5237. pp 27-52. London.*

Barndorff-Nielsen, O. E., Christiansen, C. 1988: **Erosion, Deposition and Size Distribution of Sand.** *Proc. R. Lond. A 417. pp 335-352.*

Brezina, J., 1979: **Particle Size and Settling Rate Distribution of Sand-sized Materials.** *2nd European symposium on particle characterisation (PARTEC). Nürnberg, 24-26 sep. 1979.*

Bryant, E., 1984: **Sediment Characteristics of some Eastern Australian Foreshores.** *Australian Geographer. Vol 16. pp 5-15.*

Chafetz, H. S., Kocurek, G., 1981: **Coarsening Upward Sequences in Beach Cusp Accumulations.** *J. Sed. Pet. Vol 51. No 4. pp 1157-1161.*

Christiansen, C., Hartmann, D., 1988a: **SAHARA: A Package of PC-Computer Programs for Estimating both Log-Hyberbolic Grain-Size Parameters and Standart Moments.** *Computer and Geoscienses. Vol 14. No 5. pp 557-625.*

Christiansen, C., Hartmann, D., 1988b: **On Using the Log-hyperbolic Distribution to Describe the Textural Characteristics of Aeolean Sediments - Discussions.** *J. Sed. Pet. Vol 58. No 1. pp 159-160.*

Dalsgaard, K., Sørensen, M., 1985: **A Method of Calibrating Sieves.** *Proc. Int. Workshop Phys. Blown Sand. Aarhus. Vol 3. pp 587-607.*

Fieller, N. R. J., Gilbertson, D. D., Olbricht, W., 1984: **A New Method for Environmental Analysis of Particle Size Distribution Data from Shoreline Sediments.** *Nature. Vol 311.*

Flemming, N. C., 1964: **Tank Experiments on the Sorting of Beach Material during Cusp Formation.** *J. Sed. Pet. Vol 34. pp 112-122.*

Folk, R. L., Ward, W. C., 1957: **Brazos River Bar: A Study in the Significance of Grainsize Parameters.** *J. Sed. Pet. Vol 27. No 1, pp 3-26.*

Ghionis, G., Ferentinos, G., 1988: **Geomorphological Studies in the Gulf of Kyparissia**. *National Technial University of Athens. 2nd European Workshop on Coastal Zones. Sept. 26-30 Loutraki, Greece.*

Gorycki, M. A., 1973: **Sheetflood Structure: Mechanism of Beach Cusp Formation and Related Phenomena.** *J. Geol. Vol 81. pp 109-117.*

Griffiths, J. C., 1967: **Scientific Methods in Analysis of Sediments.** *McGraw-Hill.*

Guza, R. T., Bowen, A. J., 1981: **On the Amplitude of Beach Cusps.** *J. Geoph. Res. Vol 86. No C5. pp 4125-4132.*

Hartmann, D., 1988: **Coastal Sands of the Southern and central Part of the Mediterranean Coasts of Israel - Reflections of Dynamic Sorting processes.** *Unpubl. Ph. D. Thesis. Dept. of Eath Sciences. University of Aarhus.*

Hansen, F., 1974: **Problemer med anvendelse af $^4\sqrt{2}$-skala på Bagnold diagrammer.** *Skrifter i Fysisk Geografi. Nr 7. Dept. of Earth Sciences. University of Aarhus.* (In Danish).

Kennedy, S. K., Ehrlich, R., Kana, T. W., 1981: **The Non-Normal Distribution of Intermittent Suspension Sediments Below Breaking Waves.** *J. Sed. Pet. Vol 51. No 4. pp 1103-1108.*

Komar, P. D., 1976: **Beach Processes and Sedimentation.** *Prentice-Hall, New Jersey. pp 429.*

Kuenen, Ph., 1948: **The Formation of Beachcusps.** *J. Geol. Vol 56. pp 34-40.*

Kristensen, P. J., 1991: **Beach Cusp Granulometry.** *Unpubl. M.Sc. thesis. Dept. of Earth Sciences. University of Aarhus.* (In Danish).

Lund-Hansen, L. C., Oehmig, R., 1992: **Comparing Sieve and Settling Analyses of Beach, Lake and Eolean Sediment using Log-Hyperbolic Distribution Parameters.** *Marine Geology.* (In press)

Pannagos et al., 1976: **Grainsize Parameters and Environmental Fields of Beach Sands from the Coast of the Western Peloponnese, Greece.** *Ann. Géol. Hellen. Vol 28. pp 275-298.* (In Greek).

Pyökäri, M., 1982: **Breaching of a Beach Ridge and the Formation of Beach Cusps.** *Canadian Geographer XXXVI: 4. pp 332-348.*

Reed, W. E., Le Fever, R., Moir, G. J., 1975: **Depositional Environment Interpretation from Settling-Velocity (PSI) Distributions.** *Geol. Soc. Am. Bull. Vol 86. pp 1321-1328.*

Russels, R. J., McIntire, W. G., 1965: **Beach Cusps.** *Geol. Soc. Am. Bull. Vol 76. pp 307-20.*

Sengupta, S., Veenstra, H. J., 1968: **On Sieving and Settling Techniques for Sand Analysis.** *Sedimentology. Vol 11. pp 83-98.*

Wright, L. D., Short, A. D., 1984: **Morphodynamik Variability of Surf Zones and Beaches: a Synthesis** *Marine Geology. Vol 56. No 1/4. pp 93-118.*

SEDIMENTOLOGICAL AND GEOCHEMICAL CHARACTERISTICS OF THE CARBONATIC BEACHES OF THE GULF OF OROSEI (EAST-CENTRAL SARDINIA)

A. Cristini, F. Di Gregorio & C. Ferrara
Dipartimento di Scienze della Terra
Università di Cagliari
Via Trentino 51, 09100 Cagliari, Italy

Summary
Results of research on the geo-environmental characteristics of beaches in the Gulf of Orosei in east-central Sardinia are illustrated. Moving from north to south, the geomorphological characteristics and sedimentological, minero-petrographic and geochemical parameters of sediments making up the single beaches are described in an attempt to discover possible correlations between sedimentological parameters and heavy metals read as tracers in different conditions of beach formation, both from mixed (granitic, metamorphic, carbonatic) and exclusively carbonatic petrographic provinces. Grain size distributions point to the unimodality of most of the sediments, with localized modes in determined grain-size intervals depending on the geomorphological and hydrodynamic characteristics at the sampling points. In the first five stations, from N to S, the mode is in the ambit of coarse sands, with fairly evident tails of gravel. In samples 6 to 11, where the carbonatic component is prevalent and contributions come mainly from detritis and overhanging cliffs, the mode shifts to the ambit of gravels, with some heavily polymodal samples. Samples 12 to 14 fall within the ambit of medium sands and gravel tails. Results emerging from the sedimentological analysis are compared to those obtained from the minero-petrographic and geochemical investigation. Using the fluorescence technique, each sample was analyzed for the principal elements (Al, Si, Ti, Mn, Fe) characterizing the sedimentation environments, while metal concentrations (Cr Co, Ni, Cu, Zn, Rb, Sr, Cd, Ba, Pb), investigated as possible tracers, were determined by means of spectrophotometry (ICP, AA) after aqua regia solubilitation.

1. Geomorphological organization of the coast

The Gulf of Orosei (Figure 1), included on Sheet 208 "Dorgali" of the 1:100 000 map of Italy, is situated in east central Sardinia. The coastal phase considered in the present work goes from the settled area of Cala Gonone to the Cape of Monte Santo. Geologically speaking, the entire coastal arch is closed in by thick banks of Mesozoic, carbonatic rocks of chemical and biogenic nature. Stratigraphically, from bottom to top, the succession is represented by dolostones and brown dolomitic limestones of the lower-Dogger

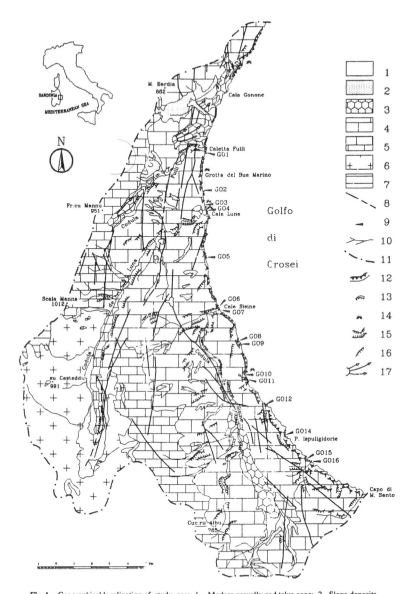

Fig. 1 - Geographical localization of study area. 1 - Modern gravelly and talus cone; 2 - Slope deposits and talus cone; 3 - Olivine basalts (*Plio-Quaternary*); 4 - Boundstones and oolitic limestones (*upper Malm*); 5 - Calcarenitic or oolitic limestones, dolostones and brown dolomitic limestone (*lower Malm-Dogger*); 6 - Granites and microgranites, felsitic porphyry of Hercynic igneous activity; 7 - Metasandstones and metasiltites; 8 - Faults and fractures; 9 - Sampling point; 10 - Drainage patterns; 11 - Watersheds; 12 - Cliff coast; 13 - Sandy or gravelly-sandy shores; 14 - Coastal caverns; 15 - Canyons; 16 - Edges; 17 - Basalt lava emission centers and flows directions.

Malm followed by stratified brown limestones, sometimes calcarenitic or oolitic, of the lower Malm, and by biogenic limestones of compact bioherms and oolitic limestones of the upper Malm (AMADESI et al., 1959; DIENI & MASSARI, 1970, 1971) which outcrop extensively over the entire curve of the Gulf. The structural set-up is determined by a series of monoclinal faults arranged in an arch-like pattern which determine the state of the Gulf. These faults dip at a mean slope of 25 to 30° (198) towards E, NE and SE in correspondence to the Tyrrhenian Sea, where the collapse of a part of the carbonatic assises took place. The faults are separated from one another or are broken on the inside by a system of fractures with a prevalently N-S trend to the north and a NW-SE trend in the southern part (CHABRIER, 1970; ASSORGIA et al., 1974). Along some of these faults, or where they intersect, the emission of olivinic basalts has taken place (LAURO, 1975) and these have led to the formation of more or less extended plateaux, some of which coalescing near the coast (e.g. Cala Gonone) or stretching out parallel to it (e.g. San Pietro). Just south of Cala Gonone, an extended fault of eboulis ordonnees, minute and well-sorted, weathering from Monte Bardia, terminates on the coast with a steep scarp. Geomorphologically, the Gulf is composed of a slightly pronounced bay closed in by high cliffs carved into the thick Mesozoic carbonatic assises and interrupted, only sporadically, by small sandy or sandy-pebbly beaches, usually situated at the outlets to the sea of water course beds known as "codule" (e.g. Codula de Fuili, Codula de Luna, Codula de Sisine). Besides these beaches, which are found in correspondence to the main "codulas" and which sometimes have the heads of their basins set to the west of the carbonatic complex (e.g. Codula di Luna) in Paleozoic, granitic (normal, medium grain granites, microgranites and pegmatitic granites), metamorphic (quartzites, crystalline schists and phyllites and rocks in dikes (aplites, porphyries, quartz, etc.), there are also locally present other small, extremely attractive beaches at the base of rocky cliffs and fed by their disintegration or the detritic faults above them.

Along the coasts characterized by cliffs and inside the numerous coastal caves, among which the one known as the "Grotta del Bue Marino", which is exploited as a tourist attraction, now-inactive notches representing coastlines going back to the Riss-interglacial period (CAROBENE, 1972; 1978) are clearly recognizable (ASSORGIA et al., 1968) and raised to a height of about ten meters above sea level. It appears that inside some caves (e.g. those of Cala Luna), outliers of marine and continental Upper Pleistocene deposits are to be found together with forms of erosion (conches and holes left by lithophagous organisms), which clearly show up four fossil coastlines (ASSORGIA, 1968; OZER & ULZEGA, 1980).

Isolated outliers of eolian sands, slightly reddened and in cross beddings, which can be attributed to the Würmian, are set here and there in the cliffs and overthrust the Pleistocene deposits. One of these (GO 9) was sampled and analyzed, with sampling carried out from N to S along the water line, with the exception of samples GO 9 of the Würmian eolian sands, GO 8 taken higher up on the beach and GO 12 taken from a submerged beach (-3 m) inside a coastal cave.

Samples underwent sedimentological and geochemical analyses, the latter by means of fluorescence and spectrophotometry. The sedimentary environment analyzed is prevalently composed of carbonatic sediments, although part of the sediments sampled on the beaches are to be considered mixed since their composition is also influenced by contributions from local water courses whose feed basins have different petrographic characteristics. It is therefore of interest to investigate the differences existing in the distribution and geochemical behaviour of metals, passing along the coast from sediments with a higher percentage of silicates to the north to those of a prevalently carbonatic composition to the south; consequently, it is also opportune to see if the presence of certain metals (Sr, Pb, Zn, Mn and Fe) depends on the type of diagenesis (BENCINI & TURI, 1974; DUCHI et al., 1990).

Sedimentological analyses were performed by means of dry sieving with 1/4 Φ sieves. The most significant grain-size distribution curves and textural parameters (Mz, SKI, sI)(FOLK & WARD, 1957) led to the definition of the sediments from N to S.

Chemical analyses were carried out by means of x-ray fluorescence for the principal elements and spectrophotometry (ICP) for the others.

2. Sedimentological characteristics of samples examined

From N to S, a clear distinction was observed in grain-size characteristics, which can be attributed to different contributions to the beaches analyzed. On the whole, grain-size parameter values (Figure 2) indicate sediments whose mean diameter is between granules and medium sands (-1.70<Mz<+1.70 Φ) with a sorting grade between "very good" and "poor" (0.19<sI<2.55) and negative asymmetric distribution curves (SKI) down to -0.51. Within this overall framework, samples can be divided into two groups (Fig. 3); the first, including samples from 1 to 6, is less subject to carbonatic contributions and is for the most part made up of unimodal sediments (0.25<Mz<1.57 Φ), very well sorted and with granule tails. The second group (Fig. 3) includes samples 7 to 15, which conversely have strongly negative Mz values: Mz down to -1.70 Φ (Fig. 2) owing to the prevalence of the coarse carbonatic component. In this group, an exception is represented by sample 12 which, due to its provenance (underwater beach), contains a larger amount of fine material and thus presents positive Mz values.

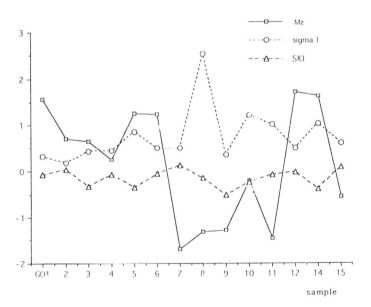

FIG.2 - Range and trend of grain size parameters (Mz, σI, SKI) of analyzed samples, from N to S.

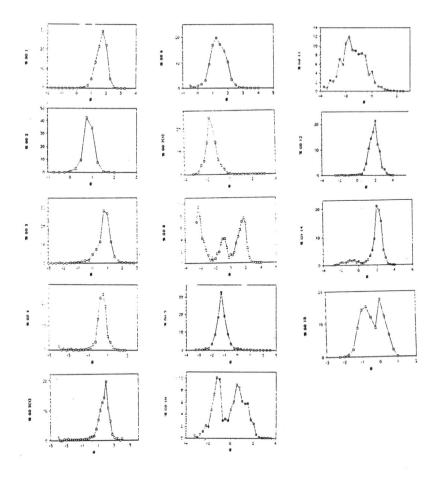

FIG. 3 - Granulometric disribution curves showing the different characteristics of samples from N to S.

3. Geochemical characteristics

From an examination of chemical data, one observes a general increase in carbonate values from N to S, with a sharp increase starting from Cala Sisine. The group of samples north of Cala Sisine is characterized by a prevalence of silicates and a lesser amount of carbonates which, on going north, increase and become the preponderant part of the sediment (samples from 7 to 15)(Fig. 4). Ti and Fe show the same behaviour. Their concentration tends to decrease as the carbonate increases, while Mn remains constant on all beaches with the exception of sample 9, taken in the Würmian eolian area set in the cliff a few meters above present sea level (Fig. 5). The trend of Mn, with its lowest value in correspondence to the paleo-beach (sample 9), shows that this element is connected with the clayey phase residue of the carbonates. Compared to Mg, Ca gradually diminishes from Punta Goloritzè (sample 15) to Cala Sisine (samples 7, 8), while the Ca/Mg ratio is constant from Cala Gonone to Lupiru (sample 6)(Fig. 6): In effect, on proceeding southwards down to Cala Sisine, the presence of dolomites predominates in the sediment, only to decrease from that point on and be replaced by calcite, in line with the geolithological characteristics of the coast.

As was to be expected, Sr is correlated with Ca, presenting some anomalies in samples 6, 10 and 12 (Fig. 7) taken in correspondence to coastal caves, perhaps owing to a local diagenetic factor depending on larger amounts of this element in the concretionary carbonatic facies.

The Zr and Sr trend from N to S shows a Zr decrease with an increase in carbonate and Sr content in the sediments (Figs 8, 9). Zr is probably connected with the siliciclastic sandy component since it is a residual mineral from the weathering of granitic and metamorphic lithologies.

Pb and Zn contents decrease more or less with the increase in carbonate content, with the exception of sample 12, which was taken from the submerged beach inside a cave at a depth of three meters (Fig. 10); this high Pb content may have some connection with the diagenetic phenomenon described for Sr, since both these elements are connected with the aragonitic phase and decrease with conversion in to the calcitic phase (DUCHI et al. 1990).

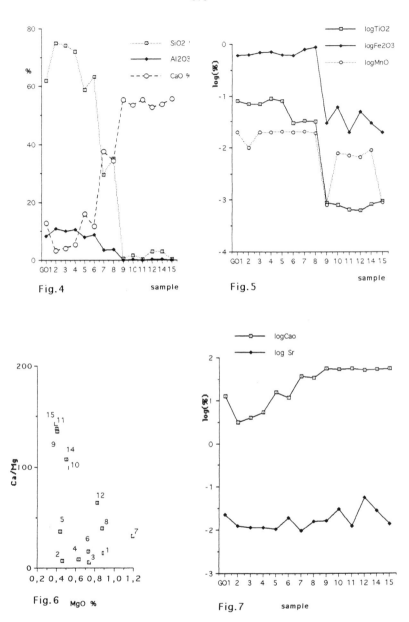

Fig. 4

Fig. 5

Fig. 6

Fig. 7

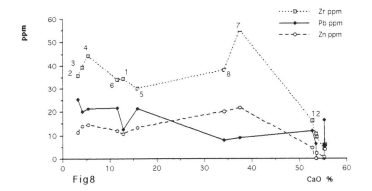
Fig 8

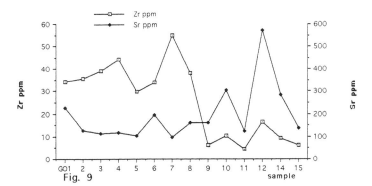
Fig. 9

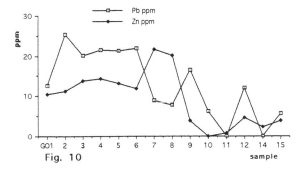
Fig. 10

4. Conclusions

From the geomorphological, sedimentological and geochemical investigations carried out, it can be assumed that at present no carbonatic sedimentation, with the production of carbonate, is taking place along the coast of the Gulf of Orosei. Its beaches appear to be fed prevalently by detritic contributions from Mesozoic cliffs and water courses, which mostly involve samples taken from the northern part of the coast, up to the Cala Sisine beach, where analyses showed higher SiO_2 contents coming from rill washing of metamorphic, granitic and dike rocks outcropping in the upper parts of the watersheds of Rio Codula de Luna and probably in the inner watershed of Rio Codula Fuili and partially of Rio Codula de Sisine.

Samples with the largest amounts of SiO_2 were those which also had mean diameters within the range of medium sands (samples 1 to 5); starting from sample 6 the trends of the two parameters (SiO_2, Mz) are similar: when sediments tend towards gravel, the SiO_2 content decreases; when the mean diameter reaches sand values it increases. Consequently, the result appears to be that sediments containing a high percentage of SiO_2 are not associated with the coarse fraction (Fig. 11).

As concerns the dependency of carbonates on mean diameter, the graph in Fig. 12 shows an increase in the content of carbonates with the increase in grain size. This may indicate that carbonatic detritis was selected from within the coarse fraction and that the fine grained calcite had been rill washed. From that it would appear that to the north the dilution with siliciclastic sediments and the higher presence in percentage of the micritic facies has led to a concentration of SiO_2 compared to CaO. To the south, carbonate was selected in coarser grain size and, in agreement with Argast and Donnelly (1987), was influenced by rill washing caused by storm waves and coastal currents, which have completely eliminated the fine fraction.

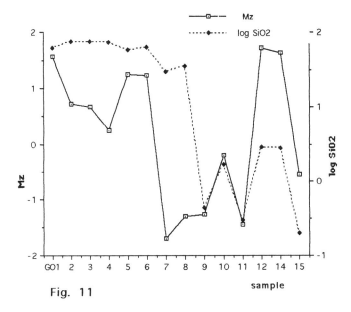

Fig. 11

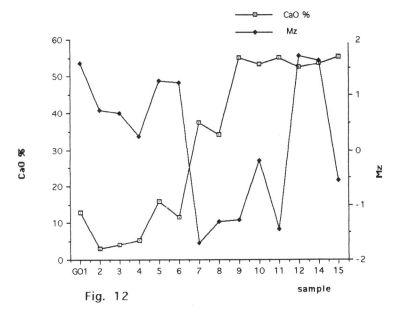

Fig. 12

REFERENCES

AMADESI E., CANTELLI C;, CARLONI G., RABBI E.(1959)- Ricerche geologiche su terreni sedimentari del foglio208 Dorgali. Giornale di Geol. di Bologna, 28, 59-87.

ARGAST S. and DONNELLY T.(1987)- The chemical discrimination of clastic sedimentary components. J.Sed.Petrol., 57, 813-823.

ASSORGIA A.(1968)- Sopra alcuni lembi di Tirreniano fossilifero in grotte costiere del Golfo di Orosei. Boll. S. S. S. N. Anno II, vol.III, 65-73.

ASSORGIA A., BENTINI L. e BIONDI P.(1974)- Caratteristiche strutturali delle assise carbonatiche del Golfo di Orosei. Mem.Soc.Geol.It. , 13(2), 209-219.

BENCINI A. e TURI A.(1974)- Mn distribution in the Mesozoic carbonate rocks from Lima Valley northern Appennines. J. Sed. Petrol. , 44, 774-782.

CAROBENE L. (1972)- Osservazioni sui solchi di battente attuali ed antichi del Golfo di Orosei in Sardegna. Boll. Soc. Geol. It., 91(3), 583-601.

CAROBENE L. (1978)- Valutazione dei movimenti recenti mediante ricerche morfologiche su falesie e grotte marine del Golfo di Orosei. Mem. Geol. It. , 19, 641-649.

CHABRIER G.(1970)-Tectonique de Socle d'age alpin en Sardeigne centro-orientale. C. R. Accad. Sc. Paris, 271(12 octobre), 1252-1255.

CRISTINI A., DI GREGORIO F., FERRARA C.(1992)- Sedimentological and geochemical characteristics of beaches on the southeastern coast of Sardinia and their dependence on source. Boll. Ocean.Teorica e Pratica, Trieste, in press.

DIENI I., MASSARI F.(1970)- Tettogenesi gravitativa di età oligocenica nella Sardegna centro-orientale. Boll. Soc. Geol. It. , 89(1), 57-64.

DIENI I., MASSARI F.(1971)- Scivolamenti gravitativi ed accumuli di frana nel quadro della morfogenesi plioquaternaria della Sardegna centro-orientale. Mem. Soc. Geol. It., 10(4), 313-345.

DUCHI V., MAZZONI S., TURI A.(1990)- Distribuzione di Zn e Pb nelle rocce carbonatiche mesozoiche della Valle di Lima (Appennino Settentrionale). Boll. Soc. Geol. It., 109, 427-436.

FOLK R.L. e WARD W.C.(1957)- Brazos River bar: a study in the significance of grain size parameters. J. Sed. Petrol, 27, 3-26.

TURI A. e PICOLLO M. e VALLERI G.(1990)- Mineralogy and origin of the carbonate beach sediments of Malta and Gozo, maltese island. Boll. Soc. Geol. It., 109, 367-374.

Describing the coastline of Europe

J.P. Doody[1], N.C. Davidson[1] & F. van der Meulen[2]

Addresses: [1] *Coastal Conservation Branch, Joint Nature Conservation Committee, Monkstone House, City Road, Peterborough, PE1 1JY, UK.* [2] *Landscape and Environmental Research Group, Unversiteit van Amsterdam, Nieuwe Prinsengracht 130, 1018 VZ Amsterdam, Netherlands.*

ABSTRACT. This paper describes three levels of methodology for describing and monitoring the European coastal resource. The approach is being developed under the co-ordination of the Science Commission of the European Union for Coastal Conservation (EUCC) and some parts of the methodology are under development for the UK by the Joint Nature Conservation Committee's Coastal Review Unit. The three levels of approach are: 1) simple inventories of habitats at a Europe-wide scale and country/regional coastline descriptions; 2) standardised review methodologies using structured computer databases for more detailed assessments of the coastal resource and the human activities affecting it; and 3) detailed surveillance and monitoring studies (using remote sensing and GIS technology) focusing on selected key areas of coast leading to development of a monitoring system for the European coastal zone. Further collaborators are sought for contributions to each part of this programme.

INTRODUCTION

Fundamental to any consideration of the requirements of nature conservation is a knowledge of the resource and the impact of man's activities on it. Only when this range of information is available is it possible then to identify clearly the range and significance of key habitats and species and to determine policy requirements for their protection. The protection of such habitats and species forms a major planks of conservation policy.

For many years vegetation scientists have developed classification systems for vegetation. This work was led by European phytosociologists such as Braun-Blanquet and Tuxen. More recently in Great Britain, a survey and classification of natural vegetation (the National Vegetation Classification) has been undertaken by Lancaster University and is being published in the series *British Plant Communities* by Cambridge University Press (Rodwell 1991a, 1991b). These and other classification exercises, including those associated with geomorphological features including erosional/depositional characteristics have provided a framework for survey, but there little systematic and Europe-wide work available.

The studies funded during the 1980s by the Council of Europe are a notable exceptions. Those of relevance to coastlines and the wider coastal zone include the reports on saltmarshes (Dijkema 1984), dunes and shoreline vegetation (Gehu 1985) and marine benthos (Mitchell 1987). Even here, however, only Dijkema (1984) includes information on sites. So an overview of both the resource and the best representative sites is not available for the majority of coastal habitats. Such an overview is an important prerequisite for any selection of sites to form part of a Europe wide series of protected coastal areas such as in meeting the requirements of the EC Directive on the Conservation of Habitats and Species.

It is not possible in this short paper to review or list all the published work and many current initiatives that contribute information towards our understanding of the European coastal resource and its conservation. Instead this paper describes the approach being promoted by the Science Commission of the European Union for Coastal Conservation (EUCC) and also some of the related work of the UK Joint Nature Conservation Committee (JNCC), the body with U.K. responsibility for European and international matters affecting nature conservation. The EUCC and JNCC work aims to provide a framework within which the selection of representative sites can be chosen, a means of collecting and collating information at varying levels of detail on the resource and human impacts, and a basis for further survey and monitoring.

EUROPEAN UNION FOR COASTAL CONSERVATION (EUCC)

The European Union for Coastal Conservation is primarily concerned with promoting the conservation of coastal landscapes, habitats and species. Its principle objectives are:

- to provide a source of information and advice on coastal conservation issues; and

- to aid the process of policy formulation in the coastal zone, at local, national and international levels.

To this end it provides a forum for the co-ordination and communication of ideas and experience of members and organisations concerned with nature conservation on the coastlines of Europe. In addition it provides advice on the implications of policy development in the wider coastal environment, in particular in relation to the management requirements of sites and coastal conservation strategies at all levels.

The Science Commission works within the European Union for Coastal Conservation and was established at the EUCC congress in Galway, Ireland in 1991. Its principle objective is to support the work of the Union by providing information and advice on the scientific aspects of coastal conservation. To this end the Science Commission is concerned with three main activities :

- providing information on the coastal resource of Europe (both the natural environment and human impacts upon it);
- facilitating contact and collaboration between scientists across Europe (including survey and monitoring work for both habitats and species); and
- identifying research programmes and needs (including those of management in the coastal zone).

This paper focuses on the work being developed on just the provision of information of Europe's coastal resource. It describes three levels of approach which can be generalised as:

- **inventories** - simple collation of existing information on topics on a broad geographical scale, or consistent collation of data on particular sites or regions;

- **reviews** - more detailed collection and collation of resource data; and

- **surveillance & monitoring** - detailed studies, including time-series analysis and resource management appraisals.

It is important to note that, whilst these divisions provide a convenient way of considering and describing the spectrum of approaches embodied in the EUCC concept, there is in reality no clear distinction between all parts of the approach and information collated at one level generally forms part also of the work at other levels. For example detailed collection of data can be compiled both to present fine-scale resource management guidance and, by extraction of simple attributes, to produce basic inventories.

INVENTORIES OF EXISTING INFORMATION

Selecting sites as representative examples of particular natural formations or species concentrations is an important part of any nature conservation strategy. This is now being reinforced in a European context by the recent adoption of the EC habitats and species Directive (*Council Directive 92/43/EEC on the conservation of natural habitats and of wild fauna and flora*). Fulfilling the requirements for site selection under this directive requires a detailed knowledge of the habitats for each country. At the same time it is important that areas being selected by each country as internationally important (for designation as Special Areas of Conservation - SACs) can be set into their European context. Whilst the work of "Corine" has started the process through classifying zones of coastal erosion and by collecting site data, no overall description of the European coast, its habitats and species, yet exists.

Describing this nature conservation resource for the whole of the European coast looks to be a daunting prospect, but a structured approach eases the task. A considerable amount of relevant information on coastal habitats, sites and species concentrations for individual countries does already exist. Much of this key information is, however, unpublished or available only in documents with a restricted distribution. As a first stage in the preparation of a comprehensive description of the coastline of Europe the EUCC is co-ordinating reviews of the available data on habitats and sites. This is being developed in two ways: **habitat inventories** and **country/regional descriptions.**

Habitat inventories

Habitat inventories provide an overview of the location and importance of the main coastal habitats in Europe. The recently completed *Sand Dune Inventory of Europe* (Doody 1991) provides an illustration of the approach which includes the collection and collation of existing descriptive data in published form and from summary contributions from individuals with specialist knowledge of a particular area. Following this model a synthesis of similar information is planned for each main coastal habitat across Europe. Following the example of the sand dune inventory these inventories will each be compiled by an editor collating information supplied by a habitat specialist in each country. Each habitat specialist is responsible for providing information on:

- the location and size of the total resource;
- the location and size of important sites;
- the identification of other nature conservation interest;
- a summary of impacts and management problems; and
- conservation requirements.

It is planned as a first phase to encourage the production of inventories in the following habitats: **sand and mud flats; sand dunes** (now published as Doody 1991); **saltmarshes; shingle structures and beaches; lagoons; coastal grazing marshes** and **sea cliffs**. Other habitats may also be included in the series.

Describing the Europe-wide description of coastal habitats is of course one key component of resource descriptions covering the wide range of wildlife features that contribute towards the overall nature conservation resource. There is potential to extend the approach to other features. Descriptions for some of these other features have already been made at various geographical scales. A well-known example is the distribution of the migrant and wintering waterfowl for which the coastlines of western Europe are of great international importance. Detailed distributions have been described at the level of a country, e.g. United Kingdom (Prater 1981), a major wetland such as the international Wadden Sea (Smit & Wolff 1981), and in a North Sea context (Davidson 1990). In addition these distributions have been set into their European context to aid compliance with EC directives (Stroud, Mudge & Pienkowski 1990) as well as more broadly, covering an entire migratory flyway involving Europe (e.g. Smit & Piersma 1989) or even world-wide for some species (e.g. Piersma & Davidson 1992). Similar contextual descriptions could and should be made for many other features of coastal wildlife.

Country/regional descriptions

In order to set the individual habitat data in context within a country, it is necessary also to compile descriptions of the coastlines of individual countries. To this end a coastal specialist is being identified for as many countries as possible who will have the responsible for providing data on:

- Length and type of coastal formations;
- Distribution of habitat and species concentrations;
- Protected sites (for nature conservation);
- Coastal status (e.g. built up; protected by sea bank, dike etc.; natural coast, eroding, accreting); and
- Conservation problems.

There are many different existing sources of these categories of information in different countries but these are generally scattered through a wide variety of published and unpublished sources. Our aim here is to promote the generation, derived from these existing sources and specialist knowledge, of a broadly comparable view of the coastal resource and its status in each country. Together these will help to provide the vital overview within which the more detailed reviews needed to identify sites for designation under the Habitats Directive can take place. Some regional descriptions, with varying amounts of detail, have also been compiled in support of other governmental commitments. An example is the *Directory of the North Sea coastal margin* (Doody et al. 1991), covering the North Sea coastline of the UK and compiled as input to the Ministerial Conferences on the North Sea.

Country coastal specialists willing to participate in this programme have already been identified for eight European countries and more offers of collaboration are being sought. It is intended that these will be published as a series by the EUCC over a period of 2-3 years.

THE UK COASTAL REVIEW UNIT

The inventory approaches to describing the coastline as outlined above provide a relatively rapid approach to the provision of background information to help identify the main habitats and sites important for nature conservation. The approach is not, however, developed from systematic methods of data collection and hence is not amenable to storage and analysis using computer technology. Nor can the descriptions, except in the very broadest terms, help identify issues which require conservation policies.

A more detailed and systematic approach is being adopted by the UK Joint Nature Conservation Committee, which has established a Coastal Review Unit (CRU) for the UK.

This will include a data storage and retrieval system for information on coastal habitats and sites and on the impact of man on them, designed to assist in coastal zone conservation work. The detailed data model focuses initially on the UK coastal resource, but the methodology under development has much wider relevance to Europe-wide and international shores.

The CRU system under development builds on the methodology for data collection and synthesis established during the earlier *Estuaries Review*, begun in 1988. This has already produced a comprehensive assessment of the British estuarine resource, its wildlife, its conservation status and the many human activities and their impacts on the resource (Davidson *et al.* 1991). Much of this assessment was made through the collation of existing information, but at a more detailed level than the Europe-wide and national coastal descriptions outline above. In addition, further data collection was initiated where pre-existing information was inadequate.

An important feature of the Coastal Review Unit view is its development of an information service capable of covering the whole breadth of the coastal zone, from hard rocky shores to extensive soft tidal flats; from subtidal marine systems through tidally inundated saltmarshes to wholly terrestrial maritime habitats such as sand dunes. A second key part of the approach has been data collection on the whole range of human activities taking place on the coast as a baseline to identifying the context for locations and activities of actual or potential conflict or impact.

The roles of the CRU are thus to co-ordinate and collate coastal data, so as to help set UK standards for the information underpinning coastal conservation science, and to collaborate in the development of data handling methods to achieve effective information flow. The CRU needs to be able to handle coastal zone data from many places and at many scales, from information on individual sites to countrywide surveys, and to package this to provide consistent UK resource data.

We anticipate that coastal information flow will be achieved by establishing links between the great variety of existing datasets held on coastal zone conservation; by managing some of these datasets, for instance on some coastal habitats; by developing protocols for surveillance of the state of coastal zone wildlife; and by identifying and promoting the filling of gaps in our current knowledge.

Collating and linking datasets is of course not the end-point of the process. The CRU is developing a variety of products that will make coastal information available in useful forms for coastal conservation and management. Products will include the publication of nation-wide reviews and inventories of features and sites, electronic databases, an information and advice service for coastal zone data.

Developing the CRU information service is a potentially substantial and complex task, even just for the UK. There are, for example, many relevant existing computerised datasets held within the JNCC, as well as others managed by other conservation agencies and many other organisations. Such information varies greatly in various attributes, notably its geographical coverage, level of detail, structure and extent of computerisation and type of information system (e.g. database, GIS etc.) utilised. An appropriate and effective data model must thus be developed with the flexibility to respond effectively to a variety of information sources and needs. To this end the CRU is currently undertaking a wide-ranging information systems review. This will focus on data handling for the UK, but will also help greatly in planning for Europe-wide (and wider) links.

There are already various products being produced using the existing resources and the methodology available to the CRU, some in collaboration with other national conservation agencies. These include:

- UK-wide distributional analyses of human activities on estuaries;

- the scale and pattern of water-based recreational activities on British estuaries;

- national resource reviews for various features of all or part of the British coastal zone. These are based on resource survey work at various levels of detail, and include a detailed mapped vegetation survey of shingle structures (Sneddon in press), and similarly detailed surveys of sand dune vegetation in England, Scotland and Wales (Dargie & Radley in prep.).

- other resource reviews covering a wider range of topics but more restricted geographical scope, such as the *Directory of the North Sea Coastal Margin* (Doody *et al.* 1991), compiled for the UK Government's Department of the Environment (see also above).

The CRU is nearing completion of another inventory project, *An inventory of UK estuaries* (Buck 1992). This publication will give a summary of location, wildlife, conservation status and human activities of each of the 163 UK estuaries is designed to be widely available for use as background to CZM plans and as a baseline for resource monitoring. The British volumes of this inventory are well advanced, and Northern Ireland volume of this inventory is now being developed. This last volume is a good example of a collaboration between statutory and voluntary conservation bodies, in this case the JNCC, the Department of the Environment (NI) and the RSPB.

This type of summary inventory (here on a site-by-site basis) compiled from the more detailed data holdings of the Coastal Review work illustrates an important point about our data handling methodology: that complex datasets can generate both detailed analyses and simpler inventory listings akin to the detail contained in habitat or country/regional inventories prepared through our 'first stage' approach. This 'double-stage' value emphasises the utility of our flexible database approach in developing the CRU.

The CRU approach has been supported recently by the 1992 *Odessa Protocol on international co-operation on migratory flyway research and conservation* (Wader Study Group 1992). This recommends that "*full use is made of existing relevant information gathered by simple techniques ... on site inventories of wader habitats, ... and analysis of human activities."* The Protocol, developed particularly to foster collaborative work on migratory birds between east and west Europe and Asia, but equally applicable to other coastal wildlife features and geographical areas, also *"emphasises that ... organisations with experience ... should assist others by ... assisting in establishing compatible databases."*

As part of the CRU's implementation of the Odessa Protocol we are now also considering ways of developing, in collaboration with several other international organisations, a simplified questionnaire-based inventory methodology applicable to a wide variety of countries, locations and circumstances, and to encouraging the compilation of such inventories in several areas world-wide.

SURVEILLANCE AND MONITORING

The first two levels of assessment ("inventories" and "reviews") can will provide a rapid assessment of the coastal resource, an indication of its scale and significance, and of the magnitude of human activity impacts upon it. To the extent that the second approach includes detailed data from habitat and vegetation surveys it will also provide a baseline for monitoring (including the 'ground-truthing' of remotely sensed images), and an opportunity to identify gaps in coverage. It is, however, difficult to use the results as the only basis for detailed surveillance and monitoring. For this a third, more standardised and detailed level survey is required to establish a suitable baseline, not only covering habitats and species but also appropriate to the type of activity likely to cause environmental impact. With the increased availability of new technology particularly in the fields of satellite imaging and computer Geographical Information Systems (GIS) technology a more fundamental appraisal can be undertaken.

If this 'third level exercise' is to be carried out across Europe as part of the EUCC initiative it will, given its possible scale, require the co-operation of many groups and the participation of a number of countries. A proposal is being developed, based on earlier assessments of a monitoring programme (Meulen 1990; Meulen & Janssen 1992) and is described further below. Contributions have so far been offered from the UK (detailed habitat and vegetation maps), Denmark (a survey of sand dunes) along with further collaborations with groups in Greece and France.

At the same time the ability to include new data sets will depend on the availability of funding from either a central point such as the European Commission or from contributions by participating countries. It would be over-ambitious to attempt to establish this detail of baseline survey for the whole European coastline at once. Rather our approach will be to selected areas chosen as appropriate to fill in the picture and provide information in especially sensitive locations and habitats. The University of Amsterdam, in collaboration with the Universities of Wageninen and Athens, has devised a research programme based on this approach of developing an integrated resource management tool for the European coastal zone. Several EC countries have already indicated their interest, and co-operation with the French National Forestry Commission is already underway. Further discussions are taking place with other countries and with the EC in Brussels to establish the most appropriate way of funding the work.

The project will last for several years. It aims to establish a surveillance/monitoring system applicable to the European coastal zone. This will be achieved through making detailed inventories for selected areas of the European coast, with classification units based on simple criteria which can be listed for monitoring in order to detect or predict changes in coastal habitats. This will be done by combining in a Geographical Information System (GIS) existing databases (for example information from the CORINE database) with "high level" (satellite) and "low level" (video and traditional air photos) imagery, and by integrating these with other relevant data such as thematic maps and field measurements from detailed vegetation surveys. Such surveys are for example complete or nearing completion in Britain covering several habitats (Burd 1989; Sneddon in press; Dargie & Radley in prep.).

This work will complement other data collection exercises and more detailed surveys by several groups working in coastal zone conservation and management, such as:

- recently completed surveys for GIS-based sea defence management in East Anglia in south east England (e.g. Townend 1990);

- assessment of erosion and vegetation change in Essex and Kent saltmarshes during the last 15 years - a study involving ground survey and GIS data analysis (Burd 1992); and

- several GIS-based systems developed as part of the research and management resource for the German part of the Wadden Sea (see e.g. Landesamt für den Nationalpark Schleswig-Holsteinisches Wattenmeer 1992).

Several different scales of surveillance are under consideration. Video-monitoring appears to be especially useful for monitoring highly dynamic large scale objects (mobile dunes, erosion coasts, dykes and dams during periods of storms). Air photos cover intermediate scales of up to 1:10,000 while satellite images yield information on smaller scales up to 1:100,000. Definitions of coastal types will use the habitat classification contained in the EC habitats and species directive.

Combining this detailed surveillance approach with the human activity information derived from the Coastal Review Unit methodology described above can then provide also a means of assessing priority issues for the implementation of coastal zone management strategies. The results will be integrated also with more detailed social and economic data (e.g. population, settlements, rural development, urbanisation, tourism, agriculture, harbours etc.) in order to provide for the production of an integrated resource management tool based on GIS and accessible to local as ell as national management bodies. This integration with socio-economic information will form a 'fourth level' project, details of which are yet to be finalised.

So this detailed approach is directed at establishing a basis for the surveillance and monitoring of habitats and the socio-economic interactions with them. It aims to use the experience of current local monitoring exercises e.g. at Albufeira Dunes (Majorca) and the Sefton coast (UK) to provide a series of reference sites linked to a wider-scale surveillance of the European coastal resource.

CONCLUSIONS

The future conservation of the coastline of Europe will depend on being able to identify adequately the areas and sites that are important as examples of natural and semi-natural habitats and species concentrations. This will be especially important in the context of the future designations under the EC Habitats Directive. The various levels of work described here outline one approach which can serve the two purposes:

1. providing a framework within which to assess the context for designating sites nationally and internationally; and

2. establishing a sound basis from which to consider nature conservation policy options, by combining the framework in 1. with the more detailed methodology being developed in the UK for collecting and collating existing information both on the resource and human influences upon it.

These policy options concern not just ways of protecting the important sites but also the development of sound coastal zone management strategies which takes the natural coastline and its dynamics fully into account. Further development of a standardised survey and monitoring programme across Europe, undertaken as a collaborative venture between member states, will allow proper assessment of the efficacy of management strategies as they develop. It will also provide a basis for monitoring changes which may be consequent upon global warming if this becomes a reality.

A major aim of this paper is intended to inform coastal scientists and managers about the approaches under development, and to encourage further contributions and collaborations in this programme of work. For further information and/or offers of collaboration, contact:

- Pat Doody - Inventories and Country/regional descriptions;

- Nick Davidson - UK Coastal Review Unit; and

- Frank van der Meulen - Surveillance and Monitoring Programme.

REFERENCES

Buck, A.L. **1992**. *An inventory of U.K. estuaries.* Joint Nature Conservation Committee, Peterborough.
Burd, F. H. **1989**. *The Saltmarsh Survey of Great Britain: an inventory of British saltmarshes.* Research & Survey in Nature Conservation No 17. Nature Conservancy Council, Peterborough.
Burd, F. H. **1992**. Erosion and vegetational change on the saltmarshes of Essex and north Kent between 1973 and 1988. *Research & survey in nature conservation* No. 42. Nature Conservancy Council, Peterborough.
Dargie, T. & Radley, G. in prep. *Sand dune survey of Great Britain.* JNCC, Peterborough.
Davidson, N.C. **1990**. The conservation of British North Sea estuaries. *Hydrobiologia* 195: 145-162.
Davidson, et al. **1991**. *Nature conservation and estuaries in Great Britain.* Nature Conservancy Council, Peterborough.
Doody, J.P. **1991**. *Sand dune inventory of Europe.* Joint Nature Conservation Committee, Peterborough & EUCC, Leiden.
Doody, J.P., Johnston, C. & Smith, B. (eds.) **1991**. *Directory of the North Sea coastal margin.* Consultation draft. Joint Nature Conservation Committee, Peterborough.
Dijkema, K.S. (ed.) **1984**. *Salt marshes in Europe.* European Committee for the Conservation of Nature and Natural Resources, Council of Europe, Strasbourg.
Gehu, J.M. **1985**. *European dune and shoreline vegetation.* Nature and Environment series No. 32, Council of Europe, Strasbourg.
Landesamt für Nationalpark Schleswig-Holsteinisches Wattenmeer. 1992. *Ökosystemforschung Wattenmeer. Das projekt im Überblick.* Landesamt für Nationalpark Schleswig-Holsteinisches Wattenmeer, Tönning.
Meulen, F. van der & Janssen, M. **1992**. Towards a monitoring programme for European coastal environments. In: R.W.G. Carter, T.G.F. Curtis & M. J. Sheeny-Skeffington (eds.), **Coastal Dunes**, pp 517-523. Balkema, Rotterdam.
Meulen, F. van der. **1990**. Landscape ecological impact of climate change on coastal dunes of Europe. *Eurodunes* 2(2): 47-54.
Mitchell, R. **1987**. Conservation of marine benthic biocenoses in the North Sea and the Baltic.
Piersma, T. & Davidson, N.C. (eds.). **1992**. *The migration of Knots. Wader Study Group Bulletin 64, Supplement.*
Prater, A.J. **1981**. *Estuary Birds of Britain and Ireland.* T. & A.D. Poyser, Berkhamsted.
Rodwell, J.S. **1991a**. *British Plant Communities. Volume 1: Woodlands and scrub.* Cambridge University Press, Cambridge.
Rodwell, J.S. **1991b**. *British Plant Communities. Volume 2: Mires and heaths.* Cambridge University Press, Cambridge.
Smit, C. & Wolff, W.J. **1981**. *Birds of the Wadden Sea.* Wadden Sea Working Group, Report 6. A.A. Balkema, Rotterdam.
Smit, C. & Piersma, T. **1989**. Numbers, midwinter distribution, and migration of wader populations using the East Atlantic Flyway. In: H. Boyd & J.-Y. Pirot (eds.), *Flyways and reserve networks for waterbirds. IWRB Special Report* No. 9: 24-63.
Sneddon, P. in press. *Vegetated shingle structures of Great Britain.* JNCC, Peterborough.
Stroud, D.A., Mudge, G.P. & Pienkowski, M.W. **1990**. *Protecting internationally important bird sites: a review of the EEC Special Protection Area network in Britain.* Nature Conservancy Council, Peterborough.
Tabor, A. **1991**. United Kingdom Digital Marine Atlas. *NERC News, July 1991.* Natural Environment Research Council, Swindon.
Townend, I.H. **1990**. The management of GIS for Coastal Zone Management. *Proceedings of the 1st European Conference on GIS (EGIS-90.).*
Wader Study Group. **1992**. The Odessa Protocol on international co-operation on migratory flyway research and conservation. *Wader Study Group Bulletin* 65: 10-12.

Coastal woodlands, forestry and nature conservation

R.M.H. Tekke & A.H.P.M. Salman

Introduction

The European coastal environment includes an enormous variety of natural habitats, many of which are specific to the coastal zone. But the last decades many coastal habitats have been destroyed or altered significantly by human activities. For example, about 75% of all Mediterranean sand dunes have disappeared since 1950. Next to urban and tourist development one of the main causes of this decline has been the planting of wood (Géhu, 1985, Salman & Strating, 1991, Doody, 1992).
This paper gives a short overview of the main effects of afforestations in the coastal environment, especially in sand dune areas. Some attention will be paid to current developments in international policies as well as in woodland management concepts. Finally the first results will be presented of an EUCC project concerning the types, distribution and natural values of natural, semi-natural and planted woodlands.

Coastal afforestations

In the 16th and 17th century coastal dune instability in Europe increased due to overgrazing by domestic cattle and burning. In the 18th century millions of trees were planted to stabilize sand dunes. Most of them were non-native pine species like *Pinus nigra, P. maritima* and *P. pinea*.
In the last two centuries timber production became more important leading to the introduction of other fast growing species like e.g. *Acacia cyanophylla, Eucalyptus globulus* and *Eucalyptus rostrata*. This process of increasing sand instability due to over-exploitation, followed by large scale planting of exotic trees, is still going on in the Mediterranean region (especially in Turkey).

In Europe the amount of afforested dune areas differs from country to country and from area to area. In some regions there are no tree plantations at all, while others like f.e. the Culbin Sands (Scotland) are almost completely afforested (94%, pine plantations). In Great Britain as a whole approximately 14 % of the total dune surface is afforested. It is estimated that 20% of the Atlantic and North Sea dunes are affected by afforestation. Since 1961, 29% of the Turkish dunes have been afforested (Doody, 1989 and 1992; Salman & Strating, 1991; Salman, 1992).
In this century vast areas on coastal plains and many cliff top areas have been afforested as well, e.g. in Western France and all along the Mediterranean. Eucalyptus and Pinus species are mostly used here.

Effects of afforestations on the coastal environment

Afforestations have several effects on the coastal environment. A very obvious effect is the complete change of landscape. Open dune areas with high geological and ecological diversity change into monotonous, high rising plantations while on other places natural woodlands, with a rich variety in plant - and animal wildlife and complex vegetation structure, are replaced by plantations which lack this diversity.

But also the geomorphological structure of coastal areas is affected by plantations. Trees were partly planted on bare/mobile sand or on places with scarce vegetation cover, thus affecting natural geomorphology, both patterns and processes. Besides, conifers also affect the soil in a very disruptive way. The old needles contribute to the formation of a micro-podzolic soil. This soil type is very acid and, if it has developed to its full extent, prevents natural vegetation to develop even after the pine trees have been cut (Atkinson et al., 1991). This problem is even worse for Eucalyptus stands, which make the soil very poor and unproductive through falling bark and leaves.

The natural hydrological system of coastal areas is also affected by pine and Eucalyptus plantations. A pine forest of one hectare evaporates 4 to 5 times more than one hectare of open dune vegetation (see table 1). This increased evaporation causes a drop in groundwater level. In adjacent unforested areas (Bakker et al.,1979) this will reduce winter flooding of dune slacks and other coastal wetlands, which causes the loss of characteristic vegetation and fauna in these areas. Many wetlands are completely desiccated due to afforestations. A side effect is often the invasion of birch and pine in desiccated slacks and wetlands, unwillingly increasing the size of the affected area (Rothwell, 1985; Leach & Kinnaer, 1985; van der Meulen, 1982).

Table 1: Effective precipitation (in millimetres) of different dune vegetation types (according to T.W.M. Bakker et al., 1979)

	mm
Bare dune sand	580
dry dune vegetation	390
deciduous woodland	335
dune slack vegetation	220
wet deciduous woodland [1]	220
conifer woodland	183
wet conifer woodland [1]	65

[1] "Wet" means in contact with groundwater

Except for the indirect influences of afforestations on the ecological processes through changes in hydrology, geomorphology and soil the natural vegetation is completely destroyed by planting trees. This has negative effects on the natural wildlife. Animals often depend on certain plant species (especially invertebrates) or natural vegetation structure (avifauna). In afforestations the characteristic plant species and vegetation structure of natural dunes is often completely destroyed, thus affecting the natural wildlife.

International policies

Forestry and woodland conservation policies have been a major issue for some decades. The conservation aspects of forest, and more specifically of natural tropical forests, were an important topic during the UNCED conference (Brasil, June 1992). During this conference there was a large consensus on the importance of natural woodland systems, including those in Europe, for mankind which resulted in signing a treaty which expresses this consensus.

The European Communities forestry policy is also developing. Subsidies for planting trees (on agricultural land), and especially those for planting deciduous woodlands (except Eucalyptus), have been increased this year to stimulate timber production (European Commission, 1991). The EC's forestry policy may result in further development of wood plantations in threatened coastal habitats.

But also the EC nature conservation policy is in full development. The main instrument for this policy is the Habitats Directive (European Commission, 1991), which was accepted in December 1991. This directive lists natural and semi-natural habitats and species which are of Community importance. It contains a list of species and habitats which need protection, including more than 200 habitats of which 35 coastal habitat types. All natural and semi-natural coastal woodlands are classified important within the directive.

Woodland management in natural areas

In most European countries views on woodland management in natural areas have changed, especially during the eighties. The management strategy for a particular woodland should be based on sound objectives and criteria. A good example of a coherent national approach is provided by the Dutch government policies in its Nature Policy Plan (Ministry of Agriculture, Nature management and Fisheries, 1990).

The main function for all coastal dunes in the Netherlands (and for many other areas) is nature conservation, which is elaborated as "the sustainable conservation, rehabilitation and development of nature and landscape".
Within this concept four major aspects are distinguished: ecological value, geomorphological and geological structure, cultural heritage and scenic value. The most important criteria for the ecological value are:

1. "Naturalness": most important are the spontaneous geomorphological and ecological processes in dunes and the patterns or forms (geomorphological features, vegetation, wildlife) in which they result. Important related criteria are: self-regulation (through spontaneous processes), completeness (of natural patterns and processes) and "authenticity" (processes and patterns, e.g. species which are characteristic for the natural system).
2. Diversity, e.g. of natural habitats, vegetation and species. This criterium is often used at an international scale. A related criterium is the international significance of habitats and species, which can be related to e.g. the EC-Habitats Directive.

Natural reference situation

Since nature conservation has replaced timber production as the primary function of many dune woodlands, the criteria "naturalness" and "authenticity" should be applied in woodland management as well. This makes it necessary to draw up "natural reference situations" (natural or potential vegetation types) in a particular area. For example, it can be considered which would be the natural vegetation succession and which would be the natural climax, given a combination of abiotic conditions.
Therefore it is important to identify and protect the few remaining (nearly) natural situations left, while in other, less natural situations management measures should stimulate developments towards the "natural reference". However, as "natural reference situations" are not sufficiently known or described (and especially not their dynamic/succession aspects), research on this subject is needed.

Notwithstanding the growing concern for the natural state of our woodlands and for proper woodland management, afforestation programmes are often continuing and, as already mentioned, national and EC forestry policy are currently developing as well. These policies may easily result in new afforestation in important coastal habitats. A careful planning may avoid further damage and, instead, may be helpful to improve the situation. Here as well, taking account of the "natural reference" is essential. If stabilisation and erosion control is necessary one should look carefully at all possibilities. If the instability is related e.g. to grazing one should consider the possibilities of a proper management scheme to control (over)grazing. If planting is necessary this means that only indigenous species should be used and that all forestry activities should further - or at least not jeopardize- developments towards the "natural reference".

EUCC-Eurosite programme "Coastal vegetations and forestry"

These changes in woodland management, both positive and negative regarded from a nature conservation point of view, and the questions which arose from them were the main reasons for EUCC to start a programme "Coastal vegetation and forestry", jointly with EUROSITE.

With this programme the EUCC will contribute to conservation and restoration of the natural values of dry coastal habitats, especially with respect to woodland management and forestry policies. Provisional funding of the programme by the European Commission (DG XI) has been obtained through the European Habitats Forum. Within this programme EUCC has started a pilot project "Coastal dune woodlands and forestry" in January 1992, focusing on Atlantic and North Sea coasts.

Main activities in the pilot project will be:
A. Surveying the major dune woodland types, their natural/original distribution, their present occurrence and their main characteristics.
B. The drawing up of provisional "natural reference woodland types".
C. Assessing the natural values of the various sorts of woodland (both natural/semi-natural and afforested woodlands), with special reference to the habitats and species of the EC-Habitats Directive.
D. Providing a series of recommendations for woodland management and forestry policies.

Some remarks on methodology

The project has been started with defining all terms and parameters. Important within this project is the definition of a "natural woodland". It was defined as: "a woodland originated in a natural way, flora and fauna are native and developed spontaneously. There is no influence of man" (Westhoff et al., 1969).

Of each country the following information has been gathered through literature research and questionnaires:

1. An inventory of natural, semi-natural and afforested woodland types, including:
   a. their present distribution and an estimation of cover in hectares
   b. their natural distribution (by means of literature search and their basic habitat requirements)

2. An inventory of the basic habitat requirements/conditions of the various woodland types:
   - Hydrology (dry, moist, wet)
   - Geomorphology (stabilized, semi-mobile, mobile)
   - Soil development (no, moderate, high soil development)

During the first months of the project it was necessary to specify the different parameters to determine the natural distribution of the different types. A parameter like e.g. soil development was further specified by including the PH-range and origin of the mother-material.

By means of two case-studies some preliminary results of the project will be presented. The first case study originates from the Netherlands, the second from the Province of Huelva, Andalucía, Spain.

Current status and first provisional results

The survey (A) and, to a certain extent, the drawing up of "reference types" (B) have been executed provisionally for the Netherlands, Belgium, Great Britain, southwest Spain and France. First attempts have also been made in assessing the natural values of the different types of woodland (C), which appeared to be very difficult due to lack of information. The work has concentrated on the development of a proper method which can be used all along the European coast. The first results were discussed during a workshop on "Woodland Conservation and Management", organized jointly by EUCC and Eurosite in Wissant (France), June 1992.

Case study 1: **Coastal dune woodland types and their distribution in the Netherlands**

In the Netherlands a total of 6 "natural reference" woodland types were distinguished in the Younger Dune systems (formed since 1000 AD). An example of such a reference type is the *Convallario-Quercetum dunense* association dominated by Common oak (*Quercus robur*) and, sometimes, Black poplar (*Populus nigra*).
The habitat requirements of the various reference types were determined. During this determination one specific feature appeared to be very important: the fact that the substratum was calcareous (2-10% calcium carbonate) or non-calcareous (0-2% calcium carbonate).

The natural (original) distribution and the present occurrence of *Convallario-Quercetum dunense* vegetations are shown in figure 1. Its distribution is limited to calcareous, dry (groundwater level more than 1 meter below surface), stabilized dunes with a high soil development (humus layer). Its present occurrence is limited to only three sites with a total of app. 50 ha. An additional 300 hectares of semi-natural Convallario-Quercetum woodland is also remaining, which was used as coppice in the past.

The total dune surface covered by natural woodlands in the Netherlands is 600 hectares, with an additional 600 hectares of semi-natural woodlands and at least 4500 hectares of afforestations (app. 3500 hectares coniferous and 1000 hectares deciduous afforestations).
The total coastal dune area of the Netherlands is app. 40,000 hectares. 3000 Hectares of this area would have been covered by natural woodland vegetations in a relatively undisturbed situation (Janssen & Salman, 1992).

Case study 2: **An assessment of natural values: Southwest-Spain.**

The assessment of natural values of the different woodland types is a difficult operation due to a lack of information. Detailed information on natural values is often not available. Sometimes it is available, but not related to one particular woodland type.

For the province of Huelva (Andalucia, Spain), however, it has been possible to gather information on the natural values of coastal woodland types. A first assessment could therefore be made. However, the data were not sufficient to determine overall biodiversity (total number of species) as well as "naturalness".

In the natural woodlands left in Huelva, 13 species of endemic plants (for the Iberian Peninsula) occur, 5 of which are considered rare and endangered in Spain and 2 of which are included in the EC-Habitats Directive. These species occur on just 3,6% of the total surface of coastal dune woodlands in Huelva. In the remaining semi-natural woodlands 9 endemic plant species were found even including 3 Habitats Directive species (on 17,8% of the total coastal dune woodland surface). In coastal dune afforestations however, covering 79,1% of the total coastal dune woodland surface, just 5 endemic plant species were present, of which none is included in the Habitats Directive. Although this should be regarded as a rough indication, it seems that natural and semi-natural woodlands have higher natural values than afforestations with regard to endemic, rare and endangered species.

Conclusions

After six months, the first preliminary conclusions can be drawn concerning the countries involved in the project so far. It can be concluded that:
- there is still a large variety of natural woodland types left in coastal dune areas
- in general only very small fragments of these natural types (and the derived semi-natural types) are left.

Concerning afforestations it can be concluded that:
- there are numerous "types" of afforestation due to the use of a variety of trees, many of which are exotic
- afforestations cover large areas of coastal dunes
- afforestations have replaced and destroyed natural vegetations and affected natural wildlife as well as the natural dune (geo)morphology and hydrology.

Besides it is important to note that information on coastal dune woodland is scarce especially concerning:
- the present distribution of natural and semi-natural woodlands
- natural values of the different woodland types

Natural reference types are only known by a small number of scientific experts, and only seldom to the woodland managers to which knowledge of the natural reference types is of the greatest importance. It is in the interest of both conservationists, foresters and those involved in erosion control to ensure the conservation of the remaining natural woodlands and the restoration by adapted management of the semi-natural woodlands. Informing those directly involved in coastal dune woodland management on the natural reference types and their management is therefore a necessity.

Continuation of the project

This year the EUCC will compile a dossier on the coastal dune woodlands along the European Atlantic coasts. Next year the programme will continue with Mediterranean and Baltic coasts, followed by a broader approach including other types of coastal woodlands like those on cliffs, coastal plains and river delta's.

Literature

- Atkinson, D & P.W. Sturgess (1991). Restoration of sand dune communities following deforestation. In: Perturbation and recovery of terrestrial ecosystems, London.
- Bakker, T.W.M., J.A.Klijn, F.J. van Zadelhoff (1979). Duinen en duinvalleien; een landschapsecologische studie van het Nederlandse duingebied. Pudoc, Wageningen.
- Doody, J.P., (1989). Management for nature conservation. In the proceedings of the Royal Society of Edinburgh.
- Doody, J.P., (1992). The coastal dunes of Europe. Paper presented at the Danish coastal dune seminar, Bunken, Denmark 1992.
- European Commission (1991). Habitats Directive, adopted in December 1991, not yet officially published.
- European Commission (1991). Community aid scheme for forestry measures in agriculture. Com (91) 415, Brussels.
- Gehu, J.M. (1985). European dune and foreshore vegetation. Nature and Environment series no. 32. Council of Europe, Strasbourg.
- Janssen, M.P.J.M. & A.H.P.M. Salman (1992). Duinen voor de wind. Stichting Duinbehoud, Leiden.
- Leach, S.J. & P.K. Kinnear (1985). Scrub and woodland management: Tentsmuir, National Nature Reserve. In Sand dunes and their management. Focus on Nature Conservation no. 13. Nature Conservancy Council, Peterborough.
- Ministry of Agriculture, Nature management and Fisheries, the Netherlands (1990). Nature Policy Plan.

Geographisches Institut der Universität Kiel

- Rothwell, P. (1985). Management problems on Ainsdale Sand Dunes, National Nature Reserve. In Sand dunes and their management. Focus on Nature Conservation no. 13. Nature Conservancy Council, Peterborough.
- Salman, A.H.P.M. & K.M. Strating (1991). European coastal dunes and their decline since 1950. EUCC Internal report series.
- Salman, A.H.P.M. (1992). Coastal dune woodlands and forestry. Paper presented at the workshop "Coastal woodland conservation and management", Wissant, France (June 1992).
- Van der Meulen, F. (1982). Vegetation changes and water catchment in a Dutch coastal dune area. Biological Conservation 24.
- Westhoff, V. & A.J. den Held (1979). Plantengemeenschappen in Nederland. Thieme, Zutphen.

Drs. Robert Tekke and drs. Albert Salman are working at the General Secretariat of the **European Union for Coastal Conservation**, P.O.Box 11059, 2301 EB Leiden, The Netherlands.

The **European Union for Coastal Conservation (EUCC)** is an international organization which aims to promote nature conservation and wise use of European coastal environments. The union consists of a network of more than 500 scientific experts, managers, government bodies and nature conservation organizations (in 33 countries) who cooperate to provide a cohesive approach to the protection and management of European coastal systems.

This paper has been accepted for presentation at the International Coastal Conference Kiel '92, 7-11 September 1992.

Figure 1: The natural distribution and present occurence of the vegetation type "Convallario-Quercetum dunense"

 Present occurence (total of 50 hectares)

 Natural distribution (dry, stabilized calcareous coastal dunes with high soil development)

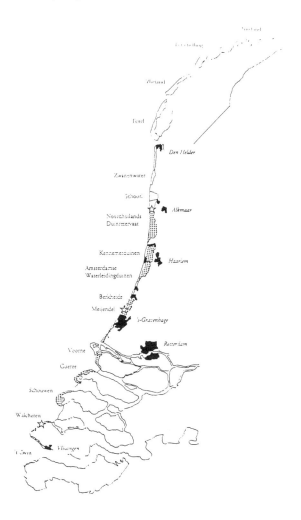

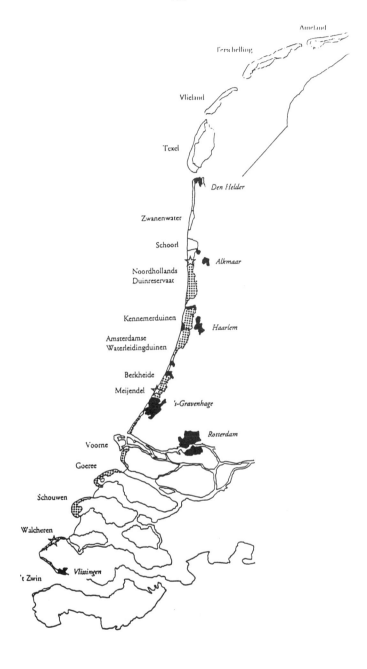

Distribution of epigaeic arthropods in dune–habitats and salt–meadows along the Baltic coast of Schleswig–Holstein

Dipl.–Biol. Claus Hoerschelmann, Forschungstelle für Ökosystemforschung und Ökotechnik, Christian–Albrechts–Universität, Kiel

Introduction

During the past decades, tourism has become an important economic factor in coastal areas of the Baltic sea. But recreational use of beaches, dunes and other forms of typical coastal landscape often leads to destruction of vegetation and a loss of habitats. The areas of dry grasslands and spit habitats have decreased to a mere 1 –3 % of their original amount in Schleswig–Holstein (HEYDEMANN 1980).

Our survey focused on the question, whether there is a typical species community of spit ecosystem habitats and which environmental variables could have influence on its composition.

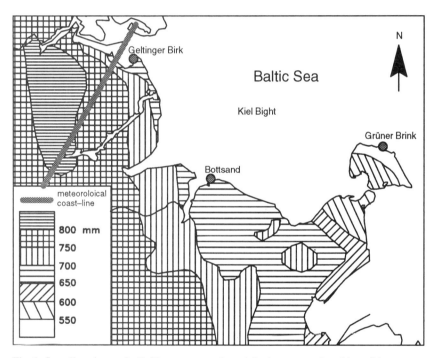

Fig. 1: Sampling sites at the Baltic coast, annual precipitation rates and position of the meteorological coast–line (BELL 1950) (SCHLENGER, PAFFEN, STEWIG, 1970, changed)

Materials and Methods

The activity–density of epigaeic arthropods (Coleoptera: Carabidae; Araneae: Dictynidae, Gnaphosidae, Clubionidae, Zoridae, Thomisidae, Salticidae, Lycosidae, Hahniidae, Theridiidae, Tetragnathidae, Linyphiidae) were assessed by pitfall catches. In each of the eleven sampling sites 3 parallels were used. Samples were taken from 1.6.88 to 23.10.89. The carabid beetles and hunting spiders are caught when hunting, spiders that build nets are caught when wandering on the ground in order to find mating partners. Most of the species of epigaeic arthropods can travel over greater distances by active flight (Carabidae) or by ballooning (spiders). In order to establish stable populations, the ecological demands of species have to be fulfilled. Often climatic and structural factors limit the existence of species in habitats. (HEYDEMANN 1956a, DUFFEY 1966, SCHAEFER 1970)

Species distribution and species composition at the sampling sites in relation to environmental variables were calculated by a canonical correspondence analysis (CANOCO). Ordination offers a way to plot centroids of species distribution, species composition and to analyze the role of environmental variables. For calculation of matrices and analysis of factors see ter BRAAK (1986) and JONGMANN, ter BRAAK et van TONGEREN (1988).

In order to cover the whole extent of the baltic coast of Schleswig–Holstein, samples were taken from three different spit systems. The sampling sites were established in natural reserves in order to eliminate influences of touristic or agricultural use of the habitats. The natural reserve "Geltinger Birk" is located in the north–west of the coast, at the mouth of the Flensburg Förde. Samples were taken in the habitats dune, dry grassland, reed and salt–marsh. The "Bottsand" is a spit system situated at the mouth of Kiel Förde. Samples were taken in the habitats dune, dry grassland, reed and salt–marsh. The "Grüner Brink" is an older spit system at the north–eastern coast of the isle of Fehmarn, at the Fehmarn Belt. Samples were taken in the habitats dune, dry grassland and reed.

The spitsystems consist of dunes and dry grassland–habitats on sandy substrata on the seaward side of the spit and wet salt–marshes and reeds on silt sediments the leeward side. The vegetation consists mainly of grasses. The dunes are dominated by *Elymus arenarius*, *Ammophila arrenaria/ Ammocalamagrostis baltica*, *Carex arenaria* and *Agropyron repens*. The vegetation of the adjacent dry grassland mainly consists of *Festuca ovina*, *Festuca rubra*, *Poa pratensis*, *Carex arenaria*, moss and lichens. The reeds and salt–marshes are characterized by wet soils and frequent but irregular flooding. *Phragmites australis* reeds with *Festuca rubra*, *Holcus lanatus* and *Agrostis stolonifera* border to salt–marshes. The vegetation is dominated by *Juncus gerardi*. *Festuca rubra*, *Glaux maritima* and *Plantago maritima* can be found frequently. *Bolboschoenus maritimus* and *Puccinellia maritima* are found at the mere edges of the salt–marshes. Salt–accumulation caused by evaporation of seawater is frequent in salt–marshes and reeds. Cl– concentrations of 35‰ were measured. Because of the grass–type vegetation in the sampling sites, vegetation–density was chosen as a measure for vegetation structure. The highest densities of vegetation are found at the salt–marsh and reed habitats, followed by the dry grassland at Flensburg Förde and the dune habitats. The vegetation at the dry grassland habitats at Kiel Förde and at Fehmarn Belt is comparatively sparse.

Results

The analysis is based on 25 536 individuals from 206 species, 146 species of spiders and 60 carabid species. The majority of species is not restricted to coastal habitats, but can be found in various open grassland biotops inland.

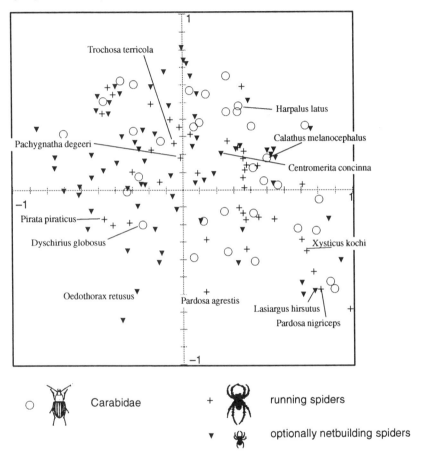

Fig. 2: CCA – scatter–diagram of epigaeic arthropods in coastal habitats of the Baltic Sea in Schleswig–Holstein based on activity–density data in sampling period 1.6.88 – 23.10.89. The more often species are found in a sample, the closer their centroids are placed together. The Eigenvektor of the first axis is 0.3667, the Eigenvektor of the second axis is 0.318.

Rather than describing each single species, the community structure is analyzed. The scatter diagram shows the centroids of distribution of the species in coastal biotops of the Baltic in Schleswig–Holstein.

The centroids of species with moist to wet habitat preferences are plotted on the left hand side of the diagram, species with xeric habitat preferences can be found on the right hand side. The spiders *Pirata piraticus, Pardosa agrestis, Oedothorax retusus* and the carabid beetle *Dyschirius globosus* are frequent in the habitats salt–marsh and reed. The Lyniphiid spider *Centromerita concinna* predominates the community in dune habitats during winter months, while the carabid beetles *Harpalus latus* and *Calathus melanocephalus* are found here during the summer.

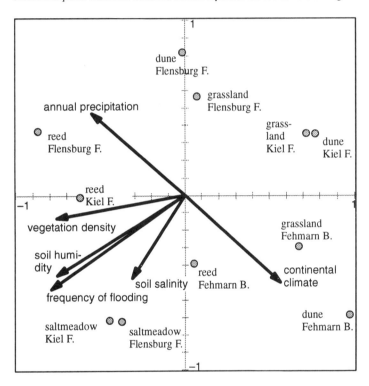

Fig 3: CCA–biplot of the sampling sites dune, dry grassland, reed and salt–marsh at the sampling areas at Flensburg Förde, Kiel Förde and Fehmarn Belt with environmental variables influencing the species composition and community structure of epigaeic arthropods based on activity–density data in sampling period 1.6.88 – 23.10.89. The vectors describe the direction and relative weight of the environmental variables in concern of species distribution. The Eigenvektor of the first axis is 0.3667, the Eigenvektor of the second axis is 0.318.

The analysis of species communities at the sampling sites shows a distinct pattern. There are several environmental variables which seem to influence species distribution and community structure in habitats of the Baltic coast in Schleswig–Holstein. A clear distinction can be made between communities of moist and xeric habitats in each of the sampling areas. Depending on the humidity of the substrata, the vegetation density as a parameter of habitat structure changes and causes differences in community–structure (SCHAEFER 1970). Differences in community structures of similar habitat–types at the Baltic coast could be explained by a gradient of climate. The climate of the north–western part of the Baltic coast of Schleswig–Holstein is influenced by the North–Sea, leading to higher amounts of precipitation and comparatively moderate winters, whereas the subcontinental climate of the south–eastern coast part of the coast leads to fewer precipitations and warmer summers. The distance of the sampling–sites to the meteorological coast–line (BELL 1950) was used to describe this climate gradient.

This climatic gradient shows covariance to distribution and community structure of epigaeic arthropods in the dryer, sandy habitats dune and dry grassland. Xerophilious species like *Pardosa nigriceps* and *Xysticus kochi* are mainly found in the dry habitats at Fehmarn Belt. The species *Lasiargus hirsutus* has its western limit of distribution on the isle of Fehmarn. The communities of habitats on substrata with higher water contents show less response to a climatic gradient. The communities of salt-marshes consist of few, mainly halotolerant species, (*Oedothorax retusus, Pardosa agrestis* and *Pirata piraticus*) that are not limited in their distribution by a gradient of macro–climate.

The fact of climatic influence on dispersal and distribution of organisms is well known so far. Since organisms of coastal habitats have little chance to refresh and stabilize their populations by immigration of individuals from inland habitats due to anthropogenic use of dry grassland areas, protection efforts should lead to a number of natural reserves of sufficient size (HEYDEMANN 1981) and in small distances from each other, functioning as skipping–stones, in addition to already existing reserve areas.

Literature

BELL, G. (1950): Die periodischen Änderungen der meterologischen Elemente IV.– Das Wetter in Schleswig–Holstein Nr.11

DEUTSCHER WETTERDIENST (Hrsg.) (1967): Klima–Atlas von Schleswig–Holstein, Hamburg und Bremen.– Offenbach, 73 S.

DUFFEY, E. (1966): Spider ecology and habitat structure.–Senck.biol., 47, 1, S. 45–49

FREUDE, H., HARDE, K. W., LOHSE, G. A. (1976): Die Käfer Mitteleuropas. Bd. 2, Adephaga 1, Krefeld, 302 S.

HEYDEMANN, B. (1956a): Die Biotopstruktur als Raumwiderstand und Raumfülle für die Tierwelt.– Verh. Dtsch. Zool. Ges. (Leipzig), S. 332–347

HEYDEMANN, B. (1981): Zur Frage der Flächengröße von Biotopbeständen für den Arten und Ökosystemschutz.– Jb. f. Naturschutz und Landschaftspflege 31, S.21–51

HOERSCHELMANN, C. (1989): Ökologisch-faunistische Untersuchungen der Verteilung von epigäischen Arthropoda (Araneae; Carabidae) in ausgewählten Strandwallbiotopen der schleswig-holsteinischen Ostseeküste.- Diploma Thesis Christian-Albrechts-Universität, Kiel

KNÜLLE, W. (1953): Zur Ökologie der Spinnen an Ufern und Küsten.- Z. Morph. Ökol. Tiere 42, S.117-158

RAABE, E.-W. (1973): Über die Belastung des Badestrandes am Bottsand.- Kieler Notizen z. Pflanzenkde. i. S.H. 5, H.3/4, S. 49-67

ROBERTS, M.J. (1985) The spiders of Great Britain and Ireland.- Bd.1-3 Harley books, London, 229 S., 204 S., 256 S.

SCHAEFER, M. (1970): Einfluß der Raumstruktur in Landschaften der Meeresküste auf das Verteilungsmuster der Tierwelt.- Zool. Jb. Bd. 97, S. 55-124

ter BRAAK, C.J.F. (1986a): Canonical correspondence analysis: a new eigenvector technique for multivariate direct gradientanalysis.- Ecology 67, S. 1167-1179

ter BRAAK, C.J.F. (1987b): CANOCO - a FORTRAN program for canonical community ordinantion by [partial] [detrended][canonical] correspondence analysis, principal components analysis and redundancy analysis (version 2.1).- ITI-TNO, Wageningen, 95 S.

ter BRAAK, C.J.F. (1987c): Ordination.- in JONGMANN, R.H.G., ter BRAAK, C.J.F. et van TONGEREN, O.F.R. (1987): Data analysis in community and landscape ecology.- Pudoc, Wageningen, S. 91-169

Author's adress:
Dipl. Biol. Claus Hoerschelmann
Forschungstelle für Ökosystemforschung und Ökotechnik
Christian-Abrechts-Universität
Biologiezentrum
Olshausenstr. 40
W-2300 Kiel 1

Biogeochemistry of semiclosed bays of the Baltic Sea and human factor of sedimentation

Dr. Olegas Pustelnikovas, Institute of Geography, Vilnius
(Lithuania)

The problems of Baltic Sea development prognosis under the conditions of human influence and increasing life necessity for the sea require its complex investigations. The shallow water zones of the baltic proper and areas of semiclosed bays (lagoons) as areas of high bioprodoctivity and recreation potential or perspective to the mass development of tourism in this case are most important.

Biogeochemical analysis of sedimentation processes in space and time allows to investigate their hydro-geochemical, dynamical and lithological factors through the prism of microorganism and hydrobiont activity. Which of the processes is at present predominating - natural or anthropogenic? A number of works determining the role of anthropogenic load in sedimentation processes of the Baltic Sea or analysing the natural background of this process often lack of complex approach (1-4 etc). Without taking into consideration natural relationship between elements in different links of marine ecosystem organic matter (OM), heavy metals (HM), hydrocarbons (HC), biogenic elements in certain ecological interpretations are rendered among anthropogenic admixtures.

Nevertheless, there exist in marine environment different objects (plants, hydrobionts, soils, peats, genetic deposits and rocks), concentrations of elements or compounds as a result of natural sorbtion, hydrological processes (upwelling and zones of convergence), fluid flows from entrails of the earth to the bottom surface, submarine discharge of ground waters from aquifers of different age, bigscale fluctuations in space and time and other insufficiently investigated phenomena. Often in sites of these natural factors manifestation of certain anomalies can be observed. They represent the elements of natural processes of the Baltic Sea development with different paleo- and neoecolgical variations (5-8). In (9-12) different stages of analysis of natural and human factors relations are reflected. The fluids drifts and sources of submarine discharge of the underground water in (13) was analysed.

To determinate accurately the anthropogenic share of heavy metals on the ground of generalized results is still impossible (4). Similar doubts on distribution of oil products are expressed in (14).

This report generalized the investigations, carried out in 1973-90 in the Finnish and Riga Bays, in the lagoon Kursiu Marios and in the Baltic Proper. The various links of the Baltic ecosystem and, at first, the geochemical barriers: river-sea, coast-sea, biota-water-bottom, upper layer (till 1m) of bottom sediment was investigated. Variety of recent sedimentation conditions has determined the biological stage and variety of its rates. It's very important to prognosize the development of the various areas, first of all, the shallow water bays and lagoons of the Baltic basin.

The methods stage investigation of various forms of HM and HC, often represented as oil products of human genesis, in different links of marine ecosystem enabled us to closely approach the problem of priority of either natural or anthropogenic factor in sedimentation process.

Concentration of HC in the water of SE Baltic is of seasonal character with maximum in spring and autumn (15). Biodegradation of n-alkanes leads to reduction of HC content near the bottom by 30-40% and alkane distribution indicate biogenic priority of HC. Anthropogenic HC in the water do not usually exceed 15% of their total amount and only in 6% of the samples fresh oil pollution was detected.

Maximum concentrations of HC in the surface layer of bottom sediments were observed in aleurite and pelite silts as most fine dispersed sediments. The maximum in the low molecular field of the spectrum, presence of polyolephines and tritherpenoids indicate natural origin of those HC which get into bottom with phytodetrite as a constituent part of bottom does not exceed 10% of sum total of HC. HC of transitional type are observed in 21% of the samples.

The SW slope of the Baltic syncline, within the limits of which the investigated aquatory is situated, is characterized by many zones of tectonic breaks related to the areas of fluid drifts from sedimentary deep layer. Anomalies of composition of HC in bottom sediments are observed in those zones. They are connected with a natural percolation of oil (petrogenic HC) from oil-gas-bearing structures. Petrogenic HC are found in 5% of investigated samples. 20% of oil tars found on the Lithuanian coast of the Baltic Sea can be related to natural HC. Content of HC in the water of the Kursiu Marios lagoon is by 1.5-2 times lower and in the bottom sediments more considerably higher than in the Baltic Sea. Nevertheless, in the subzones of geochemical barrier (mouth of the Nemunas-Klaipeda Port) it sharply increases. In such cases there is evident oil pollution (antropogenic HC). But in general biogenic HC in the water of the Kursiu Marios lagoon make 70% (in the sediments-80%) of the sum total. Thus, biogenic petrogenic, technogenic and mixed genetical components are present in the composition of water and bottom HC. The data obtained indicate the low level of anthropogenic background in different links of marine ecosystem (17). Biogenic HC prevail in 63%, oil pollution of different degrees - about 32%, natural HC 5% of investigated samples of bottom sediments.

Analysis of the upper (0-3 cm) layer of bottom sediments, dated 20-500 years, show great differences of the content of Mn, Zn, Pb, Cu, Ni, Co and Cd. The most recent sediments are richer in these elements by 1.5-3.5 times if compared with the older ones. These variations vary with every element as they play different roles in sedimentation cycle. The greatest differences are observed in the barrier zone - on the boundary of phases of sedimentogenesis and early diagenesis of bottom sediments, i.e. they are related to the phases of natural sedimentation processes. It is proved by the data obtained by the methods of various stage of desintegration of sediments.

When crystal grid of silicates was completely destroyed the content of Pb in all samples was by 1.9-4.9, Cd - 1.9-27.5, Cu - 1.5-2.6 times higher than in case of incomplete destruction also partly indicate recent mineral formations of technogenic character. They make only 10-50% of the amount of elements observed by the method of complete desintegration of crystal grid in silicates. It serves as evidence of negligible role anthropogenic component in formation of chemical composition of sediments in the phase of sedimentogenesis. The peculiarities of lithologo-geochemical structures of bottom sediments layer shows about the same. Those pecularities determined the minimum content of elements in rough sediments, maximum - in fine dispersed aleurite-pelite and pelite silts. Predominance of natural origin of the investigated elements is also proved by comparison of obtained values with clarks. Our values as a rule do not exceed the content of these elements in corresponding natural objects of lithosphere.

"Blotted" character of anomalous contents of elements (Fe, Mn, Zn, Cu) in crater zones of fluid drift areas is an indicator of substance migration in deep flows or submarine discharge of ground waters. Such assumption is confirmed by the data on bottom sediments in Gdansk and South Gotland sector. Anomalously high content of elements is related exactly with coagulated, rich in gas, cracked dark-gray silts and gray clays. Geochemical anomalies in bottom sediment layer and silty water column, which are conditioned by submarine discharge of ground waters on the example of Riga Bay are observed. Those zones are in close relationship with geologo-structural, tectonic and geomorphological peculiarities of the bay. In addition to geochemical evidence of such relationship percolation of ground water are proven by the presence of erosite-mineral which is formed at the expense of sulphides of Fe related to the outlets of sulphate-bearing ground waters.

Analysis of the upper (0-12) layer of bottom sediments of the last 50-200 years of sedimentation in Finnish Bay by the method of neutron activation has shown the absence of distinct changes in three phases of bottom sediments transformation in the process of sedimentogenesis and early diagenesis. The upper horizon of this layer (0-3 cm) represents the last (20-50 years) phase of sedimentation under the conditions of human activity, medium (3-6 cm) - the previous one (50-100 years), its sediments are in the stage of sedimentogenesis, lower (6-12 cm) - is represented by the sediments formed 100-200 years (and more) ago and now being in the stage of early diagenesis. The content of elements in the latter was taken as a background value in determining the share of possible human factor. Comparison of content 16 elements in all mentioned horizons has shown that it has not become higher during the mentioned period under the conditions of anthropogenesis. Therefore, the process of accumulation of substance and elements must be related to the conditions and rates of sedimentation which used to be, more intensive, where as recent processes do not exert substantial influence.

The prevailling natural character of surface sediments in the Kursiu Marios lagoon and Gdansk Bay is proven by stage analysis of forms of Mn and Cr (18). Stability of relationship between lithogenic and capable of reaction form is evident actually during the whole geological history.

The huge rates of sedimentation in the lagoon Kursio Marios (2.9-3.6 m/1000 years) determined by the method 210-Pb, speed up its eutrophication, change its area. This, perhaps, can explain the processes of active abrasion of Kursio Nerija spit shores in Morskoje (Sarkuva) settlement (up to 1.7 m/year). According to our calculations at the mean depth of 3.8 m the depression of the lagoon will be filled up, with sediments and macrophytes in less than 1500 year.

Our analysis-generalization of the data on biochemistry of anthropogenic admixtures (chlororganic pesticides and biphenils) has shown differnt rates of their accumulation in various links of ecosystem (8). Nevertheless a distinct relationship between their distribution and the elements of natural-geochemical a geographical nature can be observed: their maximum concentrations are found in finedispersed sediments and sites of pollution. An extremely important role of microbiocoenosis and hydrobionts must be pointed out in elimination of contaminants and reversion of elements from one form into another is clearly reflected in the time-special scale of their distribution where life-activity of macro- and microorganisms ceases in sediment layer 10-20 cm beneath bottom surface embracing age period of 100 years in average.

The upper layer (up to 0.3-0.5) of bottom sediments as a rule is rich in many elements and compounds of natural and human genesis. Many researchers connect it with the beginning of industrialization and development of rural economy in the 19th century and its intensive development today.

In our opinion such point of view is not absolutely true. Analysis of recent sedimentation rates by the method of 210-Pb, investigation of paleoecological conditions, going deeper into every link of sedimention process (mobilization, migration, differentation, accumulation, deposition, diagenesis, etc.), investigation of sedimentary matter variations in the final stage of sedimentary matter variations in the final stage of sedimentation under the impact of microorganisms, comparison of variation sedimentation with global cycles and cycles of long-term changes lead us to a conclusion that it is not the anthropogenic factor, that determines abundance of elements in the sediments, but natural process of sedimentation with its variability in space-time scale. Thus, we should not speak about the beginning of human invasion in the 19th century but about the end of the phase of active processes of sedimentogenesis and diagenesis in sediments of that age. The similar results of geochemical interpretations, are seen in the works of Prof. Ralf O. Hallberg (19). Therefore, beneath that thickness the process of deposition of sediments has started.

I want to complete my report with the words of Sabastian G. Gerlach (14). He says that if it turns out that the observed changes in the Baltic Sea are of natural character, where anthropogenic factors are of secondary importance, then it will be possible to repudiate many measures of cleaning the runoff waters which are used in the Baltic States.

At present a new cycle of freshening of the Baltic Sea has started with all ecological consequences.

References

1. Baltic Marine Environment Protection Commission-Helsinki Commission, 1990. Second periodic assessment of the state of the Marine Environment of the Baltic Sea, 1984-1988; Background Document. - Proc.Nr.35B: 432.

2. Gudelis V. and Emelyanov E. (eds), 1976 - Geology of the Baltic Sea. - Vilnius, "Mokslas", 386 (in Russian).

3. Emelyanov E. and Lukashin V.(eds), 1986. Geochmistry of the Baltic Sea sedimentation process, - Moskow, Nauka: 230 (in Russian).

4. Davidson I., Savtchuk O. (eds) 1989. - Main tendencies of the ecosystem evolution. - Leningrad, Gidrometeoizdat: 262 (in Russian).

5. Antonow A. 1987. - Large-scale changes of hydrological regime of the Baltic Sea and their influence on catches of commercial fish. - Leningrad, Gidrometeoizdat: 248.

6. Matthäus W., 1990. Langzeittrends und Veränderungen oceanologischer Parameter während der gegenwärtigen Stagnationsperiode im Tiefenwasser der Zentralen Ostsee. - Fisch.-Forsch., Rostock, Nr.28, 3, 25-34.

7. Järvekülg A., 1979. Bottom fauna of the south-eastern part of the Baltic Sea. - Tallin, "Valgus", p.381.

8. Pustelnikov O. (ed.) 1992. Biogeochemistry of semi-closed bays of the B.S. and human factor of sedimentation. Vilnius, "Academia", (in print).

9. Gudelis V., Pustelnikov O. (eds). - 1983. Biogeochemistry of the K.M. lagoon . - Vilnius, p.149.

10. Pustelnikov O., Nesterova M. (eds). 1984. Environmental influence of a black oil spill of the Baltic Sea. - Vilnius, p.150.

11. Gershanowitch D. (ed.) 1984. Outlines on Baltic Sea bioproductivity, vol 1. - Moskow, p.390.

12. Emelyanov E., Wypych K. (eds) 1987. Sedimentation processes in Gdansk basin (Baltic Sea). - Moscow, p.273.

13. Geodekian A., Trotsiuk V., Blashtchishin A. (eds) 1990. Geoacoustic and gas / lithogeochemical studies of the Baltic Sea. - Moskow, p.163.

14. Sebastian G. Gerlach, 1981. Marine polution. Diagnosis and therapy. - Springer - Verlag, Berlin-Heidelberg-New York-Tokyo, p.263.

15. Zareckas S., Pustelnikow O. 1988. Natural and anthropogenic hydrocarbons in the south-eastern part of the Baltic Sea. - Mater. of 2d Pacific Ocean symposium of marine sciences, Vladivostok, p.90-91.

16. Pustelnikov O., Zareckas S., 1988. Monitoring of hydrocarbons in the barrier zone r. Nemunas-Kursio Marios lagoon-Baltic Sea. - Ibidem, p.72-73.

17. Pustelnikow O., 1990. Anthropogenic factor, processes of natural sedimentation and solving of ecological problems of near-shore basin. - Inform. LitNIINTI, Nr.90-25, ser. 87.19.03.

18. Lukasev V. (ed.) 1986. Geochemical differentation of elements in marine and continental media. - Minsk, "Nauka i technika", p.208.

19. Hallberg R.O., 1976. A geochemical method of investigation of paleoredox conditions in sediments. - Ambio Spec. Rep., Nr.4, p.139-147.

The "Sea Level Rise" Problem: An Assessment of Methods and Data

H.-P. Plag*
Institut für Geophysik,
Christian-Albrechts-Universität zu Kiel,
Olshausenstr. 40-60,
D-W-2300 Kiel, Germany

Abstract

The rapidly increasing environmental change affected by mankind has greatly stimulated research. In order to separate natural changes of the Earth system from anthropogenic ones, a large number of studies have utilized historical global observations. Among the possible effects of anthropogenic climate change, a predicted rise in global sea level is considered to be of high importance. Thus, quite a number of recent studies have focussed on detecting the "global sea level rise" or even an acceleration of this trend. However, the results are not conclusive, though most of these studies have been based on a single global data set of coastal tide gauge data provided by the Permanent Service for Mean Sea Level (PSMSL).

A detailed discussion of the mathematical background of the methods used to arrive at estimates of the climate-related signal in relative sea level reveals the assumptions inherent in this process, and it is shown that these assumptions are invalid. The sea level data used in these studies is a coastal data set, and therefore, the resulting trend estimates cannot be more than coastal estimates, which are not necessarily climate-related. Furthermore, the data set is suffering from three severe limitations, prohibiting even the determination of a global coastal trend estimate: (1) the geographical distribution of reliable tide gauge stations is rather uneven with pronounced concentrations in some areas of the northern hemisphere (Europe, North America, Japan), and much fewer stations on the southern hemisphere where particularly few stations are located in Africa and in Antarctica; (2) the number of stations recording simultaneously at any time is far less than the total number of stations with the maximum available data from the interval between 1958 and 1989; (3) the number of long records is extremely small and almost all of them originate from a few regions of the northern hemisphere.

Nevertheless, it is noted that the data are providing valuable information concerning global and regional ocean and atmosphere-ocean dynamics, and some examples are discussed.

---
*Tel.: +49-431-880-1425, Fax: +49-431-880-4432

1 Introduction

During the last decade, the mounting consciousness of rapid global change brought about by the ever growing number of human beings has led to realizing the danger inherent in the anthropogenic rearrangement of the global environment currently taking place or being expected for the near future. For example, it is expected that the increasing emission of anthropogenic Greenhouse gases into the atmosphere will eventually result in a new global pattern of climate and weather (see e.g. Houghton et al., 1990). Another example is the currently rapid reduction of the biodiversity, a process which is thought to seriously reduce the stability of the environment, even leading to the risk of mass extinctions (Ryan, 1992).

The relation between activities of the anthroposphere and changes in the state of the other spheres, today has become one of the most important challenges to science. For meeting this challenge, the traditional separation of science into distinct disciplines will have to be overcome, and a "Geophysiology" of the Earth (Watson and Lovelock, 1983; Lovelock, 1988; Krumbein, 1983; Krumbein and Schellnhuber, 1990, 1992) not unlike to the human medicine (Schellnhuber and von Bloh, 1992) will be needed. In this context, the Earth is considered as a living natural body with the biosphere playing an active role in determining environmental conditions and shaping the Earth's surface (Krumbein and Schellnhuber, 1992), instead of passively adopting to changing environmental conditions.

Most of the direct effects currently expected to be the result of recent human activities are masked by "natural" (i.e. non-anthropogenic) variations of our environment. It is hoped for, that the knowledge of past changes will eventually lead to a better understanding of these natural fluctuations which the observable parameters of the Earth may experience and thus will enable us to detect the anthropogenic changes. Therefore, a number of recent publications related to global change have been trying to utilize historical data to extract information about climatic and environmental changes within the last hundred years or so.

Most of these papers have suffered from serious and unavoidable problems being due to the spatial and temporal scales involved in global studies and to the deficiencies of the historical data in relation to these scales. However, some of the assumptions inherent in the principal approach of these studies are to be criticized, too. The sketch in Fig. 1 illustrates this approach.

Besides the influences i_j from outside (mostly solar radiative input), the state of the Earth depends on a large number of different (internal) parameters p_j, such as the composition of the atmosphere, the chemical elements available, the number of different species in the biosphere, the number of individuals in a species. Investigating the changes in the state of the system normally builds upon analysing the changes in those parameters thought to be most indicative of changes in the environmental conditions. In detecting climate changes, the global temperature is one of these presumably sensitive parameters. Another parameter being the goal of several recent studies is the mass and volume of the ocean water, which is linked to climate changes. However, none of these parameters is directly observable but rather has to be deduced from local observations. For this purpose, the local observations have to be corrected for any inhomogeneities or errors of measurement and they have to be filtered both spatially and temporally. Disturbing

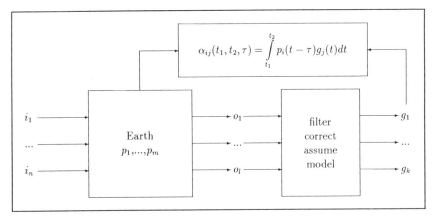

Figure 1: *The principal approach of global change research utilizing historical data.*
The conditions on the Earth being exposed to some forces i_j from outside are thought to depend on a large set of parameters p_j, be they directly observable or not. Some parameters o_j, not necessarily out of the set of p_j, are directly observable and have been monitored during the past decades. In order to deduce global parameters expected to have indicative power concerning the state of the Earth, the data have to be corrected for measurement errors, and temporally and spatially filtered. Furthermore, some assumptions have to be made concerning disturbing effects which are modelled and eliminated. Finally, global parameters g_j are determined, and they are compared to model output or expectations (a step not included in the sketch), or correlations with some of the p_i's are sought (see text).

effects are either removed by using other data or by model results, and several assumptions are required in the process of deducing the desired parameters.

Finally, a correlation is sought between the resulting global parameter g_i and other parameters p_j or inputs i_k of the general type

$$\alpha_{ij}(t_1,t_2,\tau) = \int_{t_1}^{t_2} p_i(t-\tau) g_j(t) dt \tag{1}$$

with $[t_1, t_2]$ being the time interval with observations and τ being a lag time between the changes in p_i and corresponding changes in g_j.

A typical example is the proposed correlation between CO_2 increase and global temperature (e.g. Schönwiese et al., 1990). In this last step, a most critical implicit assumption about the nature of the Earth is being made, namely the assumption that the Earth is a nearly linear system.

Most of the historical data sets used in global change studies including surface temperature, sea surface temperature, and mean sea level are handicapped by an extremely uneven spatial distribution of the sampling points, which furthermore is strongly time dependent. In the present study, the global data set of mean sea level is used to illustrate in detail the problems arising from the data deficiencies and the approches taken.

The climate–sea level link is mainly due to three processes:

(1) Changes in the sea water temperature cause volumetric expansion, with the quantitative effect on sea level depending upon the magnitude and spatial distribution (including depth) of the temperature changes;

(2) mass exchanges between cryosphere, atmosphere or terrestrial water storages and the ocean result in volume changes of the sea water;

(3) variations in the surface loading due to ice, ocean and atmosphere deform the solid Earth, resulting in changes of the height of the land and the Geoid.

However, the detection of the climate-related signal in sea level is complicated by a number of additional factors affecting relative sea level. On time scales of decades to a hundred years, vertical motion of the land is among the important ones (e.g. Douglas, 1991; for a full account of the factors affecting sea level, see e.g. Devoy, 1987).

The measurement of the relative sea level poses some additional problems, and particularly the definition and stability of the reference system for the sea level changes is most intriguing (Carter et al., 1989). However, we will not consider any measurement-related problems here, and assume ideal measurements, instead.

Within the last decade, quite a number of papers have been aiming at detecting the so called "eustatic sea level rise" (a term not uniquely defined, see e.g. Mörner, 1976; Devoy, 1987) which is closely associated with the volume change of the sea water. Extensive reviews of these papers can be found e.g. in Pirazzoli (1989), Emery and Aubrey (1991), and a short summary including a list of the respective papers is given in Section 3, following the discussion by Gröger and Plag (1992).

Most of the recent studies of the sea level rise are based on essentially the same monthly mean sea level data set derived from coastal tide gauge data. In the next section, we will outline some characteristics of this unique global sea level data set. More extensive discussions of this set of data can be found in Barnett (1983b), Pugh et al. (1987), Woodworth (1991), and Emery and Aubrey (1991).

2 The Mean Sea Level Data Base

The presently available data on sea level changes stems from basically three different sources, depending on the time scales considered. For very long time scales, stratigraphic information and especially seismic stratigraphy (Vail et al., 1977) is used to deduce variations in sea level. For time scales of kyrs, morphological features such as ancient strand lines or beaches together with other sea level related indicators are used to determine the position of relative sea level (van der Plasche, 1986; Pirazzoli, 1991). Finally, within the last 200 yrs or so, a growing number of continuously recording tide gauges are used to monitor the variations of sea level in relation to a benchmark on land (coastal tide gauges) or in relation to the sea floor (ocean bottom tide gauges).

In the present study, we will concentrate on the relative sea level data derived from coastal tide gauge records. Since 1933, the monthly mean sea level values extracted from these records are collected by the "Permanent Service for Mean Sea Level" in Bidston, UK (Spencer and Woodworth, 1991). This global data set is unique and it has been the basis for nearly all of the studies mention in the next section.

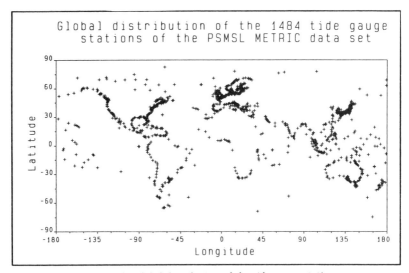

Figure 2: *The global distribution of the tide gauge stations.*
Records collected in the 'Metric' data set of the PSMSL. Note the fairly good representation of all coast lines.

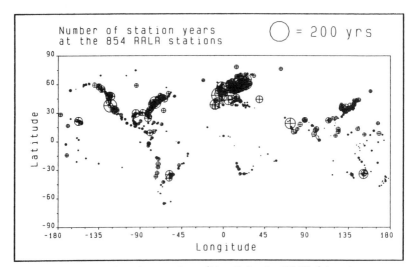

Figure 3: *Distribution of record length for the RRLR data set.*
Note that nearly all long records are located in three regions of the northern hemisphere.

The data are distributed in two different data sets, namely the so-called 'Metric' file, containing all available data directly as supplied by the national authorities (see Fig. 2), and the 'Revised Local Reference (RLR) data set, which contains a considerably smaller number of records having passed a consistency check (Woodworth et al., 1990). In the RLR set, each record is reduced to a common datum, making use of the benchmark history supplied by the national authorities. The second data set is recommended as basis for long-period sea level changes and global trend analyses (Spencer and Woodworth, 1991).

Though the records included in the RLR set have been thoroughly checked by the PSMSL (Woodworth et al., 1990), in our own studies some of the records were classified as unreliable (Gröger and Plag, 1992), with the irregularities mainly due to earthquakes or other known causes (Woodworth, 1992). Therefore, the basis for our studies is a 'revised' RLR set, denoted as RRLR.

For the determination of a global trend, especially the fairly long records and their spatial distribution is of importance. Using the length of a record as radius of a circle, the global distribution of station years is visualized in Fig. 3, revealing a distinguished unevenness of the spatial distribution, with nearly all long records clustering at the US, European, and Japanese Coasts.

The quality of a global sea level curve is depending on the number of available simultaneous recordings. Plotting this number as function of time (Fig. 4) reveals the strong temporal variation in this parameter. Only in the last four decades a fairly large number of observations is available.

Due to the presence of decadal to interdecadal sea level fluctuations (Sturges, 1987; Plag, 1988), the local trends strongly depend on the record lengths. The scatter of local trends decreases rapidly with increasing record length (Fig. 5), and especially for records of less than 20 yrs, the local trend scatter is unrealistically large. But even between 20 and 60 station years, a slight decrease of the scatter is obvious.

Consequently, the spatial distribution of local trends strongly depends on the lower limit of station years required for a record to be included (Fig. 6). Most of the larger trends are associated with either tectonic (e.g. west coast of Alaska, Emery and Aubry, 1986) or anthropogenic (e.g. Galveston; Hicks et al., 1983) causes. Particularly for higher latitudes on the northern hemispheres, the trend pattern is determined by the post-glacial deformations (Peltier and Tushingham, 1989), emphasizing the importance of the vertical motion of the land.

For a lower limit of station years of e.g. 40 yrs, there is no longer a global distribution of stations. As we will see below, this none-global distribution, among other reasons, prohibits the detection of a global climate-related sea level signal.

However, there is still important information to be extracted from this historical data set. Thus, using the temporally most comprehensive subset, Gröger and Plag (1992) detected a decadal to interdecadal fluctuation of global scale (Fig. 7), indicating some kind of climate-related teleconnection, or even pointing to the part, which the oceans may have in exciting the decadal changes in the length of day (Hide and Dickey, 1989).

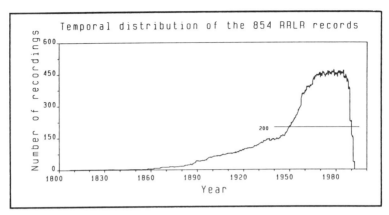

Figure 4: *The temporal distribution of the recordings available in the RRLR data set.* Note the comparably short interval between 1958 and 1989 with a large and nearly constant number of simultaneous recordings.

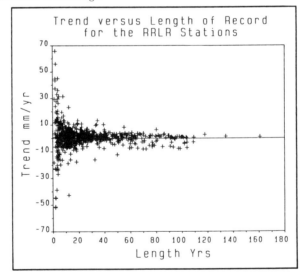

Figure 5: *Local trends as function of the record length.*
Data base is the RRLR set. Note the decrease in trend scatter as record length increases. The strong tendency to negative values for record lengths of 40 to 100 yrs is due to the dominance of the fennoscandian stations, located in an area of distinguished post-glacial uplift.

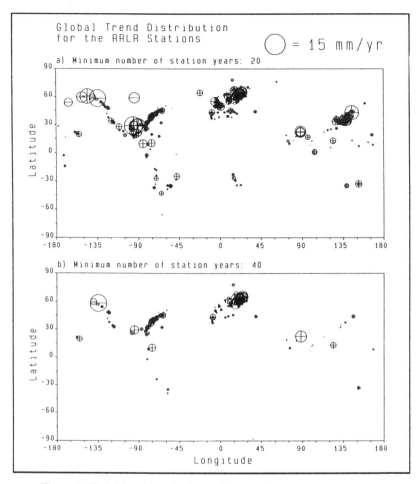

Figure 6: *Global trend distribution as function of minimum record length.*
Note the dramatic reduction of stations when using 40 yrs instead of 20 yrs as minimum record length.

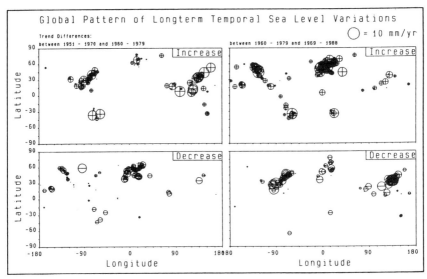

Figure 7: *Global pattern of decadal to interdecadal sea level fluctuations.* Plotted are temporal variations of the apparent trend. Data are from the time interval with the largest number of simultaneous recordings (see Fig. 4). Note the nearly inverted pattern in the right diagrams compared to the left ones, indicating a large scale east-west sea level fluctuation (adopted from Gröger and Plag, 1992).

Plag and Pomrehn (1992) used the spatially most comprehensive regional subset, i.e. the North Sea and Baltic Sea data, to study the spectral distribution of sea level variance in the 1 to 5 year period band (Fig. 8), revealing regionally coherent step-like changes occuring at least three times between 1900 and 1985. The dates of these steps coincident with similar sudden changes of other geophysical parameters such as arctic winter temperature, global temperature, or even polar motion. These changes in the sea level variance distribution are attributed to sudden changes in the atmospheric circulation over the respective oceanic regions, indicating step-like climate changes instead of gradual changes. Furthermore, they are in accord with recent findings from sediment studies, indicating rapid climatic changes in the past (for ref. see Plag and Pomrehn, 1992).

This information derived from sea level studies may constitute a very important piece of evidence concerning the nature of climate changes and the underlying mechanisms. The presence of such sudden changes renders the assumption of a more or less linear system (see Fig. 1) rather insecure, and we no longer should expect a correlation of the kind given in equation (1) to hold.

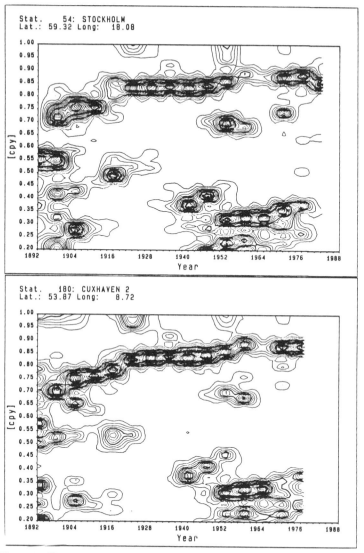

Figure 8: *Temporal variations of the spectral energy of sea level fluctuations.*
The frequency band considered is between 0.2 and 1 cpy corresponding to the 1 to 5 yrs period band. The annual variation is removed. For each time step, the variance of the frequency band has been scaled to the range [0,1]. Plotted are two typical stations in the North and Baltic Sea. Note the step like changes around 1920, 1956 and 1980 (adopted from Plag and Pomrehn, 1992).

3 Previous Trend Determinations

Most of the studies aiming at the detection of the current global sea level rise are compiled in Tab. 1, which is adopted from Gröger and Plag (1992). Though most of the more recent studies are based on the same data set described in the previous section, the scatter in the sea level trends estimates is rather large. Taking into account the given uncertainties, the full range of the results is between nearly zero and above 3 mm/yr, with some authors even doubting at all any sea level rise within the last hundred years.

To understand this scatter, Gröger and Plag separate the determination of a global trend in sea level into three main steps:

1. **Data selection:** There is no common, objective way of selecting the reliable tide gauge records from the data set. All authors set a lower limit for a record to be included in the analyses, but there is no agreement about this length, and it varies between 20 years and 60 years. Furthermore, some authors exclude stations located in tectonically active regions or in uplift areas from their analyses (see Gröger and Plag, 1992, for the respective references).

2. **Evaluation of the global trend:** Global sea level trend estimates may be obtained by either determining a trend from a global sea level curve or by determining an average of local trends. In most studies, the second alternative is chosen, and local trends are determined by linear regression. The global estimate is then deduced from the local trend values by either simply averaging the values, by weighting the local values with the ocean area they are thought to represent, or by establishing several oceanic regions, and using regionally averaged sea level trends for the determination of a global trend value.

   As a third alternative, in some of the analyses empirical orthogonal functions are used to compute a global trend.

3. **Corrections:** In order to detect the climate-related sea level trend, all disturbing factors have to be removed. Considering the vertical motion of the solid Earth to be the most important perturbation, some authors corrected the trends of single tide gauge records for the effects of tectonic or isostatic land movements. Tectonic movements are corrected by using geological data about crustal movements within the last kyrs, while geophysical models are used to calculate the vertical movement of the land associated with postglacial rebound.

Before we present a mathematical analysis of the second step in the next section, let us have a quick look at the first and last step.

Concerning the data selection, it should be noted that the number of stations actually used in the analyses is rather small with a highly non-global coverage, as was shown in the previous section (see Fig. 6).

Both attempts to correct the data for vertical land movements suffer from some limitations. Geological data are available only for some stations or coastal regions, thus further reducing the density of the useful sea level records or introducing data inhomogeneities.

Table 1: *Studies of global sea level rise based on tide gauge data.*
Note that most studies are based on the PSMSL data set. Nevertheless, the trend estimates scatter by more than a factor of two (from Gröger and Plag, 1992).

Author	Data Source	No. of Stations		Global Sea Level Rise in mm/yr	Methods
Gutenberg (1941)	AdOP		71	1.1 ± 0.8	RA
Fairbridge and Krebs (1962)	?		?	1.2	SA
Lisitzin (1974)	?		6	1.1 ± 0.4	SA
Emery (1980)	PSMSL		247	3.0	SA
Gornitz et al. (1982)	PSMSL	a)	193	1.2	RA
		b)	86	1.0	RA + GD
Barnett (1983)	PSMSL		9	1.5 ± 0.2	EOF
				1.8 ± 0.2	EOF
Barnett (1984)	PSMSL +NORPAX		155	1.4 ± 0.1	EOF + RA
				2.3 ± 0.2	EOF + RA
Gornitz and Lebedeff (1987)	PSMSL	a)	286	0.6 ± 0.4	SA
		b)	231	1.7 ± 0.3	SA
		c)	130	1.2 ± 0.3	SA + GD
				1.0 ± 0.1	RA + GD
Peltier and Tushingham (1989; 1991)	PSMSL		40	2.4 ± 0.9	EOF + GM
Trupin and Wahr (1990)	PSMSL	a)	120	1.2 ± 0.1	SA + GM
		b)	97	1.6 ± 0.1	SA + GM
		c)	84	1.75 ± 0.1	SA + GM
Douglas (1991)	PSMSL		21	1.8 ± 0.1	RA + GM

AdOP = Association d' Océanographie Physique, Liverpool
PSMSL = Permanent Service for Mean Sea Level, Bidston Observatory, Birkenhead
NORPAX = North Pacific Experiment

SA = mean of all station trends
RA = regional averages used
EOF = empirical orthogonal functions
GD = correction by geological data
GM = correction by geophysical model

In addition, some of the geological records contain large uncertainties, which are carried into the trend determinations.

Concerning the modelling of postglacial deformations, serious limitations arises from at least two of the ingredients, namely the viscoelastic Earth model used (which up to now is a spherically symmetric model with a simple depth dependent rheology, see e.g. Tushingham and Peltier, 1991) and the ice load history inducing the deformations of the Earth. In addition, the geological data used to validate the models have some inherent uncertainties. Consequently, the effect of correcting post-glacial rebound by different authors maybe contradicting, as pointed out by Gröger and Plag (1992).

4 Global Sea Level Trend Determination

Due to the problems and uncertainties outlined in the previous sections, several authors are quite skeptical about the the possibility of determining a global sea level rise from the existing data (Barnett, 1984; Pirazzoli, 1986; Emery and Aubrey, 1991; Dobrovolski, 1992). In what follows, we will show that it is indeed impossible to detect the climatically induced signal in relative sea level given the present data situation.

The global relative sea level curve g(t) can be represented as

$$g(t) = \frac{1}{O(t)} \int_{O(t)} r(t,s) ds \qquad (2)$$

where
t : time
s : a point on the ocean surface
$O(t)$: the time-dependent ocean surface
$r(t,s)$: the time-dependent height of local relative sea level.

The integration has to be carried out over the complete ocean surface, which, in principle, is time dependent. The sea level height has to be refered to a reasonably defined reference system, which, as we already mentioned in the introduction, is one of the main deficiencies of the global data set.

If we want to determine the changes in volume of the sea water, we have to eliminate the vertical motion $h(t,s)$ of the coasts and the sea bottom, neglecting for the moment any horizontal motion of the ocean boundary. Thus, we arrive at

$$g(t) = \frac{1}{O} \int_{O} (r(t,s) - h(t,s)) ds, \qquad (3)$$

where we reasonably assume the sea surface to be constant over the time scales considered. Since we are interested in the global trend, we may write

$$\dot{g}(t) = \frac{1}{O} \int_{O} (\dot{r}(t,s) - \dot{h}(t,s)) ds. \qquad (4)$$

with the dots denoting time derivatives. An assumption generally applied is the existence of a constant (linear) trend g_l at each location in both sea level and land movements, resulting in

$$g_l = \frac{1}{O}\int_O (r_l(s) - h_l(s))ds. \tag{5}$$

where both r_l and h_l depend on position but not on time. Finally, the integral in (5) is rudely approximated by

$$g_l = \frac{1}{n}\sum_{j=1}^{n}(r_{l_j} - h_{l_j}) \tag{6}$$

with the summation carried out over all selected tide gauge stations. As we see from Tab. 1, the number n of the selected stations may be as small as 6!

Inherent in this procedure are the following critical assumptions:

(1) *The global relative sea level curve is determined by the coastal sea level*, i.e.

$$\frac{1}{\partial O}\int_{\partial O} r(t,s)ds - \frac{1}{O}\int_O r(t,s)ds \approx 0, \tag{7}$$

with ∂O denoting the complete coast line.

(2) *The vertical motion of the coasts and the land below the oceans is determined by the coastal vertical motion*, i.e.

$$\frac{1}{\partial O}\int_{\partial O} h(t,s)ds - \frac{1}{O}\int_O h(t,s)ds \approx 0. \tag{8}$$

(3) *The local relative sea level contains at most a linear term but no higher degree terms*, i.e.

$$\lim_{T\to\infty} q(T,s) = c(s) \tag{9}$$

with c(s) being a function of position only, and

$$q(T,s) = \frac{2}{T^2}\int_0^T r(t,s)dt, \tag{10}$$

where we have assumed the time origin to be at the begin of the interval considered. Restricting ourselfs to time scales of centuries, the assumption is weakened to

$$q(T,s) - c_c(s) \approx 0 \qquad T_m(s) \leq T \leq T_M(s) \tag{11}$$

where the mimimum and maximum length of the time interval, T_m and T_M respectively, may depend on the location and where $c_c(s)$ again is a function of position only, which, however, may depend on the century considered.

The validity of these assumptions must be doubted for the following reasons:

(1) The dynamic ocean surface topography of up to 2 m is temporally changing with variations of several cm over decades (Sturges, 1987, Plag, 1988), and especially off-coast these changes are spatially variant (Roemmich, 1990). This rules out assumption (1).

(2) The vertical motion of the coasts is globally not representative for the vertical motion of the sea floor far away from the coast, as can be expected from the pattern of global tectonics. Modern space techniques reveal rather large vertical motion of e.g. ocean islands such as Hawaii (Kolenkiewicz et al., 1992) of up to 2 cm/yr, which might be representative for large areas, particularly around Mid Ocean Ridges. This rules out assumption (2).

(3) The linear trend assumption is clearly not justified for very long time scales. However, there might be intermediate time scales, for which $q(T,s)$ is nearly constant, and it is generally assumed that a T_m of several decades suffices (between 20 and 60 yrs, see section 3.). Nevertheless, significant sea level fluctuations exist on time scales between yrs and kyrs (see e.g. Douglas, 1991) rendering assumption (3) invalid.

The correction of the effect of the presently ongoing land uplift due to post-glacial rebound is in a slightly better position than indicated in the above analysis: The sea level equation as derived by Farrell and Clark (1976) can be used to directly calculate the sea level changes due to this effect at any location and time. However, the important incredients, namely the rheological Earth model and the ice load history are still rather simplified and uncertain, respectively, and consequently, the present day post-glacial sea level changes calculated with these models have a large degree of uncertainty.

The problems involved in e.g. determining the global temperature from historical data are rather similar to those outlined above. Therefore, the global values determined in this way (i.e. anything coming out of the right box in Fig. 1) may well be far away from the true global values.

Besides these problems in determining the global parameters, we also have to question the value of any such global parameter expressed by a single number. Both, global temperature and the sea water volume have remained fairly constant over most of the Earth's history. Nevertheless, large dramatic changes of regional conditions have taken place influencing environmental condition and the ecosystems far more than an change in global parameters could.

5 Conclusions

Coastal relative sea level changes can not be used to deduce the climate-related signal in sea level, i.e. changes of the oceanic water mass and volume. Coastal obseravtions of the relative sea level do not give sufficient information to evaluate the integral over the ocean surface required to determine a global sea level curve or a global trend.

Furthermore, the climate-related part of the global change in relative sea level is masked by a number of other influences affecting relative sea level, and these, too, may not be determined from coastal observations alone.

Therefore, from the existing historical sea level data, the determination of a relative sea level curve or trend is only possible confined to coastal regions. However, since such an average is not necessarily related to climate changes, a global average does not contain any information being worth the effort of calculating it.

Furthermore, regional averages taken along connected parts of a coast, may be of more importance for climate change impact studies than a global value (see e.g. Gornitz, 1991; Pirazzoli, 1991).

Nevertheless, the existing data set can be utilized to obtain some important pieces in the puzzle of global ocean and atmosphere-ocean dynamics. Thus, the global scale of decadal to interdecadal fluctuations as shown by Gröger and Plag (1992) may be another hint to the importance of teleconnections in the climate system.

Large scale step-like changes in the sea surface fluctuations attributed to the effect of atmospheric circulations on the ocean surface, and the coincidence of these step-like changes with similar changes e.g. in temperature, are firm evidence for the non-linear nature of the climate system. (Plag and Pomrehn, 1992).

Final, we may conclude that the existing sea level data set is a valuable source of information, provided we ask the right questions. Asking this data set for the global climate-related sea level signal is definitely the wrong question.

Acknowledgements

This work was carried out within the framework of two projects sponsored by the German Research Society, namely Zs-4/5-1,2 and Zs-4/9-1. The sea level data were kindly provided by the "Permanent Service for Mean Sea Level", Bidston Observatory, UK. The author is indebted to M. Gröger and V. Rautenberg for carefully and constructively revising the manuscript.

References

Barnett, T.P., 1983a: Possible changes in global sea level and their causes. *Climate Change*, **5**; 15–38.

Barnett, T.P., 1984: The estimation of "global" sea level change: a problem of uniqueness. *J. Geophys. Res.*, **89**; 7980–7988.

Carter, W.E., Aubrey, D.G., Baker, T.F., Boucher, C., LeProvost, C., Pugh, D.T., Peltier, W.R., Zumberge, M., Rapp, R.H., Schutz, R.E., Emery, K.O., Enfield, D.B., 1989: Geodetic fixing of tide gauge bench marks. Woods Hole Oceanographic Institution Technical Report WHOI-89-31, 44p.

Devoy, R.J.N., 1987: Introduction: First Principles and the Scope of Sea-surface Studies. In: R.J.N. Devoy (Editor): Sea Surface Studies: A Global View. Croom Helm, London, 1–32.

Dickson, R.R., Meincke, J., Malmberg, S.A. and Lee, A.J.: The "Great Salinity Anomaly" in the northern North Atlantic, 1968-1982. *Prog. Oceanog.*, **20**, 103–151.

Dobrovolski, S.G., 1992: Global Climatic Changes in Water and Heat Transfer - Accumulation Processes. Elsevier, Amsterdam, Developments in Atmospheric Science, Vol. 21., 265p.

Douglas, B.C., 1991: Global Sea Level Rise. *J. Geophys. Res.*, **96**; 6981–6992.

Ellsaesser, H.W., MacCracken, M.C., Walton, J.J., and Grotch, S.L., 1986: Global climatic trends as revealed by the recorded data. *Rev. Geophys.*, **24**; 745–792.

Emery, K.O., 1980: Relative sea levels from tide-gauge records. *Proc. Natl. Acad. Sci. USA*, **77**; 6968–6972.

Emery, K.O. and Aubrey, D.G., 1986: Relative sea level from tide gauge records of western North America. *J. Geophys. Res.*, **91**; 13941–13953.

Emery, K.O., and Aubrey, D.G., 1991: Sea Levels, Land Levels and Tide Gauges. Springer, Berlin, 237p.

Farrell, W.E. and Clark, J.A., 1976: On postglacial sea level. *Geophys. J. R. astro. Soc.*, **46**; 647–667.

Gornitz, V., Lededeff, S., and Hansen, J., 1982: Global Sea Level Trend in the Past Century. *Science*, **215**; 1611–1614.

Gornitz, V. and Lebedeff. S., 1987: Global sea-level changes during the past century. In: D. Nummedal, O.H. Pilkey, and J.D. Howard (Editors), 1987: Sea Level Fluctuations and Coastal Evolution, Soc. Econ. Palaeontol. Miner., Spec. Publ. 41; 3–16.

Gornitz, V., White, T.W., and Cushman, R.M., 1991: Vulnerability of the U.S. to future sea level rise. Coastal Zone '91, Proceedings of the 7th Symposium on Coastal & Ocean Management, ASCE/Long Beach, July 8-12, 1991, 2354–2368.

Gröger, M., and Plag, H.-P., 1992: Estimation of a Global Sea Level Trend: Limitations from the Structure of the PSMSL Global Sea Level Data Set. *Submitted to: Global and Planetary Change.*

Gutenberg, B., 1941: Changes in sea level, postglacial uplift, and mobility of the Earth's interior. *Bull. Geol. Soc. Amer.*, **52**; 721–772.

Hide, R. and Dickey, J.O., 1990: Earth's variable rotation. JPL Geodesy and Geophysics preprint No. 205, subm. to *Science*, 46 pp.

Houghton, J.T., Jenkins, G.J., and Ephraums, J.J., (Eds.), 1990: Climate Change, the IPCC Scientific Assessment. Cambridge University Press, Cambridge, 1990, 364pp.

Kolenkiewicz, R., Smith, D.E., Pavlis, E.C., Porrence, M.H., 1992: Direct estimation of vertical variations of satellite laser tracking sites from the SL8 LAGEOS Analysis. *Annales Geophysicae*, Supplement I to Vol. 10, C111.

Krumbein, W.E. (ed.), 1983: Microbial Geochemistry. Blackwell, Oxford, p. 330.

Krumbein, W.E., and Schellnhuber, H.-J., 1990: Geophysiology of carbonates as a function of bioplanets. In: Ittekot, V., Kempe, S., and Michaelis, W. (eds.): Facets of Modern Biogeochemistry, pp 5-22, Springer-Verlag, Heidelberg.

Krumbein, W.E., and Schellnhuber, H.-J., 1992: Geophysology of mineral deposits - a model for a biological driving force of global changes through Earth history. *Terra Nova*, **4**, 351-362, Global Change Sepc. Issue, ed. by F.-C. Wezel.

Levitus, S., 1989: Interpentadal variability of temperature and salinity in the deep North Atlantic. *J. Geophys. Res.*, **94**; 16125.

Lisitzin, E., 1974: Sea Level Changes. Elsevier Oceanography Series, No. 8, Elsevier Scientific Publishing Company, Amsterdam, 286 pp.

Lovelock, J.E., 1988: The Ages of Gaia. Norton, New York.

Peltier, W.R. and Tushingham, A.M., 1989: Global sea level rise and the Greenhouse effect: Might they be connected? *Science*, **244**; 806–810.

Mörner, N.-A., 1976: Eustacy and geoid changes. *J. Geol.*, **84**, 123-151.

Peltier, W.R. and Tushingham, A.M., 1989: Influence of Glacial Isostatic Adjustment on Tide Gauge Measurements of Secular Sea Level Change. *J. Geophys. Res.*, **96**, 6779-6796.

Pirazzoli, P.A., 1986: Secular Trends of Relative Sea-Level (RSL) Changes Indicated by Tide-Gauge Records. *J. Coastal Res.*, Spec. Issue 1; 1–26.

Pirazzoli, P.A., 1989a: Present and near-future global sea-level changes. *Palaeogeogr. Palaeoclimatol. Palaeoecol. (Global and Planetary Change Section)*, **75**; 241–258.

Pirazzoli, P.A., 1989b: Recent sea-level changes in the North Atlantic. In: D. B. Scott et al. (eds.): Late Quaternary Sea-Level Correlation and Applications. Kluwer Academic Publishers, 153-167.

Pirazzoli, P.A., 1991a: World Atlas of Holocene Sea-Level Changes. Elsevier Oceanography Series, 58, Elsevier Science Publishers B.V., Amsterdam, 300 pp.

Pirazzoli, P.A., 1991b: Possible defenses against a sea-level rise in the Venice area, Italy. *J. Coastal Res.*, **7**; 231–248.

Plag, H.-P., 1988: A Regional Study of Norwegian Coastal Long-Period Sea-Level Variations and Their Causes with Special Emphasis on the Pole Tide. Berliner Geowissenschaftliche Abhandlungen, Reihe B, Band 14, Verlag von Dietrich Reimer, Berlin, 175 pp.

Plag, H.-P., and Pomrehn, W., 1992: Evidence for rapid climatic changes of the Earth system. Submitted to *Nature*.

Plassche, O. van de, 1986: Introduction. In: O. van de Plassche (Editor): Sea-Level Research: A Manual For the Collection and Evaluation of Data. Geo Books, Norwich, 1–26.

Pugh, D.T., Spencer, N.E., and Woodworth, P.L., 1987: Data Holdings of the Permanent Service for Mean Sea Level. Permanent Service for Mean Sea Level, Bidston, Birkenhead, 156p.

Ryan, J.C., 1992: In Worldwatch Institute: State of the World 1992. W.W. Norton and Company, New York. Roemmisch, 1990:

Schellnhuber, H.-J., and von Bloh, W., 1992: Homöostase und Katastrophe: Ein geophysiologischer Zugang zur Klimawirkung. In: Schellnhuber, H.-J., and Sterr, H. (eds.): Klimaänderung und Küste: Einblick ins Treibhaus. Springer-verlag, Heidelberg, in press.

Schönwiese, C.-D., Birrong, W., Schneider, U., Stähler, U., and Ullrich, R., 1990: Statistische Analyse des Zusammenhangs säkularer Klimaschwankungen mit externen Einflusgrösen und Zirkulationsparametern unter besonderer Berücksichtigung des Treibhausproblems. Berichte des Instituts für Meteorologie und Geophysik der Universität Frankfurt/Main, Nr. 84, 260p.

Spencer, N.E. and Woodworth, P.L., 1991: Data Holdings of the Permanent Service for Mean Sea Level (January 1991). Bidston Observatory, Birkenhead, 136 pp.

Sturges, W., 1987: Large-scale coherence of sea level at very low frequencies. *J. Phys. Oceanogr.*, **17**, 2084–2094.

Trupin, A. and Wahr, J., 1990: Spectroscopic analysis of global tide gauge sea level data. *Geophys. J. Int.*, **100**; 441–453.

Vail, P.R., Mitchum, R.M., Todd, R.G., Widmier, J.M., Thompson, S., Sangree, J.B., Bubb, J.N., and Hatlelid, W.G., 1977: Seismic Stratigraphy and Global Changes of Sea Level. In: Payton, C.E. (ed.): Seismic Stratigraphy - applications to hydrocarbon exploration. The American Association of Petroleum Geologists, Tulsa, Oklahoma, 49-212.

Watson, A.J., and Lovelock, J.E., 1983: Biological homeostasis of the global environment. *Tellus*, **35B**, 284–389.

Woodworth, P.L., 1991: The Permanent Service for Mean Sea Level and the Global Sea Level Observing System. *J. Coastal Res.*, **7**, 699-710.

Woodworth, P.L., 1992: personal communication.

Woodworth, P.L., Spencer, N.E., and Alcock, G., 1990: On the availability of European mean sea level data. *International Hydrographic Review, Monaco*, **67**; 131–146.

Vulnerability of the coasts of Germany due to Impacts of Climate Change: Analyses and Research Demands

Sterr, H.

Potential risks of global change

Climate is quickly changing on a world-wide scale as a result of man's production and emission of greenhouse gases and substances which chemically transform the atmosphere. This appears to be a fact, now widely accepted among the scientific community and beyond. With special consideration of the coastal zones the following adjustments of the earth's system to this external input are generally assumed by climatologists or predicted by climate models (GCMs):

- by 2100 the surface temperatures over land and particularly over ocean areas will be 1,5 to 4,5 (most likely 3) degrees Celsius higher than now.

- in consequence of the global warming trend the sea level will rise world-wide at rate greatly accelerated compared to that of the past decades. Sea level-rising trends of 30-100 cm over the next 100 years are most commonly predicted, with a best estimate usually indicating a 66 cm rise (IPCC, 1990).

- Other climate parameters relevant to the coastal system, such as wind direction and intensities, storm frequencies, precipitation patterns or others, will be affected on a regional scale; their impacts on coasts could strongly vary from place to place and cannot yet be specified, neither regarding specific qualitative changes nor intensities or trends.

Given this mixture of more precise and less clear assumptions it is not surprising that most coastal scientists and managers attribute (and have done so in the past) the highest risks for the world's coastal zones to a globally accelerating rise of sea level (ASLR). Such a view has been reflected by the vast number of studies dealing with the ASLR phenomenon as well as its possible impacts (see contribution of HILLEN et.al. 1993, in this volume) (BARTH & TITUS, 1984; KELLETAT, 1990; SCHELLNHUBER & STERR, 1993). In this rather general and somewhat simplisitic approach two aspects are often overlooked or disregarded:

1. a reference sea level on a global scale does not really exist, because the relative heights and movements of land and water surfaces vary greatly from one coastal area of the world to another. Hence a global trend for ASLR cannot truly be detected or determined and thus should not be used for considering risk potentials (see contribution of PLAG, 1993 this volume) (GRÖGER & PLAG, 1992).

2. The very recent scientific discussion of (global) sea level adjustments to climate change is quite controversial. It is presently strongly disputed as to whether global warming will lead to a rise of the oceans' surfaces at all or whether perhaps an increased build-up of ice masses in Antarctica and Greenland might (over)compensate for this effect (SCHNEIDER, 1992). The latest calculations from improved ocean-coupled climate models are somewhat in line with this argument, indicating a rather moderate (average) sea level rise of only about 15 cm for the next 100 years (Spiegel, 1992).

Attention is drawn to the two aspects above so as to call for some precaution when dealing with the assessment of coastal risks or when applying impact scenarios to vulnerability analyses and related coastal studies. However, I do not mean to suggest that - because of these aspects - the potential hazards for the coastal zones are less than previously described or that they might be negligible. Instead, one might argue that it could be wise not to confine such studies to sea level rise as the predominant risk factor but to pay more attention to the various other impact parameters which are likely to affect coastlines all over the world: e.g. long-term changes in wave climates, tidal ranges, sediment budgets, storminess, bio-production changes, migration of habitats and the like.
Commonly, these variations of physical and biological process-response mechanisms are more difficult to detect and to determine than sea level changes. However, they are closely interrelated in many ways and often combine to yield a wide spectrum of effects on the coastal system as a whole, including both the natural and the human parts.

In order to find out and understand such consequences or possible hazards of climate/environmental change a new research program called **Climate Change and the Coast** was established in Germany in 1991. It is the top-priority theme in a new line of climate and impact research which was recently started jointly by the German Federal Government and the authorities of the five coastal States.

Physiography of the Coastal Zone

The coastal landscape of Germany is divided into two rather different units, the North Sea Coast to the West and the Baltic Coast to the East. They are separated by German-Danish-Peninsula which averages about 100 km in width.
The total coast length, including islands, estuaries, Förden (low-rimmed fiords) and Bodden (nearly enclosed bays), comprises 3160 km; nearly two thirds of it, 2005 km, are along the Baltic Sea, about one third, 1155 km, lies along the North Sea. Although the largest portion, perhaps as much as 90 %, are very low-lying coastal segments backed by marshes, beach ridge plains or low dunes, the specific physiographic and ecological character varies greatly along the German coastline, mainly from west (North Sea) to east (Baltic Sea) but also between outer (exposed) and inner (sheltered) coastal sections. With regard to the evaluation of climate-related impacts the following features of or differences between the various coastal systems are worth to be noted:

North Sea coast:

- generally low hinterland, mostly artificially-drained and reclaimed marshland accumulated during late-Holocene times;

- a series of funnel-shaped estuaries at the mouths of the four major rivers (Ems, Weser, Elbe, Eider) which dissect the southern and eastern rim of the German Bight;

- a chain of islands accompanying the mainland coast at a distance of some 10 to 15 km; some of these are made from dune sand, others from marshland while a few contain a core of Pleistocene till;

- a wide intra-tidal area between mainland and islands, the so-called Wadden Sea, which was formed by mid- to late Holocene sediment accumulation; it is very rich in biological life and productivity and contains a large number of precious ecotones;

- characteristic coastal features are: a very gently dipping foreshore, wide beaches in front of large dunes, widespread saltwater marshes with gradual transition zones towards the wadden flats;

- main sources for coastal sediments are rivers, sea bottom and organic matter produced in the intra-tidal zone;

- a meso-tidal hydrographic environment with tidal range from 1,5 m to 4,0 m and a generally high-energy wave climate;

- recurrent storm floods as a result of combined tidal effects and westerly wind storms;

- a high-salinity (25 to 30 permille) environment prevails along the open parts of the coast changing to moderate-salinity water conditions within the estuaries;

- large and deep tidal channels cut across the Wadden flats (usually in the gaps between the islands) through which the exchange of water, sediments, nutrients, biomass etc. takes place between coast, tidal zone and open sea.

- almost the whole length of the mainland coast is rimmed by high dikes and other solid protective structures and so are some of the low or exposed island sections.

Baltic Sea coast:

- the hinterland consists partly of hilly terrain formed in late-Pleistocene glacio-fluvial materials (cliffed coasts) and partly of low terrain, underlain by young marsh and peat deposits and recently covered by beach or dune sediments;

- no large rivers drain into the German section of the Baltic;
- inlets (Förden, Bodden) are inundated preexisting depressions cut by ice or meltwater prior to the post-glacial transgression;

- tidal range is less than 0,2 m in all of the Baltic and no intra-tidal zone exists seaward of the mean water line;

- the hydrodynamic system is dominated by wind-driven waves and currents in a narrow zone close to the shore;

- sediment sources are abrading cliffs of till or sand as well as the nearshore abrasion platform;

- longshore sediment transport lead to widespread formation of spits which results in reshaping and smoothing of an originally very irregular bay-and-headland coast: enclosure of bays, Förden and Bodden.

- predominant direction of shore-parallel sediment transport is determined by local fetch conditions and varies greatly with coastal exposure;

- characteristic coastal features are earth cliffs, narrow beaches, submarine sandbars, beach ridge plains and small (low) dunes;

- storm surges caused by easterly winds ocasionally create considerable water level set-ups and high-energy wave condition along east-facing coastal segments;

- brackish water conditions generally prevail: salinity is medium (15-12 permille) along open parts of the coast, decreasing considerably within enclosed Förden or Bodden to as low as 3-4 permille;

- (brackish water) marshes and wide belts of phragmites prevail along sheltered coastal segments with low salinity and low wave energy input;

- a wide variety of ecosystems and habitats occur on the exposed coasts but especially within the semi-enclosed embayments; within these water quality has been affected greatly in recent years by human influx;

- relatively short sections of the Baltic coasts have been protected by dikes or solid structures and the natural process pattern still prevails along most parts.

From this comparative description of the two German coastal areas it becomes clear that they may be quite differently affected by or react to climate-induced changes in their specific hydrographic, morphologic and ecologic regimes. Therefore, a regional evaluation of the coastal systems´ vulnerability is most precise and efficient when focusing on each of these areas individually while using the same basic assumptions and scenarios for system changes. Although not very much is yet known about possible impact-response mechanisms within the landscape-ecological part of the coastal system and even less about adjustments in the socio-economic part, an attempt is made here to sum up a number of the key aspects as we view them today.

Possible impacts to the coastal system induced by or related to climate change

Analysis and interpretation of effects resulting from climate change is a difficult task, for various reasons (GROET & de BOER, 1990). In a strictly systematic approach one would like to differentiate between primary changes and secondary, tertiary or higher order change reactions. Very often, though, this is neither possible nor meaningful, as demonstrated by a simple example:

a primary climate change such as a *rise of air/water temperatures* causes some direct effects (e.g. on marine biologic activity, salinity etc.) and stimulates a significant secondary change: *rise of sea level*; this again causes flooding as direct result but also leads to an *increase of wave energy* (= tertiary change) in the nearshore zone which - as forth-order impact - enhances *shoreline erosion*; loss of dune areas and habitats may be an even higher-order consequence in this chain reaction, yet the destruction of dunes could just as easily reflect a response to a change of the regional wind regime or to an increase in recreational activities (= man-made environmental change).

The complex feedback mechanisms between input parameters and their response processes thus prevent a detailed, step-by-step analysis of the coastal system and its individual components. So far, in our limited understanding of such mechanisms we are confined to a "best-guess" estimation of some first-order variations of major climate factors as well as to a still quite speculative appreciation of resulting impacts. Tab. 1 attempts to summarize the most likely or obvious relationships between climate-induced changes (column 1) and subsequent adjustments (column 3) within the German coastal zone and its typical landscape subsystems (column 2).

In addition, some man-made factors of environmental change are also included because they closely interact with the climate-dependent impacts. The list of the aspects given in Tab.1 is by no means complete and may lack some very essential parameters, but respective coastal research is only at the starting point (SCHELLNHUBER & STERR, 1993).
Generally, more is known about those aspects referring to the hydrodynamic-morphodynamic part of the system than about the ecologic and socio-economic components. Previous field studies provide some insight into key parameters such as sea level, tidal cycle, storm surges, wave climate, sediment balance, erosion trends, salinity, sea ice and others.

1. Regional Sea level Rise

   Along the North Sea coast "average" sea level rise (at mean tidal level) is estimated to average 15 cm since 1900 and the rates recorded from Baltic stations are similar (GRASSL, 1991). At several North Sea gauges the mean high-tide water level has shown a faster rise of 25-30 cm in this century (LASSEN & SIEFERT, 1991). According to PLAG, 1993 (this volume) the reliability of the German tidal gauge records is somewhat uncertain and the recordings are not included in the PMSL data sets. Based on various ocean-coupled climate models (general circulation models-GMC) the predicted sea level rise to be expected in the North Sea region for the next 100 years under the "business-as-usual" scenario greatly exceeds the global average rate (ENQUETE-COMMISSION, 1990; IPCC, 1990) and may be as much as 60-100 cm by the year 2100. However, for reasons mentioned above, these forecasts could turn out to exceed actual trends and perhaps will be corrected in later model-runs when more data about the Greenland ice sheet are available. Nevertheless, hazards related specifically to future sea level rise are likely to be among the most serious threats to German and many other coastal zones (GORNITZ, 1991; IPCC, 1992).

2. Tidal cycle

   The tidal cycle and amplitude shows an irregular pattern and long-term trend in the German Bight (North Sea), since 1910. Especially during the last three decades (since 1965) the trends of mean high-tide vs. mean low-tide levels ran opposite to each other which has caused a considerable increase in tidal range (JENSEN et.al., 1988). The change in tidal amplitude is most marked within the estuaries of Elbe and Weser but also extends across the Waddensea and the southern and western fringe of the North Sea (KUNZ, 1993). While the physical processes causing the change in the tidal regime are not very well understood (to a certain extent they may be man-made by artificial deepening of shipping channels, building large dams across tidal flats etc.) it appears that these changes are of greatest significance, because a greater volume of water is moved through the same surface area in the same time period. Increase in tidal amplitudes are likely to bear effects, for instance, on tidal current and sediment transport velocities; morphodynamic processes in coastal compartments; erosion and accretion patterns as well as sediment budgets; the whole hydrological, hydrochemical and biological cycles within the estuaries; on stability and salinity in the seawater-groundwater-interface and on many other parameters.

3. Storm events, storm surges, floods

   With temperature gradients increasing between low-latitude and high-latitude regions meteorological conditions will probably experience a greater variability, in particular with respect to wind regimes, extreme storm and rainfall probability (ENQUETE-KOMMISSION, 1991). This, in combination with sea level rise and increasing tidal range will lead to a statistically significant higher frequency of peak water tables over the next decades. Although a marked cyclicity is obvious in the frequency distribution of North Sea storm surges in the period 1900-1990 (KUNZ, 1993, fig.7), the last thirty years clearly show a significantly higher frequency level than the earlier period. This holds true in a similar fashion for the residence time at peak water levels (Führböter, 1988). While the causes for these water set-up events cannot (yet) be clearly related to climatic factors, chances for a continuation or enhancement of observed trends appear to be high - thus greatly magnifying the risk for flooding, storm-wave action and related effects/damages. Similar trends of storm recurrence and intensities as well as frequencies of extreme water level set-ups abd set-downs (which - like the former - also bear significant morphologic effects) are to be observed in the Baltic Sea in recent years (STERR, 1993) - thus making a dependance on changing climatic conditions very likely.

4. Nearshore hydrodynamics, sedimentary balance and shoreline trends

It is clear that the three hydrographic parameters: sea level, tidal cycle and storminess will most likely act in conjunction with each other and cause mutual reinforcement and enhancement of both processes and related impacts. This is particularly important with respect to nearshore wave climate and current velocities which in turn are responsible for sediment transport and local characteristics, coastal cell budget, erosion-accretion patterns, shoreline migration and many other aspects of littoral morphology and ecology. From the above considerations it appears that we'll have to expect a general inforcement of wave energy input in the nearshore zone as well as an increase in currents induced by waves, wind force or tides. As a result, intensified sediment redistribution and export will lead to erosion trends both on the shallow sea bottom and on the backshore (MISDORP, et.al., 1990). Indeed, an acceleration of shoreline and nearshore erosion seems to have begun both along Baltic and North Sea coasts some 20 years ago (STERR, 1990; EHLERS, 1989). This trend of increased sediment mobilization is reflected by phenomena such as nearshore bar migration, lowering of catchment divides on the tidal flats, downcutting and headward erosion of tidal channels, accelerated beach and cliff retreat and others. Naturally, changes in wave enery input, the distribution of sediment patterns etc. may have numerous effects on the marine and litoral biology, but details are not well understood at this point.

5. Salinity and sea ice cover

Salinity - by determining the sea water density and thus affecting currents and bottom circulation - is an important parameter of the "sea climate" (Backhaus, 1993). Also coupled positively with the salt content is sea ice formation. The Baltic, as an enclosed sea with a low salinity and a very restricted exchange of water with the North Sea through narrow outlets, is particlarly sensitive to changes in these oceanographic parameters while their variations in the North Sea basin should be comparatively minimal. Under the assumptions of rising sea level, increased storminess and wave dynamics, as given above, an increased and more frequent influx of saline North Sea waters into the Baltic Basin should be expected. This along with a trend towards warmer winter temperatures could significantly decrease sea ice formation, especially within the shallow Bodden inlets. Physically, lack of ice likely enhances erosion trends due to longer exposure of coasts to winter waves; on the other hand, salinity and sea ice changes will have great effects on the litoral and marine ecosystem with marked shifts in species as well as habitats - although exact biologic reactions are not yet known.

Along the low-lying coastal sections of North and Baltic Sea salt water influx might be inforced along the seawater-groundwater-interface as a result of higher and more frequent sea level-setups. Salinity may thus become a problematic parameter both seaward and landward of the coastline.

Vulnerability of Coastal Compartments

Given the above assumptions concerning hydrologic and other important parametric adjustments to climate change, a wide variety of impacts are to be expected on the physical, biological and socio-economical system of the North Sea and Baltic coastal zones. Both the impacts and the responding effects observed in a particular compartment of the coast together determine the vulnerability of this area, as defined by *IPCC Common Methodology* (IPCC 1992). In Germany a detailed case study is commonly prepared according to these latest (updated) IPCC standards. Therefore, a reliable quantitative data base is not yet available at present on which assessment of coastal vulnerability can be based. On the other hand, initial, still somewhat tentative estimations on how individual parts of the coastal system (e.g. of the hydrologic, morphologic, biologic or economic regime) might respond, have recently been presented by an interdisciplinary group of experts (SCHELLNHUBER & STERR 1993). Their findings along with information and data collected by the author since 1987 allow to present a preliminary overview on the sensitivity and vulnerability of specific compartments of the coastal landscape. For the two coastal areas the following landscape elements are considered:

North Sea coast	Baltic Coast
dune islands & sandbanks	sublitoral nearshore zone
tidal flats / Wadden	high coast, cliffs
salt marshes	beaches, beach plains, spits
mainland coast	narrow inlets / Bodden
estuaries	phragmites-dominated wetland

North Sea - Dune Islands, Sandbanks

Dune-covered islands and its "embryonic" counterparts, i.e. the offshore sandbanks, are highly dynamic and mobile components of the North Sea coast. The East Frisian Islands, aligned in a west-east direction, are known to be steadily migrating both eastward and land-(=south)ward under the influence of the continuing late-Holocene sea level rise. In past decades parts of these islands were stabilized by protection works such as seawalls, revetments, groins etc. Accelerated sea level rise and increased wave- and storm-energy input contribute to even greater mobility while other effects such as salt water intrusion into the islands´ groundwater reservoir may be considerable. The future trend to be expected includes the following impacts:
- beach erosion; steepening of the foreshore slope;
- dune erosion along exposed, unprotected sections;
- dune migration; loss of dune vegetation and habitats;

- salt water infiltration into dune base effecting both ecosystem stability and freshwater supply;
- increased sensitivity to external stress (trampling, freshwater pumping etc.);
- increased functional deprivation: loss of ecological as well as recreational functions;
- depreciated economic revenues from reduced use; loss of island settlements and infrastructure.

In short, the majority of the dune islands may be facing gradual diminuation and, in the long run, destruction, if existing morphodynamic trends are acerbated further by climate-change impacts and/or by structural interference with acting natural processes. Artificial fixation of sand by local protective works not only inhibits the natural reconstruction of beaches and dunes but also alters the equilibrium sediment balance in the surrounding Wadden Sea.

The outer sands, lying between some of the islands on the seaward fringe of the Wadden Sea, are affected even more strongly by the changing hydrodynamic conditions. Over the past 2 to 3 decades they have been migrating landward at an accelerated rate, ranging from 5 to 50 m/year. They could suffer from erosion and perhaps destruction mainly by fortification of extreme storm-flood conditions.

North Sea - Wadden Sea Tidal Flats

By area the largest part of the coastal system, the Wadden will likely suffer from massive morphologic and ecologic adjustments to climate-related and man-made impacts. Recent studies of the Wadden system indicate the following trends of change (KÖSTER 1993):
- the outer edge of the tidal flats is retreating landward as a result of intensified wave-energy input and subsequent erosion;
- erosion of tidal channels, both downward and headward, is accelerated;
- channel drainage divides, i.e. large portions of the tidal flats, are lowered by erosion or sediment redistribution;
- the mud-flat areas covered by organic substances are diminished and commonly shifted landward; the sand-flat areas which are biologically less productive are extending;
- all of the tidal flats are subjected to longer submergence than previously, causing a complete disruption of the equilibrium system of moisture, temperature and bio-productivity on the Wadden surface (REISE 1993);
- extensive disturbance of ecologic functions especially with respect to fish, shellfish and bird wildlife (major birth and breeding area in North Sea) is to be expected;
- long-term deterioration of fish catch could occur (von WESTERNHAGEN 1993);
- protective function of tidal flats for the mainland coast (natural wave-breaking system) is being lost, particularly around the headward ends of large channels;
- a considerable increase in maintenance cost for mainland or island protection structures is likely; in places, set-backs of structures and abandonment of agricultural land may be necessary.

In short, the Wadden area will be smaller and generally more fragile, losing parts of its ecologic and economic functions and values.

North Sea - Salt Marshes

Salt marshes fringe the landward rim of the tidal flats where the Wadden surface has grown above mean high-tide water level. They are not only the most delicate of the North Sea litoral ecotones but also most precious as plant & wildlife habitat, as natural "protective structure" in front of the mainland cost and, last not least, as filter for nutrients and pollutants. Salt marshes are, however, greatly threatened both by intensified hydraulic impacts on their seaward side and by artificially set boundaries (dikes, dams, seawalls) on their landward side which - in combining effects - leads to massive horizontal and vertical erosion. Therefore, they represent the most vulnerable component of this coast and the accelerated rate of their reduction gives rise to great concern (WWF 1991), especially with regard to the unique botanic characteristics of this habitate and its zoologic functions (DIERSSEN 1993). In economic terms, disappearance of the salt marshes would necessitate a considerable investment in maintaining/protecting existing defense structures (IPCC 1992).

North Sea - Mainland Coast

Along the German North Sea coast only small portions are preserved in a quasi-natural state while most of the marshlands have been surrounded by dikes, artificially drained and subsequently used for farming. From the impacts affecting - under the given hydrologic scenario - the islands, sandbanks, tidal flats and salt marshes seaward of the dikes it may be concluded that stability of extisting protection structures will sooner or later be endangered; thus readjustments in the "defense line" and in the traditional approach of moving further seaward through land reclamation measures will eventually be required. At present the following conclusions appear plausible:

- the current protection installations along most coastal sections are likely to withstand increased sea level rise and storm-surge impacts for some time;
- their projected life-time will, however, be considerably shortened by the observed or expected trends of erosion on tidal flats, salt marshes and within tidal channels (KÖSTER 1993);
- exposure of structures to erosional trends, i.e. the length of vulnerable shoreline, will generally increase, thus raising costs for maintenence or fortification measures sizeably in upcoming years;
- unprotected sections where previously accumulation prevailed, such as along the dune-covered edge of Eiderstedt peninsula (North Frisia region), will be subject to erosion as well;
- due to drainage of the reclaimed land wide areas behind the dikes are now lying below high-tide level which puts them at risk in case of dike failure and also makes them highly vulnerable to salt water infiltration at high water levels;
- in sparsely populated areas where agricultural use is seriously disturbed by such effects it may be economically and ecologically sound to abolish protection measures in the mid- to long-term future;

All in all, the efforts will have to be increased substantially to maintain full and safe protection of the mainland coast and hinterland along the North Sea, meaning that reassessment of priority needs should be carried out in a coastal management plan with respect to user interests, nature preservation aspects and economic expenditures. A redesign of the present use of coastal areas behind the dikes, e.g. converting some agricultural properties to "free land" with a chance of turning into natural-state wetland, could reduce the overall vulnerability and protective costs along the mainland coast.

North Sea - Estuaries

Four larger estuaries enter into the German part of the North Sea at the mouths of Eider, Elbe, Weser and Ems Rivers. These are the 'pathways' through which the influence of the sea is allowed to extend far inland. Along all of the German North Sea coast the highest tidal ranges are thus found at the interior of these estuaries (e.g. in the Weser Estuary Bremen, now registering a record high of 4,1 m, experienced at least 3 m less tidal range about 100 years ago). In conjunction with their intensive use as shipping routes and locations for industrial production, these tidal inlets have been strongly modified by man, especially by dredging and/or damming. In consequence, in- and outflow patterns, tidal range, water chemistry or other hydrographic parameters - already altered artificially to a great extent - are bound to be even more seriously affected by the climate-induced impacts discussed here (KUNZ 1993). Some typical phenomena may be observed within the estuaries as a result of this partly man-made and partly quasi-natural trend:

Rising mean sea level pushes the brackish water front further inland; increased tidal oscillations will lead to stronger tidal displacements of the estuarine water bodies; both of these are likely to increase current velocities and, subsequently, erosion or sedimentation cycles. Embankment erosion has been observed to occur more rapidly lately, giving rise to flood protection problems while inland migration of the "salinity front" alters habitat conditions for aquatic, littoral and lowland biocenoses and puts additional salinity pressure on surrounding groundwater aquifers. Moreover, from proposed regional climate-change scenarios increase in freshwater inflow during winter times should be expected (GRASSL, 1991, 1993), causing additional drainage problems in areas adjacent to the estuaries.

The two largest coastal cities of Germany, Hamburg and Bremen are situated at the most flood-prone localities within Elbe and Weser estuary, respectively, where rising high tides levels and peak influx of freshwater from upstream areas might add up to new record water marks in upcoming years. It was thus recently reported by the media (Hamburger Abendblatt 1992) that installation of sufficient flood protection for the city of Hamburg alone would cost close to 800 Mio DM in the upcoming 2-3 decades, thus giving an estimation of the vulnerability of the estuarine coastal system.

Baltic - Nearshore Zone

Compared to the coastal compartments discussed above the nearshore zone - extending from the mean water mark to approximately the 10 m isobath - appears to be less vulnerable, at least from a human perspective. In terms of economic values "only" some harbor structures (e.g. jetties), military installations and protection works reaching seaward in shallow water depths will probably be affected by likely hydro- and morphodynamic adjustments to climate-change impacts. The latter, however, are thought to be quite considerable in the nearshore zone as a result of higher storm frequencies, sea level rise, shorter recurrence periods for critical water level set-ups and set-downs, general increase in wind-induced wave energy input and, finally, of declining occurrence of sea ice (STERR 1993).

While we may expect that the submarine bottom profile - dominated by an abrasion platform with overlying longshore bars - remains basically in equilibrium, abrasion and transport processes will generally be enhanced. An increase in the vertical abrasion rate on the wave-cut platform in conjunction with an accelerated seaward export of sand due to higher current velocities will render offshore structures to greater instability. This process could be greatly enhanced, if the reconstruction of bars after a storm, which under present conditions requires up to 2 months, is hindered by shorter storm recurrence periods. In this case the equlibrium bottom profile will be lost, causing more massive structural failures or damage.

Baltic - Cliffed Coast

Basically, cliffed segments are less sensitive and vulnerable to coastal erosion or related climate impacts than are low-lying parts of the coast. The cliffs of the southwestern Baltic region, however, are made of weakly consolidated morainal material, with locally high contents of loose sandy components; therefore, they are not very resistant against basal wave attack or down-slope mass movement processes.
Long-term field observations and measurements along the cliff sections (between about 1870 and now) show a marked increase in rates of cliff retreat in the second half of this century, in particular since about 1960 (STERR 1990). For the state of Schleswig-Holstein, cliff retreat rates were found 30 to 50 % higher in this period than in previous decades and similar trends have been observed further to the east (STERR 1993; STERR & GURWELL 1991). Part of this acceleration of erosion processes is attributed to the combined effects of sea level rise trends (about 2mm/year) and an increase in storminess and (wind-induced) water set-up frequency. On the other hand, the installation of armed structures in the vicinity of the cliffs may also contribute to this observed trend, but to an unknown extent.
Although along the Baltict the loss of land from retreat of cliffed coastal segments is not excessive and - for agricultural land - is considered bearable to the land owners, there are a number of settlements and infrastructural installations (recreational, military, traffic etc.) which are threatened by accelerated cliff retreat trends. However, cliff protection measures are often considered to yield more negative than positive results (e.g. depletion of the coastal sedimentary budget) and thus, by judging the overall coast-benefit ratio, are not first choice of coastal managers. The vulnerability of cliffed coasts might therefore be greater if existing trends increase further.

Baltic - Beach plains and Spits

The erosional trends indicated for cliffs appear even more pronounced along the low-lying coastal sections of the Baltic which consist of beaches, beach-ridge plains or spits. About 75% of these areas are seen to undergo erosion at present time (STERR & GURWELL 1991) and most of this erosional work is attributed to the stimulation of nearshore hydro- and morphodynamic processes induced by climatic effects (man-made influence from building coastal structures should not be neglected, though). Susceptibility to such processes (e.g. flooding, storm-wave action) and thus the degree of vulnerability varies according to coastal exposition and topography. It gradually increases from narrow (pebbly) beaches to wide (sandy) beaches, to beach-ridge plains (partly covered by dunes) and, finally, to single or multiple spit systems. The highest risk potential is seen to occur along the freely growing spits which are subject not only to wave attack along their exposed flanks but also from overwash mechanisms after water table set-up beyond a critical level (commonly 1,5 to 2,2 m above mean).
A further rise in sea level in combination with an increase in storm surge frequency and in nearshore wave energy will probably lead to the breaching of some spits and/or the partial reduction of beach plains and their specific littoral ccotones (salt grassland, dunes, beach ridge habitats etc.) (STERR et.al. 1993).
On the other hand, many low-lying coastal segments of the Baltic have - in the course of this century - been used quite extensively, mainly as recreational, agricultural or military sites. Most of these areas were therefore protected by (low, earth-fill) dikes which often were built upon the foremost beach ridge and are now at risk from overwash, breaching or basal erosion (STERR 1990). For these low coastal plains flooding hazards may become a very serious problem, if - as current trends indicate - the recurrence period for critical water tables (>1,5m above mean), related to easterly storms, is significantly shortened. As along the North Sea coast it is mainly through the failure of (landward) dynamic re-adjustment of beaches,

berms and dunes that the vulnerability of these coastal segments is acerbated.

Baltic - Wetlands and Inlets

The inner, protected portions of Baltic coastal inlets (Förden or Bodden) are affected mainly by the changes which are initiated along the exposed coastal segments, as discussed above. These inlet coasts are widely preserved in a near-natural state, characterized by phragmites-dominated wetlands and alternating cliff sections, and are lacking intensive use or man-made infrastructure. The climate-induced adjustments are therefore expected to occur here primarily within the litoral-marine ecosystems, stimulated or initiated by the following morpho-hydrological impact scenarios:

- "stormier" conditions, inducing more flooding events and higher wave-energies in the inlets´ interior;
- more frequent and stronger exchange of water masses in and out of the inlets, raising the low salinity of Bodden waters;
- breaching or reduction of protective spits and bars, leading to additional influx of wave energies for sediment transportation and erosion within the Boddens; and
- warmer air/water temperatures in winter, preventing frequent coverage of inlets by sea ice.

Together, these impacts both fortify and extend the periods of strong wave attack in the shallow inlets causing destruction/retreat of phragmites wetland areas and revival of cliff erosion. Increase in flooding periodicities alone, without storm wave action, in turn would lead to a gradual landward shift of litoral vegetation zones and their destruction would be significant only in coastal sections backed by cliffs. On the other hand, the changes listed above is bound to increase transport and turbulence within water and sediment columns which have been highly burdened by man-made trophication in recent years. Thus, oxygen supply in the water-sediment-interface should be revived by these processes which could help to improve the highly deteriorated living conditions for benthic flora and fauna in the Bodden. It must be kept in mind, however, that the buffer and filter functions of these inlets might be lost, if water masses are enabled to circulate more freely between inlets and the open sea (SCHIEWER, 1993) while the water quality outside the inlets would be deteriorating.

Economic vulnerability generally is lower along the inlet-wetland coasts than along other coastal compartments, mainly originating from "secondary" adjustments following ecologic changes (e.g. suitability of inlets for aquaculture etc.)

Research Demands

The overview given in this paper regarding the sensibility of coastal systems and coastal areas toward climate-related impacts along North Sea and Baltic coasts makes clear that many of the expected changes are not yet very well understood. Therefore the German Federal Ministry of Research and the responsible ministries of the five coastal states have launched a research program, which deals with issues of coastal vulnerability to climate-change impacts. Following a truely interdisciplinary and "top-down" approach this program intends to deal with physical as well as ecological and socio-economical aspects of coastal systems' adjustments to impacts such as sea level rise, increased storminess etc. Public acceptance of findings and coastal management proposals should be good, if straightforward assessments of (economic) risks and sound cost/benefit analyses for measures to divert these risks will be presented.

Summary

Germany has coasts both with the North Sea and the Baltic Sea which differ greatly in morphologic appearance, hydrographic conditions and ecological as well as socio-economic structure. The low-lying North Sea coast to the west is very vulnerable to (accelerated) sea level rise and to a possible increase in storm surge magnitude and frequency in the German Bight. The high risk potential along this coast is associated with a general magnification of erosional processes, extension and shifts of tidal channels, loss of wadden and dune areas, dike failures and resulting inundation and, last but not least, setback of nearshore land use structures (tourism, settlements, agriculture etc.). Thus, for this coast detailed vulerability assessments and coast-benefit analyses with regard to protection or adaption strategies are required which follow plausible sea level rise and climate change scenarios.

The Baltic Sea coast to the east displays a wide range of natural landforms and ecological characteristics. In the absence of extensive coastal protection works this coastal region is susceptible to increasing erosion rates along cliffs, beaches and dunes but also to flooding of low-lying sections and river mouths as a result of accelerated SLR and changing storm patterns. In addition to the high vulnerabibility of protecting spits, which when flooded or breached would render Bodden interiors to higher wave energies, increased warming trends here might have strong effects on salinity and sea ice cover. These factors in turn might have considerable effects on flora and fauna of the Bodden margins, thus changing a precious biological habitat. Currently large-scale research projects are prepared to estimate climate induced impacts and management options in both the North Sea and the Baltic coastal zones of Germany.

Literature:

Barth, M. & Titus, J. (eds.) (1984): Greenhouse effect and sea level rise: a challenge for this generation. Van Norstrand Reinhold Co.

Boer, M.M. & de Groet R.S. (eds.) (1990): Landscape-ecological impact of climate change. 1990. IOS Press. Amsterdam.

Dierssen, K. (1993): Possible effects of global change on the vegetation development in tidal salt marshes. In: Schellnhuber & Sterr (eds.): The Changing Coast. 189-195, Springer Heidelberg

Ehlers, J. (1988): The Morphodynamics of the Wadden Sea. Rotterdam

Enquete-Kommission "Schutz der Erdatmosphäre" (Hrsg.): Klimaänderung gefährdet globale Entwicklung. 1992. Economica-Verlag.

Führböter, A. (1988): Changes of the tidal water levels at the German North Sea coast. Proc. 6th Intl. Wadden Sea Symposium, List/Sylt

Gornitz, V. (1991): Global coastal hazards from future sea level rise. Paleogeography, Paleoclimatology, Paleoecology, 89, 379-398.

Grassl, H. (1991): Sea level rise - short review. In: WWF Internl. (ed.). The Common Future of the Wadden Sea. Technical Report. Husum. p. 79-86.

Gröger, M. & Plag, H.P. (1993): Estimations of a global sea level trend: limitations from the structure of the PSMSL global sea level data set. Global and Planetary Change

Hillen, R., Bijlsma L. & Misdorp, R. (1993) Towards sustainable development of coastal zones. Report on the activities of the Coastal Zone management Subgroup of IPCC. (this volume).

IPCC (1990): Climate Change. The IPCC Scientific Assessment (Houghton & Ephraums eds.), Cambridge University Press, 364 pp.

IPCC-Coastal Management Subgroup (1992): Climate Change and the Rising Challenge of the Sea. Report on Margarita Island Workshop (IPCC-CZMS update-report).

Kelletat, D (1990): Meeresspiegelanstieg und Küstengefährdung. Geographische Rundschau, 12/1990, 648-652.

Kunz, H. (1993): Effects of climate change on the hydrography of coastal waters and related impacts on water management. In: Schellnhuber, H.-J. & Sterr, H. (eds.) (1993). Die Küste im Wandel. Auswirkungen von Klimaänderungen auf den deutschen Küstenraum. p. 97-126, Springer, Heidelberg (in press)

Lassen, H. & Siefert, W. (1991): Mittlere Tidewasserstände in der südöstlichen Nordsee - säkularer Trend und Verhältnisse um 1980. Die Küste, 52, 120-132.

Misdorp, R., Steyaert, F., Hallie, F., de Ronde, J. (1990): Climate change, sea level rise and morphological development in the Dutch Wadden Sea, a marine wetland. In J.Beukema et. al. (eds.) Expected effects of climate change on marine coastal ecosystems. Kluwer Publ., Dordrecht.

Plag, H.P. (1993): The "sea level rise" problem: an assessment of methods and data. (this volume)

Reise, K. (1993): Dim future for the tidal flat ecotone of the Wadden Sea. In: Schellnhuber & Sterr (eds.): The Changing Coast. 223-232, Springer, Heidelberg

Schellnhuber, H.-J. & Sterr, H. (eds.) (1993). Die Küste im Wandel (The Changing Coast). Auswirkungen von Klimaänderungen auf den deutschen Küstenraum. 400 p., Springer Verlag, Heidelberg (in press)

Schirmer, M. (1993): Climatic changes and their consequences for the coastal zone: impacts on estuaries. In: Schellnhuber & Sterr (eds.): The Changing Coast. 244-259, Springer

Schiewer, U. (1993): Stability conditions in microbial foodwebs in ecosystems of brackish water. In: Schellnhuber & Sterr (eds.): The Changing Coast. 244-259, Springer

Schneider, S. (1992): Will sea levels rise or fall? Nature, 356, 11-12

der Spiegel 37/1992: Interview mit Prof. Hasselmann

Sterr, H. (1990). Comparative coastal erosion studies in the Federal Republic of Germany. Journal of Coastal Research, Spec. Issue #9, 821-837, (Proc. Intl. Skagen Symposium, 1990, Skagen, DK)

Sterr, H. & Gurwell, B. (1991). Coastal problems and coastal research along the Baltic coast of the former German Democratic Republic. in: Brückner H. & Radtke U. (Hrsg.) (1991) Von der Norsee bis zum Indischen Ozean. Steiner Verlag, 47-64.

Sterr H. (1993a): The influence of climatic variations on the morphodynamic system of the German Baltic coast. In: Schellnhuber & Sterr (eds.) Die Küste im Wandel. p. 153-173, Springer Verlag, Heidelberg (in press)

Sterr, H. (1993b): Geomorphological Hazards along the Coasts of the FRG (North Sea and Baltic). In: Embleton, C. (ed.): National IGU-Reports on Rapid Geomorphological Hazards. (in press)

Sterr, H., Janke, W. & Kliewe, H. (1993): Holocene formation of the coast of Mecklenburg-Vorpommern and its future development with respect to climate change. In: Schellnhuber & Sterr (eds.) The changing coast. p. 137-152, Springer, Heidelberg

Westernhagen, H. (1993): Possible influences of climate on the ichtyofauna of the North Sea. In: Schellnhuber & Sterr (eds.): The Changing Coast. 212-222, Springer, Heidelberg

climatic changes	areas of climate impacts	impacts/effects of climate change
increase of long-term mean air and water temparature	active cliffs	change of hydrodynamic and hydrochemical characteristics
change of seasonal temperature and precipitation pattern	beaches, beach ridge systems spits and barriers	(accelerated) rise of sealevel
change of predominant wind direction and wind velocity	coastal wetlands and low-lying backshore areas	increasing tidal range
increase of extreme events (storms, rainfloods etc.)	tidal flats	increase of storm surges
change of the regional "ocean climate" (currents etc.)	salt marshes	increase of wave energy, bottom currents & other hydrodynamics
	litoral margins of Bodden, fiords and enclosed embayments	change in sediment transport patterns and sediment budgets
	nearshore zone	change in seasonal river runoff
	tidal estuaries	increase or decrease of near-shore salinity
changes of environmental factors: (mainly man-made)	coastal dunes	temporal or permanent rise in groundwater level(s)
		decrease of sea-ice formation
increase of nutrients in water/soil	deltas	increase in shoreline/cliff erosion
increase of toxic substances in air, soil, rivers and sea	coral reefs	change of sediment composition and aereal distribution
intensive use of the coastal zone	mangrove swamps	increase of photosynthesis
overuse/decrease of sediments	sea-ice dominated coasts	reduction of floral/faunal species
construction of barriers inhibiting morphodynamic/ecological processes		increase of eutrophication in coastal waters
over-exploitation of natural/ ecological resources		zonal migration or destruction of habitats (dunes, salt marshes)
		loss of ecological or protective function of coastal system elements

Tab. 1

Detailed Holocene sea level curve, Northern Denmark
William F. Tanner, Florida State Univ., Tallahassee, Fla., U.S.A.

Abstract

Low-energy beach ridges in extreme north Denmark have yielded a history of sea level change since about 11,500-12,000 B.P., with details every 40-50 years (C-14 years). Results to date include six grain-size moment measures and other para-meters for 154 samples in one system (one per ridge) in a full sequence from oldest (almost 8,000 B.P.) to the modern beach (eastern edge), and similar data from an older system. Kurtosis and the Sixth Moment Measure identify small sea level changes (1-5 m), but show no storms. The field method precludes collecting from dunes; detailed analysis shows that no eolian sand is present in the samples. Accretion was rapid relative to isostatic rebound, so uplift had no role in shaping the grain size distribution.

The resulting sea level curve shows in detail the amount, direction, timing and rate of sea level changes associated with the Little Ice Age (last 800 years). It also matches, back to about 3,200 B.P., curves obtained from beach ridge systems (U.S.A.) which do not go back any earlier.

Granulometry and ridge spacing show the early Holocene rise, the Mid-Holocene high stand (+ 2 m, about 6,800-5,000 B.P.), the subsequent Mid-Holocene low (-3 to -4 m, about 4,500-3,200 B.P.), and various rises and falls (1 to 3 m) since then. In addition, construction of a single beach ridge depends on sand supply rate and on couplets of very small sea level change, commonly at 40-50 year intervals, but not on storms.

The history deduced here shows a steady sea level rise (warming trend) for the last 250-300 years: recovery from the Little Ice Age. This cannot have been due to industrialization.

Beach ridge studies on the northeastern coast of the Gulf of Mexico show a reversal of wave approach and prevailing winds, 250-300 years ago. This means northward shift of the boundary between westerlies (to the north) and easterlies (to the south), hence warming in the last 250-300 years.

Primary Study Area

The towns of Skagen, Hirtshals, Tversted, Fredrikshavn and Jerup are located on the Skagen Spit in far northern Denmark (fig.1). This spit in-cludes three large beach ridge systems. One of the systems, on the high-energy northwest coast (Atlantic Ocean/North Sea), is in good part dune covered; it is not suitable for the kind of analysis reported here.

The other two systems represent low-energy beach conditions (on the coast of the water body to the east, between Denmark and Sweden; together they cover about 11,500-12,000 C-14 years. The older of these (Tversted system) formed during the interval (roughly) 12,000-8,000 B.P. The younger (Jerup system) was built since then. The two low-energy beach ridge systems contain little eolian sand, and are suitable for detailed analysis.

These ridges are of the low-energy, swash-built type (Tanner 1987) but have a recognizable settling-lag component (Tanner and Demirpolat 1988).

A large relict depression and the north-flowing creek that drains into it separate the Tversted and Jerup systems. The depression and creek show up clearly on topographic maps and on vertical aerial photos. The depression was formed during a short time

when growth of the Skagen spit, toward the northeast (enlargement of the high energy system) was faster than the low-energy waves to the southeast could fill in the bay.
The Jerup and Tversted systems have been sampled in detail, one sample for each ridge. The Jerup system is represented by 154 samples, including the modern low-energy beach to the east, with data points 50-51 years apart. Tversted system ridges are more difficult to identify, but there are about 90 ridges (field criteria; not photo count), some 40 years apart.
Details of procedure, hydrodynamic logic and background data are given elsewhere (Tanner 1992-a, 1992-b). Each field sample was roughly 50 grams.

Granulometry

The Jerup system data set includes granulometric measurements (first six grain-size moment measures plus many other parameters) for 154 samples, each of which was collected (with few exceptions) from the midheight point on the seaward face of a ridge, in a continuous sequence from oldest to youngest. Sieving was done for 30 minutes in a quarter-phi nest of screens.
The Kurtosis and the Sixth Moment Measure, given as moving averages, reflect accurately small sea level changes (such as 1-5 m), and are almost always indicators of fair weather rather than storm conditions (Tanner 1991-a, 1991-b, 1992-a, 1992-b).
This statement applies only to low-to-moderate energy beach ridges made of quartz sand and having straight and parallel map patterns (or nearly so), and is not appropriate for the high-energy dune-covered ridges on the northwestern coast or for highly curved deposits at the tip of a spit.
The choice of field area prevents collecting from storm deposits, and the sampling method precludes collecting from dunes; statistical treatment confirms that both of these precautions were successful. The rapid rate of depositional progradation (approximately 1 m per year), relative to isostatic rebound (close to 2.2 mm per year), provides that uplift had no important role in shaping the grain size distribution.
Both the Kurtosis (inverted; fig.2) and the Sixth Moment Measure show small sea level changes, typically in the range 1-5 m. These changes have occurred on several different time scales, but generally covered two or more centuries, up to more than 1,000 years.

The Little Ice Age (last 700-800 years; Lamb 1981) shows up well in the Kurtosis data, which defines a sea level drop of 1-2 m, beginning about 1,200 A.D., and starting to recover about 1,700 A.D. The Kurtosis and other parameters identify the environment of deposition, correctly, as a low-energy beach with minimum eolian influence and no evidence of storms. Changes in the Kurtosis, with time (dK/dt), show sea level changes, and give the direction of change and the approximate magnitude.
Mean size in the Jerup system decreased with time from 0.26 mm to 0.17 mm; two subtle reversals do not make important departures from that trend. The result may illustrate the selective transport model of May (1973).
The Standard Deviation shows relatively little variability and does not reflect most sea level events, but does record the Little Ice Age. This parameter had an average value, for the first 15 ridges in the system, of 0.424, and returned to an average value of 0.393 in the last few centuries, but through most of the last 4,000 years was close to 0.30, with excursions of the 15-item average to as low as 0.289 (and individual values to 0.262). All of these numbers indicate excellent sorting, and suggest a beach.
The Skewness varied between -1.365 and 3.581, with positive numbers indicating a

larger number of fine classes than the Gaussian distribution would show (hence settling). Negative Skewness became more common in the latter part, indicating a long-term decrease in settling with time.

The Settling Index is defined as SI = K - 10.61 * exp(-3.272*StdDev). Where it is large, settling is important. Where settling is relatively unimportant, SI is close to zero and may be negative. In the Jerup system, the Settling Index, plotted vs time, has a pattern somewhat like that of Kurtosis, but in general shows a decrease with time: settling became less important, and wave work more important, as time passed. The Settling Index is commonly less variable, sample to sample, than is the Kurtosis.

Tversted system ridges are in general finer grained than Jerup system ridges, but also represent a low-to-moderate-energy beach setting with an important settling component. An unusually high ridge in this system stands more than three meters above the adjacent swales; it is near the older edge of the system, and is here dubbed the "Old Ridge" (but not "oldest"). A pit about four meters deep had been dug in the southern end of this ridge, providing an excellent exposure of bedding.

Despite an eolian appearance in the field and on the air photos, the "Old Ridge" contains only low-angle swash bedding and horizontal bedding. Twelve different granulometric tests were applied to a suite of nine samples from this one locality; they indicate low-to-moderate energy beach, plus a settling component. Wind work was minor, probably about what one finds on almost any modern beach face, but the ridge is clearly not a dune. No other ridge in the system is tall enough to be a dune, except near the far northern end, where wind-blown sand from the northwest coast (Atlantic Ocean; high energy) has migrated landward.

The "Old Ridge" is thought to be about 11,500 years old, for reasons given elsewhere in this report.

Ridge Spacing

The beach ridges in the Jerup system can be classified, by using the Kurtosis, as indicating either high sea level (low K), or low sea level (high K). In other study areas, such as on St. Vincent Island, Fla. (Tanner 1992-a 1992-b), the high sea level ridges occur in distinctive sets of 5 to 20 ridges, which stand a meter or so above adjacent low sets. The latter are easy to identify, in the most obvious cases, because they stand low enough to be partly flooded, they show up differently on the air photos, they have a different type of vegetation, and they have unique Kurtosis values. They also have different map spacing, ridge crest to ridge crest.

Some of these criteria cannot be used in the Jerup area, for example because slow glacio-isostatic rebound has raised the entire system far enough that low parts cannot now be flooded by sea water. However, the "low" sets (high K) have typical ridge-to-ridge spacing of about 35 m, and "high" sets (low K) have a spacing well above the system average of 47 m.

Tversted system ridges likewise have variations in spacing, but the system average is only about 40 m.

Therefore sea level history from the granulometry can be confirmed by a study of ridge spacing (in general, but not in detail, because spacing data vary with the set, not with the ridge).

Dates: Jerup System

Carbon-14 dates were obtained from nine precisely-located samples collected by Poul Hauerbach, in the high energy beach ridge system on the northwestern coast; they have been replotted from his report in a height-vs-age diagram (Tanner 1990-a). These nine dates represent beach ridge swale peats, now exposed on the wave-cut cliff, going back to 5,535 years ago. The high energy swale peats are better for dating purposes than Jerup system swale peats, because the former are exposed on a modern cliff, where one can be sure that the oldest peat in any one swale is actually sampled. This cannot be done in the bore holes in the Jerup system, thus creating an elevation error of a few meters (several thousand years).

Hauerbach's collecting sites are only a few kilometers north of the Jerup system, and therefore elevations for the individual dates can be transferred to the latter, without much loss of precision.

Because sea level has both risen and fallen significantly in the 5,500 years covered by the dates, the height-vs-age plot shows an S-shaped curve. The last 3000 year segment of this plot matches the sea level curve from St. Vincent Island, Fla. and elsewhere in North and South America. Therefore we:

1. Assign dates to Jerup system ridges, in the last 3,000 years, on the basis of dated elevations and the sequence in which they occur.
2. Calculate the growth rate for that part of the Jerup system. The average is about 0.93 m/yr of growth, or about 50.5 years per ridge.
3. Note that in other areas the accretion rate tends to be almost con-stant, as long as there are no major changes in the geometry of the system (Tanner 1990-b). This requirement is met near Jerup, so we extrapolate the new calendar back to the oldest ridge in this one system.
4. Note that (with constant growth rate assumed) the oldest ridge in the system is about 7780 years old.
5. Because the oldest ridge is a few hundred meters east of the relict depression, adopt a round number (maximum) of 8,000 years.
6. Do not extrapolate any of these numbers across the depression and into the Tversted system, which has certain differences in character and therefore must be dated in some other way.

These projected dates match the general time of the Mid-Holocene high stand (about 6,800-5,000 B.P.), and the Mid-Holocene low position (about 4,500-3,200 B.P.). This agrees with (a) Granulometric data from Jerup, (b) Hauerbach's original curve prior to 3,000 B.P., (c) Isolated dates produced by other workers (e.g., Forman 1990, from Spitsbergen), and (d) Other data from elsewhere (Tanner 1991-c). Hauerbach was not able to obtain samples close enough to each other and young enough to show the Little Ice Age.

These dates can be used, also, to determine the local rate of glacio-isostatic rebound. It has been close to 2.2 mm/yr during the last few thousand years, but apparently has had a decreasing value, hence plots best on a semi-log diagram over a longer time span.

Kurtosis (inverted and smoothed) from Jerup system ridges has been plotted on the new calendar (fig.2). This work requires only the local granulometry, local topographic map information, and Hauerbach's C-14 dates, and does not depend on data from other areas or sources.

Dates: Tversted System

Radiometric dates have not been obtained within the Tversted ridge system. An older limiting date is 13,000 B.P. (Petersen 1990), but it does not represent initiation of the beach ridge plain; the latter must be more recent, perhaps about 12,000 B.P. The youngest ridges cannot be younger than the relict depression and the creek which drains northward into it: some 8,000-8,500 B.P. The Tversted system must fit within this interval.

The closely-dated Swedish sea level curve of Svensson (1989) covers the period 12,000-8,300 B.P.; sea level changes at about 9,600 (rise) and 9,200 (fall) are matched by events in Tversted system ridge history very well.

The rebound history of the Tversted-Jerup area, using a realistic semi-log plot as the starting point, can be projected backward into time, to see where (if anywhere) it matches other published sea-level histories. The best match for the older part of the Tversted system is 11,000-12,500 B.P.

If the oldest ridge considered in the present study (88-90 ridges) is estimated to be 11,500 years old, and the youngest Tversted ridge 8,000 years old, then the average interval between ridges was 40 years. This is subject to a small adjustment, to somewhere between 35 and 45 years.

The growth rate appears to have been 0.95-1.1 m/year along the transverse profile that was studied, and slightly larger farther north. This range includes the most common values for quartz sand ridges elsewhere, and permits ages of individual ridges to be estimated by interpolation.

Results

The Jerup sea level curve shows in detail, with points about every 50.5 years, the amount, direction, timing and rate of sea level changes associated with the Little Ice Age (last 700-800 years). It also matches, back to about 3,200 B.P., sea level curves obtained in the U.S.A. from beach ridge systems having no uplift but shorter history. And it matches other dated evidence in the U.S., from river deltas, back to about 6,000 B.P. (Tanner 1991-c).

The mid-Holocene high sea level, about 6,800-5,000 B.P., and the subsequent Mid-Holocene low (-3 m to -4 m), about 4,500-3,200 B.P., are obvious. Several rises and falls in the interval 3,000-1,000 B.P. can be seen. It is also clear, from work on the Tversted system, that there were sea level fluctuations in early Holocene time as well.

Spectral analysis, auto-covariance and other kinds of statistical treatment have not produced satisfactory evidence of periodicity for the full suite of Jerup samples. However, several types of behavior show up:

1. 1,000 years or longer: The Mid-Holocene high stand (2-2.5 m above the present position), the subsequent Mid-Holocene low (3-4 m below present sea level), and the oscillating curve of the interval 3,000-1,000 B.P.
2. The Little Ice Age, with a steady sea level drop from about 1,200-1,300 A.D. to perhaps 1,500, followed by a rise beginning about 1,700 A.D. and perhaps still continuing.

3. Individual fluctuations of a meter or so, typically only 100-300 years long. These show up clearly even after smoothing of the data.
4. The individual ridges themselves, about 50-51 years apart in the Jerup system and perhaps 40 years apart in the Tversted system, appear to represent pairs of sea level changes of 5-40 cm, and provide the only obvious periodicities in the study area.
5. A periodicity of approximately 15 ridges (750 years), in the latest 3,000 years. The auto-correlation "r" is 0.75, but the peak on the correlogram is broad rather than sharply-defined, and might be placed almost anywhere from 14 to 17 ridges.

There is a hint of asymmetry in the data: some of the sea level drops have been faster than the corresponding rises. Exceptions have been evenly balanced, and did not have the reverse asymmetry.
Maximum rates of change appear to have been about 1-1.5 cm/yr, for those changes that persisted for 100 years or longer.

Wind Pattern Reversal

Dog Island, on the northeastern coast of the Gulf of Mexico, has two beach ridge systems, one at each tip. The system at the northeastern tip has not been growing in historical times; instead, the southwestern system has been active, in keeping with the fact that the prevailing easterly winds in the area now drive waves which transport sand along the beach toward the southwest (Tanner and Spicola 1986). The northeastern system was completed before the first southwestern ridge was deposited (Tanner 1992-b).

How is this 180-degree reversal of transport direction to be explained? The northeastern ridges were built by waves from the west, driven by prevailing winds from the west (like the wind pattern over the North American continent, to the north). The younger (southwestern) ridges were built by waves from the east, driven by the prevailing winds from the east which blow across the Gulf of Mexico today. A south-to-north shift of the boundary between these two wind systems, by a distance of only 100-300 kilometers, would cause this reversal.

Dog Island beach ridge patterns and granulometry show that this important event took place about 200-250 years ago, and was associated with a small sea level rise (about 1 m). It is here taken as part of the warming trend known from written European history of the latter part of the Little Ice Age (Lamb 1981).

A similar record may be preserved on St. Joseph peninsula, about 70 Km southwest of Dog Island, but the timing appears to be quite different, and appears to reflect an earlier warming event (about 1,200-1,500 B.P.).

Other clear-cut reversals of shore-parallel transport are not now known, perhaps for any of three reasons:

(a) Not many sand islands are located in the narrow belt where a small south-north shift in the prevailing wind pattern might take place,
(b) Not many sand islands are oriented properly for this change to be preserved in beach ridge history, and
(c) Perhaps all the right islands have not been studied yet.

The Jerup ridges are not oriented properly, and are not located at the right latitude, for this kind of reversal to show up.

Oversize oxbow lakes from the southwestern part of the U.S. (Melton 1938, Tanner 1992-c) show that there have been spectacular changes in precipitation, in that region, in the last 1,200 years. Some of these lakes are as much as six times as wide as the modern river channel. Because the oversize oxbows are numerous on any one floodplain, they cannot be evidence for stream piracy. Because they are found as far south as the Mexican border, they cannot represent flow of glacial meltwater. Along with Mississippi River delta dates, they confirm parts of the history deduced here.

Useful Guides

Several useful lemmas are given here.
1. Where sand beach ridges are straight or gently curved, and more-or-less parallel, without gaps in the sequence or important changes in ridge geometry, parameters such as ridge spacing, accretion rate (seaward growth; m/yr) and ridge growth rate (yrs/ridge) tend to be almost constant (the standard deviation is numerically small). This means that extrapolation from one part of the record, to another where time control is not as good, is reasonably reliable.
2. The time interval between ridges (T; yrs) can be calculated from the accretion rate (R; m/yr) and the spacing between ridges (S; in m): $T = S/R$.
3. Short-term sea level change (dL; in m) is the difference between ridge height above the adjacent swale (without dune decoration; h; in m) and run-up height (RU; in m): $dL = h - RU$.
4. Run-up height is the product of wave height (estimated from smoothed inverted kurtosis) and the run-up factor (perhaps 1.5, but typically between 1.2 and 2.0): $RU = H * RUF$.
5. A simple relationship between Kurtosis and wave height is available: K about 3 indicates high wave energy, $K > 3.5$ indicates low wave energy, and $K > 6$ indicates low energy plus settling. If a more precise relationship is desired, then a special local calibration must be undertaken.
6. Smoothed Kurtosis, over 100 years or more, can be read in terms of amount of sea level change.
7. The simple fact that these swash-built sand ridges exist, indicates small couplets of sea level change (5-40 cm). These changes do indeed have a strong periodicity.
8. Where spacing differs markedly from an older part of the ridge sequence, to a younger part, there may have been a change in the offshore sand supply rate. These changes commonly take place in abrupt steps, rather than smoothly, and reflect a change from one ridge periodicity, to another periodicity.

Other Areas

Beach ridge systems have been studied in more than 30 areas, on three continents. In Germany, immediately south of the glacio-isostatic rebound region that covers essentially all of Denmark, several beach ridge systems are well-developed. Three of these have been inspected in the field: those at the Geltinger Birck, Schleimünde and Heiligenhafen. Their histories have been relatively short, but they do show evidence of recovery from the Little Ice Age. They cannot include mid-Holocene or older ridges, because these areas have not been subject to glacio-isostatic rebound, as has the Skagen spit in Denmark.

The Darß beach ridge plain may go back to about 3,000-3,600 years ago, but it is not now known whether or not Darß ridges contain too much eolian sand for the kind of analysis done in the Skagen area. In any event, it does not now appear that Darß ridges date far enough back to confirm the early and middle parts of the sea-level history deduced in northern Denmark.

Sandy beach ridge systems older than 3,000-3,500 B.P. have not been studied elsewhere, although gravelly ridges are known in various areas of Pleistocene glaciation (e.g., Hillaire-Marcel et al 1979).

Temperatures

Long-term mean temperature changes during the Little Ice Age were 0.8° to 1.5° C (Lamb 1981). These numbers can be placed on the plot of smoothed Kurtosis vs time, for the Jerup area, to provide a tentative temperature scale. The values adopted here are 0.45°C per unit change in Kurtosis (conservative), and 0.55°C per unit change in Kurtosis. These tentative scales, when applied to the Holocene sequence of Kurtosis values, provide the following estimates: Mid-Holocene High stand, 0.7°-1.0°C warmer than at present; Mid-Holocene Low stand, 1.5°-1.85°C cooler than at present; maximum drop in the late Holocene, 1.6°-2.0°C (in about 150 years); and maximum rise in the late Holocene, 1.6°-2.0°C (in about 250 years).

Winter-time ocean surface water temperature near the northern tip of Denmark has been reported to be about 3.5°C (Gerasimov 1964 p. 45). Lamb (1981) estimated that in the Little Ice Age, ocean water off of southern Norway was 5°C colder than now, which would give a winter average of about -1.5°C (not cold enough for sea water to freeze). However, during the Little Ice Age, very shallow water in Albaek Bay (east of Jerup) probably was cold enough to make near-shore ice in winter-time; and during the Mid-Holocene Low stand, winter water temperature in the area probably dropped to -4°C, which is indeed cold enough to develop an ice cover, possibly all the way to Sweden.

Conclusions

Sea level changes fit into several categories (Tanner 1992-a):

I. Big; 50-130 m; 10,000-100,000 years; e.g. Riss-Würm time.
II. Small; 1-6 meters; few centuries; such as in the Holocene.
III. Very small; 5-40 cm; every few decades.

Construction of a single beach ridge depends on the sand supply rate and on a couplet of very small sea level changes (commonly 40-50 years apart), but not on storms.

The lack of clear persistent periodicities exceeding 50 years, in the detailed record over Holocene time, indicates that there is not one single control, or even only a very few controls: the interactions among many geological, meteorological, astronomical and oceanographic causes and feedbacks are numerous and complicated. Therefore we are unable to forecast what sea level will do, in the future, by examining only one or two factors, no matter how logical the relationships among them may be.

The latest trend, at Jerup, is a sea level rise, presumably due to warming. This trend started 250-300 years ago, and there is no evidence in the data that it is over. The primary cause was not anthropogenic.

From the sea level curve, now available from three continents, one might well infer that the next important change will be cooling and a concomitant sea level drop: the ocean surface now stands quite high (although not a record) in the recorded range from Holocene time. If this turns out to be correct, we may want all the artificial warming that we can get.

Forecasts that we are now moving into the preliminary phase of a new ice age, much like the Riss-Würm (Wisconsinan), do not consider the very high variability in the record for Holocene time.0 Even heavily smoothed curves of sea level change show more than a score of reversals in this time interval. Some of the trends lasted 300-500 years, and half of them represented cooling. Therefore a consistent cooling record even as much as 500 years long does not predict a new ice age.

The problem we have in making long-range forecasts is that the zig-zag geometry of the detailed sea level curve, through all of Holocene time, does not provide a convenient base in relation to which departures can be evaluated.

References

Forman, S.L., 1990. Post-glacial relative sea level history of northwestern Spitsbergen, Svalbard. Bull. Geological Society of America, vol. 102 p. 1580-1590.

Gerasimov, I.P., 1964. Physico-Geographical Atlas of the World. (In Russian.) Acad. of Sciecne of the U.S.S.R., Moscow; 298 p.

Hillaire-Marcel, C., J.-S.Vincent, and others, 1979. Holocene stratigraphy and sea level changes in Southeastern Hudson Bay, Canada (Guide book, Hudson Bay field meeting). The Quebec Association for the Study of the Quaternary, Quebec, Canada; 177 p.

Lamb, H.H., 1981. An approach to the study of the development of climate and its impact on human affairs. Climate and History; T.M.Wigley, M.J. Ingram and G.Farmer, eds.; Cambridge Univ. Press; p.291-309.

May, James P., 1973. Selective transport of heavy minerals by shoaling waves. Sedimentology v.20 p. 203-212.

Melton, F.A., 1938. Underfit meanders of floodplain streams. Proceedings, Geological Society of America (for 1937), p. 324.

Petersen, K.S., 1990. On the geological setting of the marine deposits during the last 15000 years in the Skagen area. Jour. Coastal Research, Special Issue 9, pp. 660-675.

Svensson, N.-O., 1989. Late Weichselian and early Holocene shore displacement in the Central Baltic, based on stratigraphical and morphological records from eastern Smaaland and Gotland, Sweden. Ph.D. dissertation, Lundqua Thesis Series No. 25, Lund University, Lund, Sweden; 195 p.

Tanner, W.F., 1987. Spatial and temporal factors controlling overtopping of coastal ridges. Pp. 241-248 in: Flood Hydrology; V.P.Singh, ed.; D. Reidel Publishing Co., Dordrecht, The Netherlands.

Tanner, W.F., 1990. Origin of barrier islands on sandy coasts. Trans. Gulf Coast Assoc. of Geological Societies, v.40, p.819-824.

Tanner, W.F., 1990. Mean sea level change vs isostasy near Jerup, Denmark. Pp. 31-40 in: 9th Symp. on Coastal Sedimentology, Florida State Univ., Tallahassee, Fla., U.S.A.

Tanner, W.F., 1991. Suite statistics: The hydrodynamic evolution of the sediment pool. P.225-236, in: Principles, methods and application of particle size analysis; J.P.M.Syvitski, ed.; Cambridge Univ. Press.

Tanner, W.F., 1991. Application of suite statistics to stratigraphy and sea-level changes. P.283-292, in: Principles, methods and application of particle size analysis; J.P.M.Syvitski, ed.; Cambridge Univ. Press.

Tanner, W.F., 1991. The "Gulf of Mexico" sea level curve and river delta history. Trans., Gulf Coast Assoc. of Geological Socs, v.41, p.583-589.

Tanner, W.F., 1992. 3000 years of sea level change. Bull., American Meteorological Soc., v.73 p.297-303.

Tanner, W.F., 1992. Late Holocene sea-level changes from grain-size data: evidence from the Gulf of Mexico. The Holocene, v.2, p.164-170.

Tanner, W.F., 1992. Oversize oxbows: tentative dates, effects and risks. Trans., Gulf Coast Assoc. of Geological Societies, v.42, in press.

Tanner, W.F., and S.Demirpolat, 1988. New beach ridge type: severely limited fetch, very shallow water. Trans. Gulf Coast Assoc. of Geological Societies, v.38, p. 367-373.

Tanner, W.F., and J.Spicola, 1986. The asymmetrical "a-b-c..." model. P. 369-387, in: Proc. Iceland Coastal & River Symp., ed. by G. Sigbyarnarson; N.E.A., Reykjavik, Iceland; 387 p.

Captions

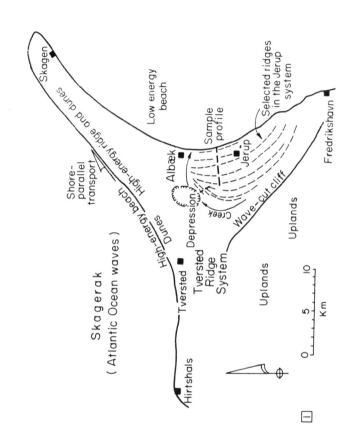

Fig. 1. Map of the study area, extreme northern Denmark. A few selected ridges are shown in the Jerup low-energy system, none in the older Tversted low-energy system. The relic depression is about 8,000-8,300 years old (C-14 years). The creek south of the depression is the boundary between the two beach ridge systems.

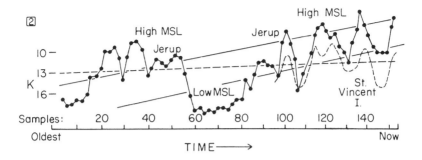

Fig. 2. Inverted Kurtosis vs time, for the Jerup system (continuous line, with filled circles), and for St. Vincent Island, Fla., U.S.A. (dashed line). The Kurtosis scale on the left is for the Jerup system only. The data points for the Jerup curve are moving averages. They have been combined two at a time, so that only 74 points (filled circles) appear, representing smoothed values of 154 samples. Sample numbers are given: the time spacing is 50-51 years for the data, 101 years for the plotted points.

MODELLING HOLOCENE SEA LEVELS IN THE IRISH AND CELTIC SEAS

Robin Wingfield, Coastal Geology Group, British Geological Survey, Keyworth, Nottingham, NG12 5GG, U.K.

Introduction

The time range considered covers the end-Weichselian and early-Holocene from some 11 to 7 thousand years before present (ka BP). South-west of Britain changes of sea level relative to present mean sea level (pmsl) are found of over 100 m (J.D.Scourse *pers.comm.*, 1992). In contrast since some 5 ka BP changes have been limited to a few metres (Shennan, 1989); and over the last 300 years covered by tide gauge records to less than 0.5 m a century (Gornitz & Lebedeff, 1987).

The Irish and Celtic seas (Fig.1) lie in a sector south-west of Scotland where the last local, major, ice-sheet developed in Late Weichselian times from about 30 to 10 ka BP. At its maximum at about 18 ka BP this ice-sheet covered most of Britain and Ireland, the Irish Sea and the north Celtic Sea to 51 degrees north. Since the other major ice sheets which developed at about the same time in Scandanavia, Iceland and Greenland were distant from the sector south-west of Scotland, the simple modelling of the glacio-isostatic effects of ice used here allows only for the effects of Scotland-centred ice; such modelling in the North Sea would have to also model the effects of Scandanavia-centred ice.

Method

The Courtmacsherry Raised Beach on the coasts of both Ireland and Britain about the Celtic Sea and southern Irish Sea is attributed a "Last Interglacial" age (Synge, 1985; Warren, 1985). This beach lies at 4 to 6 m above pmsl, and its preservation suggests that in the region considered here since the last interglacial from 132 to 120 ka BP (Chen *et al*, 1990) that the changes of sea level by tens of metres shown by onshore (Eyles and McCabe, 1989) and offshore evidence (*q.v.*) were effected by cyclic controls of sea level which have recycled to approximately the same conditions. Permanent changes of the land surface by neotectonic effects (subsidence, compaction, warping, folding and fault displacement) might have occurred, but on a scale too small to be significant to sea level changes over this time-span.

Three significant cyclic controls on sea level are identified: glacio-eustasy, glacio-isostasy and hydro-isostasy. These are separately described within the applied models:

(i) GLACIO-EUSTASY. Worldwide changes of sea level are controlled by changes in the amount of ice stored on land. Fig.2 shows a glacio-eustatic curve established by coral reef terraces in Barbados since 17 ka BP (Fairbanks, 1989). This curve is applied directly in the model. It can be used in reverse, if the superimposed effects of other controls can be assessed, to allow estimation of the glacio-eustatic component and consequently identify the age of a specific feature.

(ii) GLACIO-ISOSTASY. A mass imposed on the earth's litho-sphere sufficiently large to cause its rigid strength to be exceeded will result in depression of the lithosphere. Boulton (1990) suggested that such depression occurs by the displacement by flow of mass in the sub-lithospheric, asthenosphere (weak layer) at depths of more than 100 km. An effect of the depression of the lithosphere, either caused directly by its flexure or by the buoyancy of the radially displaced mass, is to produce a zone of uplifted lithosphere over thicker astheonsphere, the forebulge, around the depressed area; and as the imposed ice-sheet enlarges the glacio-isostatic effects dilate such that the forebulge advances outwards as a wave (Boulton, 1990). The simplest possible, purely geometric, model is applied here. The Scotland-centred ice is considered as a point-loading centred on Rannoch Moor (Fig.1) producing a conical depression in a flat-surface, which is surrounded by an annular forebulge of triangular radial cross-section. If it is assumed that the volume of the modelled glacio-isostatic depression is equal to that of the forebulge, the parameters of this crude model are controlled only by the (assumed unchanging) angle of slope of the conical surfaces. This angle is established empirically . The simple model is dilated and contracted to mimic ice-sheet growth and wastage. The expansion and contraction of the glacio-isostatic effects lag corresponding ice-sheet changes by several thousand years, as is seen in the continuing glacio-isostatic effects in Scotland and Finland respectively 10 and 8 ka after the disappearence of ice-sheets.

(iii) HYDRO-ISOSTASY. Both the flooding (loading) and withdrawal (unloading) of the sea relative to the present level will produce a further crustal isostatic effect due to the adding or subtracting of the water mass. It is assumed that this effect also requires a finite time scale to operate, and that it does not respond to short period changes such as tidal cycles. Hydro-isostatic effects on the crust act to increase relatively higher sea levels and further decrease relatively lower sea levels. In each case the full hydro-isostatic enhancement applied only to areas where the entire change of water level produced by glacio-eustatic and glacio-isostatic controls was imposed, and the hydro-isostatic enhancement decreased to zero across areas of lesser water-level changes. With relatively higher sea levels the hydro-isostatic enhancement was zero at the limit of flooding (= contemporary coast) and increased to the full at the present coastline and across present sea areas. In contrast with relatively lower sea levels the hydro-isostatic enhancement was zero at the present coastline and for present inland areas and increased to the full at the contemporary coastline and for areas seaward of it. If dH is the relative change of sea level from pmsl modelled for the glacio-eustatic +/- the glacio-isostatic effects; then taking the average densities of sea-water and the lithosphere to be respectively 1 and 3, the effect of water loading is dH x 1/3 x 0.5 (Turcotte and Schubert, 1982):

Hydro-isostatic enhancement = dH x $(6^{-1} + 6^{-2} + 6^{-3} .... 6^{-n})$

.... enhancement = dH x 0.2

Other cyclic controls of former sea levels, such as the gravititional attraction of the ice, are considered to be small enough to be neglected, while the significant, but varying, effects on levels of the tidal ranges, sedimentation and erosion, are allowed for using available or inferred data.

The simple models described for the three principal controls - glacio-eustasy, glacio-isostasy and hydro-isostasy - can be combined and adjusted to fit the available evidence of former sea levels during Lateglacial-Postglacial times. This empirical testing is not described in detail here. Offshore evidence of former sea levels is crucial since the combined models show that only where glacio-isostatic depression exceeded glacio-eustatic fall did relative sea level rise above pmsl. Offshore sea level data has become available across the north-west European continental shelf in the last ten years but is still sparse. The model shows that the lowest relative sea levels occur at the position of the peak of the forebulge (= height forebulge + glacio-eustatic fall). If the empirical values of the largest falls of sea level as radial distance from Rannoch Moor are plotted against relative fall below pmsl, a graph can be constructed, **representing the locus of the forebulge peak.** The graph of the locus accomodates these values in the form of a sine wave showing the passage outward (southward through the Irish and Celtic seas) to a maximum (in the southern Bay of Biscay), and the passage inward (northward through the Irish and Celtic seas) back to Scotland. This graph was solved by inspection to provide a best fit angle for the modelled glacio-isostatic cones (with an inclination of 25 m over 100 km). By isolating the glacio-eustatic components dates were assigned. Fig.3 shows graphs of sea level changes derived by this simple, albeit crude, analysis for points down the coastline of Europe from Scotland to northern Spain covering the 11 to 7 ka BP interval of forebulge return north and its concurrent diminution.

Results

The graphs of Fig.3 can be applied to the known bathymetry of the Irish and Celtic seas to present the coastal dispositions in 4-D as a series of time slices (Fig.4). Allowances have been made for subsequent changes effected by deposition or erosion.

The locus of the forebulge peak was located for empirical testing as follows:

(i) pmsl - 135 to -140 m in central Celtic Sea at about 50°N 8° 30'W shown by moribund tidal sand ridge crests at this depth (Pantin & Evans, 1984; Scourse *et al.*, 1990). The deduced date is 11.0 ka BP.

(ii) pmsl -115 to -120 m in north Celtic Sea at 51° 20'N 06° 15'W shown by deposits with diagnostic intertidal biota at pmsl - 123 m in a core. These deposits are underlain by cold water marine deposits with amino-acid ratios indicating an 11 ka BP age, and overlain by temperate marine deposits with amino-acid ratios indicating 8.9 ka BP age (J.D.Scourse, *pers. comm.* 1992). The modelled date is 10.5 ka BP, when at this site (Fig.4) a long bay is shown with inferred (bayhead) tidal ranges of 10 to 14 m, similar to tidal ranges in the present upper Bristol Channel (Fig.1). Thus these intertidal deposits might have formed up to 7 m below contemporary msl.

(iii) pmsl - ca.80 m in west Irish Sea at 53° 25'N 05° 30'W indicated by scour hollows in present depths of water from pmsl -90 down to - 185 m about the narrow strait (Fig.4). The inferred age is 9.25 ka BP. Similar scour hollows are found at 52° 50'N 05° 15'W, which may mark an earlier strait formed by submergence of the last land bridge to Ireland. Such tidal scour hollows in matched pairs at either end of narrow, shallow and high tidal volume, straits are described among the Japanese islands by Mogi (1979) even cut in lithified basement rocks.

(iv) Out of area evidence from the North Sea supports the placing of the arc marked "-51-10" in the map for 9.0 ka BP in Fig.4. Along this arc to the north-east in the Geordie Trough (Fig.1) D.Long *pers.comm.* (1992) reports that intertidal deposits were cored at pmsl - 60 m depth. Organic clasts in the intertidal deposits were dated by ^{14}C acclerated mass spectrometry at 9.0 +/- 0.2 ka BP. These deposits are underlain by marine deposits and truncated up by an erosion surface overlain by marine sands.

(v) pmsl ca. -40 m between Ireland and Islay (Fig.1) at 55° 35'N 06° 45'W. This fall would produce a narrow straits (Fig.4). The deduced ages are between 8.5 and 8.0 ka BP. No land bridge would have formed as postulated by Devoy (1985). Tidal scour cauldrons (Mogi, 1979) mark these modelled straits with present depths of pmsl - 60 to -110 m.

The Fig.3 graphs bear close comparison with sea level changes over a similar interval reported adjacent to the Laurentian ice-sheet in New England (Shipp *et al*, 1991) and eastern Canada (Shaw and Forbes, 1990, 1992).The sequence of changes modelled in 4-D in Fig.4 correlates with a range of Lateglacial and Early Holocene observations:

(a) TIDAL SAND RIDGES in the Celtic Sea are presently moribund. Such ridges formed in shallow seas in the tidal streams parallel to contemporary coasts with crestal levels at or just below contemporary sea levels. Crestal levels of these ridges are at progressively lesser present water depths northwards. The northern extent of these ridges is where a land-bridge is modelled to have formed to Ireland at 10.5 ka BP, with the inference that former tidal sand ridges northwards were destroyed by this emergence.

(b) A LAND-BRIDGE INTO SOUTH-EAST IRELAND is modelled from 10.5 to 9.5 ka BP. This interval included at most some 700 years of the Holocene from 10.2 ka BP (Mörner, 1976) and would provide a route for emigration into Ireland in temperate conditions of land plants and animals as required by studies of the contemporary biota (Mitchell, 1972, 1990; Preece et al., 1986). The land-bridge may have been open in temperate conditions for the full 1 ka modelled, if provisional results from a new Greenland ice-core (Johnsen et al, 1992) are confirmed.

(c) A LATER LAND-BRIDGE FROM BRITAIN TO THE ISLE OF MAN is modelled from about 9.75 to 8.25 ka BP. Regression to produce temperate water, backbeach and shallow marine deposits unconformably over deepwater, boreal marine sediments in this area was inferred by Pantin (1977, 1978) to account for cored sequences taken in present 40 m water depths. A land-bridge here, which postdated the last land-bridge to Ireland was suggested by Mitchell (1990) to explain the differences between the biotas of Ireland, the Isle of Man and Britain. It is notable that the last submergence of this land-bridge is modelled by a purely glacio-eustatic rise before 8.0 ka BP south of the limit forebulge, at the same time (+/- 0.2 ka) that the last land-bridge between Britain and mainland Europe would have been submerged by the same rise (after Jelgersma, 1979, with allowances for hydro-isostatic enhancement and tidal ranges).

(d) SHALLOW WATER TIDAL SCOUR FEATURES, attributed to the passages of very large tidal volumes through restricted straits, are modelled between 9.5 and 8.0 ka BP in the west Irish Sea and between Ireland and Islay as noted, and also north-west of the Isle of Man (Fig.1). The latter evidence takes the form of scour moats about rock pinnacles with 60 to 70 m least present water depths, which evidently stood as islets or shoals when scour hollows formed about them in the surrounding sediments up to 40 m deeper. Similar features are not found where narrow straits are modelled between 11 and 10.5 ka BP in the north Celtic Sea before the emergence of the land-bridge to Ireland. It may be that these earlier straits were blocked by sea-ice, restricting the tidal currents, during the end-Lateglacial, Younger Dryas Stadial.

(e) THE RATES OF SEA LEVEL RISE OR FALL MODELLED nowhere exceed 5 m in 100 years and any point passing from subaerial to submarine conditions, or *vice versa*, remained in the intertidal zone for at least this period if average tidal ranges for the western British Isles operated. Any such point was therefore exposed to upwards of 70,000 diurnal tidal cycles and a lengthier period in the surf zone, and relict features and sediments were almost entirely reworked by coastal processes.

CONCLUSIONS

The model presented is acknowledged to be crude, but it provides a simple and adjustable method which may be refined to accomodate the slowly expanding data-set of former sea level evidence.

The sequence of time-slices presented for the Irish and Celtic seas fits the geomorphological and stratigraphic evidence found offshore. Relatively higher former sea levels modelled for 11 ka BP (Fig.4) and earlier are in line with the levels on land of glacimarine deposits widely reported about the Irish Sea (Eyles and McCabe, 1989).

ACKNOWLEDGEMENTS

The use of data obtained by the British Geological Survey continental shelf mapping programme and bathymetric surveys by the Hydrographic Office, Taunton, UK, is acknowledged. Particular thanks are given to D.Long and J.D.Scourse for permissions to refer to their unpublished research. This paper is published with the permission of the Director, British Geological Survey (NERC).

REFERENCES

Boulton, G.S. 1990. Sedimentary and sea level changes during glacial cycles and their control on glacimarine facies architecture. In Doweswell, J.A. and Scourse, J.D. (eds) *Glacimarine Environments: Processes and Sediments*. Geological Society Special Publication, **53**: 15-52.

Chen, J.H., Curran, H.A., White, B. and Wasserburg, G.J. 1991. Precise chronology of the last interglacial period. ^{234}U-^{230}Th data from fossil coral reefs in the Bahamas. *Geological Society of America Bulletin*, **103**: 82-97.

Devoy, R.J. 1985. The problem of a Late Quaternary landbridge between Britain and Ireland. *Quaternary Science Reviews*, **4**: 43-58.

Eyles, N. and McCabe, A.M. 1989. The Late Devensian (<22,000 BP) Irish Sea Basin: the sedimentary record of a collapsed ice sheet margin. *Quaternary Science Reviews*, **8**: 307-351,

Fairbanks, R.G. 1989. A 17,000-year glacio-eustatic sea level record: influence of glacial melting rates on the Younger Dryas event and deep-ocean circulation. *Nature*, **342**: 637-642.

Gornitz, V. and Lebedeff, S. 1987. Global sea-level changes during the past century.In Nummedal, D., Pilkey, O.H. and Howard, J.D. (eds) *Sea-level Fluctuation and Coastal Evolution*. Society of Economic Palaeontologists and Mineralogists, Special Publication **41**: 3-16.

Jelgersma, S. 1979. Sea-level change in the North Sea basin. In Oele, E. Schütterhelm, R.T.E. and Wiggers, A.J. (eds) *The Quaternary History of the North Sea*. Acta Univ.Ups.Symp. Univ.Ups. Annum Quingentesimum Celebrantis: **2**, Uppsala: 233-248.

Johnsen, S.J., Clausen, H.B., Dansgaard, W., Fuhrer, K., Gundestrup, N., Hammer, C.U., Iversen, P., Jouzel, J., Stauffer, B. and Steffensen, J.P. 1992. Irregular glacial interstadials in a new Greenland ice-core. *Nature*, **359**: 311-313.

Mitchell, G.F. 1972. The Pleistocene history of the Irish Sea: second approximation. *Scientific Proceedings of the Royal Dublin Society*, **A4**: 181-199.

Mitchell, G.F. 1990. *The Shell Guide to reading the Irish landscape*. Michael Joseph Country House, Dublin. 228 pp.

Mogi, A. 1979. *An atlas of the sea floor around Japan, aspects of submarine geomorphology*. University of Tokyo,Tokyo:1-96.

Mörner, N.-A. 1976. The Pleistocene/Holocene boundary: proposed boundary stratotype in Gothenburg, Sweden. *Boreas*, **5**: 193-275.

Pantin, H.M. 1977. Quaternary sediments from the northern Irish Sea. In Kidson, C. and Tooley, M.J. (eds) *The Quaternary History of the Irish Sea*. Geological Journal Special Issue **7**, Seel House Press, Liverpool: 27-54.

Pantin, H.M. 1978. Quaternary sediments from the north-east Irish Sea, Isle of Man to Cumbria. *Bulletin of the Geological Survey of Great Britain*, **64**: 1-43.

Pantin, H.M. and Evans, C.D.R. 1984. The Quaternary history of the central and southwestern Celtic Sea. *Marine Geology*, **57**: 259-293.

Preece, R.C., Coxon, P. and Robinson, J.E. 1986. New biostratigraphic evidence of the Post-glacial colonization of Ireland and for Mesolithic forest disturbance. *Journal of Biogeography*, **13**: 487-509.

Scourse, J.D., Austin, W.E.N., Bateman, R.M., Catt, J.A., Evans, C.D.R., Robinson, J.E. and Young, J.R. 1990. Sedimentology and micropalaeontology of glacimarine sediments from the Central and Southwestern Celtic Sea. In Doweswell, J.A. and Scourse, J.D. (eds) *Glacimarine Environments: Processes and Sediments*. Geological Society Special Publication **53**: 329-347.

Shaw, J. and Forbes, D.L. 1990. Relative sea-level change and coastal response, north-east Newfoundland. *Journal of Coastal Research*, **6**: 641-660.

Shaw, J. and Forbes, D.L. 1992. Barriers, barrier platforms, and spillover deposits in St.George's Bay, Newfoundland: Paraglacial sedimentation on the flanks of a deep coastal basin. *Marine Geology*, **105**: 119-140.

Shennan, I. 1989. Holocene sea-level changes and crustal movements in the North Sea region: an experiment with regional eustasy. Scott, D.B, Pirazzoli, P.A. and Honig, C.A. (eds) *Late Quaternary Sea-Level Correlation and Application*. Kluwer Academic Publications, Dordrecht: 1-25.

Shipp, R.C., Belkap, D.F. and Kelley, J.T. 1991. Seismostratigraphic and geomorphic evidence for a post-glacial sea-level lowstand in the northern Gulf of Maine. *Journal of Coastal Research*, **7**: 341-364.

Synge, F.M. 1985. Coastal evolution. In Edwards, K.J. and Warren, W.P. (eds) *The Quaternary History of Ireland*, Academic Press, London: 115-131.

Turcote, D.L. and Schubert, G. 1982. *Geodynamics: applications of continuum physics to geological problems*. John Wiley and Sons, New York: 450 pp.

Warren, W.P. 1985. Stratigraphy. In Edwards, K.J. and Warren, W.P. (eds) *The Quaternary History of Ireland*, Academic Press, London: 39-65.

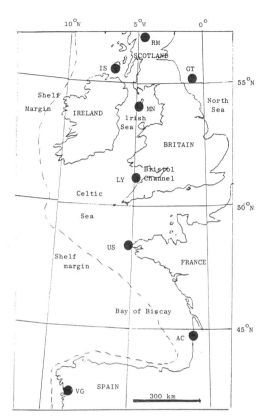

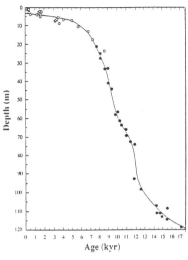

Fig.2. A Barbados sea level curve for the last 17,000 years based on dating of submerged corals.
After Fairbanks, 1989.

Fig.1   LOCATION MAP

Codes :
RM	Rannoch Moor	LY	Lundy
GT	Geordie Trough	US	Ushant
IS	Islay	AC	Arcachon
MN	Isle of Man	VG	Vigo

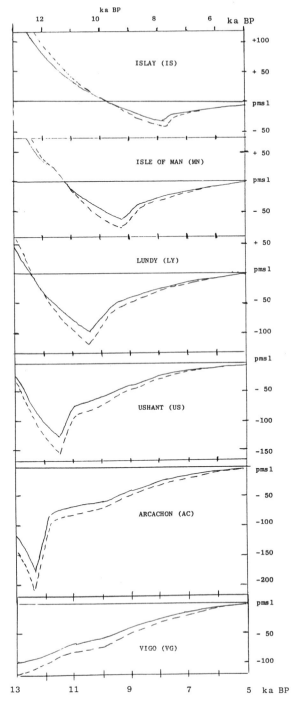

Fig.3. MODELLED GRAPHS.

Modelled contemporary sea levels relative to present mean sea level (pmsl) at six localities from southwest Scotland to northwest Spain (see Fig.1).

The y-coordinates are in METRES above (+) and below (−) pmsl (= 0).

The x-coordinates are in 1000's of years (ka) before present (BP).

Solid lines ——— are the modelled combined glacio-eustatic and glacio-isostatic effects.

Dashed lines ----- are the levels caused by full hydro-isostatic enhancement.

Fig.4.1   For title and key see Fig.4.2

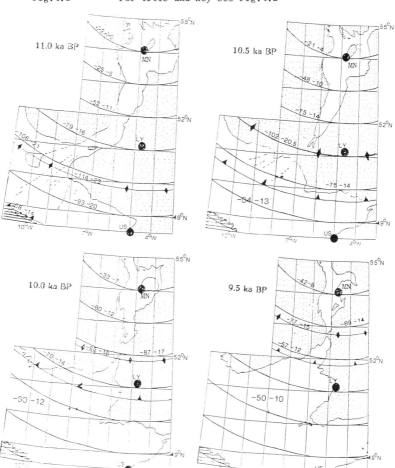

Fig.4.1 & 2. PALAEOCOASTLINES IN THE IRISH AND CELTIC SEAS, 11 to 7 KA BP.

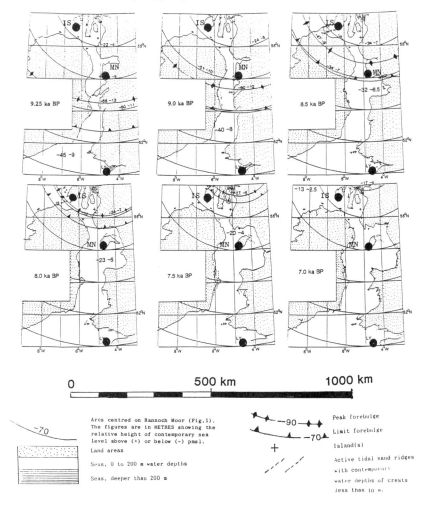

BRAMBATI A.*, AMORE C.**, GIUFFRIDA E.**, RANDAZZO G.**

RELATIONSHIP BETWEEN THE PORT STRUCTURES AND COASTAL DYNAMICS IN THE GULF OF GELA (SICILY - ITALY)

*   Istituto di Geologia - Università di Trieste
** Istituto Policattedra di Oceanologia e Paleoecologia - Università di Catania

INTRODUCTION

The Gulf of Gela, which is a physiographic area of the central - southern sicilian coast, stretching from Capo S.Angelo near Licata to Capo Scalambri near Punta Braccetto, was studied from the point of view of its evolutive tendencies. The following subjects were analized in this study - the morphologic, bathymetric and currentometric parameters, solid fluvial load, textural and composition features of the sediments on beaches both above and below sea level.
Thus a general picture of the Gulf's situation was drawn up showing it to be subject to a strong impact from civil, agricultural and industrial activities. This study can now be used to determine and plan the best solutions and practical applications to protect and conserve the area.

THE GULF OF GELA

The Gulf of Gela is 68 km long, of which 60 % is beach; just over 8% is occupied by port structures, roads and seashore walkways and 4% by dumped material from the Gela industrial installations. The offshore seabeds are slightly steep and reach the 100 m isobath about 16 km from the coast (Fig. 1).
From a morphological point of view, in strict correlation to the lithology, high coastlines and cliffs are most frequent in the western sector and in particular in Licata, Falconara, Manfria, Montelungo and Gela where, due to the presence of pelitic successions, the processes of erosion and landslides are more intense. Low rocky coastlines with sandy coves are found in the eastern sector and are most frequent between Punta Zafaglione and Punta Braccetto. There are low sandy coastlines in the central - eastern sector between Licata and Castello di Falconara, East of Monte di Poggio Lungo, at Torre Manfria, at Montelungo and so

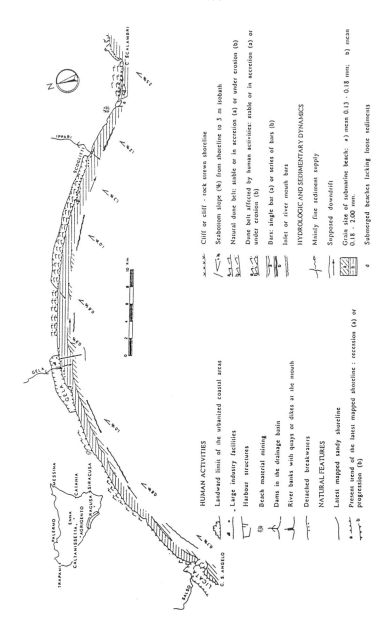

Fig. 1: Evolutive trend of the Gulf of Gela (after AMORE et Alii, 1988 a, b, c)

almost uninterrupted from Gela to Punta Zafaglione. It is rare to see evidence of the dune bars, near Macconi di S. Lucia and Cammarana, because they have been flattened and destroyed to make way for greenhouses, or have been totally altered to building or industrial ends.

From a geological point of view, in the sector to the West of Manfria, there is a succession of Tortonian clays, upper Miocene Gessoso-Solfifera series, middle Pliocene Trubi, middle Pliocene clays and Pleistocene ruditic - arenitic sediments; in the sector to the East, the clay - sandy - calcarenitic Pleistocene - Holocene successions prevalently outcrop.

From a hydrological point of view (Fig. 2), watercourses of both a fluvial and torrential nature flowing into the Gulf of Gela have their mouths blocked off by bars, which are occasionally open following heavy intense rains.

The Gulf of Gela covers an area of about 4000 km^2, with a solid load (calculated using Gavrilovic's formula) of about 4,700,000 m^3/year, of which 97 % has a pelitic component and the > 550 µ fraction is practically absent. This means that having reached the sea, most of the transported material is kept in suspension and carried out to sea, while only 3 % (that is 140,000 m^3) contributes to the littoral nourishment, a figure which is so modest as to be insufficient and ininfluential in terms of sedimentological input.

The composition of the solid load varies according to the watercourses: the Salso and Dirillo have a high percentage of garnet,

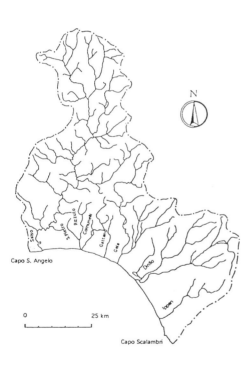

Fig. 2: Rivers flowing into the Gulf of Gela

zircon and rutile, but a low percentage of augite and diopside; the Gela shows an abundance of zircon and rutile, a good frequency of augite, diopside, epidotes and tremolite - actinolite, and a low percentage of garnet; the Ippari has an abundance of zircon, rutile and green hornblend and a low percentage of augite, diopside and garnet.

From an anemological point of view, winds from the NW and SE prevail, but as far as the specific meteomarine conditions of the shore are concerned, the winds in the III quadrant are of particular importance as they are responsible for the most destructive effects.

The surface currents generally flow in an easterly direction, with a coastal dynamic which has been mostly altered in time by a series of man - made structures both on land and at sea.

Among the principle examples of these latter there is the blocking off of the Dirillo River in 1965, the construction of the port - harbour of Gela in 1954 and the port-island in 1962, and later the setting up of a series of parallel and radiating barriers to protect urban and industrial installations in Scoglitti, Gela and Licata.

VARIATIONS IN THE COASTLINE AND BOTTOMS

To define the evolutive trend of the littoral (Fig. 3), the coastlines and isobaths for the years 1928, 1931, 1940 and 1966 were compared with those of June 1982 and October 1988.

After a period of secular progression, at the beginning of the Sixties there was a trend - inversion towards erosive processes with high points of 300 m at the mouth of the Salso River, 60 m at Punta delle due Rocche, 30 - 40 m at Gela, 20 - 30 m east of the port structures and to the right of the mouth of the Gela River, 20 m between Dirillo and Punta Zafaglione and 50 m to the right of the Ippari River.

The stretches at Montelungo, Punta Zafaglione, Porto di Scoglitti and East of Punta Braccetto are stable or only weakly eroding; there are advances of about 100 m West of the Port of Gela, of 110 - 120 m near the S.Lucia port, and also West of the mouth of the Dirillo River, near the port of Scoglitti and between the port of Scoglitti and Punta Braccetto.

When analysing the littoral in more detail, the critical points which are at the limits of irreversibility are found in Licata, West of the Castello di Falconara, in Manfria, Montelungo,

Macchitella, Gela, Contrada Macconi, between the mouth of the Dirillo River and Punta Zafaglione and in Scoglitti.

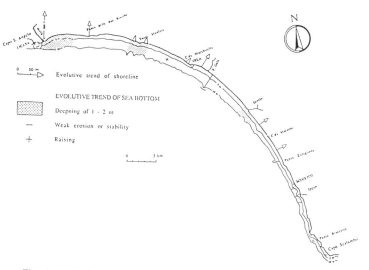

Fig. 3: Evolutive trend of the littoral of the Gulf of Gela

In Licata the coastal structures have suffered from landslides even along the cliffs, this occurred downstream of the port in an area with a littoral drift towards the eastern quadrants; at the Castello di Falconara, dumping has begun in a vain attempt to widen, reinforce and stabilize the dunes; between Capo di Manfria and Torre di Manfria the coasts are in a condition of high instability; at the Lido di Manfria the dune bar is threatened by the various tourist installations and at Montelungo there is very little of the original beach left; the dunes at Macchitella, which have been compromised by levelling and dumping, are destined to be eroded away by the sea and destroyed by the coastal road in costruction; the port - island in Gela is subject to silting and in the downstream tract, locally defended by rough rocky cliff systems, the dunes are being flattened for agricultural use, the road along the beach is threatened by the sea and the outer wharfs of the Priolo canal are undermined to the base; the Contrada Macconi dunes and those in the stretch between Dirillo and Punta Zafaglione have suffered the strong impact determined by an intense urbanization, by the contruction of a road along the beach and the development of greenhouses, structures which are now threatened by extremely violent erosion processes; lastly, to the east of Scoglitti, the situation is aggravated by an intense urbanization which extends down to the waterline and by the building of greenhouses with relative dumping and filling material of a strongly eroding beach.

The erosive processes of the coastal strip above the waterline are also to be seen in the corresponding sea bottoms; depths of 1-2 m have been measured along the shore and offshore between Licata and Punta delle due Rocche, of 1.5 m on the shoreline between Punta delle due Rocche and Torre di Manfria, of 1-1.5 m offshore between Torre di Manfria and the port - harbour of Gela, and of 1 m offshore of the mouth of the Dirillo River. Stability or just weak erosion of the sea bottom exists in the tracts between S.Lucia and Dirillo, Punta Zafaglione and Ippari, and also East of Punta Braccetto; sedimentation of a little over a metre have been found just offshore between Torre di Manfria and the port - harbour of Gela, West of the Lanterna, and also at Punta Braccetto. The beaches above sea level agree with these conditions.

TEXTURAL CHARACTERISTICS OF THE SEDIMENTS

The distribution of the sediments is here synthesized through the analysis of some statistical parameters, both for beaches above and below sea level.

On the *raised beaches* the sediments are bimodal (with 230-390 µ and 165-195 µ), typical of a mixture between alluvial deposits, and aeolian deposits (Fig. 1).

The median values show a discontinuous trend with sharp increases ascribed to fluvial origins (Salso, Cantigaglione, Ippari. La Volpe) or to particular coastal erosion processes (Castello di Falconara, Scoglitti, Macchitella, etc.) indicative of a local supply input.

However, it is to be noted that at first, the values tend to decrease from Licata almost to Dirillo, and then to increase slightly up to the mouth of the Ippari. In fact, there is a regional dispersion which means that between Dirillo and La Volpe there is a zone where the two West and East dispersive systems converge, with cycles which affect very localized areas characterized by sources with their own feedings (landslides or cliff erosion).

The sorting of sediments is worse East of Dirillo, a factor which should be related probably to the wave action which carries a strong transversal element above all in this area where calcareous sandstones and marls are found and which tend to feed landslides.

The *shore - face* is generally of fine sand at the 10 m isobath, with advances West of Punta Braccetto, due to the availability of coarser material deriving from fluvial solid loads, and with withdrawls in the Licata area due to the fine solid loads from the Salso River.

The sediments present a horizontal gradation with the coarsest mode (250-390 µ) up to the 1-2 m bathymetric, above all near the Castello di Falconara, and in the Dirillo area, which should be linked to analogous sediments along the waterline and caused by wave motion. The extension of the sediments to the 2-3 m bathymetric of the 165-195 µ mode West of Licata and near Montelungo, Macchitella and Punta Braccetto, which should be caused by the sands coming from dune bars of the raised beaches.

ORIGIN AND DISPERSION OF SEDIMENTS

The study of the origin and dispersion of sediments was carried out by means of analysis of mineralogic associations and refraction of wave motion, by identifying the sources of beach feeding and the relative influenced areas as well as the direction and extension of the littoral drift along the shore.

Regarding the mineralogical associations, the carbonates of the raised beaches gave values of over 60 % at Licata, of 45 - 55 % as far as Scoglitti, and of 50 - 60 % as far as Punta Braccetto. In the shoreface it was over 60 % at Licata; as far as Punta delle due Rocche, there are values of 50 - 60 % up to the - 6 m isobath and progressively higher values further out to sea; values of 55 - 60 % are found between the Castello di Falconara and the Canale S. Lucia and right as far as Punta Braccetto up to the 17 - 20 m isobath.

To heavy minerals, the sediments present the following mineralogical associations (Fig. 4):
- with augite, typical of the stretch to the West of the port of

Licata, characterized by a notable presence of augite and diopside, by very abundant rutile and zircon, by frequent green hornblend and low garnet values;
- with augite and garnet, typical of the central Gulf zone, with a predominance of garnet, augite and diopside and with no significant variations in other minerals;
- with augite, typical of the eastern stretch, with very high values of augite and diopside, lower levels of garnet and low levels of rutile and zircon.

Therefore there does not therefore seem to be a direct correlation between fluvial sediments and beach sediments, so the sources of the various mineralogical associations have still to be identified in terms of the destruction of the coast which is mineralogically monotonous with high levels of hematite, magnetite and sulphides.

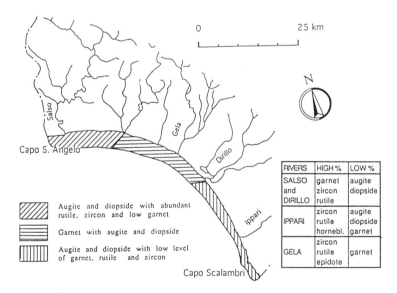

Fig. 4: Distribution of some heavy minerals present in the fluvial solid load and in the sea bottom

On the basis of the wave motion. the regional littoral drift mostly occurs in an easterly direction carried by the seas in the III quadrant, in particular for the seas from the 240°. In this case the dispersion occurs from the Salso as far as the T. La Volpe - Punta Zafaglione stretch, where a transversal drift takes over, which due to the numerous bays gives a locally easterly dispersion. The seas from the 210° also originate an easterly dispersion, as far as the Dirillo River, beyond which the westerly dispersion becomes increasingly more relevant. The seas from the 180° give a prevalently westerly drift as far as the Comunelli, beyond which a transversal drift is set up with local divergences caused by coves and promontories, and the seas from the 150° favour a westerly drift which weakens at Falconara where a slight transversal drift sets in.

The coastal hydrodynamics are therefore characterized by the increase of a transversal drift to the East, and by the contraposition of longitudinal dispersive systems towards the East for the seas from the 240° - 210°, and towards the West for the seas from the 180° - 150°.

Bearing in mind, therefore, the wind prevalence from the III quadrant and the notable frequency of those from the II quadrant, the stretch Dirillo - La Volpe constitutes the zone of sediment dispersion convergence to the East in the western sector, and to the West in the eastern one with local opposing directions in correspondence with bays and promontories.

CONCLUSIONS

With regards to the coastal environment, action has always been taken with the target of facing and solving the problems without taking into consideration induced phenomena and with absolutely no regional planning of the relations between hydrographic basins and littoral, between land - based operations and those along the coast and between maritime operations and their reflexions on bordering littorals.

The coastal devastation occurred after the Sixties coinciding with the building boom, but generalized and exasperated instabilities, and very strong erosion in the last twenty years, have led to a reduction of about 1,700,000 m^2 of beaches, also linked to the striking decrease in the solid fluvial load which can no longer manage to compensate for the longitudinal subtraction on the part of the wave motion and currents.

Summing up, it can be said that:

- apart from localized and exceptional situations, the whole littoral along the Gulf of Gela is under erosion both above and below water level, as a consequence of a general and radical process which in certain areas shows signs of being irreversible;
- the eroded sediments from the beach are generally fine ones and they are carried out from the currents towards the open sea;
- there is not relation between solid fluvial load and the beach sediments;
- the only source for the sedimentological balance of the Gulf of Gela is the sediment eroded from cliffs and low rocky coast;
- the regional drift is to the East in the western sector and to the West in the eastern sector, with a transversal drift which becomes ever stronger going towards the East.

It is thus apparent that some intervention is necessary in terms of revising those operations already carried out in the upper part of the river system, in relation to the induced effects on the littoral, and also in terms of coordinated planning of the use and development of the littoral strip; it is furthermore necessary to urgently provide for the realization of a scheme which will tend to stop the present erosion and start reconstruction processes to recuperate the littoral, these latter needing to be fast or even by means of artificial nourishment.

The problem of evaluating and choosing which stretches of cliff to sacrifice to erosion and use as sources of terrigenous material for the most seriously threatened littoral strips is still to be faced. As these operations should produce a progressive stabilization and a nourishment to the beach, it is a primary condition for the return to a natural process of reconstitution for the dunes.

In this field, it must be clear that any operation which is not the result of adequate planning structures linking the continental and coastal - marine environments is totally useless. These structures are now seen in a new and modern perspective known as the *River unicum* concept, included between the apical part of the watershed and the coastal strip or the continental shelf in front of the focial mouth, which in the final analysis leads to the conclusion that in order to protect and reconstruct at sea, it is also necessary to look, above all, at the land.

REFERENCES

AMORE C., BRAMBATI A., DI GERONIMO S., FINOCCHIARO F., GIUFFRIDA E., RANDAZZO G. (1988a) - Atlante delle Spiagge Italiane, Foglio 276 (Ragusa). C.N.R., P.F. Conserv. Suolo - Sott. Dinam. litorali, Roma.

AMORE C., BRAMBATI A., DI GERONIMO S., FINOCCHIARO F., GIUFFRIDA E., RANDAZZO G. (1988b) - Atlante delle Spiagge Italiane, Foglio 275 (Scoglitti). C.N.R., P.F. Conserv. Suolo - Sott. Dinam. litorali, Roma.

AMORE C., BRAMBATI A., DI GERONIMO S., FINOCCHIARO F., GIUFFRIDA E., RANDAZZO G. (1988c) - Atlante delle Spiagge Italiane, Foglio 272 (Gela). C.N.R., P.F. Conserv. Suolo - Sott. Dinam. litorali, Roma.

MORPHOLOGICAL TYPES OF ROCKY COAST ON SOUTHEASTERN APULIA (*)

G. Mastronuzzi (**), G. Palmentola (**) & P. Sansò (**)

Introduction.

The Adriatic coastline between Torre dell'Orso and Otranto in Apulia region, has geological, lithological and morphological characters which may represent many Mediterranean coasts. The study of this coastline allowed us to reconstruct the recent marine events which formed the present coastline shape and to estimate the mean rate of Holocene moving back to its more unstable tracts.

General aspects.

The coast stretching between Torre dell'Orso and Otranto (fig. 1) is lenghthened in NW-SE direction; it is modelled on the middle-upper pliocenic sediments (Giannelli *et alii*, 1965; 1966), made by gently sloping seaward strata of fine fossiliferous calcarenites interbedded with clayed calcareous sands and, more seldom with bluish sandy clays.

The geological body is characterized by poor physical and mechanical characteristics (unit weight: 1.4-1.7 g/cmc; compressive strength: 8-35 Kg/cm^2); it is affected by a system of subvertical joints, generally strengthened by calcite. The joints are grouped in four sets with strike NNW-SSE, NE-SW and, subordinately, WNW-ESE and ENE-WSW. The critical height, i.e. the height above which the cliff is unstable, can be computed in accordance with the relations of Terzaghi (1943). It comes out to be equal to about 28 metres.

(*) This work is financially supported by MURST under the Research Project 40% "Dinamica e caratteri geoambientali degli spazi costieri". (Resp. Naz.: Prof. G. Fierro; Resp. U.O.: Prof. G. Palmentola).
(**) Dipartimento di Geologia e Geofisica, Università degli Studi, Bari (Italy)

The next strip of land from the coastline is characterized by a nearly horizontal erosion surface which holds little developed drainage pattern; this surface is riddled with a swarm of small sinkholes and several tectonic depressions (Palmentola, 1987) are elongated parallel to the coastline. Some of them hold the Laghi Alimini. The river valleys show a flat floor bordered by moderately high cliffs and are at present dry. Probably they were cut during the deep Würmian regression and filled with alluvial deposits owing to the succeeding sea level rise.

The erosion surface lowers seaward and outlines a coastline strongly influenced by a surface of rock body. The study of the direction and the mean length of rectilinear tracts in which the coastline can be divided (fig. 2) reveals that fifteen per cent is oriented along the most frequent joint set (NNW-SSE). Ten per cent of the coastline follows the joint set with strike WNE-ESE. Although this last one is less frequent than the other sets. About seventy five per cent of coastline consists of tracts with orientation nearly uniform spread along the other directions.

The coastline is exposed to the nothern winds, locally the strongest and the most frequent ones (Mastronuzzi *et alii*, 1987), with fetches reaching a length of about 725 Km. Though direct data about waves do not exist, a study of Tomasicchio *et alii*, 1987 allows us to estimate the average height of the highest one tenth of the Waves, $H_{1/10}$, about 6 metres and a period, T, about 10.5 seconds in this area during storm events. The breaking depth, computed according to the formulas proposed by Le Méhauté & Kohn, 1967, taking into account that the near-shore zone sloping at the rate of 1:20, is estimated about 8 metres.

The morphology of the rocky coasts.

The rocky coasts between Torre dell'Orso and Otranto are very different in details to a quick succession of tracts charcterized by peculiar morphological profiles (fig. 3). They allow us to recognize three basic rocky coast types, each one marked by a typical evolution. The main features of these types and their evolutionary mechanisms could be summarized as follows:

1) GENTLY SLOPING COASTS

This group includes the rocky coasts where the land surface gently dips seaward. They are generally placed between tracts of sandy shore and tracts with high cliffs. In places the gently sloping coasts rise from erosive removal of beach sand still recognizable here and there and the consequent exhumation of the erosion surface. Some coasts of this group are bordered seaward by one or more wave-cut platforms with width up to 30 metres and elevated a little above the mean sea level (MSL). The near-shore sea floor is generally rocky and very irregular with a rough flight of steps. It appears even only in tracts arising from erosion of beaches characterized by a sandy sea floor.

The rocky coasts of these group are recognized by values of Stability Index, I_s, about zero and by different values of the Breaking Index, I_b, (*1).

The retreat and evolution of these coasts occur in a peculiar way. In fact the wave energy gathers along joints and models very elongated cylindric caves (h/1>10). The increase of these caves, is generally due to the fall of the caves inner value, that often creates little natural bridges. In this case, the waves compress the air inside the cave, and exert a prevailing pneumatic action.

This evolutionary mechanism brings about the modelling of a very indented coast with many narrow inlets. These ones are placed in connection with the main joints of rocky mass and then form various angles in the general direction of these tracts of coastline, I_f and F_f, are respectively 2.068 and 2.370 tracts/Km, which are distinctive to a very indented coastline (*2).

2) CLIFF COASTS WITH SEA-FLOOR DEEPER THAN MAXIMUM WAVE-BREAKING DEPTH

This group includes the cliffs with the depth of sea-water at its foot deeper than the maximum wave breaking depth (about 8 metres). So while they produce a standing wave during calm condition, they are affected by strong breaking waves during storms.

This type of coast is defined by morphometric indexes, I_s and I_b, reaching respectively 0.53 and about 0.8 - 1 values.

The cliffs included in this group are placed generally at the main headlands that divided the coast in many little physiographic units. Their slow retreat occurs generally by the fall of blocks moving along joint planes and, subordinately, by diffuse erosion owing to the hiting action of waves.

At most stable headlands these cliffs are bordered seaward by wide wave-cut platforms, located up to three metres above the MSL on which the wave energy is dissipated. Here, the cliff face is affected by waves only where the width of platform shrinks to a few of metres. On the contrary, erosion is very strong on the surface of the platforms, especially along joints, due to the strong increase of the local turbulence.

In these tracts the coastline is rectilinear (I_f= 1.537 - 1.946;

F_f = 1.04 - 1.09 tracts/Km), and locally complicated by little inlets where little river cuts intersect the coastline.

3) CLIFF COASTS WITH SEA-FLOOR ABOVE MAXIMUM WAVE BREAKING DEPTH

This coast-type includes the cliffs that show the near-shore sea bottom gently sloping seaward and placed at a depth which is less than the maximum of the wave breaking depth. During storms the wave breaking line is placed far from the cliff foot.

The cliffs of this group are the most frequent along the studied coast. They are very high (up to 20 metres) unstable and affected in many places by rock falls initiated by severe wave undercutting. These cliffs are marked by values of I_b ranging between 0 and 0.5, and I_s included between 0.36 and 0.71. Many times, their evolution begins with the modelling of blow-holes, and then by wide subcircular bays.

These retreating cliffs are placed inside wide bays that are divided by headlands with deep foot cliffs often showing wide raised wave-cut platforms. These bays shelter some little pocket beaches during calm condition of sea and indent the coastline, here marked by values of I_f ranging from 1.456 to 1.890 anf F_f from 1.93 up to 2.27 tracts/km.

Discussion

The morphological analysis of the rocky coasts between T. dell'Orso and Otranto allows us to provide some conclusions about the recent evolution of this coastal area.

The coastline evolution occurs in different ways in function of the different cliff morphologies. Along the tracts showing gently sloping coasts the retreatment of the coastline is strongly conditioned by the geological structure, as showed by its strong indenture. The cliffs with sea-floor above maximum wave-breaking depth owing to their height are unstable and affected by mass movements; the coastline is charcterized by wide bays owing to the coalescence of rockfall scarps. At last the rocky coasts with sea-floor deeper than maximum wave-breaking depth appears stable and regular, characterized by a slow evolution.

This later type of cliffs may show at their feet raised wave-cut platforms (up to 40 metres wide) witnessing their previous modelling during relatively high sea level. The present altimetric position of these platforms (about 3 metres) and their presence inside ancient river valleys of Würmian modelling allows us to attribute the high sea level to the Holocene. During this period some tracts of the coast retreated up to 40 metres bringing about the formation of high retreating cliffs with wide wave-cut platforms at their feet.

A few of the continential deposits and travertines still covering these platforms in spite of present severe erosion mark a successive period of relitavely lower sea level prior to the present one. Some Roman docks having depths of about four metres and some Roman tombs below sea level in the vicinity of Egnatia (Vlora, 1975), North of the area under discussion allow us to assign this period to the first or second century before Christ.

In the last 2000 - 2100 years the sea level gardually reached the present position. It began to assail the sharp outer border of the platforms. The effective action of wave energy dissipation shelters the backing cliffs so they became very stable. Now they represent the most prominent headlands.

The tracts of the coast lacking of wave-cut platforms were strongly unstable during this period. There, the coastline retreated very fast and outlined bays, inlets, stacks, arches, sea-caves and high cliffs with gently sloping near-shore sea bottom.

In some places the bays with breadth up to 60 metres are placed between headlands with raised wave-cut platforms at their seaward border. This bays because of their strong resistance to wave action mark nearly the starting point of the retreat of the coastline during the rise of sea level in the last 2000-2100 years.

The mean maximum retreating rate for this coastal type is about:

$$V_{maximum\ of\ retreat} = \frac{retreat\ (metres)}{time\ (hundred\ of\ years)} = \frac{60\ m}{21\ hy} = 2.8\ m/hy$$

NOTES

(*1) - The Stability Index, I_S, is defined by Mastronuzzi *et alii*, 1992a as

$$I_S = H/Hc$$

where **H** is the full height of cliff;

Hc is the critical height, i.e. the height above which the cliff is unstable.

The Breaking Index, I_b, is defined (Mastronuzzi *et alii*, 1992a) as

$$I_b = D/Db$$

where **D** is the depth of water at cliff foot;

Db is the depth at which waves break.

(*2) - The Index of Indent, I_f, is defined according to Mastronuzzi *et alii*, 1992b as

$$I_f = L/D$$

where **L** is the real length of coastline;

D is the distance between the ends of the considered tract of coastline.

The Frequency of Indent, F_f, is defined (Mastronuzzi *et alii*, *1992b*) as

$$F_f = n/L$$

where **n**   is the number of elementary tractzs in which the coastline can be decomposed;
  L   is the real length of coastline.

ACKNOWLEDGEMENTS

The Authors thank Prof. Elias G. Abu-Saba, A & T North Carolina State University for his fruitful discussion.

REFERENCES

Giannelli L., Salvatorini G. & Tavani G. (1965) - *Notizie preliminari sulle formazioni neogeniche di Terra d'Oranto (Puglia)*. Atti Soc. Tosc. Sc. Nat., 72, 1-19.

Giannelli L., Salvatori G. & Tavani G. (1966) - *Nuove osservazioni sulle formazioni neogeniche di Terra d'Oroso (Puglia)*. Atti Soc. Tosc. Sc. Nat., 73, 613-619.

Le Méhauté B. & Kohn R.C.Y. (1967) - *On the breaking of waves arriving at an angle to the shore*. J. Hydr. Res., 5, 67-80.

Mastronuzzi G., Palmentola G. & Sansò P. (1987) - *Osservazioni sulle caratteristiche fisiografiche del Santo leccese*. Quad. Ric. Centro Studi Geot. Ing., Lecce, 11, 223-241.

Mastronuzzi G., Palmentola G. & Sansò P. (1992a) - *Some theoretic aspects of rocky coast dynamics*. I seminario sulla dinamica e caratteri geoambientali degli spazi costieri, Alassio 3-4 ottobre 1991; Boll. Ocean. Teor. Appl., in press.

Mastronuzzi G., Palmentola G. & Sansò P. (1992b) - *Esempi di caratterizzazione morfometrica di tratti del litorale roccioso della Puglia*. XXVI Congresso geografico Italiano, Genova 4-9 maggio 1992, in press.

Palmentola G. (1987) - *Lineamenti geologici e morfologici del Salento leccese*. Quad. Ric. Centro Studi Geot. Ing., Lecce, 11, 7-30.

Terzaghi K. (1943) - *Theoretical Soil Mechanism*. New York, Wiley.

Tomasicchio U. & Longo S. (1987) - *Sulla ricostruzione dell'agitazione ondosa con applicazione e confronto per il paraggio di Brindisi*. Giornale del Genio Civile, 4-5-6.

Vlora N.R. (1975) - *Considerazioni sulle variazioni delle linea di costa tra Monopoli (Bari) ed Egnazia (Brindisi)*. Pubbl. n. 2 Ist. Geografia, Fac. Magistero.

MORPHOLOGICAL TYPES OF ROCKY COAST ON SOUTHEASTERN APULIA

G. Mastronuzzi, G. Palmentola & P. Sansò

Fig. 1 - The geographical position of studied coast.
Fig. 2-3 - Morphological and morphometric aspects of studied coast.
1 - Cliff coasts with sea-floor deeper than maximum wave-breaking depth; 2 - idem with raised wave-cut platforms at foot; 3 - cliff coasts with sea-floor above maximum wave-breaking depth; 4 - gently sloping coasts; 5 - idem with raised wave-cut platforms; 6 - sea caves; 7 - pocket beaches; 8 - tracts of unstable coast by means of rock falls.
Fig. 4 - Reconstruction of Holocenic cliff evolution.
Fig. 5 - Block-diagramm showing some peculiar charcteristics of coast.

Fig.1

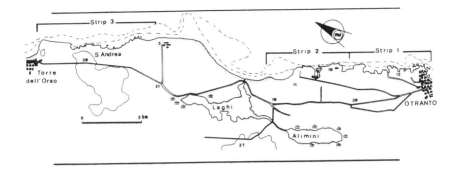

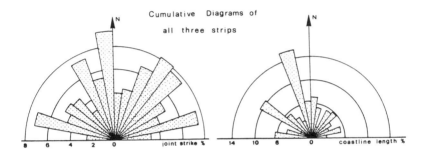

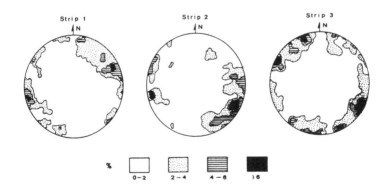

Fig. 2

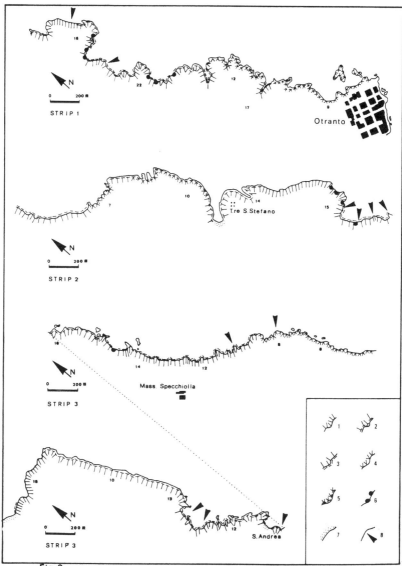

Fig. 3

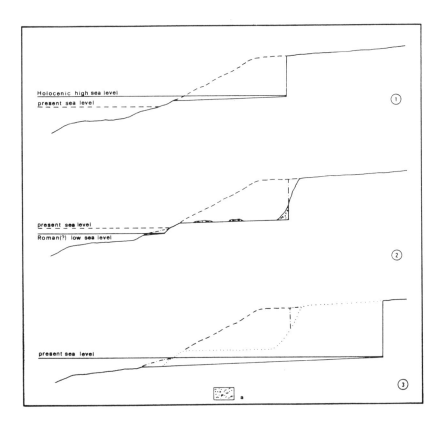

Fig. 4

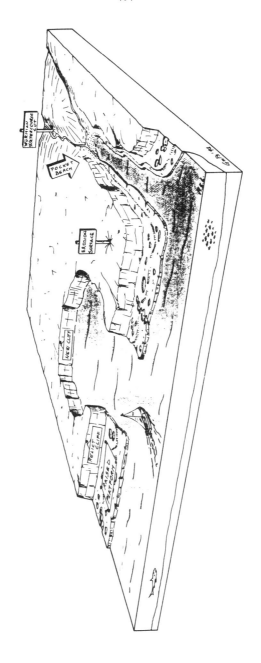

Fig. 5

List of addresses

Alveirinho Dias	J.	Instituto Hidrografico R. Trinas, 49 P - 1200 Lisboa Portugal
Alves	A.M.C.	Ciencas da Terra Universidade do Minho P - 4700 Braga Portugal
Amore	C.	Istituto Policattedra Oceanologia e Paleoecologia C. so Italia n. 55 I-95129 Catania Italy
Andrade	C.	see Freitas, M.
Anthony	E.	Université de Nice-Sophia Antipolis Dpt. de Géographie et Aménagement 98 Bd Edouard Herriot F - 06204 Nice Cedex 3 France
Basinski	T.	Institute of Hydroengineering Polish Academy of Sciences Koscierska 7 PL - 80-953 Gdansk Poland
Bijlsman	L.	see Hillen, R.
Bilgin	R.	Civil Engineering Department Karadeniz Technical University TR - 61080 Trabzon Turkey
Binderup	M.	Geological Survey of Denmark Thoravej 8 DK - 2400 Köbenhavn NV Denmark
Black	K.	University College of North Wales Marine Science Lab. Menai Bridge, Anglesey GB - Gwynedd, Wales Great Britain
Boczar-Karakiewicz	B.	INRS - Océanologie University du Québec 310, Allée des Ursulines CDN - Rimouski, Québec G5L 3A1 Canada
Boedeker	D.	I.N.A. Insel Vilm O - 2331 Lauterbach, Germany
Boldyrev	V.	SSPA "Baltberegozashita" Ribnoje 23-A GUS - 238540 Kaliningrad GUS - Russia
Brambati	A.	Instituto di Geologia Universita di Trieste I - Trieste
Bondesen	E.	Dep. of Environment, Technology and Socia Studies Postbox 260 DK - 4000 Roskilde Denmark

Caputo	C.	Universita di Roma "La Sapienza" Dipto. di Scienze Della Terra Piazzale Aldo Moro 5 I - 00185 Roma Italy
Castelli	P.	Ist. urbanistica, fac. ingegneria, Universita' di Cagliari Piazza D'armi 16 I - 09127 Cagliari Italy
Cendrero	A.	DCITTYM, Ciencias de la Tierra Fac. de Ciencias, Univ. de Cantabria Av. de los Castros, S/N E - 39005 Santander Spain
Chevalier	M.	Scheldedijk 30 B - 2730 Zwijndrecht, Belgium
Christiansen	Chr.	Dept. of Earth Science University of Aarhus Ny Munkegade, bygn. 520 DK - 8000 Aarhus C. Denmark
Christoffersen	H.	Geoscandic A/S Asumsvej 1 DK - 8530 Hjortshöj Denmark
Ciabatti	M.	Dipartimento di Discipline Geografiche e Geologico-Ambientali- Via Zamboni, 67 I - 40126 Bologna Italy
Cicco	L.	Dip. di Science Geologiche Università di Bologna I - 40126 Bologna Italy
Cieslak	A.	Institut Morski ul.Dlugi Targ 41/42 PL - 80-830 Gdansk Poland
Corradi	N.	Dipartimento di Scienza della Terra, Sezione di Geologia Corso Europa 26 I-16132 Genova Italy
Cristini	A.	Università de Cagliari Dipartimento Scienze della Terra Via Trentino 51 I - 09100 Cagliari Italy
Damm-Böcker	S.	Niedersächsisches Landesamt für Wasser u. Abfall, Küste An der Mühle 5 W - 2982 Norderney Germany
Davidson	N.	see Doody P.
De Moor	G.	Geologisch Instituut Ryksuniversiteit Gent Krijgslaan 281 B - 9000 Gent Belgium
Devapriyan	R.	Dept. of Maritime Studies University of Wales College of Cardiff GB - Cardiff CF13YP
Dewidar	Kh.	see Frihy, O.E.
Di Gregorio	F.	Università di Cagliari Dipartimento Scienze della Terra Via Trentino 51 I - 09100 Cagliari Italy

Dieckmann	R.	WSA Bremerhaven Am Alten Vorhafen 1 W - 2850 Bremerhaven 31 Germany
Doody	P.	Joint Nature Conservation Committee Monkstone House City Road GB - Peterborough PE1 1JY Great Britain
Ehlers	J.	FB5 Physiographie und Polargeographie Universität Bremen Box 330440 W - 2800 Bremen 33 Germany
Eichweber	G.	Wasser- und Schiffahrtsdirektion Nord Hindenburgufer 247 W- 2300 Kiel Germany
Eitner	V.	Niedersächsisches Landesamtes für Wasser und Abfall - Forschungsstelle Küste An der Mühle 5 W - 2980 Norderney Germany
Elmi	C.	Dip. di Scienze Geologiche Università di Bologna I - Bologna Italy
El-Fishawi	N.	Marine Geology Dep. Institute of Coastal Research 15 Faraana St. ET - El-Shalallat, 21514 Alexandria Egypt
El-Serafy	S.	Eichtalstr. 26b W - 3300 Braunschweig Germany
Fanucci	F.	Universita' Degli Studi di Urbino Instituto di Geodinamica e Sed. Via Saffi 42 I - 61029 Urbino Italy
Ferk	U.	Wittenmoor 16 W - 2000 Hamburg 54 Germany
Ferrara	C.	Università di Cagliari Dipartimento Scienze della Terra Via Trentino 51 I - 09100 Cagliari Italy
Ferreira	O.	Museu Nacional de Historia Natural (Mineralogia e Geologia) R.da Escola Politécnica, 58 P - 1200 Lisboa Portugal
Ferretti	O.	Centro Ricerche Energia Ambiente S. Teresa Casella Postale n. 316 I - 19100 La Specia Italy
Fierro	G.	see Corradi N.
Filipic	P.	Ekonomski Fakultet Split Radovanova 13 CRO - 58000 Split Croatia
Floderus	S.	Department of Physical Geography Uppsala University S - 751 Uppsala 22 Sweden

Frances	E.	DCITYM, Earth Sciences Fac. Science, Univ. of Cantabria Av. de los Castros, S/N E - 39005 Santander Spain
Fredi	P.	see Caputo, C.
Freitas	M.C.	Dep. de Geologia Faculdade de Ciencias Bloco C2 5 Piso Campo Grande P - 1700 Lisboa Portugal
Frihy	O.E.	Coastal Research Institute 15 Faraana Street ET - 21514, Alexandria Egypt
Gamito	T.	Hidrotecnica Portuguesa Rua da Guiné, Edificio HP P - 2585 Sacavem Portugal
García	M.A.	Laboratori d'Enginyeria Marítima U-P-C Gran Capitá s/n E - 08034 Barcelona Spain
Gerdes	D.	Alfred-Wegner-Institut für Polar- und Meeresforschung W - 2850 Bremerhaven Germany
Gerlach	A.	FB 7 - Biology University of Oldenburg Ammerländer Heerstr.67-69 W - 2900 Oldenburg Germany
Ghionis	G.	Departement of Earth Sciences University of Aarhus DK - Aarhus Denmark
Giuffrida	E.	see Concetto, A.
Gönnert	G.	Institute for Geography TU - Berlin Budapester Str. 46 W - 1000 Berlin 30 Germany
Gomes	C.S.F.	Departamento de Geociencias Universidade de Aveiro P - 3800 Aveiro Portugal
Granö	O.	Department of Geography University of Turku SF - 20500 Turku Finland
Gröger	M.	see Plag, H.-P.
Grotjahn	M.	see Van Bernhem, K.H.
Gurwell	B.	Abt. Küste Landesamt für Umwelt MV Schillerstr. 10 O - 2530 Warnemünde Germany
Haiden	W.	Bundesforschungsanstalt für Fischerei Triftheide 9 D - 2000 Hamburg 53 Germany
Heikkinen	O.	Department of Geography University of Oulu SF - 90570 Oulu Finland

Hennings	I.	GEOMAR Wischhofstr. 1-3	D - 2300 Kiel Germany
Hesse	K.-J.	Forschungs- und Technologiezentrum (FTZ) Werftstr. 10	D - 2242 Büsum Germany
Hillen	R.	Tidal Waters Division	Rijkswaterstaat NL - 2500 EX The Hague Netherlands
Hoerschelmann	C.	FS Ökosystemforschung u. Ökotechnik Universität Kiel Olshausenstr. 40	W - 2300 Kiel 1 Germany
Hoffmann	G.	Institut f. Ostseeforschung Seestr. 15	O - 2530 Warnemünde Germany
Hoffmann	D.	Paläontologisches Institut Universität Kiel Olshausenstr. 40	W - 2300 Kiel 1 Germany
Hofstede	J.	Abt. Hydrography Landesamt f. Wasserhaushalt und Küsten Saarbrückenstr. 38	W - 2300 Kiel 1 Germany
Houwing	E.-J.	IBN-Institute for Forestry and Nature Research/dlo	NL - 1790 AD Den Burg Netherlands
Jackson	L.A.	Gold Coast City Council	AUS - Queensland Australia
Jakobsen	P.R.	Coastal Inspektorat PO Box 100, Holjbovej 1	DK - 7620 Lemvig
Jensen	A.	see Pejrup, M.	
Jiménez	J.A.	Laboratory D'Enginyeria Maritima Departament of Hydraulics, U.P.C. Gran Capita S/N	E - 08034 Barcelona Spain
Jones	F.	see Freitas, M.	
Kaiser	R.	Niedersächsisches Landesamt für Wasser u. Abfall, Forschungsstelle Küste An der Mühle 5	W - 2982 Norderney Germany
Kamenir	Y.G.	Inst. Biol. South. Seas Nahimov Av. 2	GUS - 335000 Sevastopol Ukraine
Kelletat	D.	Institut f. Geographie (FB) Gesamthochschule Essen Postfach 103764, Universitätsstr. 5	W-4300 Essen Germany
Klatt			

Knüpling	J.	see Van Bernem, K.H.
Kocman	Asaf	Ege University Edebiyat Fakültesi Cografya Bölümü TR - 35100 Bornova-Izmir Turkey
Kohlhase	S.	Franzius Institut Universität Hannover Nienburgerstr. 4 D - 3000 Hannover Germany
Kolb	M.	GKSS, Institut für Physik Forschungszentrum Geesthacht GmbH Max-Planck-Str. W - 2054 Geesthacht Germany
Krasemann	H.L.	GKSS-Forschungszentrum Geesthacht Max-Planck-Straße W - 2054 Geesthacht Germany
Kristensen	P.J.	Departement of Earth Sciences University of Aarhus DK - Aarhus Denmark
Kunz	H.	NLWA - Forschungsstelle Küste An der Mühle 5 W - 2982 Norderney Germany
Kypraios	N.G.	National Techn. Univers. of Athens Department of Civil Engineering Pattisson 42 GR - 10682 Athens Greece
La Monica	G.B.	Dipartimento di Scienze Della Terra Universita "LA Sapienza" Piazzale Aldo Moro, 5 I - 00185 Roma Italy
Landini	B.	Dipartimento di Scienze della Terra Universita "La Sapienza" Piazzale Aldo Moro 5, I - 00185 Roma Italy
Larcher	M.	Littoral et Patrimoine S.A. 11 Cours Xavier Arnozan F - Bordeaux 33000 France
Larsen	O.F.	Civil Engineer, Consultant SEDITECH Fasanvaenget 62 DK - 6701 Esbjerg Denmark
Leatherman	S.P.	Centre for Global Change University of Maryland College Park USA - Maryland USA
Leggett	D.J.	NRA Anglian Region Kingfisher House GB - Peterbourough PE2 OZR Great Britain
Liebig	W.	Niedersächsisches Landesamt für Wasser u. Abfall, Forschungsstelle Küste An der Mühle 5 W - 2982 Norderney Germany
Lobmeyr	M.	GKSS-Forschungszentrum GmbH Max-Planck-Str. W - 2054 Geesthacht Germany

Lund-Hansen	L.-Chr.	Institute of Geology Marine, Geolog. Dept., Aarhus University Ny. Munkegade, Bygn 520 DK - 8000 Aarhus Denmark
Lupia Palmieri	E.	see Caputo, C.
Magoon	O.	Coastal Zone Foundation PO Box 279 USA - Middletown, CA 95461
Mastronuzzi	G.	Dipartimento di Geologia e Geofisica , Univ. degli Studi Campus Univesitario I - 70125 Bari
Micallef	A.	University of Malta St. Paul Street M - La Valetta Malta
Misdorp	A.	see Hillen R.
Moutzouris	C.I.	Department of Civil Engineering (NTU) National Technical Univ. of Athens Iroon Polytechniou 5 GR - 15773 Athen Greece
Müller	A.	Institut für Physik Forschungszentrum Geesthacht GmbH Max-Planck-Straße W - 2054 Geesthacht-Tesperhude Germany
Naguszewski	A.	see Boczar-Karakiewiecz, B.
Nasr	S.M.	see Frihy, O.
Nehring	S.	Institut für Meereskunde Düsternbrooker Weg 20 W - Kiel 1 Germany
Nesci	O.	see Cicco, L.
Neugeboren	L.	see Van Bernem, K.-H.
Niemeyer	H.D.	see Damm, S.
Noorbergen	H.H.S.	National Luchten Ruimtevaartlabor. National Aerospace Laboratory NLR Voorsterwey 31 NL - 1006 BM Amsterdam Netherland
Orviku	K.	Institute of Geology Estonian Academy of Sciences Bd. Estonia 7 Est - 200105 Tallinn Estonia
Palmentola	G.	see G. Mastronuzzi
Parks	J.	DYNEQS Ltd. USA - Tampa, Fl.33601 USA

Paskoff	R.	Université de Lyon 10 Squaire Saint-Floretein F - 78150 Le Chesnay France
Patzig	S.	see Van Bernem, K.-H.
Peerbolte	E.B.	see Van Overeem, J.
Pejrup	M.	Institute of Geography Copenhagen University DK - 1350 Copenhagen Denmark
Perera	N.	B/O CCD GTZ Coast Conservation Project Srilanka
Plag	H.-P.	Institut für Geophysik Universität Kiel Leibnizstr.15 W - 2300 Kiel 1 Germany
Pugliese	Fr.	Universita di Roma "La Sapienza" Dipto. di Scienze della Terra Piazzale Aldo Moro 5 I - 00185 Roma Italy
Pustelnikovas	Ol.	Institute of Geography Lithuanian Academy of Sciences Menulio 13-15 Lit - 232600 Vilnius Lithuania
Qiu Jianli		Department of Geography Hangzhou University TJ - Hangzhou 881224 P.R. China
Raey	M.El	see Nasr, S.M.
Raffi	R.	see Caputo, C.
Ragutzki	G.	see Eitner. V.
Randazzo	G.	Instituto Policattedra di Oceanologia e Paleoecologia C. so Italia N.55 I - 95129 Catania Italy
Riethmüller	R.	GKSS-Forschungszentrum GmbH Max-Planck-Str. W - 2054 Geethacht Germany
Rivas	V.	DCITTYM (Earth Science) University of Cantabria Av. de Los Carlos, S/N E - 39005 Santander Spain
Rocha	F.	Departamento de Geociencias Universidade de Aveiro Campo Universitario P - 3800 Aveiro Portugal
Roto	M.	see Granö, O.
Sach	G.	see Van Bernem K.-H.
Salman	A.H.P.M.	EUDC/ Institute for coastal Dune Conservation PO BOX 110598 NL - 2301 EB Leiden Netherlands
Sanchez-Arcilla A.		Laboratori d'Enginyeria Marítima

U-P-C Gran Capitá s/n   E - 08034 Barcelona Spain

Sansò	P.	see G. Mastronuzzi
Scharmann	L.	Geogr. Institut d. Universität Hannover Schneiderberg 50 W - 3000 Hannover 1 Germany
Schauser	U.-H.	FTZ Westküste Hafentörn W - 2242 Büsum Germany
Schröder	H.	Wasser- und Schiffahrtsdirektion Nordwest Schloßpark 9 W - 2980 Aurich Germany
Shaw	J.	Energy Mines and Resources CND - Dartmouth, Nova Scotia Canada B2Y 4A2
Simmons	S.L.	Coastal Research Unit, Science & Chemical Engineering Department GB - Mid Glamorgan CF37 1DL Great Britain
Simunovic	I.	Faculty of Economics University of Split Radovanova 13 CRO - 58000 Split Croatia
Skeie	G.	Cooperating Marine Scientists a.s Sect. of Environmental D.M. Billingstadsletta 19A N - 1361 Billingstadsletta Norway
Steijn	R.C.	see Van Overeem, J.
Sterr	H.	Geogr. Institut Christian-Albtecht-Universität Kiel Olshausenstr. 40-60 W- 2300 Kiel 1 Germany
Stieve	B.	Univ. Bremen, FB 5 Physiogeographie u. Polargeographie W - 2800 Bremen 33 Germany
Stive	M.J.F.	Delft Hydraulics PO Box 152 NL - 8300 Emmeloord
Stock	M.	WWF - Wattenmeerstelle Nevderstr. 3 W - 2250 Husum Germany
Stolk	A.	Dept. Physical Geographie University of Utrecht PO Box 80115 NL - 3508 TC Utrecht Netherland
Straube	J.	Leichtweiß-Institut für Wasserbau Abt. Hydromech. und Küstenwasserbau Beethovenstr. 51 a W - 3300 Braunschweig Germany
Strohmann	F.	see Boczar-Karakiewiecz, B.
Styczynska- Jurewicz	E.	Sw. Wojciecha 5 Str. PL - 81-347 Gdynia Poland

Suchrow	S.	see Van Bernem, K.-H.	
Suryn	T.	see Weslawski, J.M.	
Szmytkiewicz	M.	Institute of Hydroengineering Polish Academy of Sciences Koscierska 7 PL - 80-953 Gdansk Poland	
Tanner	W.F.	Geology Dept. B-160 Florida State Univ. USA - Tallahassee, Fla. 32306 USA	
Tekke	R.M.H.	European Union for Coastal Conservation PO Box 11059 NL - 2301 EB Leiden Netherland	
Townend	I. H.	c/o Sir William Halcrow & Ptnrs Ltd Burderop Park, Swindon GB - Wiltshire SN4 OQD Great Britain	
Trigo-Teixeira	A.	Hidrotecnica Portuguesa Consulting Engineers, Lisbon Rua de Guine P - 2685 Sacavem Portugal	
Uscinowicz	S.	State Geological Institute Branch of marine Geology St. Polna 62 PL - 81-740 Sopot Poland	
Valdemoro	H.	Laboratori d'Enginyeria Marítima U-P-C Gran Capitá s/n E - 08034 Barcelona Spain	
Valeur	J.	see Pejrup, M.	
Van Banning	G.K.F.M	see Van Overeem, J.	
Van Bernem	K.-H.	GKSS W - 2054 Geesthacht Germany	
Van Overeem	J.	Delft Hydraulics "De Voorst" Office NL - 8300 AD Emmeloord Netherlands	
Van Rijn		see Van Overeem, J.	
Veloso-Gomes	F.	Rua dos Poragus P - 4099 Porto Codex Portugal	
Weslawski	J.M.	Polish Academy Artic Ecology Group Institute of Oceanology PL - Sopot 81-967 Poland	
Williams	A.T.	Coastal Research Unit, Science and Chemical Engineering Department GB - Mid Glamorgan CF37 1DL Great Britain	
Wind	H.G.	University of Twente NL - 7500 AE Enschede Netherland	

Wingfield	R.T.R.	Coastal Geology British Geological Survey Keyworth GB - Nottinghamshire NG12 5GG Great Britain
Zachowicz	J.	St. Polna 62 PL - 81-740 Sopot Poland
Zajaczkowski	M.	Inst.of Oceanology, Arctic Ecology Polish Acad. Sciences PL - 81-967 Sopot Poland
Zawadzka-Kahlau	E.	Maritime Institut Abrahama 1 PL - 80307 Gdansk Poland
Zybala	J.	G.-Falke-Str. 26 W - 2000 Hamburg 13 Germany